Mechanisms and Machine Science

Volume 35

Series editor

Marco Ceccarelli, Cassino, Italy

More information about this series at http://www.springer.com/series/8779

Sébastien Briot · Wisama Khalil

Dynamics of Parallel Robots

From Rigid Bodies to Flexible Elements

Springer

Sébastien Briot
Institut de Recherche en Com. et
Cybernétique de Nantes
CNRS, IRCCyN
Nantes Cedex 3
France

Wisama Khalil
U.M.R CNRS 6597
École Centrale de Nantes, IRCCyN
Nantes Cedex 3
France

ISSN 2211-0984 ISSN 2211-0992 (electronic)
Mechanisms and Machine Science
ISBN 978-3-319-19787-6 ISBN 978-3-319-19788-3 (eBook)
DOI 10.1007/978-3-319-19788-3

Library of Congress Control Number: 2015941356

Springer Cham Heidelberg New York Dordrecht London

Printed on acid-free paper

Springer International Publishing AG Switzerland is part of Springer Science+Business Media (www.springer.com)

To our families

Preface

Dynamic models of robots play an important role in their design and control. Most publications that are meant to teach computation of dynamic models of (rigid and/or flexible) parallel robots are general works defining general equations that can be applied to constrained or closed-loop systems. However, they usually suffer from lack of the following information:

- they usually miss the fact that the Jacobian matrices used in a dynamic model to set up dynamic constraints are not so simple to compute, and no straightforward way to compute them is provided.
- most of these works do not propose efficient ways to reduce computational complexity of dynamic models. However, this reduction of complexity is crucial for obtaining models able to predict robot behavior for simulation and control, and to speed up a robot's optimal design process.
- they totally miss the facts that (i) in the presence of certain types of singularities, the dynamic models may degenerate and that (ii) this degeneracy can be avoided thanks to optimal trajectory planning.
- they do not provide experimental results to show that, even if they are complex, dynamics models of parallel robots can be very accurate.

The present book, based on material published by the two authors over the last fifteen years, aims at filling all these gaps and thus providing some tools for engineers, master and Ph.D. students dealing with the dynamics of parallel robots.

Some results given in the book were reached in collaboration with Vigen Arakelian, Nicolas Bouton, Frédéric Boyer, Etienne Dombre, Maxime Gautier, Coralie Germain, Sylvain Guégan, Ouarda Ibrahim, Philip Long, Philippe Martinet and Georges Pagis. The authors acknowledge each of them for their contributions.

The interested reader will also find within the book some links or references to free software or portions of Mathematica codes in which the presented algorithms for computing the kinematics and dynamics of some studied robots are already encoded.

The authors will be also genuinely grateful to the readers for any critical feedback on the material presented in the book and for any suggestion for its improvement.

Nantes, France
March 2015

Sébastien Briot
Wisama Khalil

Contents

Part II Dynamics of Rigid Parallel Robots

Symbols and Abbreviations

Throughout the book, vectors are represented by bold lowercase symbols and matrices by bold uppercase symbols.

Symbols

$\mathbf{a}_j$	The vector characterizing the axis of the joint j
$\bar{\mathbf{a}}_j$	The unit twist characterizing the displacement of the joint j at O_j
$\mathbf{a}^n$	A list containing the number of the intermediate frames separating the frame $\mathscr{F}_n$ from the frame $\mathscr{F}_0$
$a^n(k)$	The k-th element of the list $\mathbf{a}^n$
$a(j)$	Gives the number of the body antecedent to the body $\mathscr{B}_j$
$\alpha, \beta, \gamma, \phi, \theta, \psi$	Some angles
$\$	A unit twist
E	The kinetic energy of a system
$\mathbf{f}$	A vector of force
F	A force along one given direction
$\mathscr{F}_j$	The local frame, composed of the origin O_j and the axes x_j, y_j and z_j and attached to the body $\mathscr{B}_j$
$\mathbf{I}_{O_j}$	The inertia matrix for body $\mathscr{B}_j$, expressed at the origin of the local frame $\mathscr{F}_j$ attached to this body
$\mathbf{I}_{S_j}$	The inertia matrix for body $\mathscr{B}_j$, expressed at the center of mass of this body
$\mathbf{1}_j$	The $(j \times j)$ identity matrix
L	The Lagrangian of a system
$\boldsymbol{\lambda}$	Lagrange multipliers
l_{PQ}	The length of the segment PQ
$\mathbf{m}$	A vector of moment
M	A moment along one given direction
m_j	The mass of the body $\mathscr{B}_j$

n_{dof}	The number of degrees of freedom of the mobile platform of the parallel robot
N_{dof}	The number of degrees of freedom of the parallel robot
${}^{j}\boldsymbol{\omega}_j$	The rotational velocity of body $\mathscr{B}_j$ with respect to the fixed Galilean frame and expressed in the local frame $\mathscr{F}_j$
${}^{j}\dot{\boldsymbol{\omega}}_j$	The rotational acceleration of body $\mathscr{B}_j$ with respect to the fixed Galilean frame and expressed in the local frame $\mathscr{F}_j$
$\mathbf{q}, \dot{\mathbf{q}}, \ddot{\mathbf{q}}$	Some vectors of generalized coordinates, velocities and accelerations, respectively
${}^{i}R_j$	The (3×3) rotation matrix from the frame i to the frame j
${}^{i}\overline{R}_j$	The (6×6) augmented rotation matrix from the frame i to the frame j
${}^{j}\mathbf{r}_P$	The position of point P expressed in the local frame $\mathscr{F}_j$
${}^{i}\mathbf{r}_{PQ}$	The vector $\overrightarrow{PQ}$ expressed in a local frame $\mathscr{F}_i$
t	The time variable
$\mathbf{t}_p$	For a manipulator, the twist of its end-effector
$\dot{\mathbf{t}}_p$	For a manipulator, the acceleration of its end-effector
${}^{i}\mathbf{T}_j$	The (4×4) transformation matrix from frame $\mathscr{F}_i$ to frame $\mathscr{F}_j$
${}^{i}\overline{\mathbf{T}}_j$	The (6×6) augmented screw transformation matrix from frame $\mathscr{F}_i$ to frame $\mathscr{F}_j$
τ	An input joint effort
$\boldsymbol{\tau}$	a vector of input joint effort
U	The potential energy of a system
${}^{j}\mathbf{v}_P$	The velocity of point P with respect to the fixed Galilean frame expressed in the local frame $\mathscr{F}_j$
${}^{j}\dot{\mathbf{v}}_P$	The acceleration of point P with respect to the fixed Galilean frame expressed in the local frame $\mathscr{F}_j$
$\mathbf{w}$	A wrench, composed of a force and a moment
$\boldsymbol{x}_j, \boldsymbol{y}_j, \boldsymbol{z}_j$	The axes of the frame $\mathscr{F}_j$
$\mathbf{x}_p$	For a manipulator, the Cartesian position of its end-effector
xx_j, yy_j, zz_j	The axial moments of inertia around x_j, y_j and z_j axes, respectively, for body $\mathscr{B}_j$, expressed at the origin of the local frame $\mathscr{F}_j$
xy_j, xz_j, yz_j	The inertial cross-moments for body $\mathscr{B}_j$, expressed at the origin of the local frame $\mathscr{F}_j$
ξ	A unit wrench

Abbreviations

atan2	The four-quadrant inverse tangent function
COM	Center of mass
DDM	Direct dynamic model

DOF	Degree of freedom
FGM	Forward geometric model
FKM	Forward kinematic model
IDM	Inverse dynamic model
IGM	Inverse geometric model
IKM	Inverse kinematic model
IRCCyN	Institut de Recherche en Communications et Cybernétique de Nantes (Research Institute in Communications and Cybernetics of Nantes)
LPJTS	Leg Passive Joint Twist System
PKM	Parallel kinematic machine
PPM	Planar parallel mechanism
SPM	Spatial parallel mechanism
T&T	Tilt-and-Torsion
TPM	Translational parallel mechanism
w.r.t.	With respect to

Part I
Prerequisites

Chapter 1
Generalities on Parallel Robots

Abstract This introductory chapter presents generalities about parallel robots. After a general description and the definition of several important terms that will be used in this book, we present a general overview of the different types of parallel robots. We classify them as a function of the type of motions of their platform. We also show that parallel robots are already used for different types of applications. At the end of this chapter, we explain the reasons why we think that a book on the dynamics of parallel robots is necessary.

1.1 Introduction

Parallel robots, also called *parallel manipulators* or *parallel kinematic machines* (*PKM*), are defined in (Leinonen 1991) as robots that control the motion of their end-effectors by means of at least two kinematic chains going from the end-effector towards the fixed base.

From this definition, we see that the *PKM* are composed of different elements (Fig. 1.1):

- the (fixed) *base*, which is the fixed element of the robot
- the (moving) *platform* on which is usually mounted the *end-effector*,
- the *kinematic chains*, linking the base to the platform, and also called the robot *legs*. A leg is usually a kinematic chain of serial or tree-structure type (Figs. 1.2 and 1.3).

Parallel robots are very attractive for several applications because the manipulated load is shared by several legs of the system. Consequently, each kinematic chain carries only a fraction of the total load, which allows the creation of intrinsically more rigid robots. Such architectures also make it possible to reduce the mass of the movable links (all the actuators are mainly fixed on the base and many legs are stressed by tension/compression efforts) and, as a result, make it possible to use less powerful actuators. Such characteristics promised to create structures with high payload, high dynamic capacities and high accuracy. Nowadays, parallel robots are used for several applications, such as (the list is not exhaustive):

S. Briot and W. Khalil, *Dynamics of Parallel Robots*,
Mechanisms and Machine Science 35, DOI 10.1007/978-3-319-19788-3_1

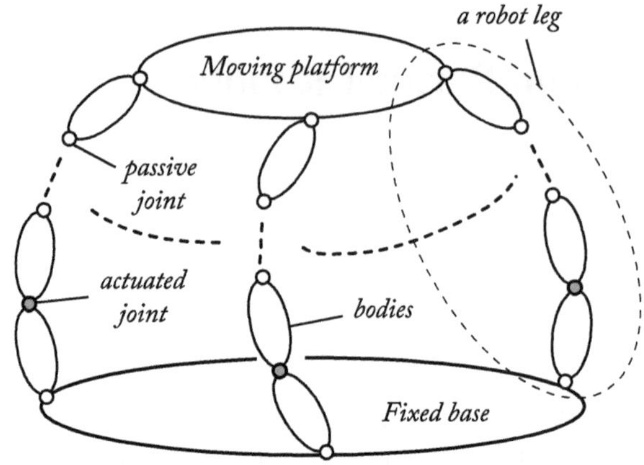

Fig. 1.1 A general parallel robot (the *gray* joints denote the actuated joints)

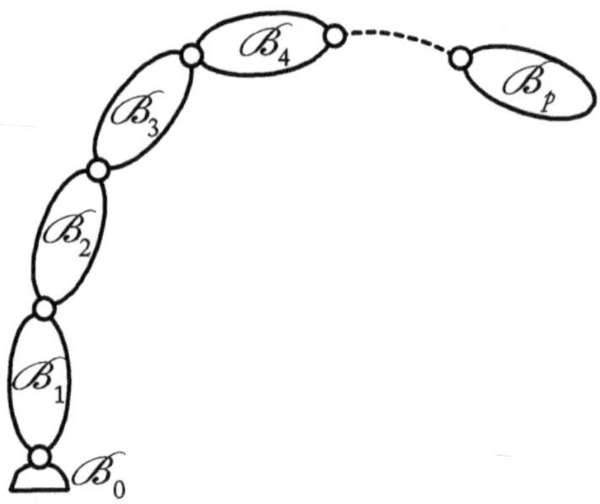

Fig. 1.2 Schematics of a serial open kinematic chain

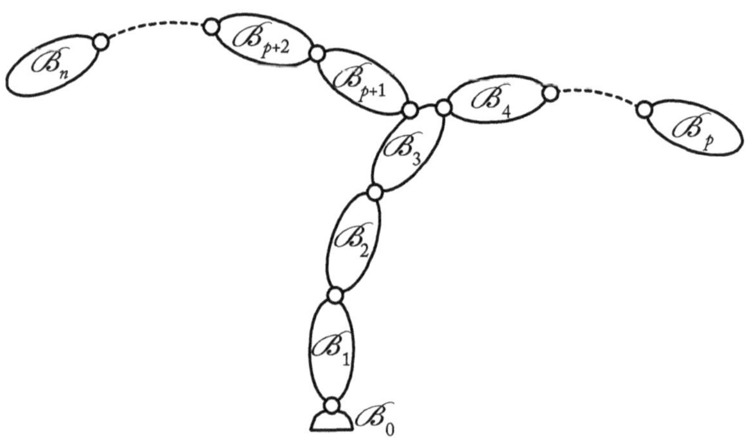

Fig. 1.3 Schematics a tree-structure open kinematic chain

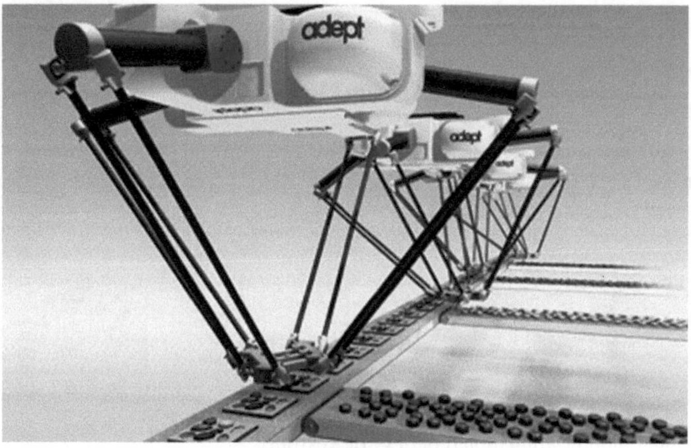

Fig. 1.4 Adept Quattro robots picking and placing chocolates (courtesy of Adept)

Fig. 1.5 The Hermes milling module (courtesy of Tecnalia)

- pick-and-place in food industry, pharmaceutical industry (Fig. 1.4),
- milling (Fig. 1.5),
- motion simulators (Fig. 1.6),
- measuring systems (measuring accuracy of some nanometers for the Delta Cube developed by the Robotic Lab. from the École Polytechnique Fédérale de Lausanne (*EPFL*)),
- micro-positioning systems,
- haptic devices (Fig. 1.7),
- medical environment (Fig. 1.8),
- accessing offshore structure (Fig. 1.9).

Fig. 1.6 A flight motion simulation (courtesy of Rexroth)

Fig. 1.7 The Oméga6 based on a Delta-like architecture (courtesy of Force Dimension)

1.2 General Definitions

Throughout this book, we define the following terms as:

- *mobility*: the mobility of a body is defined in this book as the *number and types* of independent components of its twist (rotational and translational velocity components). It is equal to six for a free body in space and three for a body in plane.
- *degree of freedom*: the number of degree of freedom (*DOF*) of a body is defined as the *number (only)* of independent components of its twist.
- *lower mobility robot*: a robot with a mobility of the platform inferior to six.
- *robot with a mobility $iTjR$*: a robot whose platform encounters i translational *DOF* and j rotational *DOF*.

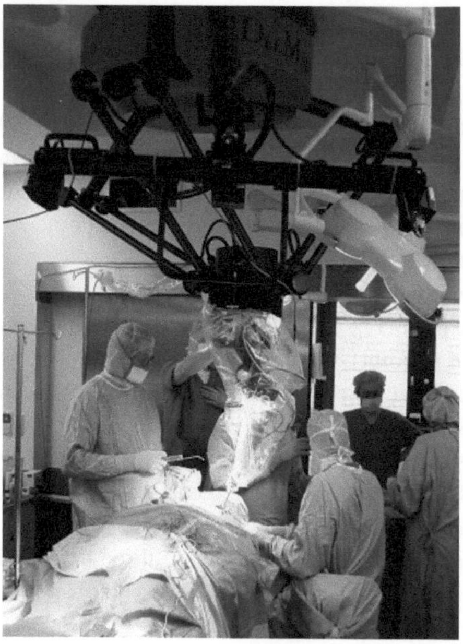

Fig. 1.8 The SurgiScope, a Delta robot carrying a 70 kg microscope (courtesy of ISIS)

Fig. 1.9 The Ampelmann platform (courtesy of Ampelmann)

- *joint*: A joint connects two successive links, thus limiting the number of degrees of freedom between them. The resulting number of degrees of freedom, m, is also called joint mobility, such that $0 \leq m < 6$. When $m = 1$, which is frequently the case in robotics, the joint is either revolute or prismatic. A complex joint with several degrees of freedom can be constructed by an equivalent combination of revolute and prismatic joints. For example, a spherical joint can be obtained by using three revolute joints whose axes intersect at a point.
- *active joint*: a joint which is actuated.
- *passive joint*: a joint which is not actuated.
- *R joint*: a revolute joint (Fig. 1.10), allowing a rotation around a given axis. If the letter is underlined ($\underline{R}$ joint), the joint is actuated. If not, it is passive.
- *P joint*: a prismatic joint (Fig. 1.11), allowing a rotation around a given axis. If the letter is underlined ($\underline{P}$ joint), the joint is actuated. If not, it is passive.
- *U joint*: a universal joint, allowing two independent rotations around two given axes. These joints are usually passive and can be represented by two R joints with orthogonal and intersecting axes.
- *S joint*: a spherical joint (Fig. 1.12), allowing three independent rotations. These joints are usually passive and can be represented by 3R joints with orthogonal and intersecting axes.
- *joint variable / coordinate*: a variable/coordinate associated to the motion of a given joint.
- *Cartesian variable / coordinate*: a variable/coordinate associated to the Cartesian position and orientation system. It is generally used to characterize the motion of the platform.
- *singularity or singular configuration*: a configuration for which the mechanism loses the ability to move along one given direction of the workspace and/or gains one uncontrollable motion. Moreover, locally, the robot mechanical performance (stiffness, accuracy, etc.) are locally decreased.

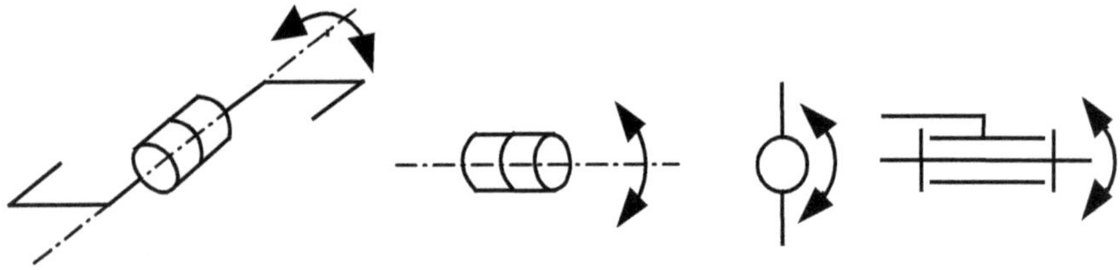

Fig. 1.10 Symbolic representation of a revolute joint

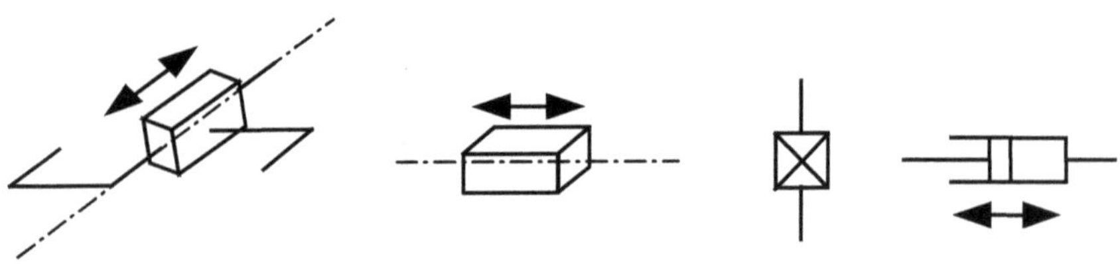

Fig. 1.11 Symbolic representation of a prismatic joint

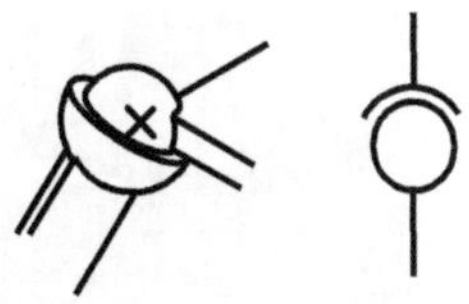

Fig. 1.12 Symbolic representation of a spherical joint

1.3 Types of *PKM* Architectures

There exists a large variety of *PKM* architectures. In a very simple approximation, if we consider that a *PKM* has six legs, and that all legs are made of six joints, the number of possible *PKM* topologies is equal to the number of possible serial legs to the power six. Therefore, it is impossible to show in this book all possible *PKM* architectures.

A large number of *PKM* architectures have been given in the books (Gogu 2008, 2009, 2010, 2012, 2014; Kong and Gosselin 2007). As no classification methods have been proposed, we group them as a function of the number and types of *DOF* of their platform.[1] We briefly introduce some of them in the following of this section.

It should be mentioned that the number of independent *DOF* of the platform of a *PKM* can be found by analyzing the rank of the parallel kinematic Jacobian matrix **A** defined in Sect. 7.3.1, when the robot is not in a singular configuration. Methods to compute the mobility of mechanisms are given in Appendix A.

1.3.1 Planar Motions of the Platform

Many *PKM* have been designed in order to be able to move their platform in a plane. We call them the Planar Parallel Manipulators (*PPM*). We can classify them into three main groups:

1. robots with 2 *DOF* able to position a point in a plane (Fig. 1.13),
2. robots with 2 *DOF* able to position a device with constant orientation in a plane (two translational *DOF* in the plane and one constrained (constant) platform orientation around the axis normal to the plane—Fig. 1.14),
3. robots with 3 *DOF* able to position a device in a plane (two translational *DOF* in the plane and one rotational *DOF* around the axis normal to the plane—Fig. 1.15).

There obviously exist other types of possible mobilities (1T1R), but they are not common.

Most of the robots of this category are planar, i.e. all their elements are constrained to move in parallel planes. However, in order to increase the stiffness of robots with

[1] Throughout this book, when we mention the number and types of *DOF* of the *PKM*, we refer to the number and types of *DOF* of its mobile platform. The number of *DOF* of the mobile platform is denoted n_{dof} while the number of *DOF* of the entire robot is denoted N_{dof}

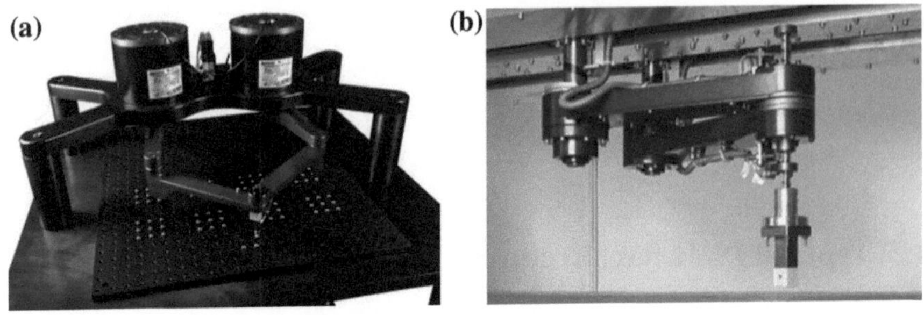

Fig. 1.13 Examples of robots with 2 *DOF* able to position a point in a plane. **a** The Dextar: a planar five-bar mechanism (*RRRRR* planar architecture) designed at ETS Montrèal (Campos et al. 2010). **b** The ParaPlacer (*PRRRP* planar architecture) from the IFW (Hesselbach et al. 2002)

Fig. 1.14 Examples of robots with 2 *DOF* able to position a device with constant orientation in a plane: Robot PacDrive Delta 2 (courtesy of Schneider Electrics)

planar motions of the platform, especially in the direction normal to the displacement plane, a recent idea was to design spatial robots able to achieve planar motions of their platform (Fig. 1.16).

1.3.2 Spatial Motions of the Platform

The large majority of *PKM* have been designed in order to be able to move their platform in the space. We call them the Spatial Parallel Manipulators (*SPM*).

The robots of this category are too numerous to mention all of them. However, we can cite:

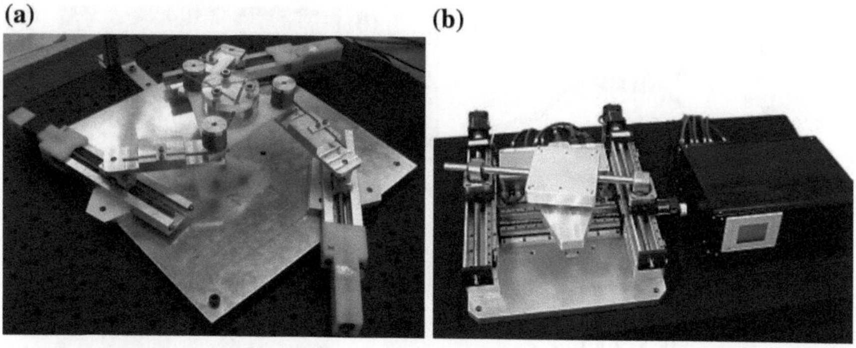

Fig. 1.15 Examples of robots with 3 *DOF* able to position a device in a plane. **a** Prototype of a 3-PRR robot (Wei and Simaan 2010). **b** A decoupled planar robot designed at ETS Montrèal (Joubair et al. 2012)

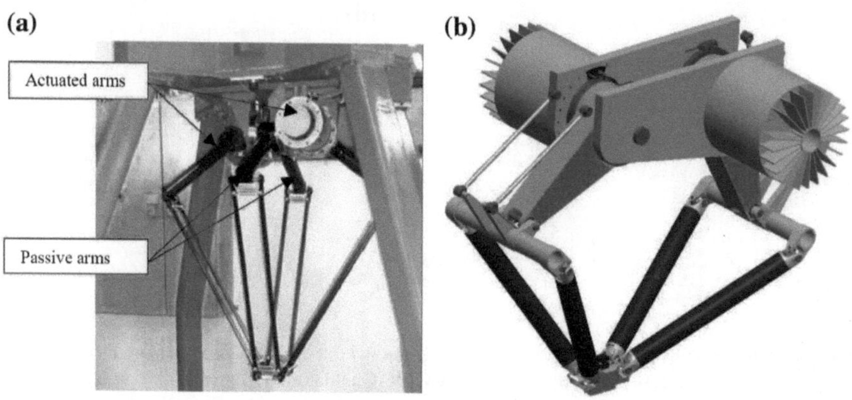

Fig. 1.16 Examples of spatial robots with 2 *DOF* able to position a device in a plane with a constant orientation of the platform. **a** The Par2 from the LIRMM, France (Pierrot et al. 2009). **b** The IRSBot-2 robot from the IRCCyN, France (Briot et al. 2012b)

- robots with three translational *DOF* (also called translational parallel manipulators (*TPM*)): among them, we can mention the Delta robot (Clavel 1990) (Fig. 1.17a), the Orthoglide (Chablat and Wenger 2003) (Fig. 1.17b), the Tripteron (Gosselin 2009) (Fig. 1.17c), etc.
- robots with three rotational *DOF* (also called spherical *PKM*): most of them allow the platform to rotate around one given fixed point (Bonev and Gosselin 2006). The most known is probably the Agile Eye (Gosselin et al. 1996) (Fig. 1.17d),
- robots with three exotic *DOF*: such types of robots have usually some *DOF* of rotation which are constrained with the *DOF* of translation [(see e.g. (Bonev 2008)]. Some of them have been designed with an additional wrist which compensates for the undesirable rotations and have found some industrial applications, especially for milling (Fig. 1.18)

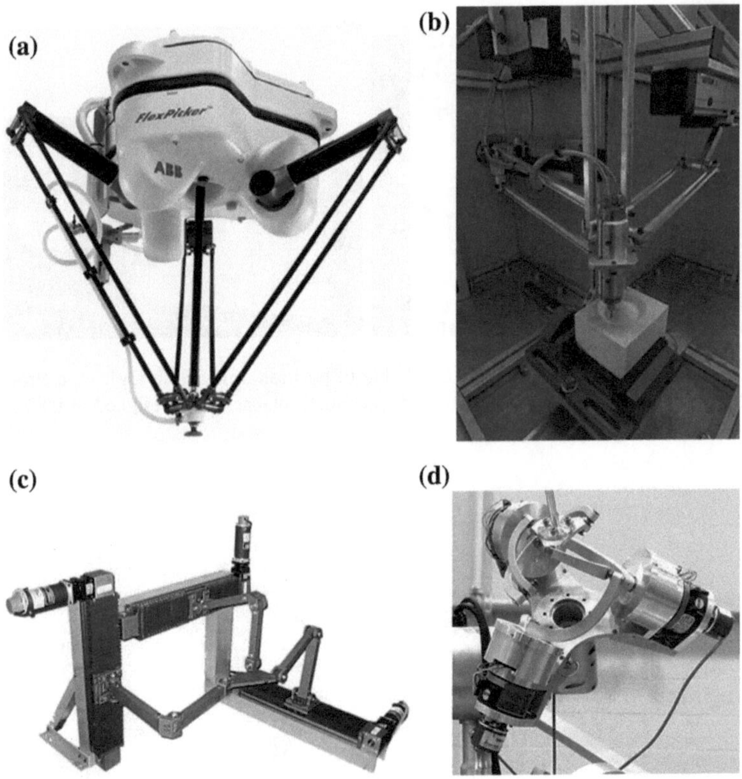

Fig. 1.17 Examples of spatial robots with 3 translational or rotational *DOF*. **a** The Delta robot by Clavel (1990), a TPM. **b** The Orthoglide of IRCCyN (Chablat and Wenger 2003). **c** The Tripteron developed by Gosselin (2009), a TPM . **d** The Agile Eye developed by Gosselin et al. (1996)

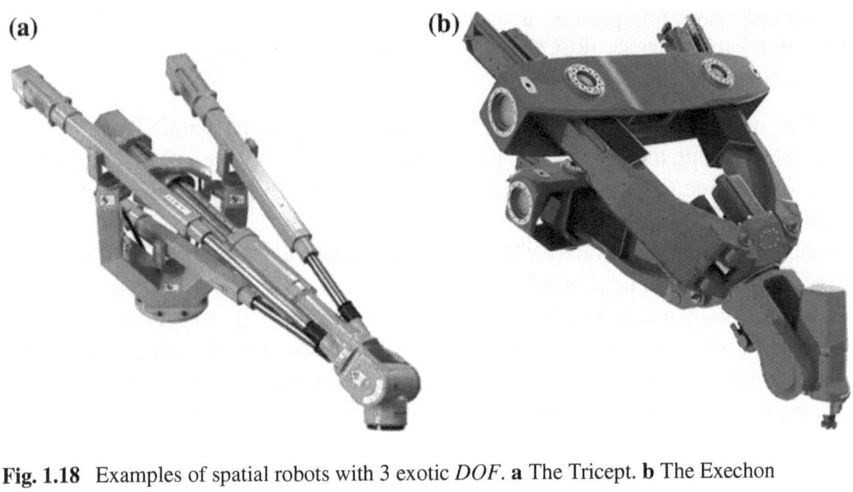

Fig. 1.18 Examples of spatial robots with 3 exotic *DOF*. **a** The Tricept. **b** The Exechon

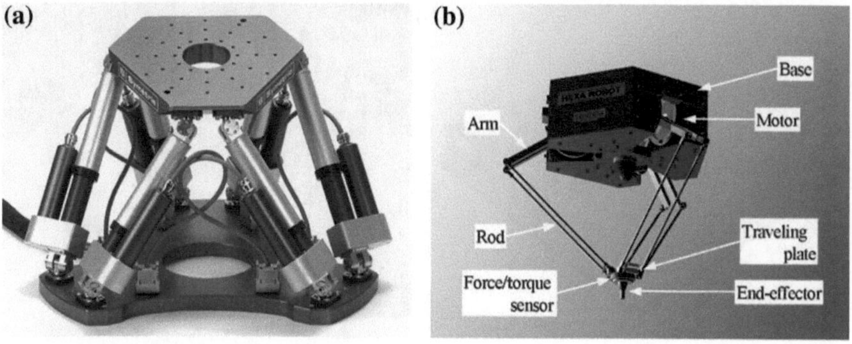

Fig. 1.19 Examples of spatial robots with 6 *DOF*. **a** A Hexapod (courtesy of Symétrie). **b** The Hexa (Pierrot et al. 1990)

- robots with three translational *DOF* and one rotational *DOF* around one given axis (also called Schönflies motion generators): they are usually used for pick-and-place operations, most often at high-speed. The most functional robot of this type is probably the Adept Quattro (Fig. 1.4)
- robots with six *DOF*: such as the Hexapod (also known as the Gough-Stewart platform—Fig. 1.19a) and the Hexa (Pierrot et al. 1990) (Fig. 1.19b).

1.3.3 Redundant PKM

Redundancy occurs when the number of active joints, n_a, is greater than the number n_{dof} of independent variables required to define the platform configuration. Redundancy in *PKM* is usually used in order to avoid their singularities which are considered as one of the main drawbacks of such robots (see Sect. 7.5).

Redundancy in parallel manipulators can be divided into two main groups:

1. *Kinematic redundancy*: in such a case, $n_a = N_{dof} > n_{dof}$,
2. *Actuation redundancy*: in such a case, $n_a > N_{dof}$.

1.3.3.1 Kinematic Redundancy

Kinematic redundancy (Ebrahimi et al. 2008) is obtained when the *total* number N_{dof} of *DOF* for the robot exceeds the number n_{dof} of independent variables necessary to define the robot's platform configuration. In such a case, we have $n_a = N_{dof} > n_{dof}$, n_a being the number of actuators. It results in an infinitude of possible solutions to the inverse kinematic problem giving the joint coordinates of the robot in terms of the platform coordinates (see Sect. 7.3.2.4). This type of redundancy occurs when extra active joints and links are added to a manipulator (Fig. 1.20a). Advantages

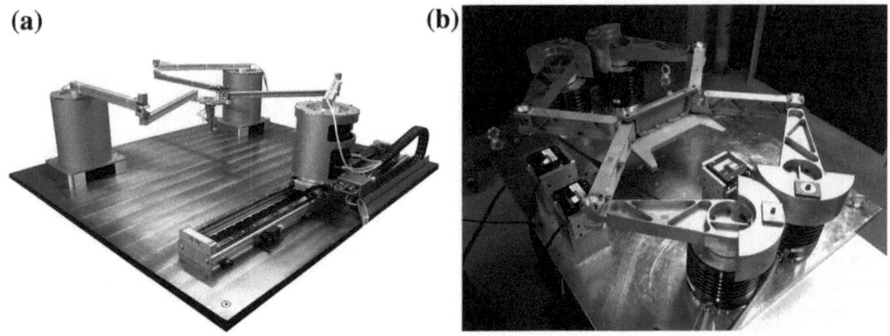

Fig. 1.20 Examples of redundant robots. **a** A kinematically redundant PPM(3-(P) RRR robot) from the institute of mechatronics systems, Germany (Kotlarski et al. 2010). **b** The DualV from the LIRMM, France (van der Wijk et al. 2011): an actuation redundant PPM(4-RRR robot)

can include larger reachable workspace, avoidance of kinematic singularities, and dexterity improvement (Ebrahimi et al. 2008).

1.3.3.2 Actuation Redundancy

Actuation redundancy occurs when the number n_a of actuators is greater than the number of robot *DOF* N_{dof}. Mathematically speaking, we have $n_a > N_{dof}$.

As a consequence,

- we cannot independently choose the active joint variables as they are constrained by n_c equations.
- there are an infinite number of possible solutions to the inverse dynamic problem (see Sect. 8.5). Internal constraint efforts may appear.
- the wrench capabilities are affected (Firmani et al. 2007) and forces of greater magnitudes can be generated.
- as an advantage, the robot workspace becomes usually free of singularity.

An example of an actuation redundant robot named the DualV is provided in Fig. 1.20b.

1.3.4 Other Types of PKM

1.3.4.1 Hybrid *PKM*

The hybrid robots are composed of serially connected parallel modules like the Logabex LX4 robot (Fig. 1.21) (Charentus and Renaud 1989) and bio-mimetic snakes robots (Chablat and Wenger 2005; Khalil et al. 2007a). The serial form of these

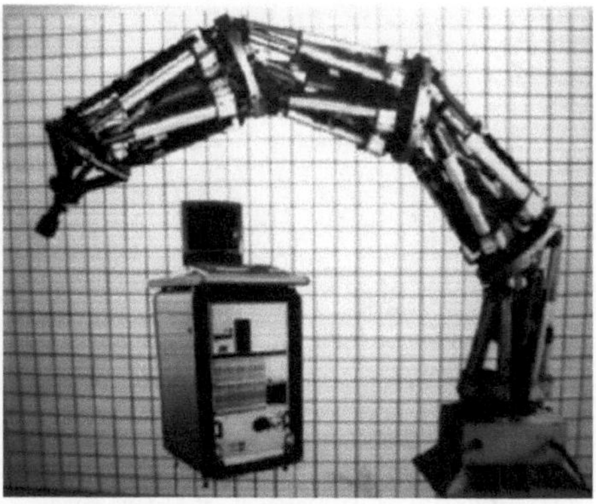

Fig. 1.21 The hybrid Logabex LX4 robot

hybrid manipulators overcomes the limited workspace of parallel manipulators and improves overall stiffness and response characteristics.

1.3.4.2 Cable-Driven Parallel Robots

Cable-driven parallel robots are quite recent types of *PKM* in which the rigid links are replaced by cables. The cables being used in tension only, it results that, for fully controlling the six *DOF* of the platform, at least seven cables must be used.

The most known example of cable-driven parallel robots is probably the Skycam. An example of cable-driven parallel robots is given in Fig. 1.22.

Please note that the dynamics of hybrid *PKM* and of cable-driven parallel robots are not investigated in the present book. However, the interested reader could investigate the works (Ibrahim and Khalil 2010; Notash and Kamalzadeh 2007).

1.4 Why a Book Dedicated to the Dynamics of Parallel Robots?

The dynamic model of robots plays an important role in their design and control. For robot design, the inverse dynamic model can be used to select the actuators (Chedmail and Gautier 1990; Potkonjak 1986), while the direct dynamic model is employed to carry out simulations for the purpose of testing the performance of the robot and to study the relative merits of possible control schemes. Regarding robot control, the inverse dynamic model is used to compute the actuator torques, which are needed to achieve a desired motion (Khalil and Dombre 2002). It is also used to identify the dynamic parameters that are necessary for both control and simulation applications.

Fig. 1.22 A cable-driven parallel robot developed at the LIRMM

Most of the works that can be used for computing the dynamic models of (rigid and/or flexible) parallel robots are general works defining general equations that can be used for constrained or closed-loop systems (Angeles 2003; Bauchau 2011; Cammarata et al. 2013; Dwivedy and Eberhard 2006a; Gallardo et al. 2003; Khalil and Dombre 2002; Khalil and Guégan 2002; Khalil and Ibrahim 2007; Moon 2007; Müller 2005; Özgür et al. 2013; Park et al. 1999; Rognant et al. 2010; Shabana 2005; Shah et al. 2013; Wang and Mills 2006; Wittbrodt et al. 2006). However, they usually suffer from the following (already mentioned above) shortage of information:

- they usually miss the fact that the Jacobian matrices used in the dynamics model to set up the dynamic constraints are not so simple to compute, and no straightforward way to compute them is provided.
- most of these works do not propose efficient ways to compute dynamic models, in terms of the reduction of the operators '$+$', '$-$', '$\times$' and '$/$' used for obtaining the expression of an individual model. However, this optimization is crucial for obtaining models able to predict robot behavior for simulation and control, and to speed up the robot optimal design process.
- they totally miss the facts that (i) in the presence of certain types of singularities, the dynamic models may degenerate and that (ii) this degeneracy can be avoided thanks to an optimal trajectory planning (optimal with respect to a criterion based on the dynamic model).

- they do not provide experimental results to show that, even if they are complex, dynamics models of parallel robots can be very accurate.

The present book aims at filling all these gaps. As a result, the book is organized as follows:

- Part I recalls some basic concepts, common to any types of robots, that are also necessary for computation of the dynamic models of *PKM*. The main Chapters of this Part contain
  - In Chap. 2, basics recalling the ways to parameterize the displacements of any body by using homogeneous transformations are given. Several types of parameters for characterizing the body orientation are shown (Euler angles, Tilt-and-Torsion parameters, Quaternions, etc.)
  - In Chap. 3, the concept of screw is introduced, in order to parameterize the velocity (twist) of a body and the efforts (wrench) exerted on it. Moreover, the expression of the acceleration of a body is derived.
  - In Chap. 4, the minimal set of parameters required for parameterizing the configuration of any type of robots is disclosed.
  - In Chap. 5, the geometric and first/second-order kinematic models of serial and tree-structure robots are investigated, because these models are useful for the computation of the kinematic relationships of *PKM*.
  - In Chap. 6, the basic dynamic principles used in this book are introduced (Lagrange equations, Newton-Euler principle, principle of virtual works).
- Part II deals with the dynamic modeling of rigid parallel robots:
  - In Chap. 7, the generic computation of the geometric and kinematic models of *PKM* is detailed. Moreover, the problem of the singularity of parallel robots is introduced.
  - In Chap. 8, the computation of the inverse and direct dynamic model of parallel robots (with and without redundancy) is treated. This computation is based on the knowledge of the dynamic models of a virtual tree structure, and the ways to optimize this model in terms of operators is shown. At the end of this Chapter, other types of models are provided (energy model and ground reaction model) and the concept of base parameters is introduced.
  - In Chap. 9, the degeneracy of the dynamic models of *PKM* is investigated. Criteria to respect around singularities in order to avoid the dynamic model degeneracy are provided.
- Part III introduces the dynamic modeling of flexible parallel robots. The goal of this Part is not to provide a complete lecture on mechanics of deformable bodies, but to show how to obtain the dynamics of a flexible *PKM* starting from basic considerations in mechanics of deformable bodies. As a result, it is organized as follows:

- In Chap. 10, the full elastodynamic model of the *PKM* is provided. For obtaining it, it is necessary to investigate the dynamics of free flexible bodies and of flexible tree structure.
- In Chap. 11, the algorithm provided in Chap. 10 is simplified in order to obtain the expressions of the stiffness and inertia matrices of *PKM*, which are necessary for the computation of the natural frequencies.

Finally, it should be mentioned that, in Parts II and III, comparative results with dynamic modeling of rigid and flexible robots and experiments are provided.

Chapter 2
Homogeneous Transformation Matrix

Abstract The transformation of frames is a fundamental concept in the modeling and programming of a robot. In this Chapter, we present a notation that allows us to describe the relationship between different frames and objects of a robotic cell. This notation, called homogeneous transformation, has been widely used in computer graphics to compute the projections and perspective transformations of an object on a screen. Currently, this is also being used extensively in robotics. We will show how the points, vectors and transformations between frames can be represented using this approach. We also make an overview of different set of parameters that are used for characterizing the orientation of a body.

2.1 Homogeneous Coordinates and Homogeneous Transformation Matrix

Let $({}^j x_P, {}^j y_P, {}^j z_P)$ be the Cartesian coordinates of an arbitrary point P with respect to the frame $\mathscr{F}_j$, which is described by the origin O_j and the axes $\boldsymbol{x}_j$, $\boldsymbol{y}_j, \boldsymbol{z}_j$ (Fig. 2.1). The homogeneous coordinates of P with respect to frame $\mathscr{F}_j$ are defined by $(w\,{}^j x_P, w\,{}^j y_P, w\,{}^j z_P, w)$, where w is a scaling factor (Newman and Sproull 1979; Roberts 1965). In robotics, w is taken to be equal to 1 (Paul 1981; Pieper 1968). Thus, we represent the homogeneous coordinates of P by the (4×1) column vector:

$$
{}^j\tilde{\mathbf{p}} = \begin{bmatrix} {}^j x_P \\ {}^j y_P \\ {}^j z_P \\ 1 \end{bmatrix}. \tag{2.1}
$$

A direction (free vector) is also represented by four components, but the fourth component is zero, indicating a vector at infinity. If the Cartesian coordinates of a unit vector $\mathbf{u}$ with respect to frame $\mathscr{F}_j$ are $({}^j u_x, {}^j u_y, {}^j u_z)$, its homogeneous coordinates will be:

S. Briot and W. Khalil, *Dynamics of Parallel Robots*,
Mechanisms and Machine Science 35, DOI 10.1007/978-3-319-19788-3_2

Fig. 2.1 Transformation of a vector

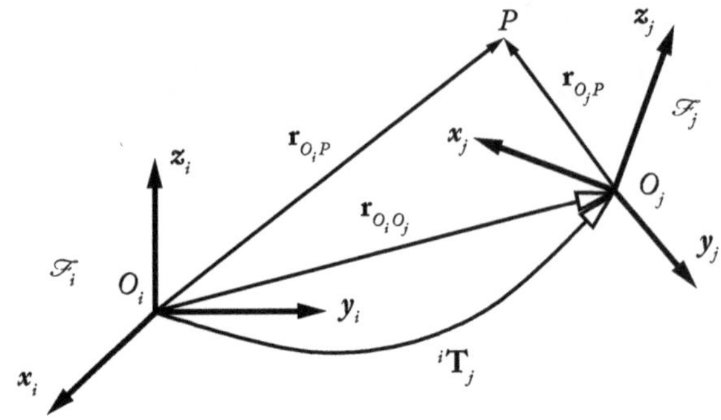

$$
{}^j\tilde{\mathbf{u}} = \begin{bmatrix} {}^ju_x \\ {}^ju_y \\ {}^ju_z \\ 0 \end{bmatrix}. \tag{2.2}
$$

The coordinates of the point P can be defined in another frame $\mathscr{F}_i$ by ${}^i\tilde{\mathbf{p}} = \left[{}^ix_P \; {}^iy_P \; {}^iz_P \; 1\right]^T$ and they can be obtained as a function of ${}^j\tilde{\mathbf{p}}$ by (Fig. 2.1):

$$
{}^i\tilde{\mathbf{p}} = {}^jx_P{}^i\tilde{\mathbf{s}}_j + {}^jy_P{}^i\tilde{\mathbf{n}}_j + {}^jz_P{}^i\tilde{\mathbf{a}}_j + {}^i\tilde{\mathbf{r}}_j = {}^i\mathbf{T}_j{}^j\tilde{\mathbf{p}} \tag{2.3}
$$

where ${}^i\mathbf{s}_j$, ${}^i\mathbf{n}_j$ and ${}^i\mathbf{a}_j$ are unit vectors directed along the $\boldsymbol{x}_j$, $\boldsymbol{y}_j$ and $\boldsymbol{z}_j$ axes with corresponding homogeneous coordinates ${}^i\tilde{\mathbf{s}}_j$, ${}^i\tilde{\mathbf{n}}_j$, ${}^i\tilde{\mathbf{a}}_j$, respectively, and are expressed in frame $\mathscr{F}_i$; ${}^i\tilde{\mathbf{r}}_j$ is the homogeneous vector representing the coordinates (parameterized by the 3D vector ${}^i\mathbf{r}_j = {}^i\mathbf{r}_{O_iO_j}$) of the origin O_j of frame $\mathscr{F}_j$ expressed in frame $\mathscr{F}_i$.

In Eq. (2.3), the matrix ${}^i\mathbf{T}_j$ allows us to calculate the coordinates of a vector ${}^j\tilde{\mathbf{p}}$ with respect to frame $\mathscr{F}_i$ in terms of its coordinates in frame $\mathscr{F}_j$. This (4×4) matrix is called the transformation matrix. It permits us to define the transformation, translation and/or rotation, of the frame $\mathscr{F}_i(O_i, \boldsymbol{x}_i, \boldsymbol{y}_i, \boldsymbol{z}_i)$ towards the frame $\mathscr{F}_j(O_j, \boldsymbol{x}_j, \boldsymbol{y}_j, \boldsymbol{z}_j)$ (Fig. 2.1) and it is represented by:

$$
{}^i\mathbf{T}_j = \left[{}^i\tilde{\mathbf{s}}_j \; {}^i\tilde{\mathbf{n}}_j \; {}^i\tilde{\mathbf{a}}_j \; {}^i\tilde{\mathbf{r}}_j\right] = \begin{bmatrix} {}^i\mathbf{R}_j & {}^i\mathbf{r}_j \\ 0 \;\; 0 \;\; 0 & 1 \end{bmatrix} \tag{2.4}
$$

where ${}^i\mathbf{R}_j$ is the rotation matrix expressing the orientation of the frame $\mathscr{F}_j$ with respect to frame $\mathscr{F}_i$ (see Sects. 2.2.2 and 2.4).

In summary:

- The matrix ${}^i\mathbf{T}_j$ represents the transformation from frame $\mathscr{F}_i$ to frame $\mathscr{F}_j$;
- The matrix ${}^i\mathbf{T}_j$ can be interpreted as representing the frame $\mathscr{F}_j$ (three orthogonal axes and an origin) with respect to frame $\mathscr{F}_i$.

2.2 Elementary Transformation Matrices

2.2.1 Transformation Matrix of a Pure Translation

A general pure translation matrix from frame $\mathcal{F}_i$ to frame $\mathcal{F}_j$ is denoted by **Trans**(a, b, c), where a, b and c denote the translation along the $\boldsymbol{x}$, $\boldsymbol{y}$ and $\boldsymbol{z}$ axes respectively, where (Fig. 2.2):

$$^i\mathbf{T}_j = \mathbf{Trans}(a,\, b,\, c) = \begin{bmatrix} 1 & 0 & 0 & a \\ 0 & 1 & 0 & b \\ 0 & 0 & 1 & c \\ 0 & 0 & 0 & 1 \end{bmatrix} = \mathbf{Trans}(\boldsymbol{x},\, a)\ \ \mathbf{Trans}(\boldsymbol{y},\, b)\ \ \mathbf{Trans}(\boldsymbol{z},\, c) \tag{2.5}$$

taking any order of the multiplication.

2.2.2 Transformation Matrices of a Rotation About the Principle Axes x, y and z

Let us consider a rotation of angle θ around the axis $\boldsymbol{x}$ and let us denote this transformation as $\mathbf{Rot}(\boldsymbol{x},\, \theta)$. From Fig. 2.3, we deduce that:

$$^i\mathbf{T}_j = \mathbf{Rot}(\boldsymbol{x},\, \theta) = \begin{bmatrix} 1 & 0 & 0 & 0 \\ 0 & \mathrm{c}_\theta & -\mathrm{s}_\theta & 0 \\ 0 & \mathrm{s}_\theta & \mathrm{c}_\theta & 0 \\ 0 & 0 & 0 & 1 \end{bmatrix} = \begin{bmatrix} & & & 0 \\ & \mathbf{rot}(\boldsymbol{x},\, \theta) & & 0 \\ & & & 0 \\ 0 & 0 & 0 & 1 \end{bmatrix} \tag{2.6}$$

where c_θ and s_θ represent $\cos\theta$ and $\sin\theta$ respectively, and $\mathbf{rot}(\boldsymbol{x},\, \theta)$ denotes the (3×3) orientation matrix.

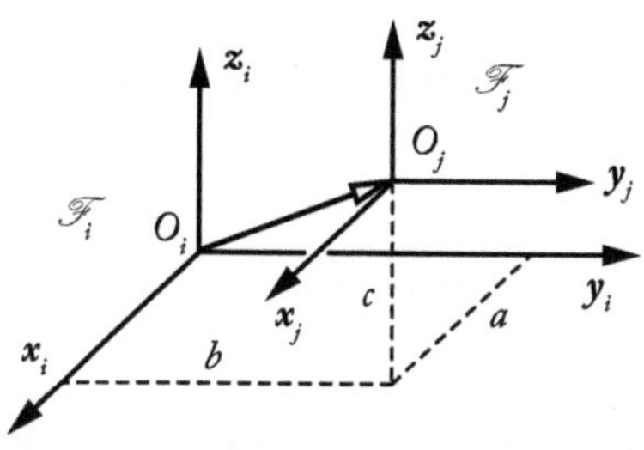

Fig. 2.2 Transformation of pure translation

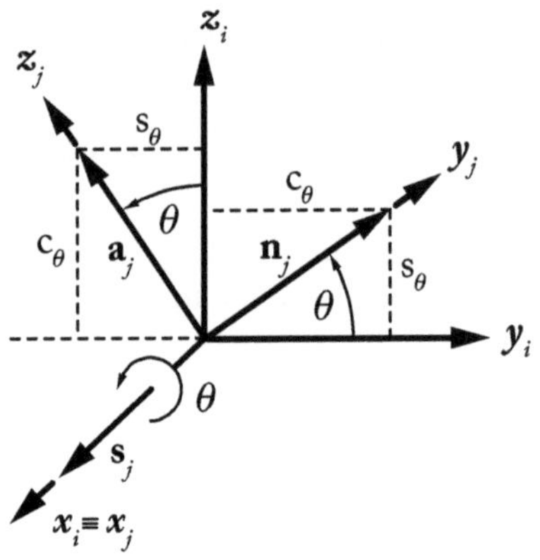

Fig. 2.3 Transformation of a pure rotation about the $\boldsymbol{x}$–axis

Similarly, the rotation of angle θ around the axis $\boldsymbol{y}$ axis is given by:

$$\mathbf{Rot}(y,\,\theta)=\begin{bmatrix} c_\theta & 0 & s_\theta & 0 \\ 0 & 1 & 0 & 0 \\ -s_\theta & 0 & c_\theta & 0 \\ 0 & 0 & 0 & 1 \end{bmatrix}=\begin{bmatrix} & & & 0 \\ & \mathbf{rot}(y,\,\theta) & & 0 \\ & & & 0 \\ 0 & 0 & 0 & 1 \end{bmatrix} \tag{2.7}$$

and the rotation of angle θ around the axis z axis is given by:

$$\mathbf{Rot}(z,\,\theta)=\begin{bmatrix} c_\theta & -s_\theta & 0 & 0 \\ s_\theta & c_\theta & 0 & 0 \\ 0 & 0 & 1 & 0 \\ 0 & 0 & 0 & 1 \end{bmatrix}=\begin{bmatrix} & & & 0 \\ & \mathbf{rot}(z,\,\theta) & & 0 \\ & & & 0 \\ 0 & 0 & 0 & 1 \end{bmatrix}. \tag{2.8}$$

2.3 Properties of Homogeneous Transformation Matrices

Before going further, we need to define the following properties of the homogeneous transformation matrices.

Property 1 *From (2.4), a transformation matrix can be written as:*

$$\mathbf{T}=\begin{bmatrix} s_x & n_x & a_x & r_x \\ s_y & n_y & a_y & r_y \\ s_z & n_z & a_z & r_z \\ 0 & 0 & 0 & 1 \end{bmatrix}=\begin{bmatrix} \mathbf{s} & \mathbf{n} & \mathbf{a} & \mathbf{r} \\ 0 & 0 & 0 & 1 \end{bmatrix}=\begin{bmatrix} & \mathbf{R} & & \mathbf{r} \\ 0 & 0 & 0 & 1 \end{bmatrix}. \tag{2.9}$$

The matrix $\mathbf{R} = \begin{bmatrix}\mathbf{s} & \mathbf{n} & \mathbf{a}\end{bmatrix}$ *represents the rotation whereas the vector* $\mathbf{r}$ *represents the translation. For a transformation of pure translation,* $\mathbf{R} = \mathbf{1}_3$ *(*$\mathbf{1}_3$ *represents the identity matrix of order 3), whereas* $\mathbf{r} = \mathbf{0}$ *for a transformation of pure rotation.*

Property 2 *The matrix* $\mathbf{R}$ *is orthogonal and its determinant is equal to 1. Consequently, its inverse is equal to its transpose:*

$$\mathbf{R}^{-1} = \mathbf{R}^T \tag{2.10}$$

where the superscript "T" indicates the transpose of the matrix.

Property 3 *The inverse of a matrix* ${}^i\mathbf{T}_j$ *is the matrix* ${}^j\mathbf{T}_i$*. Thus, to express the components of a vector* ${}^i\tilde{\mathbf{p}}_1$ *into frame* $\mathscr{F}_j$*, we write:*

$${}^j\tilde{\mathbf{p}}_1 = {}^j\mathbf{T}_i\,{}^i\tilde{\mathbf{p}}_1 \tag{2.11}$$

with:

$${}^j\mathbf{T}_i = {}^i\mathbf{T}_j^{-1}. \tag{2.12}$$

Property 4 *We can easily verify that:*

$$\mathbf{Rot}^{-1}(\boldsymbol{u},\,\theta) = \mathbf{Rot}(\boldsymbol{u},\,-\theta) = \mathbf{Rot}(-\boldsymbol{u},\,\theta) \tag{2.13}$$

$$\mathbf{Trans}^{-1}(\boldsymbol{u},\,d) = \mathbf{Trans}(\boldsymbol{u},\,-d) = \mathbf{Trans}(-\boldsymbol{u},\,d). \tag{2.14}$$

Property 5 *The inverse of a transformation matrix represented by Eq. (2.9) can be obtained as:*

$$\mathbf{T}^{-1} = \begin{bmatrix} & & & -\mathbf{s}^T\mathbf{r} \\ & \mathbf{R}^T & & -\mathbf{n}^T\mathbf{r} \\ & & & -\mathbf{a}^T\mathbf{r} \\ 0 & 0 & 0 & 1 \end{bmatrix} = \begin{bmatrix} \mathbf{R}^T & -\mathbf{R}^T\mathbf{r} \\ 0\ 0\ 0 & 1 \end{bmatrix}. \tag{2.15}$$

Property 6 *Composition of two matrices: The multiplication of two transformation matrices gives a transformation matrix:*

$$\begin{aligned} \mathbf{T}_1\mathbf{T}_2 &= \begin{bmatrix} \mathbf{R}_1 & \mathbf{r}_1 \\ 0\ 0\ 0 & 1 \end{bmatrix} \begin{bmatrix} \mathbf{R}_2 & \mathbf{r}_2 \\ 0\ 0\ 0 & 1 \end{bmatrix} \\ &= \begin{bmatrix} \mathbf{R}_1\mathbf{R}_2 & \mathbf{R}_1\mathbf{r}_2 + \mathbf{r}_1 \\ 0\ \ 0\ \ 0 & 1 \end{bmatrix}. \end{aligned} \tag{2.16}$$

In general, $\mathbf{T}_1\mathbf{T}_2 \neq \mathbf{T}_2\mathbf{T}_1$.

Property 7 *If a frame* $\mathscr{F}_0$ *is subjected to k consecutive transformations (Fig. 2.4) and if each transformation i* $(i = 1, \ldots, k)$ *is defined with respect to the current*

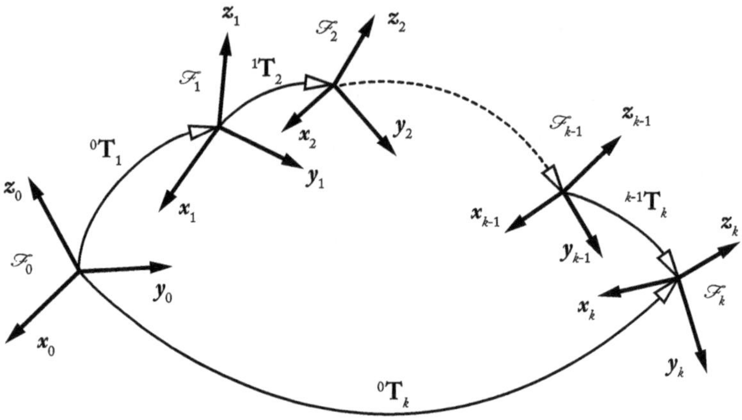

Fig. 2.4 Composition of transformations: multiplication on the right

frame $\mathscr{F}_{i-1}$, *then the transformation* ${}^0\mathbf{T}_k$ *can be deduced by multiplying all the transformations on the right as*

$$ {}^0\mathbf{T}_k = \prod_{i=1}^{k} {}^{i-1}\mathbf{T}_i = {}^0\mathbf{T}_1 \cdot {}^1\mathbf{T}_2 \cdot {}^2\mathbf{T}_3 \cdots {}^{k-1}\mathbf{T}_k. \tag{2.17} $$

Property 8 *Consecutive transformations about the same axis: We note the following properties:*

$$ \mathbf{Rot}(\boldsymbol{u},\, \theta_1)\mathbf{Rot}(\boldsymbol{u},\, \theta_2) = \mathbf{Rot}(\boldsymbol{u},\, \theta_1 + \theta_2), \tag{2.18} $$

$$ \mathbf{Rot}(\boldsymbol{u},\, \theta)\mathbf{Trans}(\boldsymbol{u},\, d) = \mathbf{Trans}(\boldsymbol{u},\, d)\mathbf{Rot}(\boldsymbol{u},\, \theta). \tag{2.19} $$

2.4 Parameterization of the General Matrices of Rotation

The orientation of a body with respect to any frame can be obtained through the use of the rotation matrix **R**. It can be calculated by using a different set of parameters. The most used representations in parallel robotics are described below.

2.4.1 Rotation About One General Axis u

The pure rotation of angle θ around any axis $\boldsymbol{u}$ parameterized by the unit vector $\mathbf{u} = [u_x\ u_y\ u_z]^T$ can be represented by (Khalil and Dombre 2002):

$$\mathbf{R} = \mathbf{rot}(\boldsymbol{u}, \theta)$$
$$= \begin{bmatrix} u_x^2(1-c_\theta)+c_\theta & u_x u_y(1-c_\theta)-u_z s_\theta & u_x u_z(1-c_\theta)+u_y s_\theta \\ u_x u_y(1-c_\theta)+u_z s_\theta & u_y^2(1-c_\theta)+c_\theta & u_y u_z(1-c_\theta)-u_x s_\theta \\ u_x u_z(1-c_\theta)-u_y s_\theta & u_y u_z(1-c_\theta)+u_x s_\theta & u_z^2(1-c_\theta)+c_\theta \end{bmatrix}. \tag{2.20}$$

Inverse problem. Let **R** be any arbitrary rotational transformation matrix such that:

$$\mathbf{R} = \begin{bmatrix} s_x & n_x & a_x \\ s_y & n_y & a_y \\ s_z & n_z & a_z \end{bmatrix}. \tag{2.21}$$

We solve the following expression for $\boldsymbol{u}$ and θ:

$$\mathbf{R} = \mathbf{rot}(\boldsymbol{u},\ \theta),\ \text{ with } 0 \le \theta \le \pi. \tag{2.22}$$

Adding the diagonal terms of Eqs. (2.20) and (2.21), we obtain:

$$c_\theta = \frac{1}{2}(s_x + n_y + a_z - 1). \tag{2.23}$$

From the off-diagonal terms, we obtain:

$$\begin{aligned} 2u_x s_\theta &= n_z - a_y \\ 2u_y s_\theta &= a_x - s_z \\ 2u_z s_\theta &= s_y - n_x \end{aligned} \tag{2.24}$$

yielding:

$$s_\theta = \frac{1}{2}\sqrt{(n_z - a_y)^2 + (a_x - s_z)^2 + (s_y - n_x)^2}. \tag{2.25}$$

From Eqs. (2.23) and (2.25), we deduce that:

$$\theta = \text{atan2}(s_\theta,\ c_\theta),\ \text{ with } 0 \le \theta \le \pi \tag{2.26}$$

where "atan2" is the four-quadrant inverse tangent function.

u_x, u_y and u_z are calculated using Eq. (2.24) if $s_\theta \neq 0$. When s_θ is small, the elements u_x, u_y and u_z cannot be determined with good accuracy by this equation. However, in the case where $c_\theta < 0$, we obtain u_x, u_y and u_z more accurately using the diagonal terms of $\mathbf{rot}(\boldsymbol{u},\ \theta)$ as follows:

$$
\begin{aligned}
u_x &= \operatorname{sign}(n_z - a_y)\sqrt{\frac{s_x - \mathrm{c}\theta}{1 - \mathrm{c}\theta}} \\
u_y &= \operatorname{sign}(a_x - s_z)\sqrt{\frac{n_y - \mathrm{c}\theta}{1 - \mathrm{c}\theta}} \\
u_z &= \operatorname{sign}(s_y - n_x)\sqrt{\frac{a_z - \mathrm{c}\theta}{1 - \mathrm{c}\theta}}
\end{aligned}
\tag{2.27}
$$

where "sign(.)" indicates the sign function of the expression between brackets, thus $\operatorname{sign}(e) = +1$ if $e > 0$, $\operatorname{sign}(e) = -1$ if $e < 0$ and $\operatorname{sign}(e) = 0$ if $e = 0$.

2.4.2 *Quaternions*

The quaternions are also called *Euler* parameters or *Olinde-Rodrigues* parameters. This is another way of parameterizing the rotation of an angle θ $(0 \le \theta \le \pi)$ about an axis $\boldsymbol{u}$. In this representation, the orientation is expressed by four parameters. We define the quaternions as:

$$
\begin{aligned}
Q_1 &= \mathrm{c}\theta/2 \\
Q_2 &= u_x \mathrm{s}\theta/2 \\
Q_3 &= u_y \mathrm{s}\theta/2 \\
Q_4 &= u_z \mathrm{s}\theta/2.
\end{aligned}
\tag{2.28}
$$

From these relations, we obtain:

$$
Q_1^2 + Q_2^2 + Q_3^2 + Q_4^2 = 1. \tag{2.29}
$$

The transformation matrix **T** is deduced from Eq. (2.20), defining $\mathbf{rot}(\boldsymbol{u}, \theta)$ (Sect. 2.4.1), after rewriting its elements as a function of Q_j. Thus, the orientation matrix is given as:

$$
\mathbf{R} = \begin{bmatrix} 2(Q_1^2 + Q_2^2) - 1 & 2(Q_2Q_3 - Q_1Q_4) & 2(Q_2Q_4 + Q_1Q_3) \\ 2(Q_2Q_3 + Q_1Q_4) & 2(Q_1^2 + Q_3^2) - 1 & 2(Q_3Q_4 - Q_1Q_2) \\ 2(Q_2Q_4 - Q_1Q_3) & 2(Q_3Q_4 + Q_1Q_2) & 2(Q_1^2 + Q_4^2) - 1 \end{bmatrix}. \tag{2.30}
$$

Inverse problem. Let us find the expression of the quaternions as functions of the direction cosines of the general matrix **R** of (2.21). Equating the elements of the diagonals of the right sides of Eqs. (2.21) and (2.30) leads to:

$$
Q_1 = \frac{1}{2}\sqrt{s_x + n_y + a_z + 1} \tag{2.31}
$$

which is always positive. If we then subtract the second and third diagonal elements from the first diagonal element, we can write after simplifying:

$$4Q_2^2 = s_x - n_y - a_z + 1. \tag{2.32}$$

This expression gives the magnitude of Q_2. For determining the sign, we consider the difference of the (3,2) and (2,3) matrix elements, which leads to:

$$4Q_1Q_2 = n_z - a_y. \tag{2.33}$$

The parameter Q_1 being always positive, the sign of Q_2 is that of $(n_z - a_y)$, which allows us to write:

$$Q_2 = \frac{1}{2}\text{sign}(n_z - a_y)\sqrt{s_x - n_y - a_z + 1}. \tag{2.34}$$

Similar reasoning for Q_3 and Q_4 gives:

$$Q_3 = \frac{1}{2}\text{sign}(a_x - s_z)\sqrt{-s_x + n_y - a_z + 1} \tag{2.35}$$

$$Q_4 = \frac{1}{2}\text{sign}(s_y - n_x)\sqrt{-s_x - n_y + a_z + 1}. \tag{2.36}$$

Contrary to Euler angles, roll-pitch-yaw angles and *T&T* angles (see next sections), quaternion representation is free of singularity. For more information on the algebra of quaternions, the reader can refer to (de Casteljau 1987).

2.4.3 Euler Angles

The orientation of frame $\mathscr{F}_k$ expressed in frame $\mathscr{F}_i$ can be determined by specifying three angles, ϕ, θ and ψ corresponding to a sequence of three rotations (Fig. 2.5).

Let us consider two intermediate frames $\mathscr{F}_j$ and $\mathscr{F}_{j'}$ defined by $\mathscr{F}_j$ $(O_i, \boldsymbol{x}_j, \boldsymbol{y}_j, \boldsymbol{z}_j)$ and $\mathscr{F}_{j'}$ $(O_{j'}, \boldsymbol{x}_{j'}, \boldsymbol{y}_{j'}, \boldsymbol{z}_{j'})$ and characterized by:

- $\boldsymbol{z}_j \equiv \boldsymbol{z}_i$ and $\boldsymbol{y}_j$ is the intersection between the two planes $(O_i, \boldsymbol{x}_i, \boldsymbol{y}_i)$ and $(O_i, \boldsymbol{x}_k, \boldsymbol{y}_k)$,
- $\boldsymbol{y}_{j'} \equiv \boldsymbol{y}_j$ and $\boldsymbol{z}_{j'} \equiv \boldsymbol{z}_k$.

Taking into account these considerations, the Euler angles are defined as:

- ϕ: *precession* angle between $\boldsymbol{y}_i$ and $\boldsymbol{y}_j$ about $\boldsymbol{z}_i \equiv \boldsymbol{z}_j$, with $0 \le \phi < 2\pi$; that angle characterizes the pure rotation of angle ϕ around $\boldsymbol{z}_i$ (see Sect. 2.2.2) that transforms the frame $\mathscr{F}_i$ into the frame $\mathscr{F}_j$;

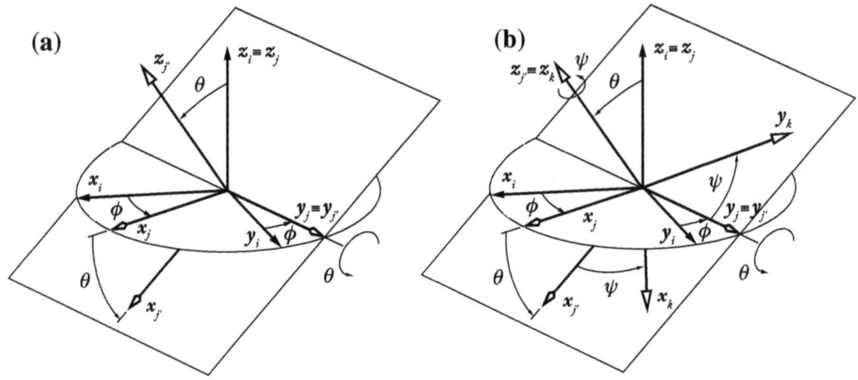

Fig. 2.5 The successive rotations that define the ZYZ Euler angles: **a** precession and nutation, **b** spin

- θ: *nutation* angle between z_i and $z_{j'}$ about $y_j \equiv y_{j'}$, with $0 \leq \theta < \pi$; that angle characterizes the pure rotation of angle θ around y_j (see Sect. 2.2.2) that transforms the frame $\mathscr{F}_j$ into the frame $\mathscr{F}_{j'}$;
- ψ: *spin* angle between $y_{j'} \equiv y_j$ and y_k about $z_{j'} \equiv z_k$, with $0 \leq \psi < 2\pi$; that angle characterizes the pure rotation of angle ψ around $z_{j'}$ (see Sect. 2.2.2) that transforms the frame $\mathscr{F}_{j'}$ into the frame $\mathscr{F}_k$.

The transformation matrix is given by:

$$\begin{aligned}\mathbf{R} &= \mathbf{rot}(z,\ \phi)\mathbf{rot}(y,\ \theta)\mathbf{rot}(z,\ \psi)\\ &= \begin{bmatrix} c_\phi c_\theta c_\psi - s_\phi s_\psi & -c_\phi c_\theta s_\psi - s_\phi c_\psi & c_\phi s_\theta \\ s_\phi c_\theta c_\psi + c_\phi s_\psi & -s_\phi c_\theta s_\psi + c_\phi c_\psi & s_\phi s_\theta \\ -s_\theta c_\psi & s_\theta s_\psi & c_\theta \end{bmatrix}.\end{aligned} \tag{2.37}$$

Inverse problem. Let us find the expression of the Euler angles as functions of the direction cosines of the general matrix **R** of (2.21). Premultiplying Eq. (2.37) by $\mathbf{rot}(z, \phi)$, we obtain:

$$\mathbf{rot}(z,\ \phi)\mathbf{R} = \mathbf{rot}(y,\ \theta)\mathbf{rot}(z,\ \psi) \tag{2.38}$$

which results in

$$\begin{bmatrix} c_\phi s_x + s_\phi s_y & c_\phi n_x + s_\phi n_y & c_\phi a_x + s_\phi a_y \\ -s_\phi s_x + c_\phi s_y & -s_\phi n_x + c_\phi n_y & -s_\phi a_x + c_\phi a_y \\ s_z & n_z & a_z \end{bmatrix} = \begin{bmatrix} c_\theta c_\psi & -c_\theta s_\psi & s_\theta \\ s_\psi & c_\psi & 0 \\ -s_\theta c_\psi & s_\theta s_\psi & c_\theta \end{bmatrix}. \tag{2.39}$$

From the elements on the second raw, third column of (2.45), we obtain:

$$-s_\phi a_x + c_\phi a_y = 0 \tag{2.40}$$

thus:

$$\phi = \text{atan2}(a_y, a_x)$$
$$\phi' = \text{atan2}(-a_y, -a_x) = \phi + \pi. \tag{2.41}$$

There is a singularity if a_x and a_y are zero. In that case, $\theta = k\pi$ $(k = 0, 1)$.

In the same way, from the elements on the first and third rows, and third column of (2.39), and then from those of the second row, first and second columns, we deduce that:

$$\theta = \text{atan2}(\text{c}_\phi a_x + \text{s}_\phi a_y, a_z)$$
$$\psi = \text{atan2}(-\text{s}_\phi s_x + \text{c}_\phi s_y, -\text{s}_\phi n_x + \text{c}_\phi n_y). \tag{2.42}$$

The described Euler angles convention is denoted as the ZYZ convention, where ZYZ denotes that we have a first rotation around z_i, then a second rotation around $\boldsymbol{y}_j$ and finally a last rotation around $\boldsymbol{z}_{j'}$. There exists in total 12 different sequences of the three rotations, and, hence, there can be 12 Euler conventions: XYZ, XZY, YXZ, YZX, ZXY, ZYX, XYX, XZX, YXY, YZY, ZXZ, and ZYZ, where the convention PQR denotes that we have a first rotation around $\boldsymbol{p}_i$–axis, then a second rotation around $\boldsymbol{q}_j$–axis and finally a last rotation around $\boldsymbol{r}_{j'}$–axis.

2.4.4 Roll-Pitch-Yaw Angles

Following the convention shown in Fig. 2.6, the angles ϕ, θ and ψ indicate roll, pitch and yaw respectively. If we suppose that the direction of motion (by analogy to the

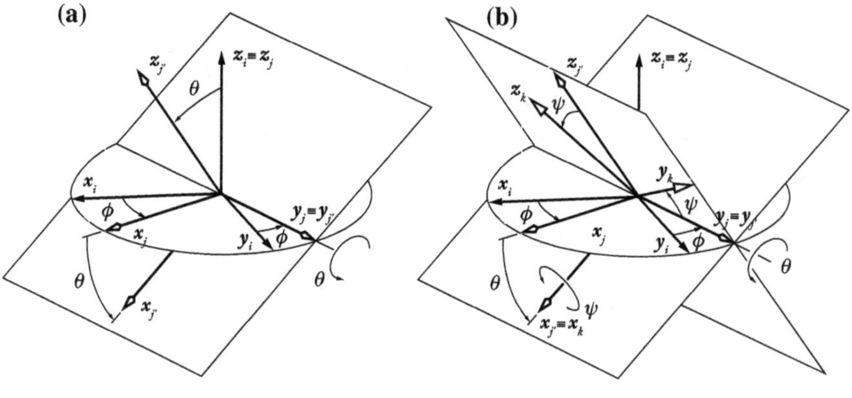

Fig. 2.6 Roll-Pitch-Yaw angles

direction along which a ship is sailing) is along the z_i axis, the transformation matrix can be written as:

$$\mathbf{R} = \mathbf{rot}(z,\ \phi)\mathbf{rot}(y,\ \theta)\mathbf{rot}(x,\ \psi)$$
$$= \begin{bmatrix} c_\phi c_\theta & c_\phi s_\theta s_\psi - s_\phi c_\psi & c_\phi s_\theta c_\psi + s_\phi s_\psi \\ s_\phi c_\theta & s_\phi s_\theta s_\psi + c_\phi c_\psi & s_\phi s_\theta c_\psi - c_\phi s_\psi \\ -s_\theta & c_\theta s_\psi & c_\theta c_\psi \end{bmatrix}. \tag{2.43}$$

This description is analogous to the ZYX Euler angle convention.

Inverse problem. Let us find the expression of the Roll-Pitch-Yaw angles as functions of the direction cosines of the general matrix $\mathbf{R}$. We use the same method discussed in the previous section. Premultiplying Eq. (2.43) by $\mathbf{rot}(z,\ \phi)$, we obtain:

$$\mathbf{rot}(z,\ \phi)\mathbf{R} = \mathbf{rot}(y,\ \theta)\mathbf{rot}(x,\ \psi) \tag{2.44}$$

which results in:

$$\begin{bmatrix} c_\phi s_x + s_\phi s_y & c_\phi n_x + s_\phi n_y & c_\phi a_x + s_\phi a_y \\ -s_\phi s_x + c_\phi s_y & -s_\phi n_x + c_\phi n_y & -s_\phi a_x + c_\phi a_y \\ s_z & n_z & a_z \end{bmatrix} = \begin{bmatrix} c_\theta & s_\theta s_\psi & s_\theta c_\psi \\ 0 & c_\psi & -s_\psi \\ -s_\theta & c_\theta s_\psi & c_\theta c_\psi \end{bmatrix}. \tag{2.45}$$

From the elements on the second raw, first column of (2.45), we obtain:

$$-s_\phi s_x + c_\phi s_y = 0 \tag{2.46}$$

thus:

$$\begin{aligned} \phi &= \text{atan2}(s_y,\ s_x) \\ \phi' &= \text{atan2}(-s_y,\ -s_x) = \phi + \pi. \end{aligned} \tag{2.47}$$

There is a singularity if s_x and s_y are zero. In that case, $\theta = \pm\pi/2$.

In the same way, from the elements on the first and third rows, and first column of (2.45), and then from those of the second row, second and third columns, we deduce that:

$$\begin{aligned} \theta &= \text{atan2}(-s_z,\ c_\phi s_x + s_\phi s_y) \\ \psi &= \text{atan2}(s_\phi a_x - c_\phi a_y,\ -s_\phi n_x + c_\phi n_y). \end{aligned} \tag{2.48}$$

2.4.5 Tilt-and-Torsion Angles

A novel three-angle orientation representation, later called the Tilt-and-Torsion (*T&T*) angles, was proposed in (Bonev and Ryu 1999). These angles were also independently introduced in (Huang et al. 1999), (Crawford et al. 1999) and (Wang 1999). They had been also proposed in (Korein 1984) under the name halfplane-deviation-twist angles. In (Bonev et al. 2002a), the advantages of the *T&T* angles in the study of spatial parallel mechanisms were further demonstrated. It was shown that there is a class of 3–*DOF* mechanisms that have always a zero torsion, that we now call zero-torsion parallel mechanisms. Furthermore, it was demonstrated in (Bonev and Gosselin 2005a) and (Bonev and Gosselin 2006) that the workspace and singularities of symmetric spherical parallel mechanisms are best analyzed using the *T&T* angles.

The *T&T* angles are defined in two stages: a tilt and a torsion. This does not, however, mean that only two angles define the *T&T* angles but simply that the axis of tilt is defined by another angle. In the first stage, illustrated in Fig. 2.7a, the body frame is tilted about a horizontal axis, $\boldsymbol{u}$, at an angle θ, referred to as the *tilt*. The axis $\boldsymbol{u}$ is defined by an angle ϕ, called the *azimuth*, which is the angle between the axes $\boldsymbol{u}$ and $\boldsymbol{y}_i$, $\boldsymbol{u}$ being at the intersection of the planes $(O_i, \boldsymbol{x}_i, \boldsymbol{y}_i)$ and $(O_i, \boldsymbol{x}_k, \boldsymbol{y}_k)$. In the second stage, illustrated in Fig. 2.7, the body frame is rotated about the body z_k axis at an angle σ, called the *torsion*.

For space limitations, we will omit the otherwise quite interesting details of the derivation process [see (Bovev et al. 2002a)], and write directly the resulting transformation matrix of the *T&T* angles, which is

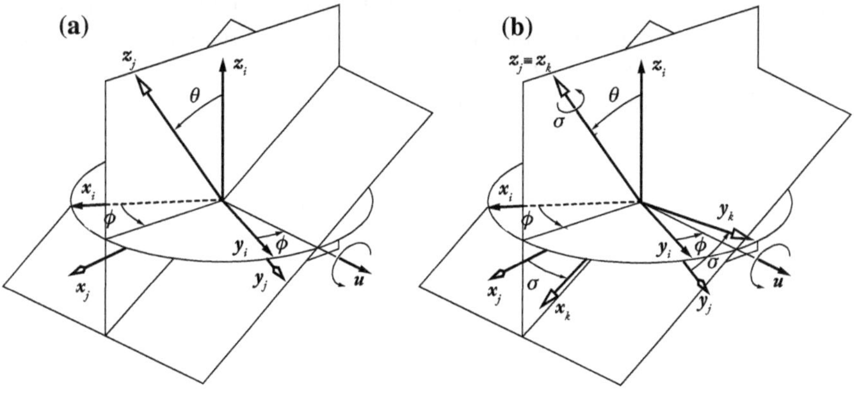

Fig. 2.7 The successive rotations of the *T&T* angles: **a** tilt, **b** torsion

$$\mathbf{R} = \mathbf{rot}(\boldsymbol{u}, \theta)\mathbf{rot}(z, \sigma)$$

$$= \begin{bmatrix} c_\phi c_\theta c_{\sigma-\phi} - s_\phi s_{\sigma-\phi} & -c_\phi c_\theta s_{\sigma-\phi} - s_\phi c_{\sigma-\phi} & c_\phi s_\theta \\ s_\phi c_\theta c_{\sigma-\phi} + c_\phi s_{\sigma-\phi} & -s_\phi c_\theta s_{\sigma-\phi} + c_\phi c_{\sigma-\phi} & s_\phi s_\theta \\ -s_\theta c_{\sigma-\phi} & s_\theta s_{\sigma-\phi} & c_\theta \end{bmatrix} \tag{2.49}$$

where $\mathbf{rot}(\boldsymbol{u}, \theta) = \mathbf{rot}(z, \phi)\mathbf{rot}(y, \theta)\mathbf{rot}(z, -\phi)$.

From the above, we see that the *T&T* angles (ϕ, θ, σ) are equivalent to the ZYZ Euler angles $(\phi, \theta, \sigma - \phi)$, i.e., the *spin* angle ψ has been replaced with $\sigma - \phi$.

Inverse problem. From the previous consideration, the inverse problem of the *T&T* angles can be solved as shown in Sect. 2.4.3, from which we find that:

$$\begin{aligned} \phi &= \text{atan2}(a_y, a_x) \text{ or } \phi = \text{atan2}(-a_y, -a_x) \\ \theta &= \text{atan2}(c_\phi a_x + s_\phi a_y, a_z) \\ \sigma &= \text{atan2}(-s_\phi s_x + c_\phi s_y, -s_\phi n_x + c_\phi n_y) + \phi. \end{aligned} \tag{2.50}$$

There is a singularity if $\theta = 0 + k\pi$ $(k = 0, 1)$.

One of the properties of three-angle orientation representation is that a given orientation can be represented by at least two triplets of angles. In our case, the triplets $\{\phi, \theta, \sigma\}$ and $\{\phi \pm \pi, -\theta, \sigma\}$ are equivalent. To avoid this and the representational singularity at $\theta = \pi$ (which is hardly achieved by any parallel mechanism), we set the ranges of the *azimuth*, *tilt*, and *torsion* as, respectively, $\phi \in (-\pi, \pi], \theta \in [0, \pi)$, and $\sigma \in (-\pi, \pi]$. Then, probably the most valuable property of the *T&T* angles is that for the ranges just defined, the angles (θ, ϕ, σ) can be represented in a cylindrical coordinate system (r, ϕ, h) through a one-to-one mapping. In other words, any orientation (except $\theta = \pi$) corresponds to a unique point within a cylinder in the cylindrical coordinate system, and vice versa. The reason is that the *T&T* representational singularity at $\theta = 0$ is of the same nature as the singularity of the cylindrical coordinate system occurring at zero-radius $(r = 0)$.

Chapter 3
Representation of Velocities and Forces/Acceleration of a Body

Abstract In this Chapter, we will use the concept of screw to describe the velocity of a body in space, as well as the forces acting on it. The definition of twists and wrenches is introduced. We also give the transformation matrices between two twists or wrenches. Finally, we define also some relations for computing the accelerations at any point of a given body.

3.1 Definition of a Screw

A vector field, $\mathbf{h}$, on $\mathbb{R}^3$ is a screw if there exists a point O_i and a vector $\boldsymbol{\Omega}$ such that for all points O_j in $\mathbb{R}^3$:

$$\mathbf{h}_j = \mathbf{h}_i + \boldsymbol{\Omega} \times \mathbf{r}_{O_i O_j} \tag{3.1}$$

where $\mathbf{h}_j$ is the vector of $\mathbf{h}$ at O_j and the symbol "$\times$" indicates the vector product. The vector $\mathbf{h}_j$ is called the moment at O_j, whereas $\boldsymbol{\Omega}$ is called the resultant of the screw.

Thus, the screw at a point O_j is well defined by the vectors $\mathbf{h}_j$ and $\boldsymbol{\Omega}$, which can be stacked in a single (6×1) vector.

3.2 Kinematic Screw (or Twist)

Since the set of velocity vectors at all the points of a body defines a screw field, the screw at a point O_i can be defined by:

- $\mathbf{v}_i$ representing the linear velocity at O_i with respect to the fixed frame $\mathscr{F}_0$, such that $\mathbf{v}_i = \dot{\mathbf{r}}_{O_0 O_j} = \left.\frac{\mathrm{d}}{\mathrm{d}t}\right|_{\mathscr{F}_0} \mathbf{r}_{O_0 O_j}$;
- $\boldsymbol{\omega}_i$ representing the angular velocity of the body with respect to the frame $\mathscr{F}_0$. It constitutes the resultant vector of the screw of the velocity vector field.

S. Briot and W. Khalil, *Dynamics of Parallel Robots*,
Mechanisms and Machine Science 35, DOI 10.1007/978-3-319-19788-3_3

Thus, the velocity $\mathbf{v}_j$ of a point O_j is calculated in terms of the velocity of the point O_i by the following equation:

$$\mathbf{v}_j = \mathbf{v}_i + \boldsymbol{\omega}_i \times \mathbf{r}_{O_i O_j}. \tag{3.2}$$

The components of $\mathbf{v}_i$ and $\boldsymbol{\omega}_i$ can be concatenated to form the kinematic screw vector $\mathbf{t}_i$, i.e.:

$$\mathbf{t}_i = \left[\mathbf{v}_i^T \ \boldsymbol{\omega}_i^T\right]^T. \tag{3.3}$$

The kinematic screw is also called the twist.

3.3 Representation of Forces and Moments (wrench)

A collection of forces and moments acting on a body can be reduced to a wrench $\mathbf{w}_i$ at point O_i, which is composed of a force $\mathbf{f}_i$ at O_i and a moment $\mathbf{m}_i$ about O_i. The moment at O_j, denoted as $\mathbf{m}_j$, can be obtained as:

$$\mathbf{m}_j = \mathbf{m}_i + \mathbf{f}_i \times \mathbf{r}_{O_i O_j}. \tag{3.4}$$

The components of $\mathbf{f}_i$ and $\mathbf{m}_i$ can be concatenated to form the wrench vector $\mathbf{w}_i$, i.e.:

$$\mathbf{w}_i = \left[\mathbf{f}_i^T \ \mathbf{m}_i^T\right]^T \tag{3.5}$$

Note that, contrary to the twist representation for which the moment of the screw (the translational velocity $\mathbf{v}_i$) is located above the resultant of the screw (the rotational velocity $\boldsymbol{\omega}_i$), we prefer to invert in (3.5) the order for the wrench representation so that the resultant $\mathbf{f}_i$ is located above the moment $\mathbf{m}_i$.

3.4 Condition of Reciprocity

A twist $\mathbf{t}^T = [\mathbf{v}^T \ \boldsymbol{\omega}^T]$ is said to be reciprocal to a wrench $\mathbf{w}^T = [\mathbf{f}^T \ \mathbf{m}^T]$ if

$$\mathbf{t}^T \ \mathbf{w} = \mathbf{v}^T \ \mathbf{f} + \boldsymbol{\omega}^T \ \mathbf{m} = 0. \tag{3.6}$$

In other words, the power induced by the wrench $\mathbf{w}$ for moving the body with a twist $\mathbf{t}$ is null.

3.5 Transformation Matrix Between Twists

Let ${}^i\mathbf{v}_i$ and ${}^i\boldsymbol{\omega}_i$ be the vectors representing a screw in O_i, origin of frame $\mathscr{F}_i$, expressed in frame $\mathscr{F}_i$. To calculate ${}^j\mathbf{v}_j$ and ${}^j\boldsymbol{\omega}_j$ representing the kinematic screw in O_j expressed in frame $\mathscr{F}_j$, we first note that:

$$\boldsymbol{\omega}_i = \boldsymbol{\omega}_j \tag{3.7}$$

$$\mathbf{v}_j = \mathbf{v}_i + \boldsymbol{\omega}_i \times \mathbf{r}_{O_iO_j}. \tag{3.8}$$

Equations (3.7) and (3.8) can be rewritten as:

$$\begin{bmatrix} \mathbf{v}_j \\ \boldsymbol{\omega}_j \end{bmatrix} = \begin{bmatrix} \mathbf{1}_3 & -\hat{\mathbf{r}}_{O_iO_j} \\ \mathbf{0}_3 & \mathbf{1}_3 \end{bmatrix} \begin{bmatrix} \mathbf{v}_i \\ \boldsymbol{\omega}_i \end{bmatrix} \tag{3.9}$$

where $\mathbf{1}_3$ and $\mathbf{0}_3$ represent the (3×3) identity matrix and zero matrix respectively and $\hat{\mathbf{r}}_{O_iO_j}$ is the cross-product matrix associated with the vector $\mathbf{r}_{O_iO_j} = \begin{bmatrix} r_x \, r_y \, r_z \end{bmatrix}^T$, i.e.

$$\hat{\mathbf{r}}_{O_iO_j} = \begin{bmatrix} 0 & -r_z & r_y \\ r_z & 0 & -r_x \\ -r_y & r_x & 0 \end{bmatrix}. \tag{3.10}$$

Projecting relation (3.9) in frame $\mathscr{F}_i$, we obtain:

$$\begin{bmatrix} {}^i\mathbf{v}_j \\ {}^i\boldsymbol{\omega}_j \end{bmatrix} = \begin{bmatrix} \mathbf{1}_3 & -{}^i\hat{\mathbf{r}}_j \\ \mathbf{0}_3 & \mathbf{1}_3 \end{bmatrix} \begin{bmatrix} {}^i\mathbf{v}_i \\ {}^i\boldsymbol{\omega}_i \end{bmatrix} \tag{3.11}$$

where ${}^i\hat{\mathbf{r}}_j = {}^i\hat{\mathbf{r}}_{O_iO_j}$

Since ${}^j\mathbf{v}_j = {}^j\mathbf{R}_i{}^i\mathbf{v}_j$ and ${}^j\boldsymbol{\omega}_j = {}^j\mathbf{R}_i{}^i\boldsymbol{\omega}_j$, Eq. (3.11) gives:

$$\begin{bmatrix} {}^j\mathbf{v}_j \\ {}^j\boldsymbol{\omega}_j \end{bmatrix} = {}^j\overline{\mathbf{T}}_i \begin{bmatrix} {}^i\mathbf{v}_i \\ {}^i\boldsymbol{\omega}_i \end{bmatrix} \tag{3.12}$$

where ${}^j\overline{\mathbf{T}}_i$ is the (6×6) transformation matrix between screws:

$${}^j\overline{\mathbf{T}}_i = \begin{bmatrix} {}^j\mathbf{R}_i & -{}^j\mathbf{R}_i{}^i\hat{\mathbf{r}}_j \\ \mathbf{0}_3 & {}^j\mathbf{R}_i \end{bmatrix}. \tag{3.13}$$

The transformation matrices between screws have the following properties:

Property 1 *Product:*

$${}^0\overline{\mathbf{T}}_j = \prod_{k=1}^{j} {}^{k-1}\overline{\mathbf{T}}_k = {}^1\overline{\mathbf{T}}_2{}^2\overline{\mathbf{T}}_3 \cdots {}^{j-1}\overline{\mathbf{T}}_j. \tag{3.14}$$

Property 2 *Inverse:*

$$ {}^{j}\overline{\mathbf{T}}_{i}^{-1} = \begin{bmatrix} {}^{i}\mathbf{R}_{j} & {}^{i}\hat{\mathbf{r}}_{j}{}^{i}\mathbf{R}_{j} \\ \mathbf{0}_{3} & {}^{i}\mathbf{R}_{j} \end{bmatrix} = {}^{i}\overline{\mathbf{T}}_{j}. \tag{3.15} $$

3.6 Transformation Matrix Between Wrenches

Similarly to what is written in the previous section, we have

$$ \begin{bmatrix} {}^{j}\mathbf{m}_{j} \\ {}^{j}\mathbf{f}_{j} \end{bmatrix} = \begin{bmatrix} {}^{j}\mathbf{R}_{i} & -{}^{j}\mathbf{R}_{i}{}^{i}\hat{\mathbf{r}}_{j} \\ \mathbf{0}_{3} & {}^{j}\mathbf{R}_{i} \end{bmatrix} \begin{bmatrix} {}^{i}\mathbf{m}_{i} \\ {}^{i}\mathbf{f}_{i} \end{bmatrix} = {}^{j}\overline{\mathbf{T}}_{i} \begin{bmatrix} {}^{i}\mathbf{m}_{i} \\ {}^{i}\mathbf{f}_{i} \end{bmatrix} \tag{3.16} $$

or also

$$ \begin{bmatrix} {}^{j}\mathbf{f}_{j} \\ {}^{j}\mathbf{m}_{j} \end{bmatrix} = \begin{bmatrix} {}^{j}\mathbf{R}_{i} & \mathbf{0}_{3} \\ -{}^{j}\mathbf{R}_{i}{}^{i}\hat{\mathbf{r}}_{j} & {}^{j}\mathbf{R}_{i} \end{bmatrix} \begin{bmatrix} {}^{i}\mathbf{f}_{i} \\ {}^{i}\mathbf{m}_{i} \end{bmatrix} \tag{3.17} $$

which, from (3.15), can be rewritten as

$$ {}^{j}\mathbf{w}_{j} = {}^{i}\overline{\mathbf{T}}_{j}^{T}\,{}^{i}\mathbf{w}_{i}. \tag{3.18} $$

3.7 Acceleration of a Body

From the differentiation with respect to time of the Eq. (3.2), we can find that

$$ \dot{\mathbf{v}}_{j} = \dot{\mathbf{v}}_{i} + \dot{\boldsymbol{\omega}}_{i} \times \mathbf{r}_{O_iO_j} + \boldsymbol{\omega}_{i} \times (\boldsymbol{\omega}_{i} \times \mathbf{r}_{O_iO_j}) \tag{3.19} $$

where $\dot{\mathbf{v}}_i$ is the linear acceleration of the point O_i, $\dot{\mathbf{v}}_j$ is the linear acceleration of the point O_j and $\dot{\boldsymbol{\omega}}_i$ is the angular acceleration of the body.

Equation (3.19) can be put under the matrix form:

$$ \begin{bmatrix} \dot{\mathbf{v}}_{j} \\ \dot{\boldsymbol{\omega}}_{j} \end{bmatrix} = \begin{bmatrix} \mathbf{1}_{3} & -\hat{\mathbf{r}}_{O_iO_j} \\ \mathbf{0}_{3} & \mathbf{1}_{3} \end{bmatrix} \begin{bmatrix} \dot{\mathbf{v}}_{i} \\ \dot{\boldsymbol{\omega}}_{i} \end{bmatrix} + \begin{bmatrix} \boldsymbol{\omega}_{i} \times (\boldsymbol{\omega}_{i} \times \mathbf{r}_{O_iO_j}) \\ \mathbf{0} \end{bmatrix}. \tag{3.20} $$

Projecting this relation in frame $\mathscr{F}_i$, we obtain:

$$ \begin{bmatrix} {}^{i}\dot{\mathbf{v}}_{j} \\ {}^{i}\dot{\boldsymbol{\omega}}_{j} \end{bmatrix} = \begin{bmatrix} \mathbf{1}_{3} & -{}^{i}\hat{\mathbf{r}}_{j} \\ \mathbf{0}_{3} & \mathbf{1}_{3} \end{bmatrix} \begin{bmatrix} {}^{i}\dot{\mathbf{v}}_{i} \\ {}^{i}\dot{\boldsymbol{\omega}}_{i} \end{bmatrix} + \begin{bmatrix} {}^{i}\boldsymbol{\omega}_{i} \times ({}^{i}\hat{\mathbf{r}}_{j}^{T}\,{}^{i}\boldsymbol{\omega}_{i}) \\ \mathbf{0} \end{bmatrix}. \tag{3.21} $$

Since ${}^j\dot{\mathbf{v}}_j = {}^j\mathbf{R}_i{}^i\dot{\mathbf{v}}_j$ and ${}^j\dot{\boldsymbol{\omega}}_j = {}^j\mathbf{R}_i{}^i\dot{\boldsymbol{\omega}}_j$, Eq. (3.21) gives:

$$
{}^j\dot{\mathbf{t}}_j = {}^j\overline{\mathbf{T}}_i{}^i\dot{\mathbf{t}}_i + {}^j\mathbf{b}_j \tag{3.22}
$$

where ${}^j\overline{\mathbf{T}}_i$ is defined at Eq. (3.13),

$$
{}^j\dot{\mathbf{t}}_j = \begin{bmatrix} {}^j\dot{\mathbf{v}}_j \\ {}^j\dot{\boldsymbol{\omega}}_j \end{bmatrix}, \quad {}^i\dot{\mathbf{t}}_i = \begin{bmatrix} {}^i\dot{\mathbf{v}}_i \\ {}^i\dot{\boldsymbol{\omega}}_i \end{bmatrix}, \tag{3.23}
$$

are acceleration vectors and

$$
{}^j\mathbf{b}_j = {}^j\overline{\mathbf{R}}_i \begin{bmatrix} {}^i\boldsymbol{\omega}_i \times ({}^i\hat{\mathbf{r}}_j^T{}^i\boldsymbol{\omega}_i) \\ \mathbf{0} \end{bmatrix} \tag{3.24}
$$

with ${}^j\overline{\mathbf{R}}_i$ the augmented rotation matrix between frames $\mathscr{F}_i$ and $\mathscr{F}_j$, i.e.

$$
{}^j\overline{\mathbf{R}}_i = \begin{bmatrix} {}^j\mathbf{R}_i & \mathbf{0}_3 \\ \mathbf{0}_3 & {}^j\mathbf{R}_i \end{bmatrix}. \tag{3.25}
$$

Chapter 4
Kinematic Description of Multibody Systems

Abstract The design and control of a robot requires the computation of some mathematical models such as the transformation models between the joint space (in which the configuration of the robot is defined) and the task space (in which the location of the end-effector is specified). These transformation models are very important since robots are controlled in the joint space, whereas tasks are defined in the task space. The modeling of robots in a systematic and automatic way requires an adequate method for the description of their structure. Several methods and notations have been proposed. The most popular among these is the Denavit-Hartenberg method. This method is developed for serial structures and presents ambiguities when applied to robots with closed or tree chains. For this reason, we will use the notation of Khalil and Kleinfinger, which gives a unified description for all mechanical articulated systems with a minimum number of parameters.

4.1 Kinematic Pairs and Joint Variables

In the approach adopted in this book, we consider joints with only one degree of freedom between the bodies, such as revolute (*R*) joints or prismatic (*P*) joints (see Sect. 1.2).

The *R* joint is parameterized by a variable denoted as θ which represents the angle of the rotation about the *R* joint axis. The *P* joint is parameterized by a variable denoted as r which represents the distance of translation along the *P* joint direction.

Using these two types of joints, it is possible to build more complex multi-*DOF* joints as long as their axes are properly arranged. For example, a spherical (*S*) joint having three rotational *DOF* may be composed of three *R* joints with linearly independent intersecting axes. However we define as a joint the connection between two and only two bodies. The *S* joint will therefore involve two virtual massless bodies that will be treated in the calculations as any physical body. However, the reader must be warned that parameterizing a *S* joint through the use of three *R* joints is equivalent to parameterizing the displacement of the considered body with three Euler angles. This parameterization is not free of representation singularities (see Sect. 2.4.3).

S. Briot and W. Khalil, *Dynamics of Parallel Robots*,
Mechanisms and Machine Science 35, DOI 10.1007/978-3-319-19788-3_4

To generalize the joint coordinates for both types of connections, we define the generalized joint coordinate q_i of the body $\mathscr{B}_i$ associated with the joint i by

$$q_i = \bar{\sigma}_i \theta_i + \sigma_i r_i \tag{4.1}$$

where

- $\sigma_i = 0$ if joint i is a R joint;
- $\sigma_i = 1$ if joint i is a P joint;
- $\bar{\sigma}_i = 1 - \sigma_i$.

To characterize the case that a body $\mathscr{B}_i$ can be rigidly attached to a body $\mathscr{B}_j$: $\sigma_i = 2$. In that case, $\dot{q}_i$ is set to 0 and $\bar{\sigma}_i$ is not defined.

4.2 Modified Denavit-Hartenberg Parameters

The modeling of robots in a systematic and automatic way requires an adequate method for the geometric description of their structure. Several methods and notations have been proposed (Craig 1986; Denavit and Hartenberg 1955; Sheth and Uicker 1971). The most popular among these is the Denavit-Hartenberg method (Denavit and Hartenberg 1955). This method is developed for serial structures and presents ambiguities when applied to robots with closed or tree chains.

The modified Denavit-Hartenberg (*MDH*) geometric parameters proposed in (Khalil and Kleinfinger 1986) allows the definition of a systematic parameterizing of the relative location of any body of the considered system. We consider here directly the general case, i.e. an open tree-structure kinematic chain as shown in Fig. 4.1. We will see how a structure with closed loops can be reduced to the study of

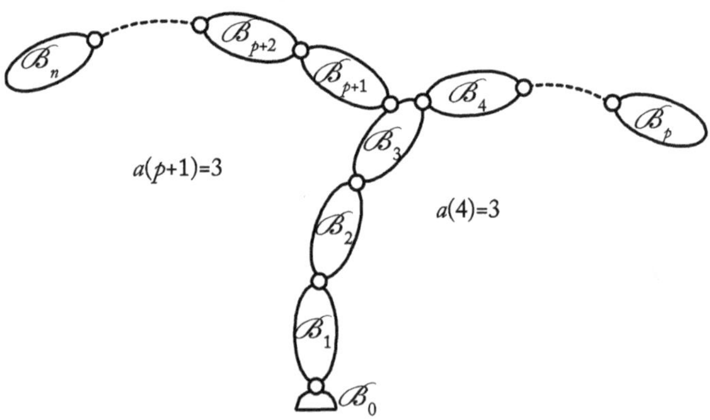

Fig. 4.1 Symbolic representation of a tree-structure open kinematic chain

an open tree-structure and we will present the additional settings required to describe structures including closed loops.

4.2.1 Parameterizing Tree-Structure Open Kinematic Chains

Let us consider a tree-structure kinematic chain consisting of $n + 1$ (physical or virtual) bodies (the body i is denoted as $\mathscr{B}_i$). That kinematic chain is composed of n joints associated with the bodies $\mathscr{B}_i$ ($i = 1, \ldots, n$) as shown in Fig. 4.1. We number each body i increasingly from the base denoted as the body $\mathscr{B}_0$. For an industrial robot fixed on the ground, the body $\mathscr{B}_0$ would be the body fixed on the ground.

We define a relationship for the antecedence between the bodies. If the body $\mathscr{B}_i$ is antecedent to body $\mathscr{B}_j$, then

$$i = a(j). \tag{4.2}$$

The numbering rules then ensure that $i < j$. By definition, in a tree-structure kinematic chain, a body can at most have one antecedent body (none in the case of the base) but several succeeding bodies. If a body has no succeeding body, this is one of the terminal bodies of the chain. A system where all the bodies have only one succeeding body is called a simple open chain. The values of $a(j)$ fully define the topology of the structure which can be seen as a graph (see Fig. 4.1).

To each body $\mathscr{B}_j$ is fixed a local $\mathscr{F}_j$ with origin O_j. The *MDH* parameterization allows to build each frame $\mathscr{F}_j$ in a straightforward way by applying the following rules for $i = 1, \ldots, n$:

- The frame $\mathscr{F}_i = (O_i, \boldsymbol{x}_i, \boldsymbol{y}_i, \boldsymbol{z}_i)$ is fixed with respect to the body $\mathscr{B}_i$;
- The axis $\boldsymbol{z}_i$ is along the axis of the joint i;
- The axis $\boldsymbol{x}_i$ is taken along the common normal between $\boldsymbol{z}_i$ and one of the succeeding joint axes, which are fixed on link $\mathscr{B}_i$. The following cases are then possible:
  - The body $\mathscr{B}_i$ is a terminal body and has no succeeding body; $\boldsymbol{x}_i$ can be set arbitrarily (as long as it remains orthogonal to $\boldsymbol{z}_i$);
  - The body $\mathscr{B}_i$, $i = a(j)$, has only one succeeding body; $\boldsymbol{x}_i$ must be then along the common perpendicular to $\boldsymbol{z}_i$ and $\boldsymbol{z}_j$;
  - The body $\mathscr{B}_i$ has several succeeding bodies; one of the succeeding bodies must then be chosen to build the axis $\boldsymbol{x}_i$. Practically, the succeeding body on which $\boldsymbol{x}_i$ is defined can be selected as the one on the path leading to the main terminal link, but this is not an obligation;
- The axis $\boldsymbol{y}_i$ is taken by the right-hand rule such that $\{\boldsymbol{x}_i, \boldsymbol{y}_i, \boldsymbol{z}_i\}$ is an orthonormal basis.

In the case where, for the body $\mathscr{B}_j$ succeeding the body $\mathscr{B}_i$ ($i = a(j)$), $\boldsymbol{x}_i$ is not orthogonal to $\boldsymbol{z}_j$, we build an additional vector $\boldsymbol{u}_j$ along the common orthogonal to $\boldsymbol{z}_i$ and $\boldsymbol{z}_j$. Note that $\boldsymbol{u}_j$ is fixed on the body $\mathscr{B}_i$.

Practically, the axes $\boldsymbol{z}_j$ of the whole bodies are first defined. Then we set the direction of the axes $\boldsymbol{x}_j$ (and $\boldsymbol{u}_k$ if needed). With the systematic definition of each body frame, it is possible to define a set of 6 parameters for each frame $\mathcal{F}_j$ which are denoted as the *MDH* parameter, that are for $i = a(j)$:

- γ_j: angle between $\boldsymbol{x}_i$ and $\boldsymbol{u}_j$ around z_i;
- b_j: distance between $\boldsymbol{x}_i$ and $\boldsymbol{u}_j$ along z_i;
- α_j: angle between z_i and z_j around $\boldsymbol{u}_j$;
- d_j: distance between z_i and z_j along $\boldsymbol{u}_j$;
- θ_j: angle between $\boldsymbol{u}_j$ and $\boldsymbol{x}_j$ around z_j;
- r_j: distance between $\boldsymbol{u}_j$ and $\boldsymbol{x}_j$ along z_j.

Those six parameters are necessary when the axis $\boldsymbol{x}_i$ of the antecedent body $\mathcal{B}_i$ is not perpendicular to z_j and when an additional vector $\boldsymbol{u}_j$ has been built. For other cases, only four parameters are necessary: α_j, d_j, θ_j and r_j. Indeed, in the last case, the vector $\boldsymbol{u}_j$ has no role since it is aligned to $\boldsymbol{x}_i$ and the *MDH* parameters becomes:

- α_j: angle between z_i and z_j around $\boldsymbol{x}_i$;
- d_j: distance between z_i and z_j along $\boldsymbol{x}_i$;
- θ_j: angle between $\boldsymbol{x}_i$ and $\boldsymbol{x}_j$ around z_j;
- r_j: distance between $\boldsymbol{x}_i$ and $\boldsymbol{x}_j$ along z_j;

with $\gamma_j = 0$ and $b_j = 0$.

Figures 4.2 and 4.3 illustrate that way of parameterizing.

In order to describe the system in a practical way, the parameters are given in a table (see Table 4.1).

Using those parameters, the homogeneous transformation matrix allowing us to transform the frame $\mathcal{F}_i$ fixed to the body $\mathcal{B}_i$ into the frame $\mathcal{F}_j$ fixed to the body $\mathcal{B}_j$ can be written as:

$$\begin{aligned}
&{}^i\mathbf{T}_j = \\
&\mathbf{Rot}(z,\gamma_j)\mathbf{Trans}(z,b_j)\mathbf{Rot}(x,\alpha_j)\mathbf{Trans}(x,d_j)\mathbf{Rot}(z,\theta_j)\mathbf{Trans}(z,r_j) \\
&= \begin{bmatrix} c_{\gamma_j}c_{\theta_j} - s_{\gamma_j}c_{\alpha_j}s_{\theta_j} & -c_{\gamma_j}s_{\theta_j} - s_{\gamma_j}c_{\alpha_j}c_{\theta_j} & s_{\gamma_j}s_{\alpha_j} & d_jc_{\gamma_j} + r_js_{\gamma_j}s_{\alpha_j} \\ s_{\gamma_j}c_{\theta_j} + c_{\gamma_j}c_{\alpha_j}s_{\theta_j} & -s_{\gamma_j}s_{\theta_j} + c_{\gamma_j}c_{\alpha_j}c_{\theta_j} & -c_{\gamma_j}s_{\alpha_j} & d_js_{\gamma_j} - r_jc_{\gamma_j}s_{\alpha_j} \\ s_{\alpha_j}s_{\theta_j} & s_{\alpha_j}c_{\theta_j} & c_{\alpha_j} & r_jc_{\alpha_j} + b_j \\ 0 & 0 & 0 & 1 \end{bmatrix} \\
&= \begin{bmatrix} & d_jc_{\gamma_j} + r_js_{\gamma_j}s_{\alpha_j} \\ {}^i\mathbf{R}_j & d_js_{\gamma_j} - r_jc_{\gamma_j}s_{\alpha_j} \\ & r_jc_{\alpha_j} + b_j \\ 0\;0\;0 & 1 \end{bmatrix}.
\end{aligned} \tag{4.3}$$

The inverse of ${}^i\mathbf{T}_j$ can be thus written as

$${}^i\mathbf{T}_j^{-1} = {}^j\mathbf{T}_i = \begin{bmatrix} & -b_js_{\alpha_j}s_{\theta_j} - d_jc_{\theta_j} \\ {}^i\mathbf{R}_j^T & -b_js_{\alpha_j}c_{\theta_j} + d_js_{\theta_j} \\ & -b_jc_{\alpha_j} - r_j \\ 0\;0\;0 & 1 \end{bmatrix}. \tag{4.4}$$

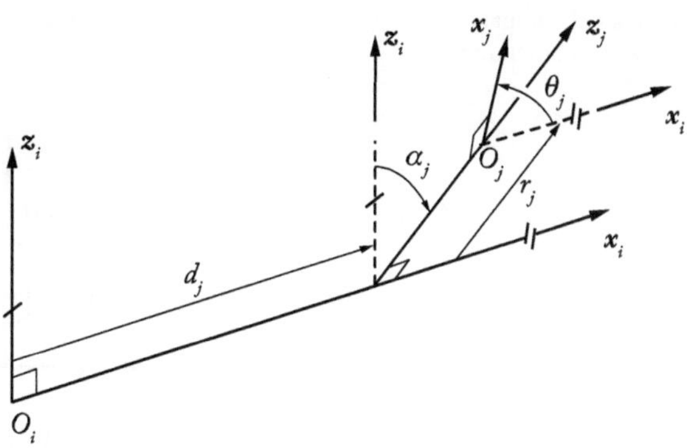

Fig. 4.2 *MDH* parameters in the case of a simple serial chain

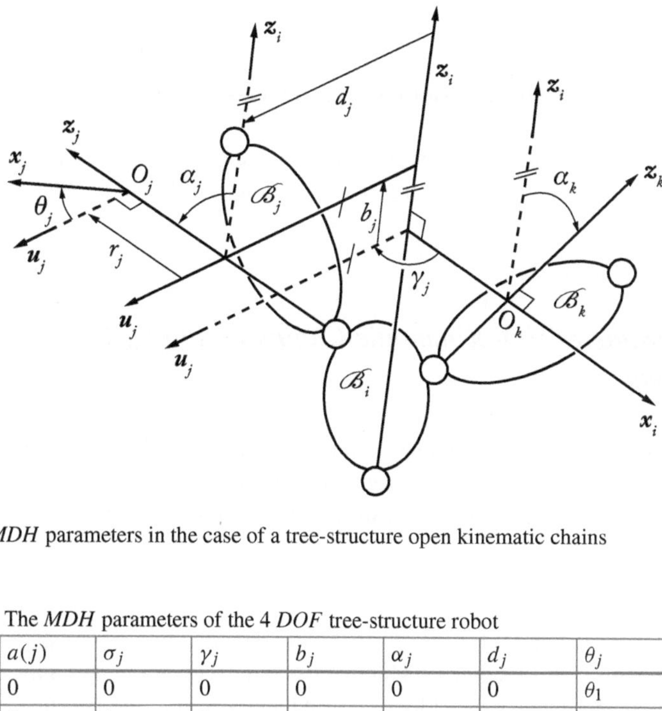

Fig. 4.3 *MDH* parameters in the case of a tree-structure open kinematic chains

Table 4.1 The *MDH* parameters of the 4 *DOF* tree-structure robot

j	$a(j)$	σ_j	γ_j	b_j	α_j	d_j	θ_j	r_j
1	0	0	0	0	0	0	θ_1	0
2	1	1	0	0	$-\pi/2$	0	0	r_2
3	2	1	0	0	$\pi/2$	0	0	r_3
4	1	0	$-\pi/2$	0	0	d_4	θ_4	r_4

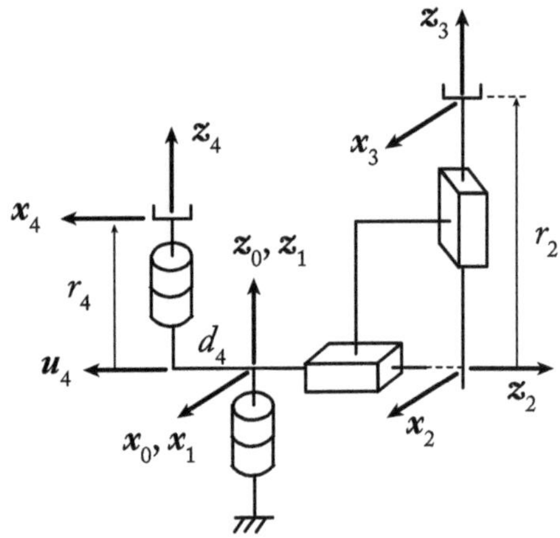

Fig. 4.4 Example of 4 *DOF* tree-structure robot

4.2.1.1 Example of *MDH* parameters for a tree-structure

Let us consider the 4 *DOF* tree-structure robot depicted in Fig. 4.4. Straightforwardly following the rules given in the present section, its *MDH* parameters are given in Table 4.1.

4.2.2 Parameterizing Kinematic Chains Including Closed Loops

As previously, the kinematic chain is composed of $n + 1$ (physical or virtual) bodies. However, now, the joint number m is greater than n. This characterizes the presence of kinematic loops in the considered structure. The number of closed loops is equal to:

$$B = m - n \tag{4.5}$$

Indeed, one single joint is only able to link two (and only two) consecutive bodies and this relation indicates that the kinematic chain necessary makes a loop, as it has more joints than bodies. Closed loops consequently lead to the fact that some joint variables depend on another due to the geometric constraints involved by the kinematic coupling. Let us denote as n_a the number of independent joint variables (that are also the active joint variables in the case of *PKM* without actuation redundancy). The remaining $m - n_a$ variables (that are also the passive joint variables in the case of *PKM* without actuation redundancy) can be obtained by solving the loop-closure equations, i.e. the equations corresponding to the kinematic constraints. We will see in Chap. 7 the ways to solve those equations.

To characterize the active and passive variables, an additional parameter is defined for each joint j of the analyzed system:

- $\mu_j = 1$ if the joint j is active;
- $\mu_j = 0$ if the joint j is passive.

Then, the study of systems including closed loops is assimilated to the study of tree-structure chain by virtually cutting each of the B loops at one joint (passive, if possible). In order to solve the loop-closure equations, it is necessary to introduce additional frames and to adapt the previous parameterization of the system as follows:

1. *Constructing an equivalent tree structure:* construct a tree structure having n joints by virtually cutting each closed chain at one of its joints (passive, if possible). Since a closed loop contains several passive joints, select the joint to be cut in such a way that the difference between the number of links of the two branches from the root of the loop to the links connected to the cut joint is as small as possible. This choice reduces the computational complexity of the dynamic model (Khalil and Kleinfinger 1986). The geometric parameters of the equivalent tree structure are determined as described in the previous section;
2. *Definition of frames on one of the links of the cut joints:* number the cut joints from $n+1$ to m. For each cut joint k, assign a frame $\mathscr{F}_k$ fixed on one of the links connected to this joint, for instance link j. The z_k axis is taken along the axis of joint k, and the $\boldsymbol{x}_k$ axis is taken along the common normal between z_k and z_j (Fig. 4.5). Let $i = a(k)$ where link i denotes the other link of joint k. The transformation matrix from frame $\mathscr{F}_i$ to frame $\mathscr{F}_k$ can be obtained as a function of the usual six (or four) geometric parameters $\gamma_k, b_k, \alpha_k, d_k, \theta_k$ and r_k, where q_k is either r_k or θ_k as usual;
3. *Definition of the cut joint frames with respect to the other link:* since frame $\mathscr{F}_k$ is fixed on link j, the transformation matrix from frame $\mathscr{F}_j$ and to frame $\mathscr{F}_k$ is constant. To avoid any confusion, this transformation will be denoted by ${}^j\mathbf{T}_{k+B}$, with B is the number of loops and $j = a(k+B)$. The frame $\mathscr{F}_{k+B}$ is aligned with frame $\mathscr{F}_k$, thus $z_{k+B} = z_k$ along the cut joint, and $\boldsymbol{x}_{k+B} = \boldsymbol{x}_k$, thus only 4 parameters non zero (at most) are needed to define ${}^j\mathbf{T}_{k+B}$. In fact γ_{k+B} and b_{k+B}

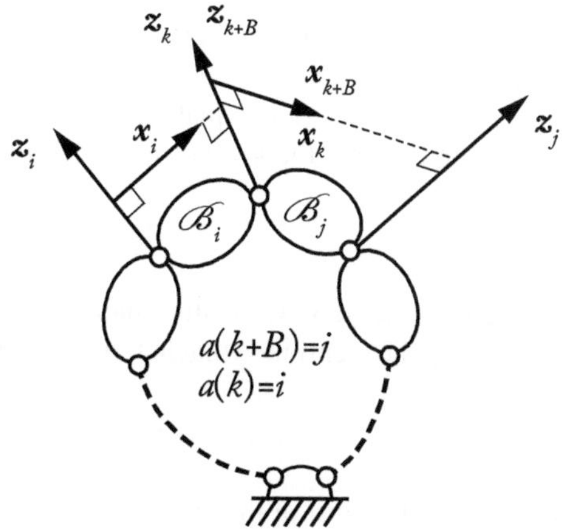

Fig. 4.5 Frames of a cut joint

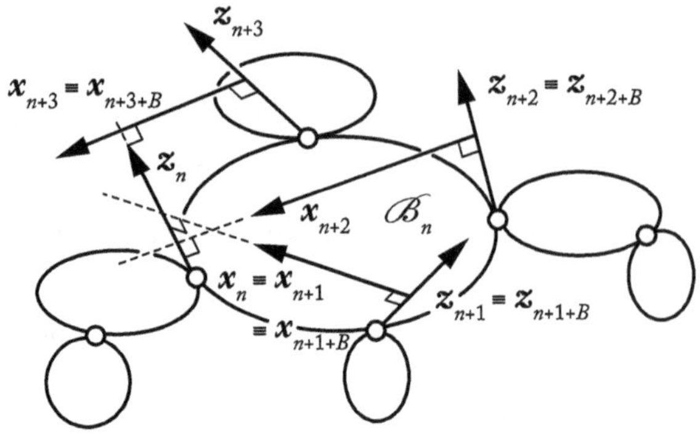

Fig. 4.6 Frames of a cut joint

will permit us to align $\boldsymbol{x}_j$ with $\boldsymbol{x}_{k+B}$ then α_{k+B} and d_{k+B} will align z_j with z_{k+B}, both θ_{k+B} and r_{k+B} are zero. The case where $\boldsymbol{x}_{k+B}$ is along $\boldsymbol{x}_j$ will lead to have also both γ_{k+B} and b_{k+B} equal to zero. Figure 4.6 shows a case where three cut joints take place on the terminal link n. Then $a(n+1+B) = a(n+2+B) = a(n+3+B) = n$.

In summary, the geometric description of a structure with closed loops is defined by an equivalent tree structure that is obtained by cutting each closed loop at one of its passive joints and by adding two aligned frames, but with different antecedent, at each cut joint. The total number of frames is equal to $n + 2B$ and the geometric parameters of the last B frames are constant.

Let us define a vector $\mathbf{q}^T = [\mathbf{q}_a^T \ \mathbf{q}_d^T \ \mathbf{q}_c^T]$ of dimension n_q in which:

- $\mathbf{q}_a$ is the vector containing the n_a active joint variables;
- $\mathbf{q}_d$ is the vector containing the $n_d = n - n_a$ passive joint variables of the equivalent tree structure;
- $\mathbf{q}_c$ is the vector containing the B variables of the cut joints. When a cut joint has several degrees of freedom (spherical, universal, ...), we can consider all of its joint variables to be belonging to $\mathbf{q}_c$.

Only the n_a active variables $\mathbf{q}_a$ are independent. Thus, there are $c = n_q - n_a$ independent constraint equations between the joint variables $\mathbf{q}$. These relations form the geometric constraint equations satisfying the closure of the loops. Since $\mathcal{F}_k$ and $\mathcal{F}_{k+B}$ are aligned, the *geometric constraint equations* for each loop can be written as:

$$ {}^{k+B}\mathbf{T}_k = \mathbf{1}_4, \ \text{for } k = n+1, \ldots, n+B \tag{4.6} $$

where $\mathbf{1}_4$ is the (4×4) identity matrix.

It should be mentioned that in terms of velocities, such equations can be expressed as:

$$ \mathbf{t}_{k+B} - \mathbf{t}_k = \mathbf{0}, \ \text{for } k = n+1, \ldots, n+B \tag{4.7} $$

where $\mathbf{t}_i$ is the twist of the body $\mathcal{B}_i$.

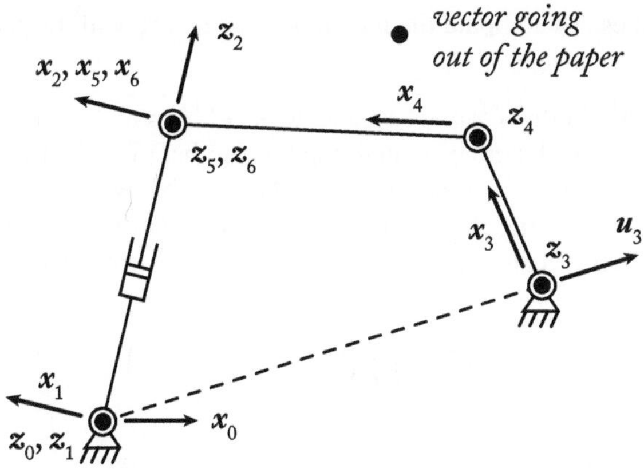

Fig. 4.7 Example of 2 *DOF* closed loop

Table 4.2 The *MDH* parameters of the 2 *DOF* closed loop

j	$a(j)$	μ_j	σ_j	γ_j	b_j	α_j	d_j	θ_j	r_j
1	0	1	0	0	0	0	0	θ_1	0
2	1	0	1	0	0	$\pi/2$	0	0	r_2
3	0	1	0	γ_3	0	0	d_3	θ_3	0
4	3	0	0	0	0	0	d_4	θ_4	0
5	4	0	0	0	0	0	d_5	θ_5	0
6	2	0	2	0	0	$-\pi/2$	0	0	0

4.2.2.1 Example of *MDH* parameters for a closed loop

Let us consider the single planar closed loop shown in Fig. 4.7, which is composed of four revolute joints and one prismatic joint. The system's mobility is equal to 2 (see Appendix A). We suppose that the active joints are the joints 1 and 3 which are fixed on the base. According to the above mentioned notations, it is possible to assign the frames of the system in the way presented in Fig. 4.7.

The geometric parameters are given in Table 4.2.

4.2.3 Computation of the Homogeneous Transformation Matrix Representing the Location of the Frame $\mathcal{F}_k$ with Respect to the Frame $\mathcal{F}_i$

Let us denote as $\mathbf{a}^k = \{0 \ \ldots \ a(a(k)) \ a(k) \ k\}$ a list containing the number of each intermediate frames separating the frame $\mathcal{F}_k$ from the frame $\mathcal{F}_0$, ordered by successive frames. To compute the location of frame $\mathscr{F}_k$ with respect to the frame $\mathscr{F}_i$, two cases can arise:

- *Case 1*: $i \in \mathbf{a}^k$; in that case, let us denote as $\mathbf{a}^{(i:k)} = \{i \ \ldots \ a(a(k)) \ a(k) \ k\}$ a sub-list of $\mathbf{a}^k$ with a length n_k containing the number of each intermediate frames separating the frame $\mathscr{F}_k$ from the frame $\mathscr{F}_i$, ordered by successive frames. The transformation allowing the computation of the position and orientation of the frame $\mathscr{F}_k$ with respect to the frame $\mathscr{F}_i$ can be deduced from (2.17) and is given by:

$$^{i}\mathbf{T}_k = \prod_{j=2}^{n_k} \left(^{a^{(i:k)}(j-1)}\mathbf{T}_{a^{(i:k)}(j)} \right) \tag{4.8}$$

  where $a^{(i:k)}(j)$ denotes the jth component of the list $\mathbf{a}^{(i:k)}$.
- *Case 2*: $i \notin \mathbf{a}^k$; in that case, we create a new list $\mathbf{a}^i = \{0 \ \ldots \ a(a(i)) \ a(i) \ i\}$ containing the number of each intermediate frames separating the frame $\mathscr{F}_i$ from the frame $\mathscr{F}_0$, ordered by successive frames. Let us denote as r_c the highest number common to both lists. Body $\mathscr{B}_{r_c}$ is thus the body where the sub-chain going from the ground to body $\mathscr{B}_k$ makes a bifurcation from the sub-chain going from the ground to body $\mathscr{B}_i$. Let us denote as $\mathbf{a}^{(r_c:k)} = \{r_c \ \ldots \ a(a(k)) \ a(k) \ k\}$ a sub-list of $\mathbf{a}^k$ with a length n_k containing the number of each intermediate frames separating the frame $\mathscr{F}_{r_c}$ from the frame $\mathscr{F}_k$, ordered by successive frames, and $\mathbf{a}^{(r_c:i)} = \{r_c \ \ldots \ a(a(i)) \ a(i) \ i\}$ a sub-list of $\mathbf{a}^i$ with a length n_i containing the number of each intermediate frames separating the frame $\mathscr{F}_{r_c}$ from the frame $\mathscr{F}_i$, ordered by successive frames. The transformation allowing the computation of the position and orientation of the frame $\mathscr{F}_k$ with respect to the frame $\mathscr{F}_i$ is here given by:

$$^{i}\mathbf{T}_k = \left(\prod_{j=2}^{n_i} \left(^{a^{(r_c:i)}(j-1)}\mathbf{T}_{a^{(r_c:i)}(j)} \right) \right)^{-1} \left(\prod_{j=2}^{n_k} \left(^{a^{(r_c:k)}(j-1)}\mathbf{T}_{a^{(r_c:k)}(j)} \right) \right) \tag{4.9}$$

  where $a^{(r_c:k)}(j)$ denotes the jth component of vector $\mathbf{a}^{(r_c:k)}$ and $a^{(r_c:i)}(j)$ denotes the jth component of vector $\mathbf{a}^{(r_c:i)}$.

The control of a robot manipulator requires fast computation of its different models. An efficient method to reduce the computation time is to generate a symbolic customized model for each specific robot. To obtain this model, we expand the matrix multiplications to transform them into scalar equations. Each element of a matrix containing at least one mathematical operation is replaced by an intermediate variable. This variable is written in the output file that contains the customized model. The elements that do not contain any operation are kept without modification. We propagate the matrix obtained in the subsequent equations. Consequently, customizing eliminates multiplications by one and zero, and additions with zero. Moreover, if the robot has two or more successive revolute joints with parallel axes, it is more interesting to replace the corresponding product of matrices by a single matrix, which is calculated using Eq. (2.18).

4.2.3.1 Example 1: Computation of the Homogeneous Transformation Matrices for a Tree-Structure Robot

Let us consider the tree-structure robot depicted in Fig. 4.4. For this robot, accordingly to the parameterization given on the picture, we have:

$$ {}^{0}\mathbf{T}_4 = {}^{0}\mathbf{T}_1\, {}^{1}\mathbf{T}_4 \tag{4.10} $$

$$ {}^{0}\mathbf{T}_3 = {}^{0}\mathbf{T}_1\, {}^{1}\mathbf{T}_2\, {}^{2}\mathbf{T}_3 \tag{4.11} $$

$$ \begin{aligned} {}^{3}\mathbf{T}_4 &= \left({}^{1}\mathbf{T}_2\, {}^{2}\mathbf{T}_3\right)^{-1}\, {}^{1}\mathbf{T}_4 \\ &= \left({}^{2}\mathbf{T}_3\right)^{-1} \left({}^{1}\mathbf{T}_2\right)^{-1}\, {}^{1}\mathbf{T}_4 \\ &= {}^{3}\mathbf{T}_2\, {}^{2}\mathbf{T}_1\, {}^{1}\mathbf{T}_4. \end{aligned} \tag{4.12} $$

4.2.3.2 Example 2: Computation of the Homogeneous Transformation Matrices for a Closed Loop

Let us consider the closed loop depicted in Fig. 4.7. For this chain, accordingly to the parameterization given on the picture, we have:

$$ {}^{0}\mathbf{T}_5 = {}^{0}\mathbf{T}_3\, {}^{3}\mathbf{T}_4\, {}^{4}\mathbf{T}_5 \tag{4.13} $$

$$ {}^{0}\mathbf{T}_6 = {}^{0}\mathbf{T}_1\, {}^{1}\mathbf{T}_2\, {}^{2}\mathbf{T}_6. \tag{4.14} $$

As the frame $\mathscr{F}_5$ is aligned to the frame $\mathscr{F}_6$, we have

$$ {}^{0}\mathbf{T}_5 = {}^{0}\mathbf{T}_6. \tag{4.15} $$

Chapter 5
Geometric, Velocity and Acceleration Analysis of Open Kinematic Chains

Abstract This chapter recalls the computation of the pose, velocity, acceleration and kinematic Jacobian matrix of any body belonging to an open (serial or tree structure) kinematic chain. The concept of singularity for serial or tree structure robots is introduced. Moreover, recursive and efficient algorithms able to decrease the computational complexity for the calculation of the velocity, Jacobian matrices and acceleration of any body are provided.

5.1 Geometric Analysis of Open Kinematic Chains

In the following of the present book, we will use the term "geometry" when speaking about the models defining the robot configuration, whereas the term "kinematics" will be used when studying velocities. This is quite unusual in English where the term "kinematics" is often used for both geometric and velocity study, however this confusion in the English terminology was pointed out by Bernard Roth during his keynote lecture at the *1987 International Conference on Advanced Robotics*. Therefore it is more appropriate to make the separation between "geometry" and "kinematics".

5.1.1 Direct Geometric Model of Open Kinematic Chains

Direct geometric model refers to the computation of the pose $\mathbf{x}_n$ of the terminal link $\mathscr{B}_n$ with respect to the base frame $\mathscr{F}_0$ in terms of the active joint coordinates $\mathbf{q}_a$. Here, body $\mathscr{B}_n$ is a terminal link but in the case of a tree-structure, other terminal links will exist. In the following, we develop computation for the body $\mathscr{B}_n$ but the generalization to other terminal links is straightforward.

For open kinematic chain (serial or tree-structure) robots, all joints are active. Thus the vector $\mathbf{q}_a$ is a vector of joint coordinates $\mathbf{q}$ of dimension m, which groups the joint coordinates of the sub-chain connecting the fixed base $\mathscr{B}_0$ to the considered terminal link $\mathscr{B}_n$.

S. Briot and W. Khalil, *Dynamics of Parallel Robots*,
Mechanisms and Machine Science 35, DOI 10.1007/978-3-319-19788-3_5

Let us denote as $\mathbf{a}^n = \{0\ l\ \ldots\ a(a(n))\ a(n)\ n\}$ a list of size s containing the number of each intermediate frame separating the frame $\mathscr{F}_n$ from the frame $\mathscr{F}_0$, ordered by successive frames and in which $a(l) = 0$. From Sect. 4.2.3 and Eq. (4.8), it turns out that:

$$^0\mathbf{T}_n = \begin{bmatrix} ^0\mathbf{R}_n & ^0\mathbf{r}_n \\ \mathbf{0} & 1 \end{bmatrix} = \prod_{k=2}^{s} \left(^{a^n(k-1)}\mathbf{T}_{a^n(k)} \right) \tag{5.1}$$

where $a^n(k)$ is the kth element of the list $\mathbf{a}^n$.

Thus, with (5.1), we can compute directly the homogeneous transformation matrix giving the relative pose of the frame $\mathscr{F}_n$ w.r.t. the frame $\mathscr{F}_0$. Vector $^0\mathbf{r}_n$ denotes the position of the frame $\mathscr{F}_n$—and of the terminal link $\mathscr{B}_n$—w.r.t. $\mathscr{F}_0$ and $^0\mathbf{R}_n$ is the orientation matrix of the frame $\mathscr{F}_n$—and of the terminal link $\mathscr{B}_n$—w.r.t. $\mathscr{F}_0$. From that orientation matrix, any set of orientation parameters can be obtained (Euler angles, Tilt and Torsion, Quaternions, etc.) by using the inverse problem methodologies given in Sect. 2.4.

5.1.2 *Inverse Geometric Model of Open Kinematic Chains*

Inverse geometric model refers to the computation of the vector of joint coordinates $\mathbf{q}$ once the pose $\mathbf{x}_n$ of the terminal link $\mathscr{B}_n$ with respect to the base frame $\mathscr{F}_0$ is known. Finding the solution of the inverse geometric model of an open kinematic chain is important for the study of *PKM* whose legs connecting the platform to the base are open kinematic chains. Indeed, if we fix the *PKM* platform pose and that we want to know the leg configuration, the inverse geometric model of each open chain constituting the legs must be solved.

Solving an inverse geometric model of an open kinematic chain could be difficult to achieve. However, methods have been developed to solve that problem in a general manner, such as the methods proposed in (Khalil and Dombre 2002; Pfurner and Husty 2010). Those methods are not detailed here, because most proposed *PKM* are made of simple legs that do not require the use of complex methods for solving their inverse geometric models. More details will be given in Sect. 7.1.

5.2 Velocity Analysis of Open Kinematic Chains

5.2.1 *Forward Kinematic Models*

The forward kinematic model of a serial or tree-structure robot manipulator gives the velocity of the terminal link $\mathbf{t}_n^T = [\mathbf{v}_n^T\ \boldsymbol{\omega}_n^T]$ in terms of the joint velocities $\dot{\mathbf{q}}$, which is defined here as a vector of dimension m and which groups all joint velocities of

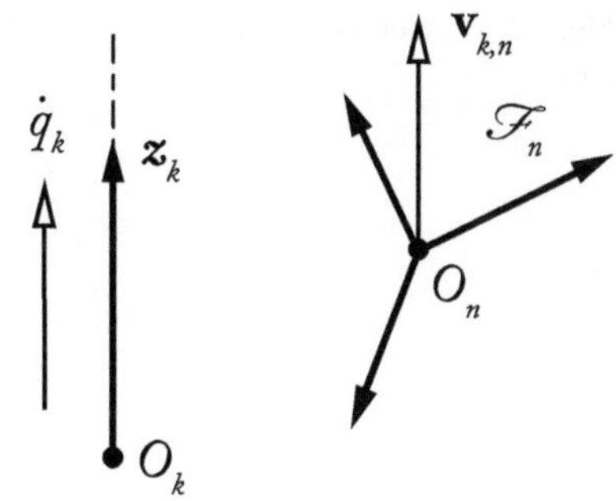

Fig. 5.1 Case of a prismatic joint

the sub-chain connecting the fixed base $\mathscr{B}_0$ to the considered terminal link $\mathscr{B}_n$. It is written as:

$$\mathbf{t}_n = \mathbf{J}_n(\mathbf{q})\dot{\mathbf{q}} \tag{5.2}$$

where $\mathbf{J}_n(\mathbf{q})$ denotes the $(6 \times m)$ kinematic Jacobian matrix. The matrix $\mathbf{J}_n(\mathbf{q})$ can be computed as follows.

The velocity $\dot{q}_k$ of joint k produces linear and angular velocities (denoted as $\mathbf{v}_{k,n}$ and $\boldsymbol{\omega}_{k,n}$ respectively) at the terminal frame $\mathscr{F}_n$. Two cases are considered:

- if joint k is prismatic (in that case, from Chap. 4, $\sigma_k = 1$, Fig. 5.1):

$$\mathbf{t}_{k,n} = \begin{bmatrix} \mathbf{v}_{k,n} \\ \boldsymbol{\omega}_{k,n} \end{bmatrix} = \begin{bmatrix} \mathbf{a}_k \\ \mathbf{0} \end{bmatrix} \dot{q}_k \tag{5.3}$$

  where $\mathbf{a}_k$ is the unit vector along the z_k axis;
- if joint k is revolute (in that case, from Chap. 4, $\sigma_k = 0$, Fig. 5.2):

$$\mathbf{t}_{k,n} = \begin{bmatrix} \mathbf{v}_{k,n} \\ \boldsymbol{\omega}_{k,n} \end{bmatrix} = \begin{bmatrix} \mathbf{a}_k \dot{q}_k \times \mathbf{r}_{O_k O_n} \\ \mathbf{a}_k \dot{q}_k \end{bmatrix} = \begin{bmatrix} \mathbf{a}_k \times \mathbf{r}_{O_k O_n} \\ \mathbf{a}_k \end{bmatrix} \dot{q}_k \tag{5.4}$$

  where $\mathbf{r}_{O_k O_n}$ denotes the position vector connecting O_k to O_n, i.e. the origins of the frames $\mathscr{F}_k$ and $\mathscr{F}_n$ linked respectively to bodies $\mathscr{B}_k$ and $\mathscr{B}_n$.

Thus, $\mathbf{v}_{k,n}$ and $\boldsymbol{\omega}_{k,n}$ can be written in the following general form:

$$\mathbf{t}_{k,n} = \begin{bmatrix} \mathbf{v}_{k,n} \\ \boldsymbol{\omega}_{k,n} \end{bmatrix} = \begin{bmatrix} \sigma_k \mathbf{a}_k + \bar{\sigma}_k (\mathbf{a}_k \times \mathbf{r}_{O_k O_n}) \\ \bar{\sigma}_k \mathbf{a}_k \end{bmatrix} \dot{q}_k = \$_k \dot{q}_k \tag{5.5}$$

where $\$_k$ is a unit twist representing the displacement of the terminal link when actuator k is moving only.

Fig. 5.2 Case of a revolute joint

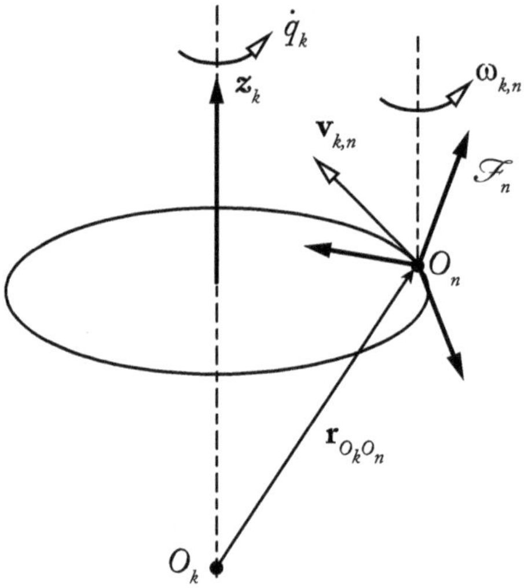

Now considering all joints of the studied chain, linear and angular velocities of the terminal frame can be written as:

$$
\begin{aligned}
\mathbf{t}_n &= \sum_{k=2}^{s} \begin{bmatrix} \mathbf{v}_{a^n(k),n} \\ \boldsymbol{\omega}_{a^n(k),n} \end{bmatrix} = \sum_{k=2}^{s} \$_{a^n(k)} \dot{q}_{a^n(k)} \\
&= \sum_{k=2}^{s} \begin{bmatrix} \sigma_{a^n(k)} \mathbf{a}_{a^n(k)} + \bar{\sigma}_{a^n(k)} (\mathbf{a}_{a^n(k)} \times \mathbf{r}_{O_{a^n(k)} O_n}) \\ \bar{\sigma}_{a^n(k)} \mathbf{a}_{a^n(k)} \end{bmatrix} \dot{q}_{a^n(k)}
\end{aligned}
\tag{5.6}
$$

where $a^n(k)$ is defined at (5.1).

Writing Eq. (5.6) in matrix form and using Eq. (5.2), we deduce that:

$$
\begin{aligned}
\mathbf{J}_n &= \begin{bmatrix} \sigma_l \mathbf{a}_l + \bar{\sigma}_l (\mathbf{a}_l \times \mathbf{r}_{O_l O_n}) & \dots & \sigma_{a(n)} \mathbf{a}_{a(n)} + \bar{\sigma}_{a(n)} (\mathbf{a}_{a(n)} \times \mathbf{r}_{O_{a(n)} O_n}) & \sigma_n \mathbf{a}_n \\ \bar{\sigma}_l \mathbf{a}_l & \dots & \bar{\sigma}_{a(n)} \mathbf{a}_{a(n)} & \bar{\sigma}_n \mathbf{a}_n \end{bmatrix} \\
&= \begin{bmatrix} \$_l & \dots & \$_{a(n)} & \$_n \end{bmatrix}.
\end{aligned}
\tag{5.7}
$$

Expressing the vectors of $\mathbf{J}_n$ with respect to frame $\mathscr{F}_i$, we obtain the $(6 \times m)$ kinematic Jacobian matrix ${}^i\mathbf{J}_n$ such that:

$$
{}^i\mathbf{t}_n = {}^i\mathbf{J}_n(\mathbf{q})\dot{\mathbf{q}}. \tag{5.8}
$$

By extension of (5.7) and (5.8), the velocity of the jth body of the chain is given by:

$$
{}^i\mathbf{t}_j = {}^i\mathbf{J}_j(\mathbf{q})\dot{\mathbf{q}} \tag{5.9}
$$

where ${}^i\mathbf{J}_j$ is the projection into the frame $\mathscr{F}_i$ of the matrix $\mathbf{J}_j$ defined by

$$\mathbf{J}_j = \begin{bmatrix} \sigma_l \mathbf{a}_l + \bar{\sigma}_l(\mathbf{a}_l \times \mathbf{r}_{O_l O_j}) & \dots & \sigma_{a(j)}\mathbf{a}_{a(j)} + \bar{\sigma}_{a(j)}(\mathbf{a}_{a(j)} \times \mathbf{r}_{O_{a(j)} O_j}) & \sigma_j \mathbf{a}_j & \mathbf{0} \dots \mathbf{0} \\ \bar{\sigma}_l \mathbf{a}_l & \dots & \bar{\sigma}_{a(j)}\mathbf{a}_{a(j)} & \bar{\sigma}_j \mathbf{a}_j & \mathbf{0} \dots \mathbf{0} \end{bmatrix}$$
$$= \left[\$_l \dots \$_{a(j)}\ \$_j\ \mathbf{0} \dots \mathbf{0}\right]. \quad (5.10)$$

5.2.2 Inverse Kinematic Models

5.2.2.1 Inverse Kinematic Model of Non-redundant Open Chains

In this section, we will deal with the inverse kinematic problems of non-redundant open kinematic chain, i.e. chains for which the size m of the vector $\dot{\mathbf{q}}$ is equal to the rank of the matrix ${}^i\mathbf{J}_n$.

If the size of $\dot{\mathbf{q}}$ is $m = 6$, the inverse kinematic model is:

$$\dot{\mathbf{q}} = {}^i\mathbf{J}_n^{-1}(\mathbf{q})\,{}^i\mathbf{t}_n \quad (5.11)$$

as long as $\mathbf{J}_n$ is not rank-deficient (see Sect. 5.2.3).

If the size of $\dot{\mathbf{q}}$ is $m < 6$, this means that the terminal body $\mathscr{B}_n$ cannot produce 6 independent motions, but only m independent motions. Let us denote as:

- ${}^i\mathbf{t}_n^r$ the vector containing the m independent coordinates of ${}^i\mathbf{t}_n$,
- ${}^i\mathbf{t}_n^r = \mathbf{D}_n\,{}^i\mathbf{t}_n$ the constraint equations between the coordinates of ${}^i\mathbf{t}_n$.

Equation (5.8) can thus be reduced to:

$${}^i\mathbf{t}_n^r = {}^i\mathbf{J}_r(\mathbf{q})\dot{\mathbf{q}} \quad (5.12)$$

and the inverse kinematic model is:

$$\dot{\mathbf{q}} = {}^i\mathbf{J}_r^{-1}(\mathbf{q})\,{}^i\mathbf{t}_n^r = {}^i\mathbf{J}_r^{-1}(\mathbf{q})\mathbf{D}_n\,{}^i\mathbf{t}_n = {}^i\mathbf{J}_{inv}(\mathbf{q})\,{}^i\mathbf{t}_n \quad (5.13)$$

where ${}^i\mathbf{J}_{inv}(\mathbf{q}) = {}^i\mathbf{J}_r^{-1}(\mathbf{q})\mathbf{D}_n$. Note once again that this expression is valid as long as ${}^i\mathbf{J}_r$ (and ${}^i\mathbf{J}_n$) is not rank-deficient (see Sect. 5.2.3).

5.2.2.2 Inverse Kinematic Model of Redundant Open Chains

For redundant open kinematic chain, the size m of the vector $\dot{\mathbf{q}}$ is strictly greater than the maximum rank of the matrix ${}^i\mathbf{J}_n$. This means that, for any terminal-link pose, there can exist a non-null vector $\dot{\mathbf{q}}_0$ such as

$$\mathbf{0} = {}^i\mathbf{J}_n(\mathbf{q})\dot{\mathbf{q}}_0. \quad (5.14)$$

Consequently, the robot inverse kinematic model can have infinite number of solutions that are given by:

$$\dot{\mathbf{q}} = {}^i\mathbf{J}_n^{+i}\mathbf{t}_n + \left(\mathbf{1}_m - {}^i\mathbf{J}_n^{+i}\mathbf{J}_n\right)\boldsymbol{\xi} \tag{5.15}$$

where, if ${}^i\mathbf{J}_n$ is of maximal rank, ${}^i\mathbf{J}_n^{+} = {}^i\mathbf{J}_n^T\left({}^i\mathbf{J}_n{}^i\mathbf{J}_n^T\right)^{-1}$ is the pseudo-inverse of the Jacobian matrix ${}^i\mathbf{J}_n$ and $\boldsymbol{\xi}$ is an arbitrary vector in the $\dot{\mathbf{q}}$ space, which, once defined, fix the solution (Khalil and Dombre 2002).

Consequently, as $\boldsymbol{\xi}$ can be defined arbitrarily, secondary performance criteria can be optimized, such as:

- minimizing the norm of the joint velocities (Whitney 1969),
- avoiding obstacles (Baillieul 1986; Maciejewski and Klein 1985),
- avoiding singular configurations (Yoshikawa 1984),
- avoiding joint limits (Fournier 1980; Klein 1984),
- minimizing driving joint torques (Baillieul et al. 1984; Hollerbach and Suh 1985).

The reader interested in the control of redundant open chains should refer to (Khalil and Dombre 2002).

5.2.3 *Inverse Kinematic Models Degeneracy/Notions of Singularity*

As mentioned in Sect. 5.2.2, the inverse geometric and kinematic models of a robot can be computed as long as the robot is not at a singular configuration (also called singularity).

Mathematically speaking, a singularity for a serial chain (or a serial sub-chain of a tree-structure robot) is defined when, at a given configuration, the rank of the kinematic Jacobian matrix $\mathbf{J}_n$ defined in (5.7) is lower than its maximal rank.

Physically speaking, and from the analysis of Eq. (5.2), there can exist one (or several, depending on the loss of rank of $\mathbf{J}_n$) non zero unit vector $\mathbf{q}_s$ which belongs to the null space of $\mathbf{J}_n$ that, *locally*, cannot produce a velocity of the end-effector, i.e.

$$\mathbf{J}_n(\mathbf{q}_n)\dot{\mathbf{q}}_s = \mathbf{0}. \tag{5.16}$$

Thus, the robot is unable to generate an arbitrary velocity along some given axes. Moreover, its inverse geometric model may have an infinity of solutions.

If $\mathbf{J}_n$ is rank-deficient, by definition, this is also the case of the square matrix $\mathbf{J}_r$ defined in (5.12). Thus, the singularity can also be characterized by:

$$\det(\mathbf{J}_r(\mathbf{q}_n)) = 0. \tag{5.17}$$

Those singularities divide the joint space in several *aspects* (Borrel 1986). The aspects are the connected regions of the joint space inside which no minor of order k extracted from the kinematic Jacobian matrix $\mathbf{J}_n$ is zero, except if this minor is zero everywhere in the joint domain. For a non-redundant robot manipulator, the only minor of order $\mathbf{J}_n$ is the kinematic Jacobian matrix itself. Therefore, the aspects are limited by the *singularity branches* (i.e. the connected components of the set of singular configurations of the joint space) and the joint limits. Consequently, they represent the maximum singularity-free regions of the joint space.

For a long time, it has been thought that the aspects also represent the uniqueness domains of the inverse geometric model solutions. Although this is indeed the case for most industrial robots with simple architectures, which are classified as *non-cuspidal* robots (El Omri 1996; Wenger 1992), the inverse geometric model of *cuspidal* robots can have several solutions in the same aspect.[1] Thus, a cuspidal robot can move from one inverse geometric model solution to another without encountering a singularity.

More details on the singularity analysis are given in Sect. 7.5.

5.2.4 Recursive Computation of Velocities and Kinematic Jacobian Matrix for Open Kinematic Chains

Let us consider two bodies $\mathscr{B}_i$ and $\mathscr{B}_j$ (where $i = a(j)$) on which are attached the frames $\mathscr{F}_i$ and $\mathscr{F}_j$, respectively. The twist of $\mathscr{B}_i$ projected in the frame $\mathscr{F}_i$ and expressed at O_i (origin of $\mathscr{F}_i$) is denoted as ${}^i\mathbf{t}_i^T = [{}^i\mathbf{v}_i^T \; {}^i\boldsymbol{\omega}_i^T]$ and the twist of $\mathscr{B}_j$ projected in the frame $\mathscr{F}_j$ and expressed at O_j (origin of $\mathscr{F}_j$) is denoted as ${}^j\mathbf{t}_j^T = [{}^j\mathbf{v}_j^T \; {}^j\boldsymbol{\omega}_j^T]$.

$\mathscr{B}_i$ being the antecedent of $\mathscr{B}_j$, we have (from Sect. 5.2.1):

$$\boldsymbol{\omega}_j = \boldsymbol{\omega}_i + \bar{\sigma}_j \mathbf{a}_j \dot{q}_j \tag{5.18}$$

$$\mathbf{v}_j = \mathbf{v}_i + \boldsymbol{\omega}_i \times \mathbf{r}_{O_i O_j} + \sigma_j \mathbf{a}_j \dot{q}_j. \tag{5.19}$$

Thus, by identification with the relation (3.12), and by projecting the equations in the frame $\mathscr{F}_j$, we have

$${}^j\mathbf{t}_j = {}^j\overline{\mathbf{T}}_i {}^i\mathbf{t}_i + {}^j\bar{\mathbf{a}}_j \dot{q}_j \tag{5.20}$$

where ${}^j\bar{\mathbf{a}}_j = \begin{bmatrix} \sigma_j \mathbf{1}_3 \\ \bar{\sigma}_j \mathbf{1}_3 \end{bmatrix} {}^j\mathbf{a}_j = \begin{bmatrix} \sigma_j {}^j\mathbf{a}_j \\ \bar{\sigma}_j {}^j\mathbf{a}_j \end{bmatrix}$, in which ${}^j\mathbf{a}_j = [0 \; 0 \; 1]^T$.

Let us now consider that those two bodies belong to a kinematic chain (serial or tree structure) composed of n bodies whose motions are described by the vector of

[1] For parallel robots, a cuspidal robot is usually considered to have several solutions of the forward geometric model in the same workspace aspect (Zein et al. 2008).

joint velocities $\dot{\mathbf{q}}$. From (5.9) and (5.13), we have ${}^i\mathbf{t}_i = {}^i\mathbf{J}_i\dot{\mathbf{q}}$ and ${}^j\mathbf{t}_j = {}^j\mathbf{J}_j\dot{\mathbf{q}}$. Thus, from (5.20), we find that:

$$
{}^j\mathbf{J}_j = {}^j\overline{\mathbf{T}}_i{}^i\mathbf{J}_i + {}^j\mathbf{A}_j \tag{5.21}
$$

with

$$
{}^j\mathbf{A}_j = \left[\mathbf{0} \ldots {}^j\bar{\mathbf{a}}_j \ldots \mathbf{0}\right] \tag{5.22}
$$

defined such that:

$$
{}^j\bar{\mathbf{a}}_j\dot{q}_j = {}^j\mathbf{A}_j\dot{\mathbf{q}}. \tag{5.23}
$$

Those recursive equations will be used for optimizing the computation of the dynamic model of the parallel robots. They are implemented in the free software OpenSYMORO (Khalil et al. 2014).

5.3 Acceleration Analysis of Open Kinematic Chains

The second-order kinematic model of a serial or tree-structure robot manipulator gives the acceleration of the terminal link $\mathcal{B}_n$ in terms of positions $\mathbf{q}_n$, velocities $\dot{\mathbf{q}}_n$ and accelerations $\ddot{\mathbf{q}}_n$ of the joints of the considered sub-chain. By differentiating Eq. (5.2) with respect to time, we obtain the following expression:

$$
\dot{\mathbf{t}}_n = \begin{bmatrix} \dot{\mathbf{v}}_n \\ \dot{\boldsymbol{\omega}}_n \end{bmatrix} = \mathbf{J}_n(\mathbf{q}_n)\ddot{\mathbf{q}}_n + \dot{\mathbf{J}}_n(\mathbf{q}_n, \dot{\mathbf{q}}_n)\dot{\mathbf{q}}_n. \tag{5.24}
$$

However, it is most efficient from the computational cost point of view to obtain $\dot{\mathbf{t}}_n$ using the recursive algorithm given below.

Let us consider once more two bodies $\mathcal{B}_i$ and $\mathcal{B}_j$ (where $i = a(j)$). The acceleration of $\mathcal{B}_i$ projected in the frame $\mathcal{F}_i$ and expressed at O_i (origin of $\mathcal{F}_i$) is denoted as ${}^i\dot{\mathbf{t}}_i^T = [{}^i\dot{\mathbf{v}}_i^T\ {}^i\dot{\boldsymbol{\omega}}_i^T]$ and the acceleration of $\mathcal{B}_j$ projected in the frame $\mathcal{F}_j$ and expressed at O_j (origin of $\mathcal{F}_j$) is denoted as ${}^j\dot{\mathbf{t}}_j^T = [{}^j\dot{\mathbf{v}}_j^T\ {}^j\dot{\boldsymbol{\omega}}_j^T]$.

$\mathcal{B}_i$ being the antecedent of $\mathcal{B}_j$, by expressing the derivative of $\boldsymbol{\omega}_j$ and $\mathbf{v}_j$ w.r.t. time, we also have (from Sect. 5.2.1 and Eq. (3.19)):

$$
\dot{\boldsymbol{\omega}}_j = \dot{\boldsymbol{\omega}}_i + \frac{\mathrm{d}}{\mathrm{d}t}\left(\bar{\sigma}_j\mathbf{a}_j\dot{q}_j\right) = \dot{\boldsymbol{\omega}}_i + \bar{\sigma}_j(\mathbf{a}_j\ddot{q}_j + \boldsymbol{\omega}_i \times \mathbf{a}_j\dot{q}_j) \tag{5.25}
$$

$$
\begin{aligned}
\dot{\mathbf{v}}_j &= \dot{\mathbf{v}}_i + \dot{\boldsymbol{\omega}}_i \times \mathbf{r}_{O_iO_j} + \boldsymbol{\omega}_i \times \frac{\mathrm{d}}{\mathrm{d}t}\left(\mathbf{r}_{O_iO_j}\right) + \frac{\mathrm{d}}{\mathrm{d}t}\left(\sigma_j\mathbf{a}_j\dot{q}_j\right) \\
&= \dot{\mathbf{v}}_i + \dot{\boldsymbol{\omega}}_i \times \mathbf{r}_{O_iO_j} + \boldsymbol{\omega}_i \times (\boldsymbol{\omega}_i \times \mathbf{r}_{O_iO_j}) + \sigma_j(\mathbf{a}_j\ddot{q}_j + 2\boldsymbol{\omega}_i \times \mathbf{a}_j\dot{q}_j)
\end{aligned} \tag{5.26}
$$

because $\frac{\mathrm{d}}{\mathrm{d}t}\left(\mathbf{r}_{O_iO_j}\right) = \boldsymbol{\omega}_i \times \mathbf{r}_{O_iO_j} + \mathbf{a}_j\dot{q}_j$ in the case of a prismatic joint and $\frac{\mathrm{d}}{\mathrm{d}t}\left(\mathbf{r}_{O_iO_j}\right) = \boldsymbol{\omega}_i \times \mathbf{r}_{O_iO_j}$ in case of revolute joint.

Expressing the vectors of (5.25) and (5.26) within frame $\mathscr{F}_j$, they can also be rewritten as:

$$ {}^j\dot{\boldsymbol{\omega}}_j = {}^j\mathbf{R}_i{}^i\dot{\boldsymbol{\omega}}_i + \bar{\sigma}_j({}^j\mathbf{a}_j\ddot{q}_j + {}^j\boldsymbol{\omega}_i \times {}^j\mathbf{a}_j\dot{q}_j) \tag{5.27} $$

$$ {}^j\dot{\mathbf{v}}_j = {}^j\mathbf{R}_i({}^i\dot{\mathbf{v}}_i + {}^i\mathbf{U}_i{}^i\mathbf{r}_j) + \sigma_j({}^j\mathbf{a}_j\ddot{q}_j + 2{}^j\boldsymbol{\omega}_i \times {}^j\mathbf{a}_j\dot{q}_j) \tag{5.28} $$

where ${}^i\mathbf{r}_j = {}^i\mathbf{r}_{O_iO_j}$ and

$$ {}^i\mathbf{U}_i = {}^i\dot{\hat{\boldsymbol{\omega}}}_i + {}^i\hat{\boldsymbol{\omega}}_i{}^i\hat{\boldsymbol{\omega}}_i \tag{5.29} $$

in which ${}^i\dot{\hat{\boldsymbol{\omega}}}_i$ is the cross-product matrix associated with ${}^i\dot{\boldsymbol{\omega}}_i$ and ${}^i\hat{\boldsymbol{\omega}}_i$ is the cross-product matrix associated with ${}^i\boldsymbol{\omega}_i$.

Eqs. (5.27)–(5.29) will be used for optimizing the computation of the dynamic model of the parallel robots. They are implemented in the free software OpenSYMORO (Khalil et al. 2014).

Finally, (5.27) and (5.28) can also be rewritten in the matrix form

$$ {}^j\dot{\mathbf{t}}_j = \begin{bmatrix} {}^j\dot{\mathbf{v}}_j \\ {}^j\dot{\boldsymbol{\omega}}_j \end{bmatrix} = {}^j\mathbf{J}_j\ddot{\mathbf{q}}_n + {}^j\overline{\mathbf{b}}_j \tag{5.30} $$

where ${}^j\mathbf{J}_j$ is given in (5.21) and

$$ {}^j\overline{\mathbf{b}}_j = {}^j\overline{\mathbf{T}}_i{}^i\overline{\mathbf{b}}_i + \begin{bmatrix} 2\sigma_j{}^j\boldsymbol{\omega}_i \times {}^j\mathbf{a}_j\dot{q}_j + {}^j\boldsymbol{\omega}_i \times ({}^j\boldsymbol{\omega}_i \times {}^j\mathbf{r}_{O_iO_j}) \\ \bar{\sigma}_j{}^j\boldsymbol{\omega}_i \times {}^j\mathbf{a}_j\dot{q}_j \end{bmatrix} \tag{5.31} $$

which can be deduced from the fact that:

$$ \begin{bmatrix} {}^j\dot{\mathbf{v}}_j \\ {}^j\dot{\boldsymbol{\omega}}_j \end{bmatrix} = {}^j\overline{\mathbf{T}}_i \begin{bmatrix} {}^i\dot{\mathbf{v}}_i \\ {}^i\dot{\boldsymbol{\omega}}_i \end{bmatrix} + \begin{bmatrix} 2\sigma_j{}^j\boldsymbol{\omega}_i \times {}^j\mathbf{a}_j\dot{q}_j + {}^j\boldsymbol{\omega}_i \times ({}^j\boldsymbol{\omega}_i \times {}^j\mathbf{r}_{O_iO_j}) \\ \bar{\sigma}_j{}^j\boldsymbol{\omega}_i \times {}^j\mathbf{a}_j\dot{q}_j \end{bmatrix} + {}^j\bar{\mathbf{a}}_j\ddot{q}_j. \tag{5.32} $$

Chapter 6
Dynamics Principles

Abstract In this chapter, some brief recalls of the main dynamics principles are made, as they will be used in the following chapters of the book. We detail the main dynamics principles: the Lagrange formulation, the Newton-Euler equations and the principle of virtual powers. We also recall the static model giving the relation between a wrench exerted on the end-effector of any type of robot and the input effort of its actuators.

6.1 The Lagrange Formulation

6.1.1 Introduction to the Lagrange Formulation

The Lagrange formulation describes the behavior of a dynamic system in terms of work and energy stored in the system. The Lagrange equations are written in the form:

$$\boldsymbol{\tau} = \frac{\mathrm{d}}{\mathrm{d}t}\left(\frac{\partial L}{\partial \dot{\mathbf{q}}}\right)^T - \left(\frac{\partial L}{\partial \mathbf{q}}\right)^T \tag{6.1}$$

where

- $\boldsymbol{\tau}$ is the vector of generalized forces applied on the system, which are equal to the input joint torques or forces,
- $\mathbf{q}$ is the vector of generalized coordinates, i.e. the vector of independent parameters that describe the configuration of the system such that, for any body $\mathscr{B}_j$ the homogeneous transformation matrix ${}^0\mathbf{T}_j$ of its corresponding frame $\mathscr{F}_j$ can be expressed as a function of $\mathbf{q}$:

$$ {}^0\mathbf{T}_j = {}^0\mathbf{T}_j(\mathbf{q}). \tag{6.2}$$

  For rigid robots, $\mathbf{q}$ is equal to the vector of active joint variables $\mathbf{q}_a$. Thus, in the case of robots without any closed loops, $\mathbf{q}$ is the vector of the joint coordinates.
- $\dot{\mathbf{q}}$ is the vector of generalized velocities, the vector of parameters that describe the velocity of any body $\mathscr{B}_j$ of the system such that we can obtain, for any body $\mathscr{B}_j$:

S. Briot and W. Khalil, *Dynamics of Parallel Robots*,
Mechanisms and Machine Science 35, DOI 10.1007/978-3-319-19788-3_6

$$\mathbf{t}_j = \mathbf{J}_j(\mathbf{q})\dot{\mathbf{q}} \tag{6.3}$$

where $\mathbf{t}_j$ is the twist of the frame $\mathcal{F}_j$,

- L is a function called the Lagragian:

$$L = E - U \tag{6.4}$$

in which E is the kinetic energy of the system and U its potential energy (due to gravity effects, deformations, etc.).

For any rigid robot, the Lagrange formulation leads to a dynamic model of the form (Khalil and Dombre 2002):

$$\boldsymbol{\tau} = \mathbf{M}(\mathbf{q})\ddot{\mathbf{q}} + \mathbf{c}(\mathbf{q}, \dot{\mathbf{q}}) \tag{6.5}$$

where $\mathbf{M}(\mathbf{q})$ is the robot *generalized inertia matrix* (also called in this book *robot inertia matrix* and defined in Sect. 6.1.2) and $\mathbf{c}(\mathbf{q}, \dot{\mathbf{q}})$ is the vector of Coriolis, centrifugal and gravity effects. This model is called the *inverse dynamic model.*

6.1.2 Computation of Kinetic Energy

The kinetic energy E_j of a body $\mathcal{B}_j$ is given by the equation:

$$E_j = \frac{1}{2}\int_{\mathcal{B}_j} \mathbf{v}_{M_j}^T \mathbf{v}_{M_j} \mathrm{d}m \tag{6.6}$$

where $\mathbf{v}_{M_j}$ is the velocity of a point M_j belonging to $\mathcal{B}_j$ that can be expressed as a function of the frame $\mathcal{F}_j$ twist $\mathbf{t}_j^T = [\mathbf{v}_j^T \ \boldsymbol{\omega}_j^T]$:

$$\mathbf{v}_{M_j} = \mathbf{v}_j + \boldsymbol{\omega}_j \times \mathbf{r}_{O_jM_j} = \begin{bmatrix} \mathbf{1}_3 & \hat{\mathbf{r}}_{O_jM_j}^T \end{bmatrix} \begin{bmatrix} \mathbf{v}_j \\ \boldsymbol{\omega}_j \end{bmatrix}. \tag{6.7}$$

Introducing (6.7) into (6.6) and expressing all equations in $\mathcal{F}_j$ leads to:

$$\begin{aligned} E_j &= \frac{1}{2}\begin{bmatrix} {}^j\mathbf{v}_j^T & {}^j\boldsymbol{\omega}_j^T \end{bmatrix} \left(\int_{\mathcal{B}_j} \begin{bmatrix} \mathbf{1}_3 \\ {}^j\hat{\mathbf{r}}_{O_jM_j} \end{bmatrix} \begin{bmatrix} \mathbf{1}_3 & {}^j\hat{\mathbf{r}}_{O_jM_j}^T \end{bmatrix} \mathrm{d}m \right) \begin{bmatrix} {}^j\mathbf{v}_j \\ {}^j\boldsymbol{\omega}_j \end{bmatrix} \\ &= \frac{1}{2}\begin{bmatrix} {}^j\mathbf{v}_j^T & {}^j\boldsymbol{\omega}_j^T \end{bmatrix} \left(\int_{\mathcal{B}_j} \begin{bmatrix} \mathbf{1}_3 & {}^j\hat{\mathbf{r}}_{O_jM_j}^T \\ {}^j\hat{\mathbf{r}}_{O_jM_j} & {}^j\hat{\mathbf{r}}_{O_jM_j} {}^j\hat{\mathbf{r}}_{O_jM_j}^T \end{bmatrix} \mathrm{d}m \right) \begin{bmatrix} {}^j\mathbf{v}_j \\ {}^j\boldsymbol{\omega}_j \end{bmatrix} \\ &= \frac{1}{2}\mathbf{t}_j^T \begin{bmatrix} m_j\mathbf{1}_3 & \widehat{\mathbf{ms}}_j^T \\ \widehat{\mathbf{ms}}_j & \mathbf{I}_{O_j} \end{bmatrix} \mathbf{t}_j = \frac{1}{2}\mathbf{t}_j^T \mathbf{M}_j \mathbf{t}_j \end{aligned} \tag{6.8}$$

in which:

- m_j is the body $\mathscr{B}_j$ mass,
- $\widehat{\mathbf{ms}}_j$ is the cross-product matrix associated with the vector $\mathbf{ms}_j = \int_{\mathscr{B}} {}^j\mathbf{r}_{O_jM_j} dm = m_j {}^j\mathbf{r}_{O_jS_j}$, with S_j the center of mass (*COM*) of the body $\mathscr{B}_j$; $\mathbf{ms}_j$ is called the vector of the first moments of inertia (also known as the vector of the static moments) and represents the mass of the body $\mathscr{B}_j$ multiplied by the position of its *COM* w.r.t. frame $\mathscr{F}_j$, i.e. $\mathbf{ms}_j = [mx_j \; my_j \; mz_j]^T$, where $mx_j = m_j x_{S_j}$, $my_j = m_j y_{S_j}$ and $mz_j = m_j z_{S_j}$,
- $\mathbf{I}_{O_j}$ is the inertia matrix of the body $\mathscr{B}_j$ at the origin of the frame $\mathscr{F}_j$ and expressed in the same frame,

$$\mathbf{I}_{O_j} = \int_{\mathscr{B}} {}^j\hat{\mathbf{r}}_{O_jM_j} {}^j\hat{\mathbf{r}}^T_{O_jM_j} dm = \begin{bmatrix} xx_j & xy_j & xz_j \\ xy_j & yy_j & yz_j \\ xz_j & yz_j & zz_j \end{bmatrix} \tag{6.9}$$

in which xx_j, yy_j, zz_j are the axial moments of inertia around $\boldsymbol{x}_j, \boldsymbol{y}_j$ and $\boldsymbol{z}_j$ axes, respectively, for body $\mathscr{B}_j$, expressed at the origin of the local frame $\mathscr{F}_j$, i.e.,

$$xx_j = \int_{\mathscr{B}} (r_y^2 + r_z^2) dm \tag{6.10}$$

$$yy_j = \int_{\mathscr{B}} (r_x^2 + r_z^2) dm \tag{6.11}$$

$$zz_j = \int_{\mathscr{B}} (r_x^2 + r_y^2) dm \tag{6.12}$$

with $\mathbf{r}^T_{O_jM_j} = [r_x \; r_y \; r_z]$, and xy_j, xz_j, yz_j are the inertial cross-moments for body $\mathscr{B}_j$, expressed at the origin of the local frame $\mathscr{F}_j$, i.e.,

$$xy_j = -\int_{\mathscr{B}} r_x r_y dm \tag{6.13}$$

$$xz_j = -\int_{\mathscr{B}} r_x r_z dm \tag{6.14}$$

$$yz_j = -\int_{\mathscr{B}} r_y r_z dm. \tag{6.15}$$

- $\mathbf{M}_j$ is the body $\mathscr{B}_j$ (6×6) generalized inertia matrix at the origin of the frame $\mathscr{F}_j$ and expressed in the same frame.

The total kinetic energy E of the system is given by:

$$E = \sum_j E_j = \frac{1}{2} \sum_j \left({}^j\mathbf{t}_j^T \mathbf{M}_j {}^j\mathbf{t}_j \right). \tag{6.16}$$

Introducing (6.3) into (6.16), it comes:

$$E = \sum_j E_j = \frac{1}{2} \dot{\mathbf{q}}^T \sum_j \left({}^j\mathbf{J}_j^T(\mathbf{q}) \mathbf{M}_j {}^j\mathbf{J}_j(\mathbf{q}) \right) \dot{\mathbf{q}} = \frac{1}{2} \dot{\mathbf{q}}^T \mathbf{M}(\mathbf{q}) \dot{\mathbf{q}} \tag{6.17}$$

where $\mathbf{M}(\mathbf{q}) = \sum_j \left({}^j\mathbf{J}_j^T(\mathbf{q}) \mathbf{M}_j {}^j\mathbf{J}_j(\mathbf{q}) \right)$ is the system inertia matrix.

6.1.3 Computation of Potential Energy

In this section, we will give the value of the potential energy due to gravity effects. Elastic potential energy is not considered as elasticity is the subject of Part III.

The potential energy of a body $\mathscr{B}_j$ due to gravity effects is given by the equation:

$$U_j = -m_j \mathbf{g}^T \mathbf{r}_{O_0 S_j} \tag{6.18}$$

where

- $\mathbf{g}$ is the vector of gravitational acceleration,
- $\mathbf{r}_{O_0 S_j}$ is the position of the *COM* of the body $\mathscr{B}_j$ w.r.t. the origin of the global frame $\mathscr{F}_0$.

Equation (6.18) can be rewritten in the base frame $\mathscr{F}_0$ as:

$$U_j = -{}^0\mathbf{g}^T \left(m_j {}^0\mathbf{r}_{O_0 O_j} + m_j {}^0\mathbf{r}_{O_j S_j} \right) = -{}^0\mathbf{g}^T \left(m_j {}^0\mathbf{r}_{O_0 O_j} + {}^0\mathbf{R}_j {}^j\mathbf{ms}_j \right) \tag{6.19}$$

or also:

$$U_j = -\left[{}^0\mathbf{g}^T \ 0 \right] {}^0\mathbf{T}_j(\mathbf{q}) \begin{bmatrix} {}^j\mathbf{ms}_j \\ m_j \end{bmatrix}. \tag{6.20}$$

Finally, the total potential energy of the system is:

$$U = \sum_j U_j = -\left[{}^0\mathbf{g}^T \ 0 \right] \sum_j \left({}^0\mathbf{T}_j(\mathbf{q}) \begin{bmatrix} {}^j\mathbf{ms}_j \\ m_j \end{bmatrix} \right). \tag{6.21}$$

6.1.4 Lagrange Equations with Constraints

For closed loop robots, such as parallel robots, the expression of kinetic and potential energies is difficult to obtain as a function of the active joint variables $\mathbf{q}_a$ and velocities $\dot{\mathbf{q}}_a$ only. Therefore, it is preferable to introduce into the vector of generalized coordinates $\mathbf{q}$ additional variables, denoted here as $\mathbf{q}_d$, that will help in obtaining kinetic and potential energies in a simpler form. Those variables will be, in the case of a parallel robot, all passive joint variables and platform Cartesian coordinates. Obviously, the variables $\mathbf{q}_d$ are not independent and can be linked to the active joint variables $\mathbf{q}_a$ through the use of constraint equations (see Sect. 7.1):

$$\mathbf{h}(\mathbf{q}_a, \mathbf{q}_d) = \mathbf{0} \tag{6.22}$$

and:

$$\mathbf{A}(\mathbf{q}_a, \mathbf{q}_d)\dot{\mathbf{q}}_d + \mathbf{B}(\mathbf{q}_a, \mathbf{q}_d)\dot{\mathbf{q}}_a = \mathbf{0} \tag{6.23}$$

where $\mathbf{A} = \left[\frac{\partial \mathbf{h}}{\partial \mathbf{q}_d}\right]$ and $\mathbf{B} = \left[\frac{\partial \mathbf{h}}{\partial \mathbf{q}_a}\right]$ are two matrices depending on $\mathbf{q}_a$ and $\mathbf{q}_d$.

Taking into account those constraints, the Lagrange equations must be rewritten by using the Lagrange multipliers $\boldsymbol{\lambda}$ (see Appendix B):

$$\begin{aligned} \boldsymbol{\tau} + \mathbf{B}^T\boldsymbol{\lambda} = \boldsymbol{\tau}_a, \text{ where } \boldsymbol{\tau}_a &= \frac{\mathrm{d}}{\mathrm{d}t}\left(\frac{\partial L}{\partial \dot{\mathbf{q}}_a}\right)^T - \left(\frac{\partial L}{\partial \mathbf{q}_a}\right)^T \\ \mathbf{A}^T\boldsymbol{\lambda} = \boldsymbol{\tau}_d, \text{ where } \boldsymbol{\tau}_d &= \frac{\mathrm{d}}{\mathrm{d}t}\left(\frac{\partial L}{\partial \dot{\mathbf{q}}_d}\right)^T - \left(\frac{\partial L}{\partial \mathbf{q}_d}\right)^T. \end{aligned} \tag{6.24}$$

We will consider two cases:

- **Case 1**: the matrix $\mathbf{A}$ is square and of full rank (this particularity will appear for computation of the kinematics of parallel robots without redundancy and with kinematic redundancy). In such a case, rewriting (6.23), we also have

$$\dot{\mathbf{q}}_d = -\mathbf{A}^{-1}\mathbf{B}\dot{\mathbf{q}}_a = \mathbf{J}\dot{\mathbf{q}}_a \tag{6.25}$$

where $\mathbf{J}$ is the Jacobian matrix relating the velocities $\dot{\mathbf{q}}_d$ to the active joint velocities $\dot{\mathbf{q}}_a$.

Eliminating $\boldsymbol{\lambda}$ from (6.24), we thus obtain

$$\begin{aligned} \boldsymbol{\lambda} &= \mathbf{A}^{-T}\boldsymbol{\tau}_d \\ \boldsymbol{\tau} &= \boldsymbol{\tau}_a - \mathbf{B}^T\mathbf{A}^{-T}\boldsymbol{\tau}_d = \boldsymbol{\tau}_a + \mathbf{J}^T\boldsymbol{\tau}_d. \end{aligned} \tag{6.26}$$

Note that (6.26) can also be rewritten as:

$$\boldsymbol{\tau} = \boldsymbol{\tau}_a + \left[\frac{\partial \mathbf{q}_d}{\partial \mathbf{q}_a}\right]^T \boldsymbol{\tau}_d. \tag{6.27}$$

Since $\mathbf{h}$ is not an explicit function in time, we may notice that

$$\left[\frac{\partial \mathbf{q}_d}{\partial \mathbf{q}_a}\right] = \left[\frac{\partial \dot{\mathbf{q}}_d}{\partial \dot{\mathbf{q}}_a}\right]. \tag{6.28}$$

- **Case 2**: the matrix $\mathbf{A}$ is rectangular with more rows than columns (this particularity will appear for computation of the kinematics of parallel robots with actuation redundancy) but the matrix $\mathbf{B}$ is square and of full rank. In such a case, rewriting (6.23), we also have

$$\dot{\mathbf{q}}_a = -\mathbf{B}^{-1}\mathbf{A}\dot{\mathbf{q}}_d = \mathbf{J}_{inv}\dot{\mathbf{q}}_d \tag{6.29}$$

where $\mathbf{J}_{inv}$ is the inverse Jacobian matrix relating the velocities $\dot{\mathbf{q}}_a$ to the active joint velocities $\dot{\mathbf{q}}_d$. Note that $\mathbf{J}_{inv}$ has more rows than columns.
Eliminating $\boldsymbol{\lambda}$ from (6.24), we now obtain

$$\begin{aligned} \boldsymbol{\lambda} &= -\mathbf{B}^{-T}(\boldsymbol{\tau} - \boldsymbol{\tau}_a) \\ \boldsymbol{\tau}_d &= -\mathbf{A}^T\mathbf{B}^{-T}(\boldsymbol{\tau} - \boldsymbol{\tau}_a) = \mathbf{J}_{inv}^T(\boldsymbol{\tau} - \boldsymbol{\tau}_a) \end{aligned} \tag{6.30}$$

which can be rewritten as

$$\mathbf{J}_{inv}^T\boldsymbol{\tau} = \mathbf{J}_{inv}^T\boldsymbol{\tau}_a + \boldsymbol{\tau}_d. \tag{6.31}$$

Equation (6.31) represents the dynamic model. The matrix $\mathbf{J}_{inv}^T$ having more columns than rows (i.e. the system (6.31) having more unknowns than equations), there is an infinity of solutions for the vector of the input efforts $\boldsymbol{\tau}$ which are all given by

$$\boldsymbol{\tau} = \boldsymbol{\tau}_a + \mathbf{J}_{inv}^{T+}\boldsymbol{\tau}_d + (\mathbf{1} - \mathbf{J}_{inv}^{T+}\mathbf{J}_{inv}^T)\boldsymbol{\eta} \tag{6.32}$$

in which $\mathbf{J}_{inv}^{T+}$ is the pseudo-inverse of the matrix $\mathbf{J}_{inv}^T$ and $\boldsymbol{\eta}$ is an arbitrary vector in the $\boldsymbol{\tau}$ space which is called the overconstraint. If $\boldsymbol{\eta} = \mathbf{0}$, we get the solution for $\boldsymbol{\tau}$ with the minimal norm.

This formalism will be used for obtaining the dynamic model of the rigid parallel robots in Part II, in which more detailed examples will be provided.

6.1.5 Dynamic Model Properties

In this section, we summarize some important properties of the dynamic model of robots:

1. The system inertia matrix $\mathbf{M}(\mathbf{q})$ is symmetric and positive definite, as well as the body $\mathscr{B}_j$ generalized inertia matrix $\mathbf{M}_j$;

2. The energy of body $\mathscr{B}_j$ is a function of $\mathbf{q}$ and $\dot{\mathbf{q}}$,
3. The kinetic energy, the potential energy and the inverse dynamic model are linear in the elements of the standard inertial parameters m_j, ${}^j\mathbf{ms}_j$ and ${}^j\mathbf{I}_{O_j}$. Grouping all those parameters in the vector $\boldsymbol{\chi}^{st}$, we can write:

$$E = \mathbf{e}(\mathbf{q}, \dot{\mathbf{q}})\boldsymbol{\chi}^{st}, \;\; U = \mathbf{u}(\mathbf{q})\boldsymbol{\chi}^{st}, \;\; \boldsymbol{\tau} = \mathbf{C}(\mathbf{q}, \dot{\mathbf{q}}, \ddot{\mathbf{q}})\boldsymbol{\chi}^{st}. \tag{6.33}$$

This property is exploited to identify the inertial parameters [(see (Briot and Gautier 2012)], to reduce the computation burden of the dynamic model, and to develop adaptive control schemes (Khalil and Dombre 2002).

6.2 The Newton-Euler Equations

The Newton-Euler (*NE*) equations allow computation of the sum of external forces $\Sigma\mathbf{f}_j$ and moments $\Sigma\mathbf{m}_{S_j}$ (including gravity effects) acting on the *COM* S_j of body $\mathscr{B}_j$ that are equal to:

$$\begin{aligned} \Sigma\mathbf{f}_j &= m_j \dot{\mathbf{v}}_{S_j} \\ \Sigma\mathbf{m}_{S_j} &= \mathbf{I}_{S_j}\dot{\boldsymbol{\omega}}_j + \boldsymbol{\omega}_j \times (\mathbf{I}_{S_j}\boldsymbol{\omega}_j) \end{aligned} \tag{6.34}$$

in which:

- $\dot{\mathbf{v}}_{S_j}$ is the acceleration of the *COM* of link $\mathscr{B}_j$,
- $\dot{\boldsymbol{\omega}}_j$ is the rotational acceleration of link $\mathscr{B}_j$,
- $\mathbf{I}_{S_j}$ is the inertia matrix of link $\mathscr{B}_j$, expressed at its *COM*.

The *NE* equations can be also expressed at the origin of the frame $\mathscr{F}_j$ attached to the body $\mathscr{B}_j$. In that case, they take the form (Khalil and Dombre 2002):

$$\begin{aligned} \Sigma\mathbf{f}_j &= m_j \dot{\mathbf{v}}_j + \dot{\boldsymbol{\omega}}_j \times \mathbf{ms}_j + \boldsymbol{\omega}_j \times (\boldsymbol{\omega}_j \times \mathbf{ms}_j) \\ \Sigma\mathbf{m}_j &= \mathbf{I}_{O_j}\dot{\boldsymbol{\omega}}_j + \mathbf{ms}_j \times \dot{\mathbf{v}}_j + \boldsymbol{\omega}_j \times (\mathbf{I}_{O_j}\boldsymbol{\omega}_j) \end{aligned} \tag{6.35}$$

in which:

- $\Sigma\mathbf{m}_j$ is the sum of external moment applied at the origin of the frame $\mathscr{F}_j$,
- $\dot{\mathbf{v}}_j$ is the acceleration of the origin of the frame $\mathscr{F}_j$,
- $\mathbf{ms}_j$ is the vector of the first moments of inertia defined in Sect. 6.1.2,
- $\mathbf{I}_{O_j}$ is the inertia matrix defined in Sect. 6.1.2.

Using the screw notation, we can rewrite these equations as

$$\Sigma\mathbf{w}_j = \begin{bmatrix} \Sigma\mathbf{f}_j \\ \Sigma\mathbf{m}_j \end{bmatrix} = \begin{bmatrix} m_j \mathbf{1}_3 & \widehat{\mathbf{ms}}_j^T \\ \widehat{\mathbf{ms}}_j & \mathbf{I}_{O_j} \end{bmatrix} \begin{bmatrix} \dot{\mathbf{v}}_j \\ \dot{\boldsymbol{\omega}}_j \end{bmatrix} + \begin{bmatrix} \boldsymbol{\omega}_j \times (\boldsymbol{\omega}_j \times \mathbf{ms}_j) \\ \boldsymbol{\omega}_j \times (\mathbf{I}_{O_j}\boldsymbol{\omega}_j) \end{bmatrix} = \mathbf{M}_j\dot{\mathbf{t}}_j + \mathbf{c}_j \tag{6.36}$$

where

- $\mathbf{M}_j$ is the (6×6) augmented inertia matrix of the body $\mathscr{B}_j$,
- $\mathbf{c}_j$ is the vector of Coriolis and centrifugal effects,
- $\dot{\mathbf{t}}_j$ is the derivative of the body $\mathscr{B}_j$ twist w.r.t. time.

The *NE* equations will be used in Part II to obtain the symbolic expression of the dynamic model of rigid parallel robots.

6.3 The Principle of Virtual Powers

The principle of virtual powers (*PVP*) (le Rond d'Alembert, 1743) states that, at equilibrium, *the power P^*_{acc} developed by the inertial effects of a body (or a system of bodies) moving with a virtual velocity $\mathbf{v}^*$ is equal to the power P^*_{ext} developed by the external forces applied on the body (or on the system of bodies) plus the power P^*_{int} developed by the internal forces applied in the body (or in the system of bodies)*. Throughout this book, all quantities followed by the superscript "*" will be considered as virtual quantities.

Let us consider a single rigid link $\mathscr{B}_j$ on which is attached the frame $\mathscr{F}_j$ (the case of flexible bodies will be considered in Part III). For that body, the *PVP* can be written as:

$$P^*_{acc} = P^*_{int} + P^*_{ext}. \tag{6.37}$$

By definition, the virtual power due to the inertial effects is equal to:

$$P^*_{acc} = \int_{\mathscr{B}_j} \dot{\mathbf{v}}^T_{M_j} \mathbf{v}^*_{M_j} \mathrm{d}m. \tag{6.38}$$

Note that $\dot{\mathbf{v}}_{M_j} dm$ are the *NE* equations applied to a particle of mass $\mathrm{d}m$.

The velocity and acceleration of point M_j can be expressed as a function of the twist $\mathbf{t}_j$ and acceleration quantities $\dot{\mathbf{t}}_j$ of the frame $\mathscr{F}_j$ through the use of (3.2) and (3.19):

$$\mathbf{v}_{M_j} = \mathbf{v}_j + \boldsymbol{\omega}_j \times \mathbf{r}_{O_j M_j} \tag{6.39}$$

$$\dot{\mathbf{v}}_{M_j} = \dot{\mathbf{v}}_j + \boldsymbol{\omega}_j \times \left(\boldsymbol{\omega}_j \times \mathbf{r}_{O_j M_j}\right) + \dot{\boldsymbol{\omega}}_j \times \mathbf{r}_{O_j M_j}. \tag{6.40}$$

Substituting (6.39) into (6.38), it turns out that:

$$P^*_{acc} = \left(\int_{\mathscr{B}_j} \dot{\mathbf{v}}_{M_j} \mathrm{d}m\right)^T \mathbf{v}^*_j + \left(\int_{\mathscr{B}_j} \mathbf{r}_{O_j M_j} \times \dot{\mathbf{v}}_{M_j} \mathrm{d}m\right)^T \boldsymbol{\omega}^*_j. \tag{6.41}$$

Adding (6.40) into (6.41) and developing the result leads to:

$$\begin{aligned} P^*_{acc} &= \left(\int_{\mathscr{B}_j} \mathbf{1}_3 \, dm \, \dot{\mathbf{v}}_j\right)^T \mathbf{v}^*_j + \left(\boldsymbol{\omega}_j \times \left(\boldsymbol{\omega}_j \times \int_{\mathscr{B}_j} \mathbf{r}_{O_jM_j} \, dm\right)\right)^T \mathbf{v}^*_j \\ &+ \left(\dot{\boldsymbol{\omega}}_j \times \int_{\mathscr{B}_j} (\mathbf{r}_{O_jM_j}) \, dm\right)^T \mathbf{v}^*_j + \left(\left(\int_{\mathscr{B}_j} \mathbf{r}_{O_jM_j} \, dm\right) \times \dot{\mathbf{v}}_j\right)^T \boldsymbol{\omega}^*_j \\ &+ \left(\boldsymbol{\omega}_j \times \left(\int_{\mathscr{B}_j} \hat{\mathbf{r}}_{O_jM_j} \hat{\mathbf{r}}^T_{O_jM_j} \, dm\right) \boldsymbol{\omega}_j\right)^T \boldsymbol{\omega}^*_j \\ &+ \left(\left(\int_{\mathscr{B}_j} \hat{\mathbf{r}}_{O_jM_j} \hat{\mathbf{r}}^T_{O_jM_j} \, dm\right) \dot{\boldsymbol{\omega}}_j\right)^T \boldsymbol{\omega}^*_j \end{aligned} \quad (6.42)$$

which can be simplified as, from (6.8),

$$\begin{aligned} P^*_{acc} &= \left(m_j \dot{\mathbf{v}}_j + \boldsymbol{\omega}_j \times \left(\boldsymbol{\omega}_j \times \mathbf{ms}_j\right) + \dot{\boldsymbol{\omega}}_j \times \mathbf{ms}_j\right)^T \mathbf{v}^*_j \\ &+ \left(\mathbf{ms}_j \times \dot{\mathbf{v}}_j + \boldsymbol{\omega}_j \times (\mathbf{I}_{O_j} \boldsymbol{\omega}_j) + \mathbf{I}_{O_j} \dot{\boldsymbol{\omega}}_j\right)^T \boldsymbol{\omega}^*_j. \end{aligned} \quad (6.43)$$

It should be mentioned that (6.43) could be rewritten as:

$$P^*_{acc} = \Sigma \mathbf{f}^T_j \mathbf{v}^*_j + \Sigma \mathbf{m}^T_j \boldsymbol{\omega}^*_j \quad (6.44)$$

where $\Sigma \mathbf{f}_j$ and $\Sigma \mathbf{m}_j$ are defined in (6.35) as the *NE* equations.

The virtual power of the external efforts can be divided into two parts,

$$P^*_{ext} = P^*_{grav} + P^*_{reac} \quad (6.45)$$

where

$$P^*_{grav} = \int_{\mathscr{B}_j} \left(\mathbf{v}^*_{M_j}\right)^T \mathbf{g} \, dm \quad (6.46)$$

is the virtual power of the gravity field $\mathbf{g}$ and P^*_{reac} is the virtual power of other external effects. In the case where two wrenches $\mathbf{w}^T_{O_j} = [\mathbf{f}^T_{O_j} \; \mathbf{m}^T_{O_j}]$ applied on point O_j and $\mathbf{w}^T_{B_j} = [\mathbf{f}^T_{B_j} \; \mathbf{m}^T_{B_j}]$ applied on point B_j act on the link,

$$P^*_{reac} = \mathbf{f}^T_{O_j} \mathbf{v}^*_j + \mathbf{m}^T_{O_j} \boldsymbol{\omega}^*_j + \mathbf{f}^T_{B_j} \mathbf{v}^*_{B_j} + \mathbf{m}^T_{B_j} \boldsymbol{\omega}^*_j. \quad (6.47)$$

Thus, introducing (6.39) in the previous expressions,

$$P^*_{grav} = \left(\int_{\mathscr{B}_j} \mathbf{g} \, dm\right)^T \mathbf{v}^*_j + \left(\int_{\Sigma_j} \mathbf{r}_{O_jM_j} \times \mathbf{g} \, dm\right)^T \boldsymbol{\omega}^*_j \quad (6.48)$$

which, from (6.8), can be rewritten as:

$$P^*_{grav} = m_j \mathbf{g}^T \mathbf{v}^*_j + \left(\widehat{\mathbf{ms}}_j \mathbf{g}\right)^T \boldsymbol{\omega}^*_j \tag{6.49}$$

and

$$P^*_{reac} = \left(\mathbf{f}_{O_j} + \mathbf{f}_{B_j}\right)^T \mathbf{v}^*_j + \left(\mathbf{m}_{O_j} + \mathbf{m}_{B_j} + \mathbf{r}_{O_j B_j} \times \mathbf{f}_{B_j}\right)^T \boldsymbol{\omega}^*_j. \tag{6.50}$$

Finally, the virtual power due to the internal effects is null for a free rigid body (no deformation and no dissipative effects), i.e. $P^*_{int} = 0$.

Thus, introducing (6.43), (6.49) and (6.50) into (6.37), and taking into account the fact that the virtual velocities $\mathbf{v}^*_j$ and $\boldsymbol{\omega}^*_j$ are independent, two equilibrium equations can be obtained:

$$\mathbf{f}_{O_j} + \mathbf{f}_{B_j} + m_j \mathbf{g} = m_j \dot{\mathbf{v}}_j + \boldsymbol{\omega}_j \times \left(\boldsymbol{\omega}_j \times \mathbf{ms}_j\right) + \dot{\boldsymbol{\omega}}_j \times \mathbf{ms}_j \tag{6.51}$$

$$\mathbf{m}_{O_j} + \mathbf{m}_{B_j} + \mathbf{r}_{O_j B_j} \times \mathbf{f}_{B_j} + \mathbf{ms}_j \times \mathbf{g} = \mathbf{ms}_j \times \dot{\mathbf{v}}_j + \boldsymbol{\omega}_j \times (\mathbf{I}_{O_j} \boldsymbol{\omega}_j) + \mathbf{I}_{O_j} \dot{\boldsymbol{\omega}}_j \tag{6.52}$$

which are equivalent to the *NE* equations of a rigid body on which are applied the gravitational effects and two wrenches $\mathbf{w}_{O_j}$ and $\mathbf{w}_{B_j}$.

The expression (6.37) can be easily extended to be used for a system of bodies.

6.4 Computation of Actuator Input Efforts Under a Wrench Exerted on the End-Effector

A well-known relation in robotics, called the static model which gives the joint torques in terms of the external wrench on the end-effector, states that:

$$\boldsymbol{\tau} = \mathbf{J}_n^T \mathbf{w}_n \tag{6.53}$$

where $\boldsymbol{\tau}$ the vector of the input efforts, $\mathbf{J}_n$ is the kinematic Jacobian matrix defined at (5.7) relating the end-effector twist to the actuated joint velocities, and $\mathbf{w}_n$ is a wrench applied on the end-effector. This formula can be demonstrated through the use of the *PVP* as follows.

Let us consider a virtual twist $\mathbf{t}^*_n$ at the end-effector which is obtained thanks to the virtual joint velocity vector $\dot{\mathbf{q}}^*_n$. In the absence of any other effects, the wrench $\mathbf{w}_n$ leads to the robot input efforts $\boldsymbol{\tau}$. So, the power conservation states that:

$$\dot{\mathbf{q}}_n^{*T} \boldsymbol{\tau} = \mathbf{t}_n^{*T} \mathbf{w}_n. \tag{6.54}$$

From (5.8), we have

$$\mathbf{t}_n^* = \mathbf{J}_n \dot{\mathbf{q}}_n^*. \tag{6.55}$$

Introducing (6.55) in (6.54) leads to:

$$\dot{\mathbf{q}}_n^{*T} \boldsymbol{\tau} = \dot{\mathbf{q}}_n^{*T} \mathbf{J}_n^T \mathbf{w}_n. \tag{6.56}$$

The virtual velocity $\dot{\mathbf{q}}_n^*$ being arbitrary, we can simplify the relation (6.56) to (6.53).

Part II
Dynamics of Rigid Parallel Robots

Chapter 7
Kinematics of Parallel Robots

Abstract The study of parallel robot dynamics cannot be done without using their kinematic relationships and studying their singularities. Therefore, in the present chapter, we investigate the computation of inverse and forward geometric and kinematic models of the usual *PKM*. The models of several robots will be solved:

- The planar five-bar mechanism,
- The planar 3–*R<u>P</u>R* planar parallel mechanism (*PPM*),
- The Orthoglide,
- The Gough-Stewart platform.

These robots have been chosen because they are typical examples found in the literature and/or their dynamic model will be defined later in the book.

7.1 Inverse Geometric Model

7.1.1 General Methodology

Let us consider a general *PKM* composed of a rigid fixed base (denoted as the element $\mathscr{B}_0$ on which is attached the global frame $\mathscr{F}_0(O, \mathbf{x}_0, \mathbf{y}_0, \mathbf{z}_0)$), a rigid moving platform (element $\mathscr{B}_p$) and n legs. Each leg is a kinematic chain (which is serial most of the time, but can also be composed of closed-loop or tree-structure sub-chains[1]) composed of bodies connected by m_i joints located at points A_{ij} (revolute, prismatic, universal, etc.—$i = 1, \ldots, n$, $j = 1, \ldots, m_i$) (Fig. 7.1—in this figure, the gray pairs denote the actuated joints). The jth link of the leg i will be denoted in what follows as the link $\mathscr{B}_{ij}$. Moreover, the joint located at point A_{ij} will be parameterized by the variable q_{ij}.

[1]At the end of the Chap. 7 on the kinematics of *PKM*, we will present the geometry and kinematic equations by considering *PKM*made of serial legs only. However, the methodology can be extended to any types of legs. Moreover, the equations of the dynamics presented in Chap. 8 are general and can be used for any types of legs made of serial, closed-loop or tree-structure chains.

S. Briot and W. Khalil, *Dynamics of Parallel Robots*,
Mechanisms and Machine Science 35, DOI 10.1007/978-3-319-19788-3_7

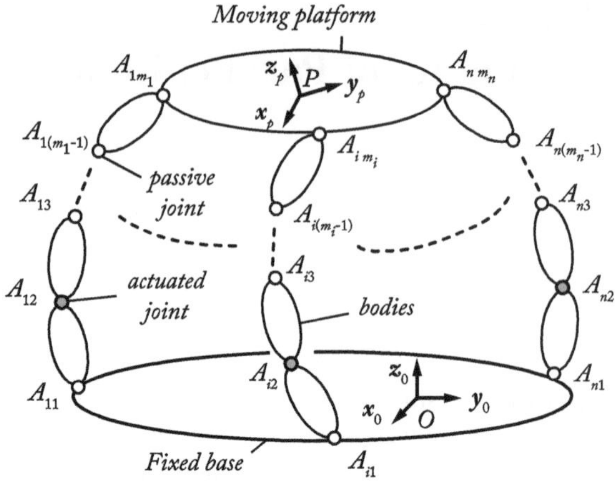

Fig. 7.1 A general parallel robot (the *gray pairs* denote the actuated joints)

The actuated variables are stacked into the vector $\mathbf{q}_a$ and the leg passive variables into the vector $\mathbf{q}_d$. The platform coordinates are denoted as $\mathbf{x}_p$; they represent the location of the frame $\mathscr{F}_p(P, \boldsymbol{x}_p, \boldsymbol{y}_p, z_p)$ attached to the end-effector w.r.t. the base frame $\mathscr{F}_0$. Only n_{dof} coordinates of $\mathbf{x}_p$ are independent, where n_{dof} is the number of *DOF* of the platform the *PKM*. They will be denoted as $\mathbf{x}$ and are related to all platform coordinates by the constraint equations:

$$\mathbf{c}_p(\mathbf{x}, \mathbf{x}_p) = \mathbf{0}. \tag{7.1}$$

The size n_a of $\mathbf{q}_a$ must be equal or superior to the number n_{dof} of *DOF* of the platform of the *PKM*. All the active and passive variables are grouped into the vector $\mathbf{q}_t$ of size $n_t = \sum_{i=1}^{n} m_i$.

The inverse geometric model (*IGM*) consists in obtaining the value of the joint coordinates $\mathbf{q}_a$ as a function of the platform coordinates $\mathbf{x}$, i.e. $\mathbf{q}_a = \mathscr{H}(\mathbf{x})$. For usual *PKM*, this problem is relatively easy to solve in a closed form solution. An additional problem, which must be solved for computing the dynamic model of the *PKM*, consists in finding all joint variables $\mathbf{q}_t$ as a function of the platform coordinates $\mathbf{x}$, i.e. $\mathbf{q}_t = \mathscr{H}_t(\mathbf{x})$.

The most used approach for solving these two problems consists in considering the *loop-closure equations* of the *PKM* through the serial chain i going from point O to point P. In that case, the location of the end-effector frame $\mathscr{F}_p$ can be obtained by considering all the displacements of the considered chain, similarly as for an open chain robot (Sect. 4.2.3).

By using the results of Sect. 4.2.3, the transformation allowing computation of the position and orientation of the frame $\mathscr{F}_p$ with respect to the frame $\mathscr{F}_0$ is given by:

$${}^0\mathbf{T}_p(\mathbf{x}) = {}^0\mathbf{T}_{i1}(q_{i1}) \prod_{j=2}^{m_i} \left({}^{i(j-1)}\mathbf{T}_{ij}(q_{ij})\right) {}^{im_i}\mathbf{T}_p, \text{ for } i = 1, \ldots, n \tag{7.2}$$

where

- ${}^0\mathbf{T}_p(\mathbf{x})$ is the homogeneous matrix defining the transformation from the base frame $\mathscr{F}_0$ to the end-effector frame $\mathscr{F}_p$,

$$
{}^0\mathbf{T}_p = \begin{bmatrix} & {}^0\mathbf{R}_p & & x \\ & & & y \\ & & & z \\ 0 & 0 & 0 & 1 \end{bmatrix}
$$

where ${}^0\mathbf{R}_p$ is the rotation matrix between the frames $\mathscr{F}_0$ and $\mathscr{F}_p$ that can be defined by any set of parameters presented in Chap. 2, and x, y and z are the translational components of $\mathbf{x}_p$ expressed in the base frame $\mathscr{F}_0$,
- ${}^0\mathbf{T}_{i1}(q_{i1})$ is the homogeneous matrix defining the transformation between the base frame $\mathscr{F}_0$ and the frame $\mathscr{F}_{i1}$ attached to the link $\mathscr{B}_{i1}$,
- ${}^{i(j-1)}\mathbf{T}_{ij}(q_{ij})$ is the homogeneous matrix defining the location between the frame $\mathscr{F}_{i(j-1)}$ attached to the link $\mathscr{B}_{i(j-1)}$ and the frame $\mathscr{F}_{ij}$ attached to the link $\mathscr{B}_{ij}$. Note that, in the case of a constant transformation, q_{ij} is a constant parameter,
- ${}^{im_i}\mathbf{T}_p$ is the homogeneous matrix defining the constant transformation between the frame $\mathscr{F}_{im_i}$ attached to the link $\mathscr{B}_{im_i}$ and the platform frame $\mathscr{F}_p$. It can be determined by the *MDH* method with only four parameters at most if $\boldsymbol{x}_{im_i}$ is defined along the common orthogonal of z_p and z_{im_i}.

All these homogeneous transformation matrices are defined in Chap. 2.

Thus, Eq. (7.2) allows us to get implicit relations between the joint and platform coordinates $\mathbf{q}_t$ and $\mathbf{x}$, which can be written in the form:

$$
\mathbf{h}(\mathbf{x}, \mathbf{q}_t) = \begin{bmatrix} \mathbf{h}_{11}(\mathbf{x}) - \mathbf{h}_{21}(\mathbf{q}_t) \\ \vdots \\ \mathbf{h}_{1n}(\mathbf{x}) - \mathbf{h}_{2n}(\mathbf{q}_t) \end{bmatrix} = \mathbf{0}. \tag{7.3}
$$

In general, each row of (7.3) is similar to the *IGM* of tree-structure robots whose solutions can give $\mathbf{q}_i$ in terms of $\mathbf{x}$, $\mathbf{q}_i$ being the joint coordinates of the sub-chain i. Many methods have been developed to solve that problem in a general manner, such as the methods proposed in Khalil and Dombre (2002), Pfurner and Husty (2010). Those methods are not recalled here, because most of proposed *PKM* are made of simple legs that do not require the use of complex methods for solving the *IGM*. Note that each row of (7.3) can be solved independently.

In most cases, the problem complexity can be reduced by eliminating $\mathbf{q}_d$ such that the active actuator coordinates $\mathbf{q}_a$ are directly linked to the moving platform coordinates $\mathbf{x}$ using a relation of the form:

$$
\mathbf{h}_p(\mathbf{x}, \mathbf{q}_a) = \mathbf{0} \tag{7.4}
$$

where $\mathbf{h}_p(\mathbf{x}, \mathbf{q}_a)$ is a vector for which we can define:

- $n_a = n_{dof}$ independent equations for parallel robots without redundancy,
- n_a equations among which only n_{dof} are independent for parallel robots with actuation redundancy (for which $n_a > n_{dof}$) (Müller 2005),
- n_{dof} independent equations for parallel robots with kinematic redundancy (for which also $n_a > n_{dof}$) (Kotlarski et al. 2010).

In a general manner, the active joint coordinates of the leg i which are grouped in the vector $\mathbf{q}_{ai}$ can be obtained by solving the reduced problem:

$$\mathbf{h}_{pi}(\mathbf{x}, \mathbf{q}_{ai}) = \mathbf{0} \tag{7.5}$$

where $\mathbf{h}_{pi}$ is a part of the vector $\mathbf{h}_p$ defined in (7.4).

Equation (7.4) is usually quite easy to solve under the form $\mathbf{q}_a = \mathscr{H}(\mathbf{x})$, as shown in the examples developed below. It is necessary to mention that the solution may not be unique. In the case when more than one solution exist, the different solutions are called the *working modes* of the robot. They correspond, for one given set of platform coordinates, to all the possible ways to assembly the mechanism legs (see the Sects. 7.1.2.1–7.1.2.5).

Finally, once the values of $\mathbf{q}_a$ are found from (7.4) as a function of $\mathbf{x}$, it is possible to introduce them into (7.3) in order to express all joint coordinates $\mathbf{q}_t$ as a function of $\mathbf{x}$. Note that sometimes, $\mathbf{q}_a$ and $\mathbf{q}_d$ can be obtained in the same step. Once again, this problem could be difficult to solve in the most general case. However, it is generally easy for usual *PKM* and, even for more complicated cases, it can now be solved using the advanced mathematical methods mentioned above (Khalil and Dombre 2002; Pfurner and Husty 2010).

In order to simplify and/or clarify the problem understanding, the following geometrical approach could be also used. The idea is to virtually cut the leg i at one given joint (generally located at the middle of the leg—but not necessarily—e.g. without loss of generality at the joint located at point A_{i2}—denoted in the following as joint A_{i2}). Then, virtually considering that the platform is fixed (and as a result the location of point $A_{i\,m_i}$) and that all joints can freely move, the configuration loci of the frame associated with joint A_{i2} (in terms of translations plus rotations) when belonging to the lower part of the leg (loci denoted as $\mathscr{C}_{i1}$—(Fig. 7.2a) and when belonging to the upper part of the leg (loci denoted as $\mathscr{C}_{i2}$—(Fig. 7.2b) can be computed. The solutions of the *IGM* are at the intersections of those two configuration loci (Fig. 7.2c).

It must be mentioned that, in usual cases, the obtained configuration loci $\mathscr{C}_{ij}$ are defined by algebraic equations. Therefore, for having an idea of the maximal number of intersection points, usual methods (such as the Bézout bounds Bézout 1764) can be used.

In the next sections we present the *IGM* of some *PKM* that are solved by analytical methods and/or by geometrical approaches.

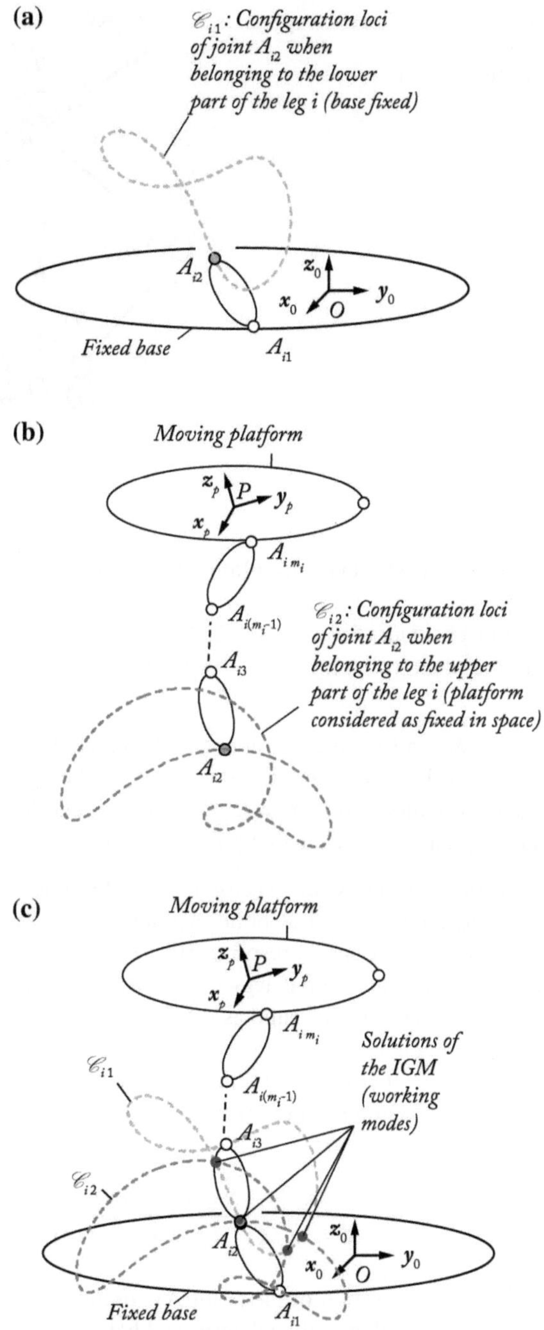

Fig. 7.2 The solutions of the *IGM* of a general *PKM*

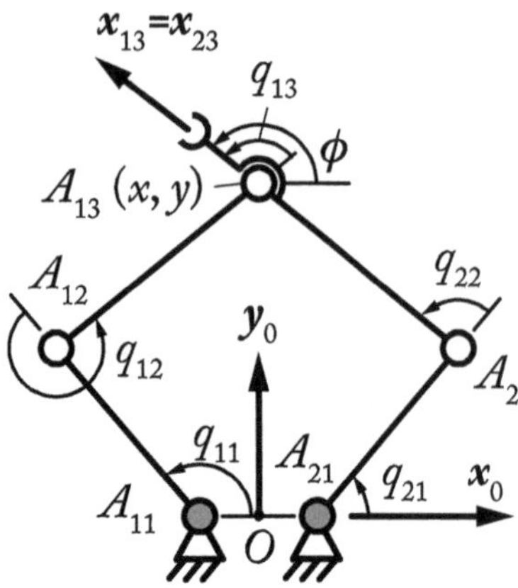

Fig. 7.3 The planar five-bar mechanism (the *gray pairs* denote the actuated joints)

7.1.2 Examples

7.1.2.1 Inverse Geometric Model of the Planar Five-Bar Mechanism

The planar five-bar mechanism (Fig. 7.3) is a 2 *DOF* parallel robot able to achieve two translations in the plane $(O, \boldsymbol{x}_0, \boldsymbol{y}_0)$ (see Appendix A) and which is composed of two legs:

- A leg composed of 3 *R* joints whose axes are parallel, directed along z_0 and located at points A_{11}, A_{12} and A_{13}, the joint located at point A_{11} being actuated, and
- A leg composed of 2 *R* joints whose axes are parallel, directed along z_0 and located at points A_{21} and A_{22}, the joint located at point A_{21} being actuated.

All other joints are passive. Thus, the vector of actuated coordinates is $\mathbf{q}_a^T = [q_{11}\ q_{21}]$. The end-effector is located at point A_{13} and its controlled coordinates along $\boldsymbol{x}_0$ and $\boldsymbol{y}_0$ are denoted as x and y, respectively.

The *MDH* parameters of the five-bar legs are shown in Table 7.1. It should be mentioned here that frame $\mathcal{F}_{23}$ is the same as the frame $\mathcal{F}_{13}$, but its antecedent is the frame $\mathcal{F}_{22}$.

As the robot has only 2 *DOF*, the orientation ϕ of the end-effector depends on the coordinates x and y. The constraint relations on the platform coordinates (7.1) can be written as:

$$\tan\phi = \frac{y - y_{A_{22}}}{x - x_{A_{22}}} = \frac{y - d_{22}\sin q_{21}}{x - d_{22}\cos q_{21} - d_{21}} \tag{7.6}$$

in which $[x_{A_{22}}\ y_{A_{22}}]^T$ is the position of the point A_{22} in the base frame. The value of q_{21} is a function of x and y that will be determined below.

From Table 7.1, the right-hand side of (7.2) can be computed. Considering in the present example that the end-effector location is parameterized by the homogeneous transformation

Table 7.1 The *MDH* parameters of the five-bar legs

ij	$a(ij)$	μ_{ij}	σ_{ij}	γ_{ij}	b_{ij}	α_{ij}	d_{ij}	θ_{ij}	r_{ij}
11	0	1	0	0	0	0	$d_{11} = -l_{OA_{11}}$	q_{11}	0
12	11	0	0	0	0	0	$d_{12} = l_{A_{11}A_{12}}$	q_{12}	0
13	12	0	0	0	0	0	$d_{13} = l_{A_{12}A_{13}}$	q_{13}	0
21	0	1	0	0	0	0	$d_{21} = l_{OA_{21}}$	q_{21}	0
22	21	0	0	0	0	0	$d_{22} = l_{A_{21}A_{22}}$	q_{22}	0
23	22	0	2	0	0	0	$d_{23} = l_{A_{22}A_{13}}$	0	0

$$
{}^0\mathbf{T}_p = \begin{bmatrix} \cos\phi & -\sin\phi & 0 & x \\ \sin\phi & \cos\phi & 0 & y \\ 0 & 0 & 1 & 0 \\ 0 & 0 & 0 & 1 \end{bmatrix}
$$

we get:

$$
\begin{bmatrix} x \\ y \end{bmatrix} = \begin{bmatrix} d_{i1} + d_{i2}\cos q_{i1} + d_{i3}\cos(q_{i1} + q_{i2}) \\ d_{i2}\sin q_{i1} + d_{i3}\sin(q_{i1} + q_{i2}) \end{bmatrix}, \text{ for } i = 1, 2 \tag{7.7}
$$

where $d_{23} = l_{A_{22}A_{13}}$, and

$$
\begin{aligned} \phi &= q_{11} + q_{12} + q_{13}, \text{ when considering leg 1} \\ \phi &= q_{21} + q_{22}, \text{ when considering leg 2.} \end{aligned} \tag{7.8}
$$

Rearranging the terms of (7.7), it can be found that

$$
\begin{bmatrix} x - d_{i1} - d_{i2}\cos q_{i1} \\ y - d_{i2}\sin q_{i1} \end{bmatrix} = \begin{bmatrix} d_{i3}\cos(q_{i1} + q_{i2}) \\ d_{i3}\sin(q_{i1} + q_{i2}) \end{bmatrix}. \tag{7.9}
$$

Then, squaring both sides of (7.9) and equating both lines of (7.8), the reduced form (7.4) of the loop-closure equations can be obtained as:

$$
\mathbf{h}_p(\mathbf{x}, \mathbf{q}_a) = \begin{bmatrix} (x - d_{11} - d_{12}\cos q_{11})^2 + (y - d_{12}\sin q_{11})^2 - d_{13}^2 \\ (x - d_{21} - d_{22}\cos q_{21})^2 + (y - d_{22}\sin q_{21})^2 - d_{23}^2 \end{bmatrix} = \mathbf{0} \tag{7.10}
$$

which can be simplified as:

$$
\mathbf{h}_p(\mathbf{x}, \mathbf{q}_a) = \begin{bmatrix} a_1\cos q_{11} + b_1\sin q_{11} + c_1 \\ a_2\cos q_{21} + b_2\sin q_{21} + c_2 \end{bmatrix} = \mathbf{0} \tag{7.11}
$$

where $a_i = -2d_{i2}\,(x - d_{i1})$, $b_i = -2d_{i2}y$, $c_i = (x - d_{i1})^2 + y^2 + d_{i2}^2 - d_{i3}^2$.

Taking into account the half-angle formula:

$$\cos q_{ij} = \frac{1-t_{ij}^2}{1+t_{ij}^2}, \ \sin q_{ij} = \frac{2t_{ij}}{1+t_{ij}^2}, \tag{7.12}$$

where $t_{ij} = \tan(q_{ij}/2)$, and introducing those expressions into (7.11), we get:

$$\mathbf{h}_p(\mathbf{x}, \mathbf{q}_a) = \begin{bmatrix} (c_1 - a_1)t_{11}^2 + 2b_1 t_{11} + (a_1 + c_1) \\ (c_2 - a_2)t_{21}^2 + 2b_2 t_{21} + (a_2 + c_2) \end{bmatrix} = \mathbf{0} \tag{7.13}$$

from which we can find that

$$t_{i1} = \frac{-b_i \pm \sqrt{b_i^2 - c_i^2 + a_i^2}}{c_i - a_i} \tag{7.14}$$

or also

$$q_{i1} = 2\tan^{-1}\left(\frac{-b_i \pm \sqrt{b_i^2 - c_i^2 + a_i^2}}{c_i - a_i}\right). \tag{7.15}$$

Note that:

- the value of q_{11} can be found from (7.7) for $i = 1$ without considering the Eq. (7.7) for $i = 2$.
- the value of q_{21} can be found from (7.7) for $i = 2$ without considering the Eq. (7.7) for $i = 1$.

This means that Eq. (7.7) can be solved independently.

In (7.15), the sign "±" correspond to the different working modes of the robot (Fig. 7.4). From a geometric point of view, solving these equations is equivalent to finding the intersection points of two circles (Fig. 7.4):

- Circle $\mathscr{C}_{i1}$ centred in A_{i1} of radius d_{i2}, which corresponds to the vertex space of the point A_{i2} when considering that it belongs to the link $A_{i1}A_{i2}$,
- Circle $\mathscr{C}_{i2}$ centred in A_{i3} (considered as fixed if the coordinates x and y are known) of radius d_{i3}, which corresponds to the vertex space of the point A_{i2} when considering that it belongs to the link $A_{i2}A_{i3}$.

The values of q_{12} and q_{22} can be obtained using (7.9) as

$$q_{i2} = \text{atan2}\,(y - d_{i2}\sin q_{i1}, x - d_{i1} - d_{i2}\cos q_{i1}) - q_{i1}, \text{ for } i = 1, 2 \tag{7.16}$$

where "atan2" is the four-quadrant inverse tangent function.

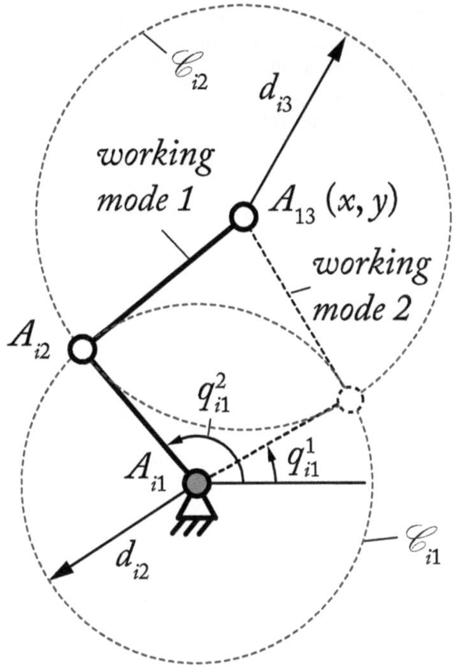

Fig. 7.4 The two working modes of the leg i of the planar five-bar mechanism

Finally, the value of q_{13} can be directly obtained by introducing (7.15) and (7.16) into the last equation of (7.13), i.e. $q_{13} = q_{21} + q_{22} - q_{11} - q_{12}$.

7.1.2.2 Inverse Geometric Model of the 3–*R*P*R* Planar Parallel Manipulator

The 3–*R*P*R* *PPM* (Fig. 7.5) is a 3 *DOF* parallel robot able to achieve two translations in the plane $(O, \boldsymbol{x}_0, \boldsymbol{y}_0)$ and one rotation around z_0 and which is composed of three legs, each leg being made of two passive *R* joints (with respective axes (A_{i1}, z_0) and (A_{i3}, z_0)) and one active *P* joint whose direction is contained in the plane $(O, \boldsymbol{x}_0, \boldsymbol{y}_0)$ and whose configuration is parameterized by the variable q_{i2} $(i = 1, \ldots, 3)$.

Thus, the vector of actuated coordinates is $\mathbf{q}_a^T = [q_{12}\ q_{22}\ q_{22}]$. The end-effector is located at point P and its coordinates along $\boldsymbol{x}_0$ and $\boldsymbol{y}_0$ are denoted as x and y, respectively. The orientation ϕ of the platform is defined as the angle between $\boldsymbol{x}_0$ and the segment $A_{13}A_{23}$.

The *MDH* parameters of the 3–*R*P*R* *PPM* legs are shown in Table 7.2. In this table, $\gamma_i = \text{atan2}(y_{OA_{i1}}, x_{OA_{i1}})$, where $x_{OA_{i1}}$ and $y_{OA_{i1}}$ are the coordinates of the points OA_{i1} expressed in the base frame $\mathscr{F}_0$.

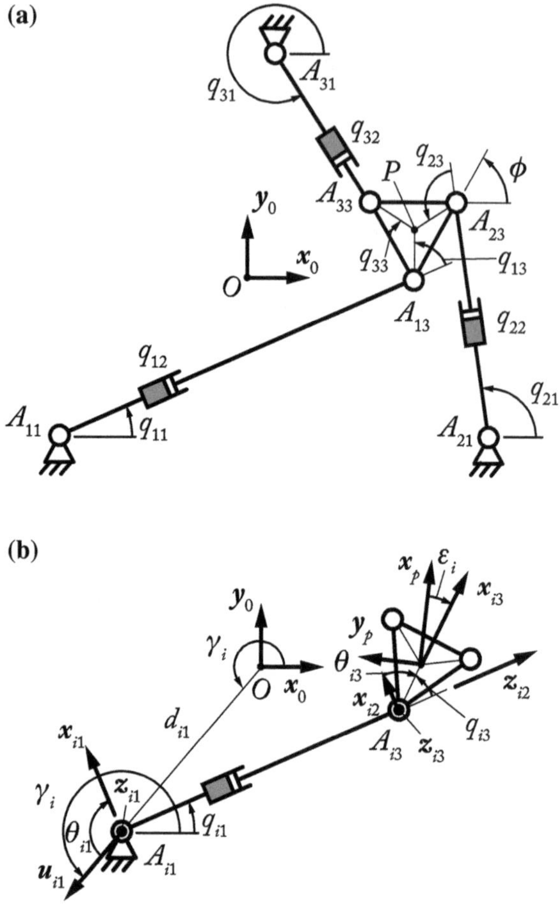

Fig. 7.5 The 3–*R<u>P</u>R PPM*. **a** Kinematic architecture (the *gray pairs* denote the actuated joints). **b** Frames associated to the *MDH* parameters

Table 7.2 The *MDH* parameters of the 3–*R<u>P</u>R PPM* legs ($i = 1, \ldots, 3$)

ij	$a(ij)$	μ_{ij}	σ_{ij}	γ_{ij}	b_{ij}	α_{ij}	d_{ij}	θ_{ij}	r_{ij}
$i1$	0	1	0	γ_i	0	0	$d_{i1} = l_{OA_{i1}}$	$q_{i1} + \gamma_i - \pi/2$	0
$i2$	$i1$	0	1	0	0	$\pi/2$	0	0	q_{i2}
$i3$	$i2$	0	0	0	0	$-\pi/2$	0	$q_{i3} - \pi/2$	0
p^*	$i3$	0	2	0	0		$d_{i4} = l_{A_{i3}P}$	$-\varepsilon_i$	0

Note '∗' the upperscript p denotes the platform frame
$\varepsilon_i = \text{atan2}({}^p y_{A_{i3}P}, {}^p x_{A_{i3}P})$, where ${}^p x_{A_{i3}P}$ and ${}^p y_{A_{i3}P}$ are the coordinates of the points A_{i3} expressed in the moving platform frame $\mathscr{F}_p$

From Table 7.2, the right-hand side of (7.2) can be computed. Considering in the present example that the end-effector location is parameterized by the homogeneous transformation

$$
{}^{0}\mathbf{T}_p = \begin{bmatrix} \cos\phi & -\sin\phi & 0 & x \\ \sin\phi & \cos\phi & 0 & y \\ 0 & 0 & 1 & 0 \\ 0 & 0 & 0 & 1 \end{bmatrix}
$$

it turns out that:

$$
\begin{bmatrix} x \\ y \\ \phi \end{bmatrix} = \begin{bmatrix} x_{OA_{i1}} + q_{i2}\cos q_{i1} + d_{i4}\cos(q_{i1}+q_{i3}) \\ y_{OA_{i1}} + q_{i2}\sin q_{i1} + d_{i4}\sin(q_{i1}+q_{i3}) \\ q_{i1} + q_{i3} - \varepsilon_i \end{bmatrix}, \text{ for } i = 1, \ldots, 3. \quad (7.17)
$$

Rearranging the terms of the two first rows of (7.17) and introducing the fact that $\phi + \varepsilon_i = q_{i1} + q_{i3}$ which is deduced from the third line, it can be found that

$$
\begin{bmatrix} x - x_{OA_{i1}} - d_{i4}\cos(\phi+\varepsilon_i) \\ y - y_{OA_{i1}} - d_{i4}\sin(\phi+\varepsilon_i) \end{bmatrix} = \begin{bmatrix} q_{i2}\cos q_{i1} \\ q_{i2}\sin q_{i1} \end{bmatrix}. \quad (7.18)
$$

Squaring both sides of (7.18) and summing the two rows, it comes that:

$$
\mathbf{h}_p(\mathbf{x}, \mathbf{q}_a) = \begin{bmatrix} \left(x_{OA_{13}} - x_{OA_{11}}\right)^2 + \left(y_{OA_{13}} - y_{OA_{11}}\right)^2 - q_{12}^2 \\ \left(x_{OA_{23}} - x_{OA_{21}}\right)^2 + \left(y_{OA_{23}} - y_{OA_{21}}\right)^2 - q_{22}^2 \\ \left(x_{OA_{33}} - x_{OA_{31}}\right)^2 + \left(y_{OA_{33}} - y_{OA_{31}}\right)^2 - q_{32}^2 \end{bmatrix} = \mathbf{0}. \quad (7.19)
$$

where $x_{OA_{i3}} = x - d_{i4}\cos(\phi+\varepsilon_i)$, $y_{OA_{i3}} = y - d_{i4}\sin(\phi+\varepsilon_i)$ are the coordinates of points OA_{i3} expressed in the base frame $\mathcal{F}_0$.

It can be finally deduced from (7.19) that, for $i = 1, \ldots, 3$:

$$
q_{i2} = \pm\sqrt{\left(x_{OA_{i3}} - x_{OA_{i1}}\right)^2 + \left(y_{OA_{i3}} - y_{OA_{i1}}\right)^2}. \quad (7.20)
$$

Finally, the passive variables can be found by:

$$
\begin{aligned} q_{i1} &= \operatorname{atan2}\left(y_{OA_{i3}} - y_{OA_{i1}}, x_{OA_{i3}} - x_{OA_{i1}}\right), \text{ if } q_{i2} \geq 0 \\ q_{i1} &= \operatorname{atan2}\left(y_{OA_{i3}} - y_{OA_{i1}}, x_{OA_{i3}} - x_{OA_{i1}}\right) + \pi, \text{ if } q_{i2} < 0 \end{aligned} \quad (7.21)
$$

and

$$
q_{i3} = \phi + \varepsilon_i - q_{i1}. \quad (7.22)
$$

In (7.20), the sign "±" correspond to the different working modes of the robot (which are, in that particular case, equivalent—Fig. 7.6). From a geometric point of view, solving these equations is equivalent to finding the distance between the points A_{i1} and A_{i3}.

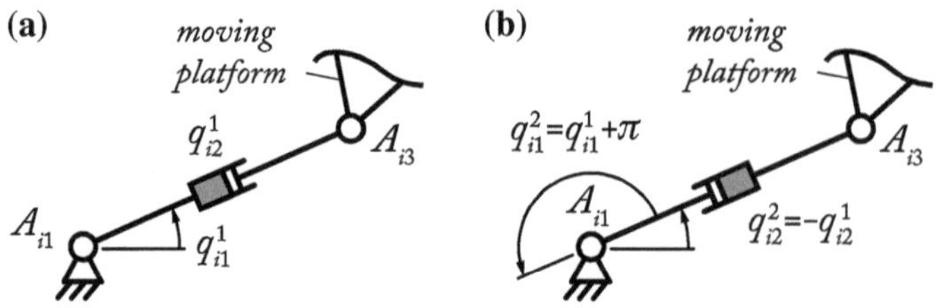

Fig. 7.6 The two working modes of the leg i of the 3–*R<u>P</u>R PPM*. **a** Working mode 1. **b** Working mode 2

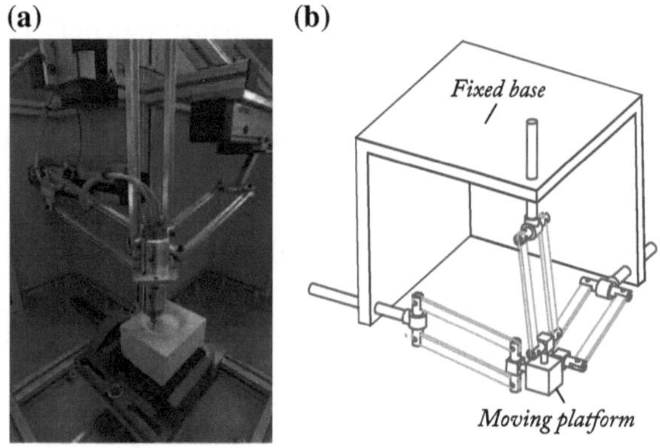

Fig. 7.7 The Orthoglide robot. **a** Prototype. **b** Kinematic chain (Courtesy of Philippe Wenger and Damien Chablat)

7.1.2.3 Inverse Geometric Model of Other Planar Parallel Manipulators

The geometrical and analytical methodologies presented in Sects. 7.1.2.1 and 7.1.2.2 can be easily extended to any types of planar parallel robots and will not be developed here. The reader is referred to Bonev (2002) for further investigations.

7.1.2.4 Inverse Geometric Model of the Orthoglide

The Orthoglide is a parallel robot composed of three identical legs (Fig. 7.7) allowing three translational *DOF* of its end-effector (parameterized by the variables x, y and z that represent respectively the translation along $\boldsymbol{x}_0, \boldsymbol{y}_0$ and $\boldsymbol{z}_0$ of the base frame $\mathscr{F}_0$). Each leg is made of one linearly actuated link (parameterized by the variables q_{i1}, $i = 1, \ldots, 3$, i.e. $\mathbf{q}_a^T = [q_{11}\; q_{21}\; q_{31}]$) linked at its extremity to a spatial parallelogram (Fig. 7.8a). The parallelogram is also attached to the mobile platform. Kinematically speaking for obtaining the inverse kinematics, and without loss of generality, each parallelogram chain can be replaced by an equivalent chain composed of two *U*joints connected by a rigid link (Fig. 7.7b).

The directions of the three linear actuators of the Orthoglide are orthogonal (Fig. 7.7b). The purpose is to create a mechanism with a workspace shape close

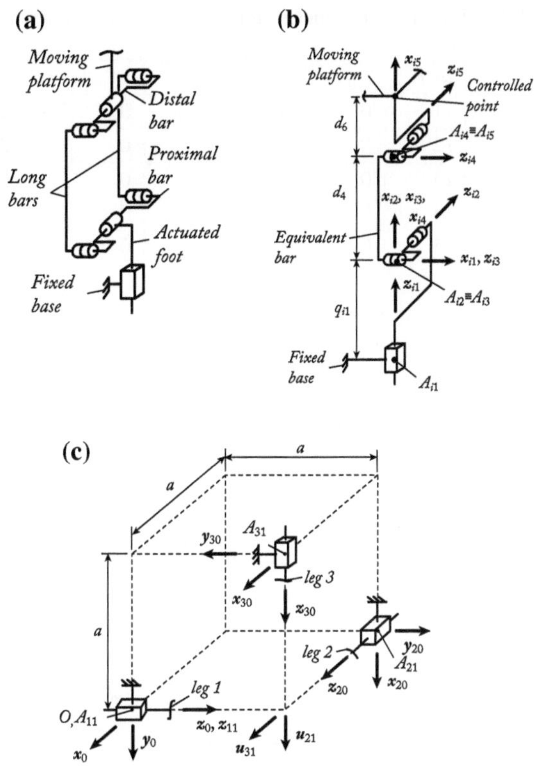

Fig. 7.8 Description of the Orthoglide kinematic chain. **a** Kinematics of one leg. **b** Equivalent kinematics of one leg. **c** Fixed base

Table 7.3 *MDH* parameters for the first frames corresponding to the first bodies of the Orthoglide legs

ij	$a(ij)$	μ_{ij}	σ_{ij}	γ_{ij}	b_{ij}	α_{ij}	d_{ij}	θ_{ij}	r_{ij}
11	0	1	1	0	0	0	0	0	q_{11}
21	0	1	1	$\pi/2$	a	$\pi/2$	0	0	$q_{21} - a$
31	0	1	1	0	a	$-\pi/2$	0	0	$q_{31} - a$

to a cube and whose behavior is close to the isotropy (Merlet 2006a) wherever it is located in its workspace (Chablat and Wenger 2003).

The leg *MDH* parameters of the equivalent kinematic chains for the legs are given in Tables 7.3 and 7.4.

From Tables 7.3 and 7.4, the right-hand side of (7.2) can be computed. Considering in the present example that the end-effector location is parameterized by the homogeneous transformation,

$$
{}^{0}\mathbf{T}_p = \begin{bmatrix} 1 & 0 & 0 & x \\ 0 & 1 & 0 & y \\ 0 & 0 & 1 & z \\ 0 & 0 & 0 & 1 \end{bmatrix}
$$

Table 7.4 *MDH* parameters for the frames corresponding to the leg i of the Orthoglide ($i = 1, \ldots, 3$)

ij	$a(ij)$	μ_{ij}	σ_{ij}	γ_{ij}	b_{ij}	α_{ij}	d_{ij}	θ_{ij}	r_{ij}
$i2$	$i1$	0	0	0	0	$-\pi/2$	0	q_{i2}	0
$i3$	$i2$	0	0	0	0	$-\pi/2$	0	q_{i3}	0
$i4$	$i3$	0	0	0	0	0	d_4	q_{i4}	0
$i5$	$i4$	0	0	0	0	$\pi/2$	0	q_{i5}	0
p	$i5$	0	2	0	0	$\alpha_{1p} = \pi/2$, $\alpha_{2p} = -\pi/2$, $\alpha_{3p} = -\pi$	d_6	0	0

it get that, for leg 1:

$$\begin{bmatrix} x \\ y \\ z \end{bmatrix} = \begin{bmatrix} d_4 \cos q_{12} \cos q_{13} + d_6 \\ -d_4 \sin q_{13} \\ q_{11} - d_4 \cos q_{13} \sin q_{12} \end{bmatrix} \tag{7.23}$$

for leg 2:

$$\begin{bmatrix} x \\ y \\ z \end{bmatrix} = \begin{bmatrix} -a + q_{21} - d_4 \cos q_{23} \sin q_{22} \\ d_4 \cos q_{22} \cos q_{23} + d_6 \\ a - d_4 \sin q_{23} \end{bmatrix} \tag{7.24}$$

and for leg 3:

$$\begin{bmatrix} x \\ y \\ z \end{bmatrix} = \begin{bmatrix} d_4 \cos q_{32} \cos q_{33} + d_6 \\ -a + q_{31} - d_4 \cos q_{33} \sin q_{32} \\ a + d_4 \sin q_{33} \end{bmatrix} \tag{7.25}$$

where a, d_4 and d_6 are geometric parameters defined in Fig. 7.8a, and, to take into account the parallelogram constraints, we have

$$\begin{bmatrix} 0 \\ 0 \end{bmatrix} = \begin{bmatrix} q_{i2} + q_{i5} \\ q_{i3} + q_{i4} \end{bmatrix} \tag{7.26}$$

Simplifying (7.23)–(7.25), it turns out that:

$$\mathbf{h}_p = \begin{bmatrix} (x - d_6)^2 + y^2 + (z - q_{11})^2 - d_4^2 \\ (x + a - q_{21})^2 + (y - d_6)^2 + (z - a)^2 - d_4^2 \\ (x - d_6)^2 + (y + a - q_{31})^2 + (z - a)^2 - d_4^2 \end{bmatrix} = \mathbf{0}. \tag{7.27}$$

Developing, each row of (7.27) leads to a polynomial of the second order in q_{i1}

$$q_{i1}^2 + c_{i1} q_{i1} + c_{i0} = 0, \text{ for } i = 1, \ldots, 3 \tag{7.28}$$

where

$$c_{11} = -2z$$
$$c_{10} = (x - d_6)^2 + y^2 + z^2 - d_4^2$$
$$c_{21} = -2(x + a)$$
$$c_{20} = (x + a)^2 + (y - d_6)^2 + (z - a)^2 - d_4^2$$
$$c_{31} = -2(y + a)$$
$$c_{30} = (x - d_6)^2 + (y + a)^2 + (z - a)^2 - d_4^2$$

from which we can find:

$$q_{i1} = \frac{-c_{i1} \pm \sqrt{c_{i1}^2 - 4c_{i0}}}{2}. \tag{7.29}$$

Finally, the passive variables can be found from (7.23) by:

$$q_{12} = -q_{15} = \operatorname{atan2}(q_{11} - z,\ x - d_6) \tag{7.30}$$
$$q_{22} = -q_{25} = \operatorname{atan2}(q_{21} - x - a,\ y - d_6) \tag{7.31}$$
$$q_{32} = -q_{35} = \operatorname{atan2}(q_{31} - y - a,\ x - d_6) \tag{7.32}$$
$$q_{13} = -q_{14} = \operatorname{atan2}(-y,\ (x - d_6)/\cos q_{12}) \tag{7.33}$$
$$q_{23} = -q_{24} = \operatorname{atan2}(-z,\ (y - d_6)/\cos q_{22}) \tag{7.34}$$
$$q_{33} = -q_{34} = \operatorname{atan2}(z,\ (x - d_6)/\cos q_{32}) \tag{7.35}$$

In (7.29), the sign "±" corresponds to the two different working modes of the robot (Fig. 7.9). From a geometric point of view, solving these equations is equivalent to finding the intersection between a line $\mathscr{L}_i$ defining the displacement of the active prismatic joints and a sphere $\mathscr{S}_i$ that represents the displacement of the point A_{i2} when the platform is fixed and the leg is virtually broken at point A_{i2} (Fig. 7.9).

7.1.2.5 Inverse Geometric Model of the Gough-Stewart Platform

The 6–$U\underline{P}S$ *PKM*, also called the Gough-Stewart platform, is a robot composed of six legs, each leg being made of a passive U joint fixed on the base, followed by an active $\underline{P}$ joint and then a passive S joint (Fig. 7.10).

The *MDH* parameters associated to the frames of Fig. 7.11 for one leg are given in Table 7.5. For simplifying the computation, the base connecting points A_{i1} are considered to all belong to the same plane $(O,\ \boldsymbol{x}_0,\ \boldsymbol{y}_0)$.

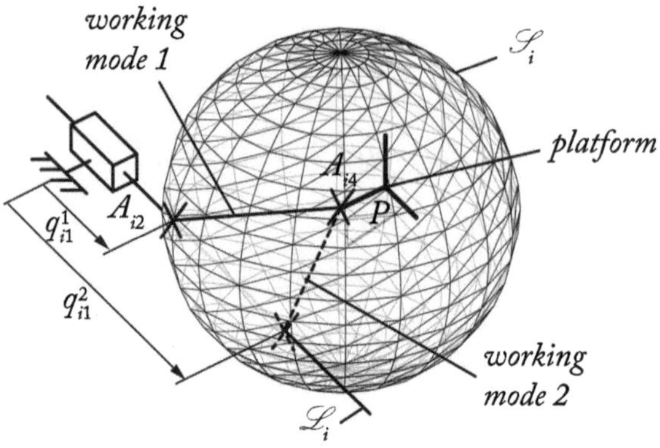

Fig. 7.9 The two working modes of the Orthoglide leg i

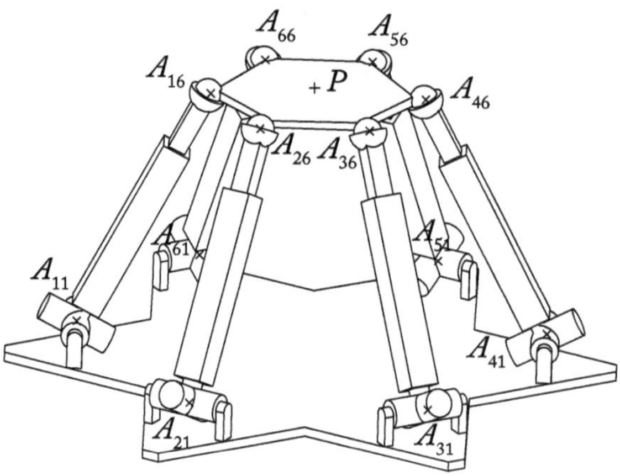

Fig. 7.10 The Gough-Stewart platform (6–$U\underline{P}S$ *SPM*)

Table 7.5 *MDH* parameters for the frames corresponding to the first three joints of the leg i of the 6–$U\underline{P}S$ *SPM*

ij	$a(ij)$	μ_{ij}	σ_{ij}	γ_{ij}	b_{ij}	α_{ij}	d_{ij}	θ_{ij}	r_{ij}
$i1$	0	0	0	γ_i	0	$-\pi/2$	d_{i1}	q_{i1}	0
$i2$	$i1$	0	0	0	0	$\pi/2$	0	q_{i2}	0
$i3$	$i2$	1	1	0	0	$\pi/2$	0	0	q_{i3}

$\gamma_i = \text{atan2}\left(y_{A_{i1}}, x_{A_{i1}}\right)$

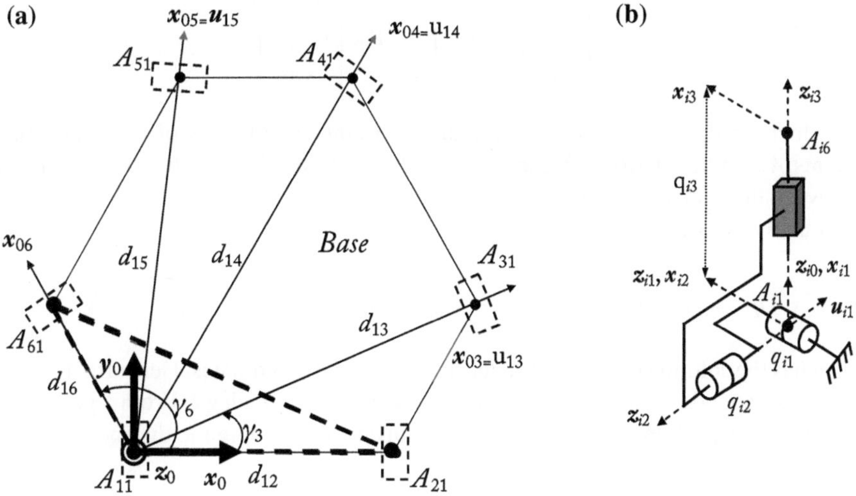

Fig. 7.11 Frames associated with the *MDH* parameters of the Gough-Stewart platform. **a** Frames associated to the base of the Gough-Stewart platform. **b** Frames associated to the leg *i*

The parameters corresponding to the *S* joint are deliberately omitted. The computation of the *S* joint coordinates is of no interest in that section as they have no effect on the dynamic model if their corresponding friction are neglected (Khalil and Ibrahim 2007). We will deliberately limit the analysis of the *IGM* of the Gough-Stewart to the computation of the active joint coordinates only (that can be obtained through the use of the translational part of (7.2) only). The computation of the passive coordinates could be a good exercise for the interested reader. In particular, the first and second joints of each leg which are needed in the dynamic model (Khalil and Ibrahim 2007).

Using the general loop-closure equations, it can be seen that the implicit equations linking the active coordinates to the platform coordinates are:

$$\mathbf{h}_p = \begin{bmatrix} (x_{A_{16}} - x_{A_{11}})^2 + (y_{A_{16}} - y_{A_{11}})^2 + (z_{A_{16}} - z_{A_{11}})^2 - q_{13}^2 \\ (x_{A_{26}} - x_{A_{21}})^2 + (y_{A_{26}} - y_{A_{21}})^2 + (z_{A_{26}} - z_{A_{21}})^2 - q_{23}^2 \\ (x_{A_{36}} - x_{A_{31}})^2 + (y_{A_{36}} - y_{A_{31}})^2 + (z_{A_{36}} - z_{A_{31}})^2 - q_{33}^2 \\ (x_{A_{46}} - x_{A_{41}})^2 + (y_{A_{46}} - y_{A_{41}})^2 + (z_{A_{46}} - z_{A_{41}})^2 - q_{43}^2 \\ (x_{A_{56}} - x_{A_{51}})^2 + (y_{A_{56}} - y_{A_{51}})^2 + (z_{A_{56}} - z_{A_{51}})^2 - q_{53}^2 \\ (x_{A_{66}} - x_{A_{61}})^2 + (y_{A_{66}} - y_{A_{61}})^2 + (z_{A_{66}} - z_{A_{61}})^2 - q_{63}^2 \end{bmatrix} = \mathbf{0} \quad (7.36)$$

where $x_{A_{i1}}$, $y_{A_{i1}}$ and $z_{A_{i1}}$ are the coordinates of the connecting points A_{i1} in the base frame $\mathscr{F}_0$ $(i = 1, \ldots, 6)$ and

$$\begin{bmatrix} x_{A_{i6}} \\ y_{A_{i6}} \\ z_{A_{i6}} \end{bmatrix} = \begin{bmatrix} x \\ y \\ z \end{bmatrix} + {}^0\mathbf{R}_p \begin{bmatrix} {}^p x_{A_{i6}} \\ {}^p y_{A_{i6}} \\ {}^p z_{A_{i6}} \end{bmatrix}$$

in which ${}^p x_{A_{i6}}$, ${}^p y_{A_{i6}}$ and ${}^p z_{A_{i6}}$ are the (constant) coordinates of the connecting points A_{i6} in the platform frame $\mathcal{F}_p$ ($i = 1, \ldots, 6$) and ${}^0\mathbf{R}_p$ is the rotation matrix between the frames $\mathcal{F}_p$ and $\mathcal{F}_0$.

Finally, we get

$$q_{i3} = \pm\sqrt{(x_{A_{i6}} - x_{A_{i1}})^2 + (y_{A_{i6}} - y_{A_{i1}})^2 + (z_{A_{i6}} - z_{A_{i1}})^2} \tag{7.37}$$

which is the solution of the *IGM* reduced to the active coordinates. Here also, the sign "±" corresponds to the two different working modes of the leg i. From a geometric point of view, solving these equations is equivalent to finding the distance between the points A_{i1} and A_{i6}.

7.2 Forward Geometric Model

7.2.1 General Methodology

The forward geometric model (*FGM*) consists in obtaining the value of platform coordinates $\mathbf{x}$ as a function of the active joint coordinates $\mathbf{q}_a$, starting from the expressions (7.3) or (7.4), i.e. to obtain $\mathbf{x} = \mathcal{H}^{-1}(\mathbf{q}_a) = \mathcal{G}(\mathbf{q}_a)$. An additional problem is to obtain the passive joint coordinates $\mathbf{q}_d$ (which are needed for the computation of the dynamic model) as a function of the active joint variables $\mathbf{q}_a$, i.e. $\mathbf{q}_d = \mathcal{G}_d(\mathbf{q}_a)$. The main idea that is usually followed is to adequately rearrange the Eq. (7.4) in order to suppress the translation parameters of the vector $\mathbf{x}$ so that a polynomial depending on the tangent, sine and/or cosine of the rotation parameters of the vector $\mathbf{x}$ can be obtained. To simplify the calculations, it is generally necessary to set the base frame origin O at one robot base anchor point (e.g. point A_{11} in Fig. 7.1) and the moving platform frame origin P at one robot platform anchor point (e.g. point A_{1m_1} in Fig. 7.1). Note that this choice can also be taken while solving the *IGM*.

In order to simplify and/or clarify understanding of the problem, the following geometrical approach could be also used. The idea is to virtually disconnect a leg from the robot moving platform, e.g. without loss of generality at the joint located at point A_{1m_1} (Fig. 7.12a). This joint will be denoted as joint A_{1m_1}. In that case, even when all the actuators are fixed, the moving platform gains one or more *DOF* and the joint A_{1m_1} can freely describe a configuration loci (translations plus rotations) denoted as $\mathcal{S}$. As joint A_{1m_1} also belongs to the first leg, it also describes, when the actuator of the leg 1 is fixed, another configuration loci denoted as $\mathcal{C}$ and called the *vertex space* of the considered leg (Fig. 7.12b). The solutions of the *FGM* are at the

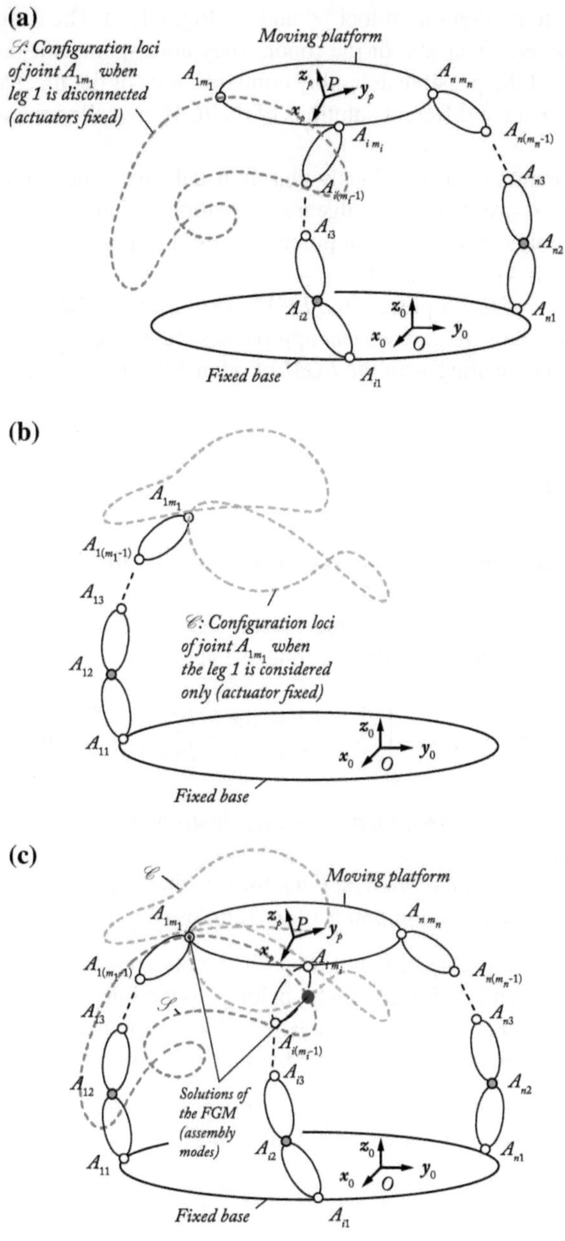

Fig. 7.12 The solutions of the *FGM* of a general *PKM*

intersection of the configuration loci $\mathscr{S}$ and $\mathscr{C}$ (Fig. 7.12c). The different solutions are called the *assembly modes* of the robot. They correspond, for one given set of active joints, to all the possible assembly configurations of the mechanism.

Let us recall that the different solutions of the *IGM* are called the *working modes* (see Sect. 7.1).

It must be mentioned as previously that, in usual cases, the obtained configuration loci $\mathscr{C}$ and $\mathscr{S}$ are defined by algebraic equations. Therefore, to get an idea of the maximal number of intersection points, the usual methods (such as the Bézout bounds Bézout 1764) can be used.

Sections 7.2.2.1–7.2.2.5 present the *FGM* of some *PKM* that are solved by analytical methods and/or by geometrical approaches. Moreover, in Sect. 7.2.2.6, a list of reference papers dealing with the *FGM* of other *SPM* is provided.

7.2.2 Examples

7.2.2.1 Forward Geometric Model of the Planar Five-Bar Mechanism

Let us consider again the five-bar mechanism presented in Sect. 7.1.2.1. Starting from (7.10) and developing the expressions, we get

$$\mathbf{h}_p(\mathbf{x}, \mathbf{q}_a) = \begin{bmatrix} x^2 + y^2 + a_1 x + b_1 y + c_1 \\ x^2 + y^2 + a_2 x + b_2 y + c_2 \end{bmatrix} = \mathbf{0} \tag{7.38}$$

where $a_i = -2(d_{i1} + d_{i2} \cos q_{i1})$, $b_i = -2 d_{i2} \sin q_{i1}$ and $c_i = (d_{i1} + d_{i2} \cos q_{i1})^2 + d_{i2}^2 \sin^2 q_{i1} - d_{i3}^2$.

From a geometric point of view, solving the two first equations of (7.38) is equivalent to finding the intersection points of two circles (Fig. 7.13):

- Circle $\mathscr{C}_1$ centered in A_{12} of radius d_{13}, which corresponds to the vertex space of the point A_{13} when considering that it belongs to the link $A_{12}A_{13}$,
- Circle $\mathscr{C}_2$ centered in A_{22} of radius d_{23}, which corresponds to the vertex space of the point A_{13} when considering that it belongs to the link $A_{22}A_{13}$.

Thus, the two robot assembly modes correspond to the intersection points of circles $\mathscr{C}_1$ and $\mathscr{C}_2$ whose expressions are, if $b_1 \neq b_2$,

$$\begin{aligned} x &= \frac{-f_2 \pm \sqrt{f_2^2 - 4 f_1 f_3}}{2 f_1} \\ y &= e_1 x + e_2 \end{aligned} \tag{7.39}$$

where

$$f_1 = 1 + e_1^2$$

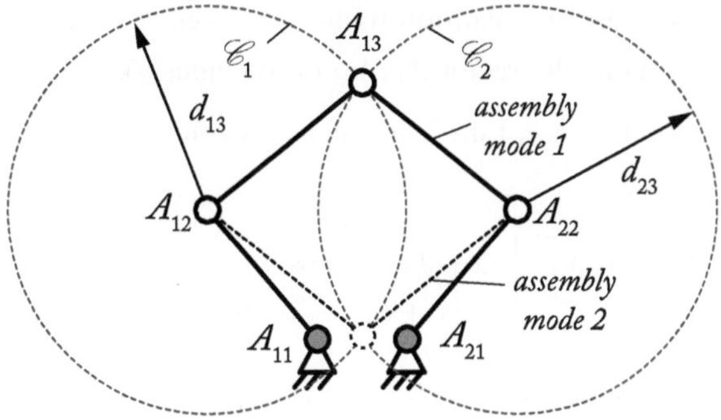

Fig. 7.13 The two assembly modes of the planar five-bar mechanism

$$
\begin{aligned}
f_2 &= 2e_1e_2 + a_1 + b_1e_1 \\
f_3 &= e_2^2 + b_1e_2 + c_1 \\
e_1 &= -d_1/d_2 \\
e_2 &= -d_3/d_2 \\
d_1 &= a_2 - a_1 \\
d_2 &= b_2 - b_1 \\
d_3 &= c_2 - c_1
\end{aligned}
\tag{7.40}
$$

or, if $b_1 = b_2$

$$
\begin{aligned}
x &= -\frac{c_2 - c_1}{a_2 - a_1} \\
y &= \frac{-b_1 \pm \sqrt{b_1^2 - 4(x^2 + a_1x + c_1)}}{2}.
\end{aligned}
\tag{7.41}
$$

Once the values of x and y are found, the expressions of the passive joint coordinates q_{12}, q_{13} and q_{22} can be obtained using the expressions given in Sect. 7.1.2.1.

7.2.2.2 Forward Geometric Model of the 3–$R\underline{P}R$ Planar Parallel Manipulator

Let us consider again the 3–$R\underline{P}R$ *PPM* presented in Sect. 7.1.2.2. The way to obtain the equations has been introduced in Gosselin et al. (1992) and this section makes only a recall of the mentioned work.

For reasons of simplicity of calculations, but without loss of generality, let us now consider that

- The base frame $\mathscr{F}_0$ is center in A_{11} (i.e. $O \equiv A_{11}$) and that axis $\boldsymbol{x}_0$ is along the vector $\overrightarrow{A_{11}A_{21}}$,

- The moving platform frame $\mathscr{F}_p$ is center in A_{31} (i.e. $P \equiv A_{31}$) and that axis $\boldsymbol{x}_p$ is along the vector $\overrightarrow{A_{13}A_{23}}$ (as previously).

In that case, Eq. (7.19) can be rewritten as

$$\mathbf{h}_p(\mathbf{x}, \mathbf{q}_a) = \begin{bmatrix} x^2 + y^2 - q_{12}^2 \\ \left(x + l_{A_{31}A_{32}} \cos\phi - l_{A_{11}A_{12}}\right)^2 + \left(x + l_{A_{31}A_{32}} \sin\phi\right)^2 - q_{22}^2 \\ \left(x + l_{A_{31}A_{33}} \cos(\phi + \varepsilon_3) - l_{A_{11}A_{13}} \cos\gamma_3\right)^2 + c - q_{32}^2 \end{bmatrix} = \mathbf{0} \tag{7.42}$$

where $c = \left(y + l_{A_{31}A_{33}} \sin(\phi + \varepsilon_3) - l_{A_{11}A_{13}} \sin\gamma_3\right)^2$ and γ_3 is defined in Sect. 7.1.2.2.

After developing, one can obtain the algebraic form

$$\mathbf{h}_p(\mathbf{x}, \mathbf{q}_a) = \begin{bmatrix} x^2 + y^2 - q_{12}^2 \\ R\,x + S\,y + Q \\ U\,x + V\,y + W \end{bmatrix} = \mathbf{0} \tag{7.43}$$

with

$$\begin{aligned}
R &= 2l_{A_{31}A_{32}} \cos\phi - 2l_{A_{11}A_{12}} \\
S &= 2l_{A_{31}A_{32}} \sin\phi \\
Q &= -2\,l_{A_{31}A_{32}}\,l_{A_{11}A_{12}} \cos\phi + l_{A_{31}A_{32}}^2 + l_{A_{11}A_{12}}^2 - q_{12}^2 + q_{22}^2 \\
U &= 2l_{A_{31}A_{33}} \cos(\phi + \varepsilon_3) - 2l_{A_{11}A_{13}} \cos\gamma_3 \\
V &= 2l_{A_{31}A_{32}} \sin(\phi + \varepsilon_3) - 2l_{A_{11}A_{13}} \sin\gamma_3 \\
W &= -2\,l_{A_{31}A_{33}}\,l_{A_{11}A_{13}} \cos(\phi + \varepsilon_3)\cos\gamma_3 - 2\,l_{A_{31}A_{33}}\,l_{A_{11}A_{13}} \sin(\phi + \varepsilon_3)\sin\gamma_3 \\
&\quad + l_{A_{31}A_{33}}^2 + l_{A_{11}A_{13}}^2 - q_{12}^2 + q_{22}^2.
\end{aligned} \tag{7.44}$$

The two last equations of (7.43) are linear in x and y and a solution that depends on ϕ is given by:

$$\begin{aligned}
x &= -\frac{-S\,W + V\,Q}{R\,V - S\,U} \\
y &= \frac{-R\,W + U\,Q}{R\,V - S\,U}
\end{aligned} \tag{7.45}$$

which is valid as long as $R\,V - S\,U \neq 0$.

Introducing (7.45) into the first row of (7.43) leads to

$$(-S\,W + V\,Q)^2 + (-R\,W + U\,Q)^2 - q_{12}^2(R\,V - S\,U) = 0 \tag{7.46}$$

which is a function that depends on $\cos\phi$ and $\sin\phi$. Taking into account that:

Fig. 7.14 Coupler surface for a 3–*R<u>P</u>R PPM* when leg 1 is disconnected and actuators of legs 2 and 3 are fixed

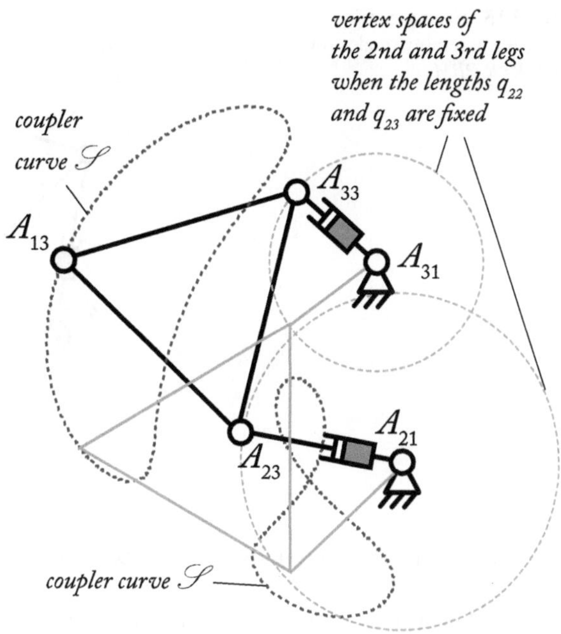

$$\cos\phi = \frac{1-t_\phi^2}{1+t_\phi^2},\ \sin\phi = \frac{2t_\phi}{1+t_\phi^2}, \tag{7.47}$$

where $t_\phi = \tan(\phi/2)$, and introducing those expressions into (7.46), it is shown in Gosselin et al. (1992) that a polynomial of degree 6 in t_ϕ can be obtained:

$$c_0 + c_1 t_\phi + c_2 t_\phi^2 + c_3 t_\phi^3 + c_4 t_\phi^4 + c_5 t_\phi^5 + c_6 t_\phi^6 = 0 \tag{7.48}$$

in which the coefficients c_i depend on the robot constant geometrical parameters and actuated variables q_{i2}. It must be mentioned that the coefficients are not given here and finding them could be a good exercise for the interested reader.

Thus, from (7.48), it can be deduced that the *FGM* can have up to 6 solutions, i.e. up to 6 assembly modes.

It is necessary to mention that the present equations miss some degenerated kinematic conditions that appear when $RV - SU = 0$. This problem is investigated in Wenger et al. (2007) and the interested reader should have a look at the mentioned paper.

From the geometric point of view, one could observe that, when the first leg of the robot is disconnected from point A_{13} and for constant values of the actuated variables q_{22} and q_{32}, the resulting mechanism is equivalent to a passive four-bar linkage whose legs can freely rotate around the *R* joints located at A_{21} and A_{31} (Fig. 7.14). As a result, the curve drawn by the point A_{13} (also called the *coupler curve*) when the

Fig. 7.15 Examples of possible assembly modes for a 3–*RPR PPM*

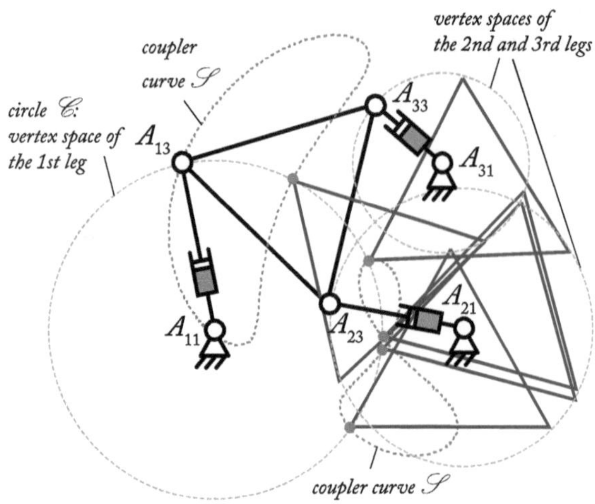

four-bar linkage is freely moving is a sextic curve $\mathscr{S}$, i.e. an algebraic curve of degree 6.

As point A_{13} also belongs to the first leg, it moves on a circle $\mathscr{C}$ centered in A_{11} and of radius q_{12} when the leg 1 actuated joint is fixed. The circle $\mathscr{C}$ is the vertex space of the leg 1. As a result, the solutions of the *FGM* of the 3–*RPR PPM* are at the intersection of the sextic curve $\mathscr{S}$ and the circle $\mathscr{C}$ and at most 6 solutions may exist (the proof is given in Merlet 2006b) (Fig. 7.15).

Once the values of x, y and ϕ are found, the expressions of the passive joint coordinates q_{i1} and q_{i3} can be obtained using the expressions given in Sect. 7.1.2.2.

7.2.2.3 Forward Geometric Model of the Other Planar Parallel Manipulators

The geometrical and analytical methodologies presented in Sects. 7.2.2.1 and 7.2.2.2 can be easily extended to other types of planar parallel robots. The reader is referred for instance to Briot et al. (2008) and Merlet (1997, 2006b) for further investigations.

7.2.2.4 Forward Geometric Model of the Orthoglide

Let us consider again the Orthoglide robot presented in Sect. 7.1.2.4. Rewriting the Eq. (7.27), we get

$$\mathbf{h}_p = \begin{bmatrix} x^2 + y^2 + z^2 + 2c_1x + 2c_2z + c_1^2 + c_2^2 - d_4^2 \\ x^2 + y^2 + z^2 + 2c_3x + 2c_4y + 2c_5z + c_3^2 + c_4^2 + c_5^2 - d_4^2 \\ x^2 + y^2 + z^2 + 2c_6x + 2c_7y + 2c_8z + c_6^2 + c_7^2 + c_8^2 - d_4^2 \end{bmatrix} = \mathbf{0} \quad (7.49)$$

where

$$\begin{aligned} c_1 &= -d_6 \\ c_2 &= -q_{11} \\ c_3 &= a - q_{21} \\ c_4 &= -d_6 \\ c_5 &= -a \\ c_6 &= -d_6 \\ c_7 &= a - q_{31} \\ c_8 &= -a. \end{aligned}$$

Solving (7.49), it can be found that

$$\begin{aligned} x &= c_9 z + c_{10} \\ y &= c_{11} z + c_{12} \\ z &= \frac{-c_{14} \pm \sqrt{c_{14}^2 - 4c_{13}c_{15}}}{2c_{13}} \end{aligned} \tag{7.50}$$

where

$$\begin{aligned} c_9 &= -(c_2 - c_5)c_7/(c_1c_4 + c_1c_7 - c_3c_7 - c_4c_6) \\ c_{10} &= -(c_1^2 + c_2^2 - c_3^2 - c_4^2 - c_5^2)c_4/(c_1c_4 + c_1c_7 - c_3c_7 - c_4c_6) \\ c_{11} &= (c_2 - c_8)(c_1 - c_6)/(c_1c_4 + c_1c_7 - c_3c_7 - c_4c_6) \end{aligned}$$

$$\begin{aligned} c_{12} &= -(c_1^2 + c_2^2 - c_6^2 - c_7^2 - c_8^2)(c_1 - c_3)/(c_1c_4 + c_1c_7 - c_3c_7 - c_4c_6) \\ c_{13} &= c_9^2 + c_{11}^2 + 1 \\ c_{14} &= 2(c_9c_{10} + c_{11}c_{12} + c_1c_9 - c_2) \\ c_{15} &= c_{10}^2 + c_{12}^2 + 2c_1c_{10} + c_1^2 + c_2^2 - d_4^2. \end{aligned}$$

The sign "$\pm$" in (7.50) corresponds to the two robot assembly modes (Fig. 7.16). Once the values of x, y and z are found, the expressions of the passive joint coordinates q_{i2}, q_{i3}, q_{i4} and q_{i5} can be found using the expressions given in Sect. 7.1.2.4.

From a geometric point of view, Eq. (7.49) are three equations of spheres $\mathscr{S}_i$ centered respectively in $O_1(0,\ 0,\ d_6+q_{11})$, $O_2(d_6-a+q_{21},\ 0,\ a)$ and $O_3(0,\ d_6-a+q_{31},\ a)$. Thus, the solutions of the *FGM* are the intersection points of those spheres (Fig. 7.16).

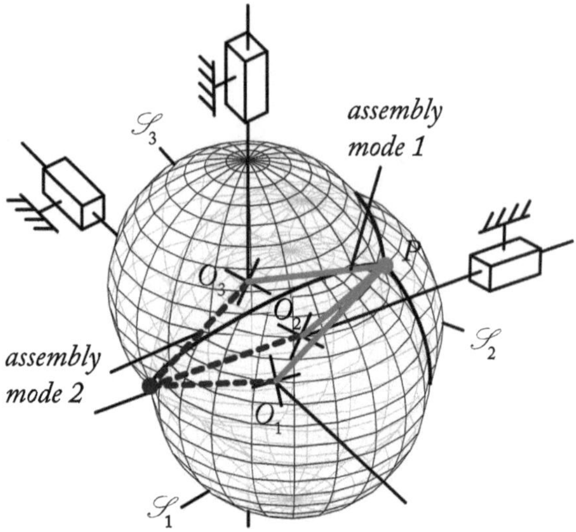

Fig. 7.16 The two assembly modes of the Orthoglide

7.2.2.5 Forward Geometric Model of the Gough-Stewart Platform

The *FGM* of the Gough-Stewart platform is probably one of the most complicated topics of the field. The ways to solve it will not be detailed here, but the aim of this section is to make brief recalls on the most relevant works concerning this problem so that the reader can have an idea of what could be interesting w.r.t. his own objectives.

In the most general case, 6–$U\underline{P}S$ *PKM* can have up to 40 assembly modes. This result was first shown in Ronga and Vust (1992), and then confirmed through different approaches proposed in Husty (1996), Lazard (1993), Mourrain (1993), Raghavan (1993) and Wampler (1996).

The first researchers who were able to give the expression of the univariate polynomial of degree 40 whose roots correspond to the assembly modes of the 6–$U\underline{P}S$ *PKM* were Husty in Husty (1996) and Wampler in Wampler (1996).

The number of solutions considerably decreases for special arrangement of the legs. For example, with the design proposed in Fig. 7.17 for which the legs 1, 2 and 3 (4 and 5, resp.) are linked to the same platform point A_{16} (A_{46}, resp.), the number of solutions is decreased to 8 and all of them can be obtained in a closed-form using the following method (Hunt and Primrose 1993; Nanua and Waldron 1991):

1. Knowing the lengths q_{13}, q_{23} and q_{33} of the legs 1, 2 and 3, compute the position of point A_{16} (that will be considered here as the platform controlled point with coordinates (x, y, z)) which is at the intersection of the three spheres centers in A_{11}, A_{21} and A_{31} of radius q_{13}, q_{23} and q_{33}, respectively; thus the translational part of the vector $\mathbf{x}$ is found,
2. Knowing the position of A_{16} the lengths q_{43} and q_{53} of the legs 4 and 5, compute the position of point A_{46} which is at the intersection of the spheres centered in A_{16} of radius $l_{A_{16}A_{46}}$ and the two spheres centered in A_{41} and A_{51} of radius

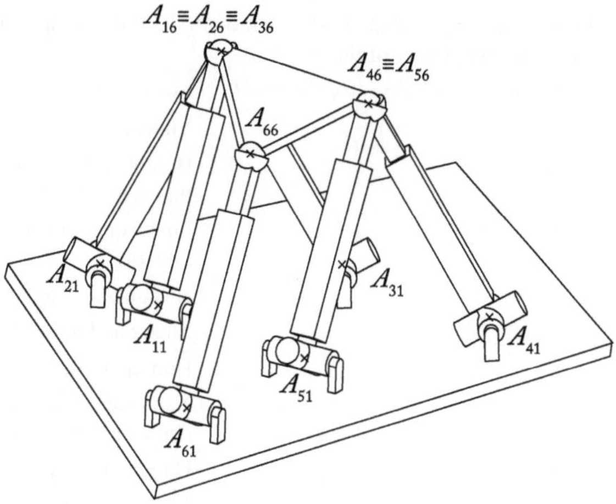

Fig. 7.17 An example of 6–$U\underline{P}S$ *PKM* with simplified *FGM*

q_{43} and q_{53}, respectively; thus two independent platform orientation parameters can be obtained while the rotation angle around the line $A_{13}A_{43}$ still cannot be computed,

3. Finally, knowing the position of A_{16} and A_{46} and the length q_{63} of the leg 6, compute the position of point A_{66} which is at the intersection of the spheres centered in A_{16}, A_{46} and A_{66} of radius $l_{A_{16}A_{66}}$, $l_{A_{46}A_{66}}$ and q_{63}, respectively; thus the rotation angle around the line $A_{16}A_{46}$ is found.

Table 7.6 proposes a (non-exhaustive) list of reference papers concerning manipulators with a special leg arrangement leading to a decrease of the number of *FGM* solutions.

7.2.2.6 Forward Geometric Model of Some Other Spatial Parallel Manipulators

It is not possible in this book to deal with the *FGM* of too many *PKM*. So, in this section, Table 7.7 presents a list of papers dealing with the *FGM* of some interesting *SPM*. It should be mentioned that this list is far from being exhaustive but only presents *SPM* quite often met in the literature.

7.2.3 Assembly Mode Selection and Numerical Methods for Solving the FGM

Previous sections have shown some methodologies to find all the possible assembly modes corresponding to a given value $\mathbf{q}_a$ of the active joint coordinates. However,

Table 7.6 List of references dealing with the *FGM* of the 6–*UPS PKM* with special leg arrangement allowing one to decrease the complexity of the *FGM*

Type of robot[a]	Number of solutions for the *FGM*	References
6–5	10	Yin and Liang (1994)
6–4	32	Hunt and Primrose (1993), Innocenti (1995)
	16	Innocenti and Parenti-Castelli (1991a)
		Husain and Waldron (1994)
		Zhang and Song (1991)
6–3	8	Hunt and Primrose (1993)
		Nanua and Waldron (1991) (robot depicted at Fig. 7.17)
5–5	24	Hunt and Primrose (1993)
5–4	32	Lin et al. (1994)
	24	Innocenti and Parenti-Castelli (1993)
		Faugère and Lazard (1995)
	16	Innocenti and Parenti-Castelli (1991b)
		Lin et al. (1994)
	8	Nair (1994)
5–3	16	Faugère and Lazard (1995)
	8	Faugère and Lazard (1995)
4–4	24	Lin et al. (1992)
	16	Lin et al. (1992)
		Innocenti and Parenti-Castelli (1992)
	8	Bruyninckx (1998)
4–3	16	Faugère and Lazard (1995)
	8	Faugère and Lazard (1995)
3–3	16	Dedieu and Norton (1990)
	8	Faugère and Lazard (1995)

[a]The symbol *m*-*n* robot corresponds to a robot with *m* attachment points on the base and *n* attachment points on the mowing platform

one point was not discussed which is: among all the possible assembly modes, how to choose the one that corresponds to the real platform pose?

This problem is still an open problem for research on *PKM*, even if some methodologies have already been proposed. To get the assembly mode knowledge, it is possible to use additional encoders mounted in the passive joints (Arai et al. 1990; Inoue et al. 1985). Such additional information can help to find the real posture of the robot, and can also help to simplify the computation of the pose. For example, if all active and passive joint coordinates of the leg k of the general robot presented in

Table 7.7 List of references dealing with the *FGM* of some interesting *SPM*

Type of *DOF*[a]	Robot name	References
2T1R	Zero-torsion *PKM*	Bonev (2008)
3T	Delta	Clavel (1989)
	Orthoglide-like family[b]	Company et al. (2002), Pashkevich et al. (2006)
	3-$U\underline{P}U$	Tsai and Joshi (2000)
	Decoupled *TPM*	Gogu (2004), Kong and Gosselin (2002)
		Carricato and Parenti-Castelli (2002)
	Pantopteron	Briot and Bonev (2009a)
3R	Agile Eye	Bonev et al. (2006)
	3–$\underline{R}SS$ wrist	Di Gregorio (2004a)
	3–$U\underline{P}U$ wrist	Di Gregorio (2003)
3 exotic *DOF*	*PKM* with 3 legs	Di Gregorio (2005)
3T1R	Quattro/Par4	Nabat (2007)
	H4, I4, I4L	Company and Pierrot (1999)
	PAMINSA	Briot et al. (2008)
	MacGill SMG	Alizadeh (2009)
	Quadrupteron	Kong and Gosselin (2011a)
	Pantopteron–4	Briot and Bonev (2010)
3T2R	5–$R\underline{P}UR$	Tale-Masouleh et al. (2011)
	Verne Machine	Kanaan et al. (2007)
3T3R	Gough-Stewart platform family	See Sect. 7.2.2.5
	Hexa family (6–$\underline{R}US$/6–$\underline{P}US$)	Same approach as for Gough-Stewart platforms
	Hexapteron	Seward and Bonev (2014)

[a]In that column, the symbol "$iTjR$" denotes that the considered mechanism has i translational *DOF* and j rotational *DOF*

[b]Orthoglide-like family regroups all 3 T robots with linear actuators mounted onto the base (whatever is their orientation) followed by passive legs of the Orthoglide type (Fig. 7.8)

Fig. 7.1 are measured, the problem remains to find the direct geometric model of a serial structure, which has a direct and unique solution (Khalil and Dombre 2002).

Another solution is to use exteroceptive sensors such as cameras. The most common approach consists of the direct observation of the end-effector pose (Espiau et al. 1992; Horaud et al. 1998; Martinet et al. 1996). However, some applications prevent visual observation of the end-effector of a parallel mechanism. For instance, it is not wise to imagine observing the end-effector of a machine-tool while it is generally not a problem to observe its legs that are most often designed with slim and rectilinear rods (Merlet 2014).

The platform pose reconstruction based on visual observation of the robot legs was proposed in Andreff et al. (2005) where vision was used to derive a visual servoing scheme based on the observation of a Gough-Stewart platform. After these preliminary works, the approach was extended to control of the robot directly in the image space by observation of the leg edges (from which the leg direction can be extracted), which has proven to exhibit better performances in terms of accuracy than the previous approach (Andreff et al. 2007). The approach was applied to several types of robots, such as the Adept Quattro and other robots of the same family (Andreff and Martinet 2006; Ozgur et al. 2011).

It is to be noticed that, for some particular cases, this last approach may not be able to give an estimation of the platform pose (Andreff and Martinet 2006), and the mapping between the leg space and Cartesian space may not be free of singularity (Briot and Martinet 2013).

Numerical methods can also be used to get information about the robot's assembly mode. The probably most efficient ones are the Interval Analysis (*IA*) (Merlet 2004) and other methodologies based on Cylindrical Algebraic Decomposition (*CAD*) (Chablat et al. 2011) that are able to give the intervals of *PKM* poses that belong to the same workspace aspect (i.e. a workspace area that is singularity-free and bounded by singularities (Merlet 2006b)). This does unfortunately not bring a 100 % guarantee that the assembly mode can be known as, for some robots, a workspace aspect can regroup several assembly modes (Zein et al. 2008).

To solve this problem, it can be interesting to solve the *FGM* iteratively. Knowing at a time t the exact pose of the *PKM*, the idea is to take advantage of the fact that the unknown current pose at time $t + \delta t$ will be close to the pose that was known at time t. Several well-known approaches that have solved non-linear numerical systems can be used. Probably, the most efficient one is the Newton-Raphson Scheme.

When it is not possible to find a closed-form solution to the constraint geometric equation (7.4) giving $\mathbf{x}$ in terms of $\mathbf{q}_a$, we can use the Newton-Raphson Scheme as follows.

Based on Eq. (7.4), we derive the following differential model, which can be used to numerically compute the variables $\mathbf{x}$ for a given $\mathbf{q}_a$:

$$\delta\mathbf{x} = \mathbf{J}_x\, \delta\mathbf{q}_a \tag{7.51}$$

where $\mathbf{J}_x$ is the robot Jacobian matrix that will be fully defined in Sect. 7.3, $\delta\mathbf{x}$ is a small increment of the platform pose $\mathbf{x}$ and $\delta\mathbf{q}_a$ is a small increment of the active joint coordinates $\mathbf{q}_a$.

To calculate the pose $\mathbf{x}_f$ corresponding to the active joint values $\mathbf{q}_{a_f}$, we use the following algorithm:

1. initialize by the current values $\mathbf{q}_{a_0}$ of $\mathbf{q}_a$ and $\mathbf{x}_0$ of $\mathbf{x}$ if known (by random if not);
2. compute the platform pose $\mathbf{x}_{k+1}$ at the iteration $k + 1$ by

$$\mathbf{x}_{k+1} = \mathbf{x}_k + \mathbf{J}_x \left(\mathbf{q}_{a_f} - \mathbf{q}_{a_k}\right) \tag{7.52}$$

where $\mathbf{q}_{a_k} = \mathscr{H}(\mathbf{x}_k)$ is the solution of the *IKM* corresponding to the platform pose value $\mathbf{x}_k$,

3. if $\left\|\mathbf{q}_{a_f} - \mathbf{q}_{a_k}\right\| < \varepsilon$, where ε is a number small enough to be fixed by the user, the algorithm can stop. If not, the second step must run again.

If the algorithm does not converge within a relatively large number of iterations, or to obtain another solution, we have to restart the calculation using a new random value of $\mathbf{x}_0$; if no convergence occurs for many different trials, it can be stated that there is no solution or that the robot is too close from a singularity (see Sect. 7.5).

More information on the convergence conditions of this algorithm can be found in Merlet (2006b).

7.3 Velocity Analysis

In this section, the kinematic relations linking the active joint velocities to the platform twist and passive joint velocities are defined and analyzed.

7.3.1 Computation of the Kinematic Constraint Relations

The kinematic relations linking the active joint velocities to the platform twist could be obtained by differentiating (7.4) w.r.t. time. However, this solution may not be computationally efficient. Therefore, we propose to use the following methodology which can take advantage of the recursive algorithms defined in Sect. 5.2.4.

Let us consider the input-output relation of the chain i (Fig. 7.1) which expresses the platform twist (which will be denoted below as ${}^0\mathbf{t}_p$, the superscript "0" denoting that the vector is given in the reference frame $\mathscr{F}_0$) as a function of all joint velocities $\dot{\mathbf{q}}_i$ for the considered chain. From (5.6) and (5.7), we have

$$ {}^0\mathbf{t}_p = {}^0\mathbf{J}_{p_i}\dot{\mathbf{q}}_i = \left[{}^0\$_{i1} \ldots {}^0\$_{i\,m_i}\right]\dot{\mathbf{q}}_i. \tag{7.53} $$

where ${}^0\$_{ik}$ is a unit twist representing the displacement of the end-effector when joint ik is moving only and m_i is the number of joints in the considered chain.[2]

Let us rewrite (7.53) by reorganizing matrix ${}^0\mathbf{J}_{p_i}$ so that we can group:

- in a sub-matrix ${}^0\$_{ia}$ the unit twists corresponding to the active joints of velocities $\dot{\mathbf{q}}_{ai}$, and
- in a sub-matrix ${}^0\$_{id}$ the unit twists corresponding to the passive joints of velocities $\dot{\mathbf{q}}_{di}$.

[2]Here, we express directly (7.53) at the center of the platform. However, all the expressions presented in this section could have also been obtained by expressing (7.53) at the last joint of the chain i, such as presented in Sect. 7.3.3.

Thus (7.53) becomes

$$ {}^0\mathbf{t}_p = \begin{bmatrix} {}^0\$_{ia} & {}^0\$_{id} \end{bmatrix} \begin{bmatrix} \dot{\mathbf{q}}_{ai} \\ \dot{\mathbf{q}}_{di} \end{bmatrix} = {}^0\$_{ia}\ \dot{\mathbf{q}}_{ai} + {}^0\$_{id}\ \dot{\mathbf{q}}_{di}. \tag{7.54} $$

To eliminate the passive joint velocities $\dot{\mathbf{q}}_{di}$ from (7.54), ${}^0\mathbf{t}_p$ can be multiplied by a wrench $\boldsymbol{\zeta}_i$ which is reciprocal to all passive joint twists ${}^0\$_{id}$ but NOT to the active joint twists ${}^0\$_{ia}$. In other words, this means that (see Sect. 3.4)

$$ \boldsymbol{\zeta}_i^T\ {}^0\$_{id} = \mathbf{0} \text{ and } \boldsymbol{\zeta}_i^T\ {}^0\$_{ia} \neq \mathbf{0} \tag{7.55} $$

$\boldsymbol{\zeta}_i$ is an actuation wrench which, if applied to the platform, can be resisted using only the actuators of the chain i. Examples of wrenches reciprocal to some usual systems of twists are given in Appendix C.

As a result, the following scalar equation can be obtained:

$$ \boldsymbol{\zeta}_i^T\ {}^0\mathbf{t}_p = (\boldsymbol{\zeta}_i^T\ {}^0\$_{ia})\ \dot{\mathbf{q}}_{ai}. \tag{7.56} $$

If the chain i has n_{ai} active joints, n_{ai} independent vectors $\boldsymbol{\zeta}_i$ can be found. Finally, considering all legs, Eq. (7.56) can be rewritten under the matrix form:

$$ \begin{bmatrix} \boldsymbol{\zeta}_1^T \\ \boldsymbol{\zeta}_2^T \\ \vdots \\ \boldsymbol{\zeta}_n^T \end{bmatrix} {}^0\mathbf{t}_p = \begin{bmatrix} \boldsymbol{\zeta}_1^T\ {}^0\$_{1a} & \mathbf{0} & \dots & \mathbf{0} \\ \mathbf{0} & \boldsymbol{\zeta}_2^T\ {}^0\$_{2a} & \dots & \mathbf{0} \\ \dots & \dots & \dots & \dots \\ \mathbf{0} & \mathbf{0} & \dots & \boldsymbol{\zeta}_n^T\ {}^0\$_{na} \end{bmatrix} \dot{\mathbf{q}}_a \tag{7.57} $$

where $\boldsymbol{\zeta}_k$ is a $(6 \times c_k)$ matrix groups all c_k *independent* unit wrenches due to the n_{ak} actuators of the kth leg. As a result, the system (7.57) is a system of $n_w = \sum_k c_k$ equations.

Thus, from (7.57), we define two matrices $\mathbf{A}$ and $\mathbf{B}$ such that

$$ \mathbf{A} = \begin{bmatrix} \boldsymbol{\zeta}_1^T \\ \boldsymbol{\zeta}_2^T \\ \dots \\ \boldsymbol{\zeta}_n^T \end{bmatrix} \tag{7.58} $$

and

$$ \mathbf{B} = -\begin{bmatrix} \boldsymbol{\zeta}_1^T\ {}^0\$_{1a} & \mathbf{0} & \dots & \mathbf{0} \\ \mathbf{0} & \boldsymbol{\zeta}_2^T\ {}^0\$_{2a} & \dots & \mathbf{0} \\ \dots & \dots & \dots & \dots \\ \mathbf{0} & \mathbf{0} & \dots & \boldsymbol{\zeta}_n^T\ {}^0\$_{na} \end{bmatrix}. \tag{7.59} $$

The matrix $\mathbf{A}$ is of dimension $(n_w \times 6)$ and the matrix $\mathbf{B}$ is of dimension $(n_w \times n_a)$. Then, we rewrite (7.57) as

$$\mathbf{A}\,{}^0\mathbf{t}_p + \mathbf{B}\dot{\mathbf{q}}_a = \mathbf{0}. \tag{7.60}$$

From (7.58), it is necessary to point out that the row i of $\mathbf{A}$ is a unit wrench which is aligned along the direction of the wrench applied by the actuator i on the platform through the leg when it is developing an input effort in a static mode of operation and in absence of any other type of external effects.

7.3.2 Kinematic Models

7.3.2.1 Condensed Form of Kinematic Constraint Relations

Let us define a vector ${}^0\mathbf{t}_r$ which groups a set of n_{dof} independent coordinates in ${}^0\mathbf{t}_p$ such that:

$${}^0\mathbf{t}_p = \boldsymbol{\Psi}_t\,{}^0\mathbf{t}_r \Leftrightarrow {}^0\mathbf{t}_r = \boldsymbol{\Psi}_t^{inv}\,{}^0\mathbf{t}_p \tag{7.61}$$

where $\boldsymbol{\Psi}_t$ is a $(6 \times n_{dof})$ matrix while $\boldsymbol{\Psi}_t^{inv}$ is a $(n_{dof} \times 6)$ matrix. Note that $\boldsymbol{\Psi}_t^{inv}$ is a constant matrix composed of 0 and 1 only while $\boldsymbol{\Psi}_t$ may not be constant (see examples in Sect. 7.3.4).

Using (7.61), we can rewrite (7.60) as:

$$\mathbf{A}_r\,{}^0\mathbf{t}_r + \mathbf{B}\dot{\mathbf{q}}_a = \mathbf{0} \tag{7.62}$$

where $\mathbf{A}_r = \mathbf{A}\,\boldsymbol{\Psi}_t$ is a $(n_w \times n_{dof})$ matrix. $\mathbf{A}_r$ is a part of the matrix $\mathbf{A}$ defined in (7.60) that relates the n_{dof} independent coordinates of the platform twist ${}^0\mathbf{t}_r$ to the actuated joints $\dot{\mathbf{q}}_a$. Thus, it has exactly the same property as $\mathbf{A}$, i.e. its row i contains n_{dof} independent coordinates of the unit wrenches which are aligned along the direction of the wrench applied by the actuator i on the platform through the legs when it is developing an input effort in a static mode of operation and in absence of any other type of external effects.

Matrices $\mathbf{A}_r$ *and* $\mathbf{B}$ *are crucial for computation of the dynamic model of the PKM.*

In a general manner, it can be proven that, for a *PKM*:

- without redundancy, the total number n_w of *independent* actuation wrenches is equal to the total number n_a of actuators and also to the number n_{dof} of independent *DOF* for the platform ($n_w = n_a = n_{dof}$), and thus matrices $\mathbf{A}_r$ and $\mathbf{B}$ are square of dimension $(n_a \times n_a)$,
- with actuation redundancy for which $n_a > n_{dof}$, the total number n_w of *independent* actuation wrenches is equal to n_a. Thus matrix $\mathbf{B}$ is a $(n_a \times n_a)$ square matrix while matrix $\mathbf{A}_r$ is rectangular of dimension $(n_a \times n_{dof})$,

- with kinematic redundancy for which $n_a > n_{dof}$, the total number n_w of *independent* actuation wrenches is equal to n_{dof}. Thus matrix $\mathbf{A}_r$ is a $(n_{dof} \times n_{dof})$ square matrix while matrix $\mathbf{B}$ is rectangular of dimension $(n_{dof} \times n_a)$.

7.3.2.2 Kinematic Models of *PKM* Without Redundancy

Matrices $\mathbf{A}_r$ and $\mathbf{B}$ being square for a *PKM* without redundancy, the forward kinematic model (*FKM*) is given by:

$$^0\mathbf{t}_r = -\mathbf{A}_r^{-1}\mathbf{B}\dot{\mathbf{q}}_a = \mathbf{J}\dot{\mathbf{q}}_a \tag{7.63}$$

$\mathbf{J} = -\mathbf{A}_r^{-1}\mathbf{B}$ is the kinematic Jacobian matrix of the *PKM*. This expression is valid as long as matrix $\mathbf{A}_r$ is not singular. The singularity conditions of matrix $\mathbf{A}_r$ are disclosed in Sect. 7.5.

The inverse kinematic model (*IKM*) is given by:

$$\dot{\mathbf{q}}_a = -\mathbf{B}^{-1}\mathbf{A}_r{}^0\mathbf{t}_r = \mathbf{J}_{inv}{}^0\mathbf{t}_r \tag{7.64}$$

where $\mathbf{J}_{inv} = -\mathbf{B}^{-1}\mathbf{A}_r$ is the inverse kinematic Jacobian matrix. For *PKM* without redundancy, $\mathbf{J}_{inv} = \mathbf{J}^{-1}$ which is not the case for other types of *PKM*. The expression (7.64) is valid as long as matrix $\mathbf{B}$ is not singular. The singularity conditions of matrix $\mathbf{B}$ are disclosed in Sect. 7.5.

Please note that $\mathbf{J}$ and $\mathbf{J}_{inv}$ are NOT the Jacobian and inverse Jacobian matrices of the robot, but the kinematic Jacobian and inverse kinematic Jacobian matrices. This is because they do not relate the actuated joint velocities $\dot{\mathbf{q}}_a$ to the derivative w.r.t. time of the platform coordinates (denoted as $\dot{\mathbf{x}}$) but to the platform twist $^0\mathbf{t}_r$. The Jacobian and inverse Jacobian matrices are defined at Sect. 7.3.2.5.

7.3.2.3 Kinematic Models of *PKM* with Actuation Redundancy

In that case, matrix $\mathbf{A}_r$ being rectangular while matrix $\mathbf{B}$ is square, the *IKM* can be defined without any problem using (7.64)—as long as matrix $\mathbf{B}$ is not singular—while the *FKM* becomes:

$$^0\mathbf{t}_r = \mathbf{J}_{inv}^{+}\dot{\mathbf{q}}_a \tag{7.65}$$

where $\mathbf{J}_{inv}^{+} = \left(\mathbf{J}_{inv}^T\mathbf{J}_{inv}\right)^{-1}\mathbf{J}_{inv}^T$ is the pseudo-inverse of the inverse kinematic Jacobian matrix $\mathbf{J}_{inv}$ defined at (7.64).

7.3.2.4 Kinematic Models of *PKM* with Kinematic Redundancy

In that case, matrix $\mathbf{B}$ being rectangular while matrix $\mathbf{A}_r$ is square, the *FKM* can be defined without any problem using (7.63)—as long as matrix $\mathbf{A}_r$ is not singular—while the *IKM* becomes:

$$\dot{\mathbf{q}}_a = \mathbf{J}^{+0}\mathbf{t}_r + \left(\mathbf{1}_{n_a} - \mathbf{J}^+\mathbf{J}\right)\boldsymbol{\xi} \tag{7.66}$$

where $\mathbf{J}^+ = \mathbf{J}^T\left(\mathbf{J}\mathbf{J}^T\right)^{-1}$ is the pseudo-inverse of the kinematic Jacobian matrix $\mathbf{J}$ defined at (7.63) and $\boldsymbol{\xi}$ is an arbitrary vector in the $\dot{\mathbf{q}}_a$ space (Khalil and Dombre 2002).

7.3.2.5 Relation Between the Platform Coordinate Derivatives with Respect to Time and the Platform Twist

In Sects. 7.1 and 7.2, we denote as $\mathbf{x}$ the n_{dof} independent platform coordinates. As a result, the vector $\dot{\mathbf{x}}$ is different from the platform twist ${}^0\mathbf{t}_p$, and corresponds only to the vector of the derivatives w.r.t. time of the independent platform coordinates.

From the differentiation w.r.t. time of (7.1), $\dot{\mathbf{x}}$ is related to $\dot{\mathbf{x}}_p$, i.e. the vector of the derivatives w.r.t. time of all platform coordinates by

$$\dot{\mathbf{x}} = \boldsymbol{\Psi}_x\dot{\mathbf{x}}_p \text{ or also } \dot{\mathbf{x}}_p = \boldsymbol{\Psi}_x^{inv}\dot{\mathbf{x}} \tag{7.67}$$

where $\boldsymbol{\Psi}_x$ and $\boldsymbol{\Psi}_x^{inv}$ are usually constant matrices but not necessarily.

$\dot{\mathbf{x}}_p$ is usually a vector composed of 6 coordinates (it can be greater than 6 if homogeneous transformation matrices, quaternions, or other less usual parameters are used to define the platform pose, such as the Study parameter (Pfurner and Husty 2010), but that case will not be treated here). Let us consider here that those 6 components are:

- $\dot{x}, \dot{y}, \dot{z}$ the three components of the translational velocity of the platform frame $\mathscr{F}_p$ w.r.t. $\mathscr{F}_0$ and expressed in $\mathscr{F}_0$, and
- $\dot{\phi}, \dot{\theta}, \dot{\psi}$ the derivatives w.r.t. time of the three ZYZ Euler angles characterizing the rotation of $\mathscr{F}_p$ w.r.t. $\mathscr{F}_0$.

Thus, $\dot{\mathbf{x}}_p^T = [\dot{x}\ \dot{y}\ \dot{z}\ \dot{\phi}\ \dot{\theta}\ \dot{\psi}]$.

In that case, the rotational velocity of the platform is given by:

$${}^0\boldsymbol{\omega}_p = \begin{bmatrix} \dot{\psi}s_\theta c_\phi - \dot{\theta}s_\phi \\ \dot{\psi}s_\theta s_\phi + \dot{\theta}c_\phi \\ \dot{\phi} + \dot{\psi}c_\theta \end{bmatrix} \tag{7.68}$$

which can be rewritten in the matrix form

$$^0\boldsymbol{\omega}_p = \begin{bmatrix} 0 & -s_\phi & s_\theta c_\phi \\ 0 & c_\phi & s_\theta s_\phi \\ 1 & 0 & c_\theta \end{bmatrix} \begin{bmatrix} \dot{\phi} \\ \dot{\theta} \\ \dot{\psi} \end{bmatrix} = \mathbf{D} \begin{bmatrix} \dot{\phi} \\ \dot{\theta} \\ \dot{\psi} \end{bmatrix}. \tag{7.69}$$

Thus, we can deduce that the platform twist $^0\mathbf{t}_p$ is equal to:

$$^0\mathbf{t}_p = \begin{bmatrix} \mathbf{1}_3 & \mathbf{0}_3 \\ \mathbf{0}_3 & \mathbf{D} \end{bmatrix} \dot{\mathbf{x}}_p = \overline{\mathbf{D}} \dot{\mathbf{x}}_p. \tag{7.70}$$

Similar equations can be found whatever are the types of parameters.

Now, left-multiplying (7.70) by the matrix $\boldsymbol{\Psi}_t^{inv}$ defined in (7.61) and introducing (7.67) in it, we get

$$^0\mathbf{t}_r = \boldsymbol{\Psi}_t^{inv} \overline{\mathbf{D}} \boldsymbol{\Psi}_x^{inv} \dot{\mathbf{x}} = \overline{\mathbf{D}}_r \dot{\mathbf{x}} \tag{7.71}$$

where $\overline{\mathbf{D}}_r = \boldsymbol{\Psi}_t^{inv} \overline{\mathbf{D}} \boldsymbol{\Psi}_x^{inv}$ is a $(n_{dof} \times n_{dof})$ invertible matrix.

7.3.2.6 Obtaining the Kinematic Models by Differentiating the Geometric Constraint Equations

If the relation (7.4) is available, the kinematic relationships can be obtained through its differentiation w.r.t. time, which leads to:

$$\mathbf{A}_d \dot{\mathbf{x}} + \mathbf{B}_d \dot{\mathbf{q}}_a = \mathbf{0} \tag{7.72}$$

where,

$$\mathbf{A}_d = \left[\frac{\partial \mathbf{h}_p}{\partial \mathbf{x}} \right] \tag{7.73}$$

and

$$\mathbf{B}_d = \left[\frac{\partial \mathbf{h}_p}{\partial \mathbf{q}_a} \right]. \tag{7.74}$$

Introducing (7.71) into (7.62), we can link the usual equations (7.72) to the kinematic relationships defined in Sect. 7.3.2.1 by

$$\mathbf{A}_r \overline{\mathbf{D}}_r \dot{\mathbf{x}} + \mathbf{B} \dot{\mathbf{q}}_a = \mathbf{0}. \tag{7.75}$$

By identification between (7.72) and (7.75), we thus obtain

$$\mathbf{A}_d = \mathbf{A}_r \overline{\mathbf{D}}_r \tag{7.76}$$

$$\mathbf{B}_d = \mathbf{B} \tag{7.77}$$

which gives the relations between the usual equations (7.72) and the kinematic constraints defined in Sect. 7.3.2.1.

Please note that, from (7.72), we can define the expression of the Jacobian matrix which is, for a mechanism without redundancy:

$$\mathbf{J}_x = -\mathbf{A}_d^{-1}\mathbf{B} \tag{7.78}$$

so that the following relations hold:

$$\dot{\mathbf{x}} = \mathbf{J}_x \dot{\mathbf{q}}_a \tag{7.79}$$

and

$$\dot{\mathbf{q}}_a = \mathbf{J}_x^{-1}\dot{\mathbf{x}}. \tag{7.80}$$

7.3.3 Computation of the Passive Joint Velocities

The computation of the passive joint velocities is necessary for the computation of the dynamic model. Therefore, the way to compute them is defined in this section.

All joint velocities could be obtained by differentiating (7.3) w.r.t. time. However, this solution may not be computationally efficient. Therefore, we propose to use the following methodology which is based on equating the twist at the terminal frame of each chain as a function of the platform twist ${}^0\mathbf{t}_p$ from one side, and as a function of the joint velocities of the chain from the other side. This method can take advantage of the recursive algorithms defined in Sect. 5.2.4.

Let us consider the chain i of the *PKM*, which is composed of m_i joints (Fig. 7.1). From (3.2), we can compute the twist of the platform at point $A_{i\,m_i}$ (that will be denoted as $\mathbf{t}_p^i$) as:

$${}^0\mathbf{t}_p^i = {}^0\mathbf{t}_p + \begin{bmatrix} {}^0\boldsymbol{\omega}_p \times {}^0\mathbf{r}_{PA_{i\,m_i}} \\ \mathbf{0} \end{bmatrix} \tag{7.81}$$

which can also be rewritten in the matrix form:

$${}^0\mathbf{t}_p^i = \begin{bmatrix} \mathbf{1}_3 & -{}^0\hat{\mathbf{r}}_{PA_{i\,m_i}} \\ \mathbf{0} & \mathbf{1}_3 \end{bmatrix} {}^0\mathbf{t}_p = \mathbf{J}_{t_i}\,{}^0\mathbf{t}_p \tag{7.82}$$

where $\mathbf{J}_{t_i}$ is a (6×6) matrix.

As the joint located at $A_{i\,m_i}$ also belongs to the chain i, its twist can be obtained by using the relation (5.8) as:

$${}^0\mathbf{t}_p^i = {}^0\mathbf{J}_{i\,m_i}\dot{\mathbf{q}}_i = \begin{bmatrix} {}^0\$_{i\,m_i}^{i1} & \cdots & {}^0\$_{i\,m_i}^{i\,m_i} \end{bmatrix} \dot{\mathbf{q}}_i \tag{7.83}$$

where $\dot{\mathbf{q}}_i$ represents the vector of all joint velocities of the chain i, ${}^0\mathbf{J}_{i\,m_i} = {}^0\overline{\mathbf{R}}_{i\,m_i}{}^{i\,m_i}\mathbf{J}_{i\,m_i}$ is the chain i kinematic Jacobian matrix of dimension $(6 \times m_i)$ and $\$^{ik}_{i\,m_i}$ is a unit twist representing the displacement of the chain tip $A_{i\,m_i}$ when joint ik is moving only. The expression of ${}^{i\,m_i}\mathbf{J}_{i\,m_i}$ can be obtained by the recursive algorithm (5.21).

Equation (7.83) can be rewritten by reorganizing the matrix ${}^0\mathbf{J}_{i\,m_i}$ so that we can group:

- in a sub-matrix ${}^0\mathbf{J}_{ai}$ the unit twists ${}^0\$^{ik}_{i\,m_i}$ corresponding to the active joints of velocities $\dot{\mathbf{q}}_{ai}$, and
- in a sub-matrix ${}^0\mathbf{J}_{di}$ the unit twists ${}^0\$^{ik}_{i\,m_i}$ corresponding to the passive joints of velocities $\dot{\mathbf{q}}_{di}$.

Thus (7.83) becomes

$$ {}^0\mathbf{t}^i_p = {}^0\mathbf{J}_{i\,m_i}\dot{\mathbf{q}}_i = \begin{bmatrix} {}^0\mathbf{J}_{ai} & {}^0\mathbf{J}_{di} \end{bmatrix} \begin{bmatrix} \dot{\mathbf{q}}_{ai} \\ \dot{\mathbf{q}}_{di} \end{bmatrix} = {}^0\mathbf{J}_{ai}\dot{\mathbf{q}}_{ai} + {}^0\mathbf{J}_{di}\dot{\mathbf{q}}_{di} \tag{7.84} $$

or also

$$ {}^0\mathbf{t}^{c\,i}_p = {}^0\mathbf{J}_{di}\dot{\mathbf{q}}_{di} \tag{7.85} $$

where ${}^0\mathbf{t}^{c\,i}_p = {}^0\mathbf{t}^i_p - {}^0\mathbf{J}_{ai}\dot{\mathbf{q}}_{ai}$.

As $\dot{\mathbf{q}}_{di}$ is a vector of dimension $n_{di} < 6$, only n_{di} components of ${}^0\mathbf{t}^{c\,i}_p$ are independent. Let us define a vector $\mathbf{t}^{c\,i}_r$ of n_{di} independent components such that

$$ \mathbf{t}^{c\,i}_r = \boldsymbol{\Psi}_{t\,i}{}^0\mathbf{t}^{c\,i}_p \tag{7.86} $$

with $\boldsymbol{\Psi}_{t\,i}$ a $(n_{di} \times 6)$ matrix. Note that usually, the matrix $\boldsymbol{\Psi}_{t\,i}$ can be found by projecting the Eq. (7.85) in the last frame (or the last one before) of the chain (see the examples in Sect. 7.3.4).

Introducing (7.61), (7.82) and (7.86) into (7.84) and (7.85) leads to:

$$ \mathbf{J}^c_{t_i}\,{}^0\mathbf{t}_r - \mathbf{J}_{tai}\dot{\mathbf{q}}_{ai} = \mathbf{J}_{tdi}\dot{\mathbf{q}}_{di} \tag{7.87} $$

with

$$ \mathbf{J}^c_{t_i} = \boldsymbol{\Psi}_{t\,i}\,\mathbf{J}_{t_i}\,\boldsymbol{\Psi}_t \tag{7.88} $$

a $(n_{di} \times n_{dof})$ matrix,

$$ \mathbf{J}_{tai} = \boldsymbol{\Psi}_{t\,i}{}^0\mathbf{J}_{ai} \tag{7.89} $$

a $(n_{di} \times n_{ai})$ matrix (n_{ai} being the number of actuated joints in the chain i), and

$$ \mathbf{J}_{tdi} = \boldsymbol{\Psi}_{t\,i}{}^0\mathbf{J}_{di} \tag{7.90} $$

a $(n_{di} \times n_{di})$ matrix.

Thus, as the platform twist ${}^0\mathbf{t}_r$ and active joint velocities $\dot{\mathbf{q}}_{ai}$ are considered known by using the equations presented in Sects. 7.3.1 and 7.3.2, the joint velocities $\dot{\mathbf{q}}_{di}$ of the chain i can be found by inverting the matrix $\mathbf{J}_{tdi}$ in (7.87). The conditions of singularity of matrix $\mathbf{J}_{tdi}$ are disclosed in Sect. 7.5.2.

Now, considering all legs, we have

$$\mathbf{J}_{td}\dot{\mathbf{q}}_d = \mathbf{J}_t\,{}^0\mathbf{t}_r - \mathbf{J}_{ta}\dot{\mathbf{q}}_a \tag{7.91}$$

with

$$\mathbf{J}_{td} = \begin{bmatrix} \mathbf{J}_{td1} & \mathbf{0}_{n_{d1}\times n_{d1}} & \cdots & \mathbf{0}_{n_{d1}\times n_{dn}} \\ \mathbf{0}_{n_{d2}\times n_{d1}} & \mathbf{J}_{td2} & \cdots & \mathbf{0}_{n_{d2}\times n_{dn}} \\ \vdots & \vdots & \ddots & \vdots \\ \mathbf{0}_{n_{dn}\times n_{d1}} & \mathbf{0}_{n_{dn}\times n_{d2}} & \cdots & \mathbf{J}_{tdn} \end{bmatrix} \tag{7.92}$$

a $(n_d \times n_d)$ matrix (n_d being the total number of passive joints).

Moreover,

$$\mathbf{J}_t = \begin{bmatrix} \mathbf{J}_{t_1}^c \\ \vdots \\ \mathbf{J}_{t_n}^c \end{bmatrix} \tag{7.93}$$

a $(n_d \times n_{dof})$ matrix,

$$\mathbf{J}_{ta} = \begin{bmatrix} \mathbf{J}_{ta1} & \mathbf{0}_{n_{d1}\times n_{a1}} & \cdots & \mathbf{0}_{n_{d1}\times n_{an}} \\ \mathbf{0}_{n_{d2}\times n_{a1}} & \mathbf{J}_{ta2} & \cdots & \mathbf{0}_{n_{d2}\times n_{an}} \\ \vdots & \vdots & \ddots & \vdots \\ \mathbf{0}_{n_{dn}\times n_{a1}} & \mathbf{0}_{n_{dn}\times n_{a2}} & \cdots & \mathbf{J}_{tan} \end{bmatrix} \tag{7.94}$$

a $(n_d \times n_a)$ matrix.

Finally, we get

$$\dot{\mathbf{q}}_d = \mathbf{J}_{td}^{-1}(\mathbf{J}_t\,{}^0\mathbf{t}_r - \mathbf{J}_{ta}\dot{\mathbf{q}}_a). \tag{7.95}$$

It is necessary to mention that the matrices $\mathbf{J}_t$, $\mathbf{J}_{ta}$ *and* $\mathbf{J}_{td}$ *are crucial for computation of the dynamic model of the PKM.*

7.3.4 Examples

In this section, we present the first-order kinematic models of some *PKM* described in the previous section.

7.3.4.1 First-Order Kinematics of the Five-Bar Mechanism

Let us consider again the five-bar mechanism presented in Sect. 7.1.2.1. For this mechanism, one can show that the relation (7.53) can be expressed as:

$$ {}^0\mathbf{t}_p = \left[{}^0\$_{11}\ {}^0\$_{12}\ {}^0\$_{13}\right]\dot{\mathbf{q}}_1, \tag{7.96}$$

$$ {}^0\mathbf{t}_p = \left[{}^0\$_{21}\ {}^0\$_{22}\right]\dot{\mathbf{q}}_2, \tag{7.97}$$

with

$$ {}^0\$_{11} = \begin{bmatrix}-d_{12}\sin q_{11} - d_{13}s_{121} & d_{12}\cos q_{11} + d_{13}c_{121} & 0 & 0 & 0 & 1\end{bmatrix}^T, \tag{7.98}$$

$$ {}^0\$_{12} = \begin{bmatrix}-d_{13}s_{121} & d_{13}c_{121} & 0 & 0 & 0 & 1\end{bmatrix}^T, \tag{7.99}$$

$$ {}^0\$_{13} = \begin{bmatrix}0 & 0 & 0 & 0 & 0 & 1\end{bmatrix}^T, \tag{7.100}$$

and

$$ {}^0\$_{21} = \begin{bmatrix}-d_{22}\sin q_{21} - d_{23}s_{122} & d_{22}\cos q_{21} + d_{23}c_{122} & 0 & 0 & 0 & 1\end{bmatrix}^T, \tag{7.101}$$

$$ {}^0\$_{22} = \begin{bmatrix}-d_{23}s_{122}\ d_{23}c_{122} & 0 & 0 & 0 & 1\end{bmatrix}^T, \tag{7.102}$$

where ${}^0\mathbf{t}_p = [\dot{x}\ \dot{y}\ 0\ 0\ 0\ \dot{\phi}]^T$, $\dot{\mathbf{q}}_1 = [\dot{q}_{11}\ \dot{q}_{12}\ \dot{q}_{13}]^T$, and $\dot{\mathbf{q}}_2 = [\dot{q}_{21}\ \dot{q}_{22}]^T$, and $c_{12i} = \cos(q_{i1} + q_{i2})$ and $s_{12i} = \sin(q_{i1} + q_{i2})$ $(i = 1, 2)$.

From (7.54) and the fact that the first revolute joint of each leg are actuated only, it turns out that:

- $\dot{\mathbf{q}}_{a1} = \dot{q}_{11}$, $\dot{\mathbf{q}}_{a2} = \dot{q}_{21}$,
- $\dot{\mathbf{q}}_{d1} = [\dot{q}_{12}\ \dot{q}_{13}]^T$, $\dot{\mathbf{q}}_{d2} = \dot{q}_{22}$,
- ${}^0\$_{a1} = {}^0\$_{11}$, ${}^0\$_{a2} = {}^0\$_{21}$,
- ${}^0\$_{d1} = [{}^0\$_{12}\ {}^0\$_{13}]$, ${}^0\$_{d2} = {}^0\$_{22}$.

The system of wrenches $\boldsymbol{\zeta}_i$ satisfying the Eq. (7.55) (and which are thus reciprocal to the passive joint twists and NOT to the active joint twists) are:

- For the leg 1, $\boldsymbol{\zeta}_1 = \begin{bmatrix}c_{121} & s_{121} & 0 & 0 & 0 & 0\end{bmatrix}^T$,
- For the leg 2, $\boldsymbol{\zeta}_2 = \begin{bmatrix}c_{122} & s_{121} & 0 & 0 & 0 & 0\end{bmatrix}^T$.

Let us note that $\boldsymbol{\zeta}_1$ is a pure force along $\overrightarrow{A_{12}A_{13}}$ reciprocal to ${}^0\$_{12}$ and ${}^0\$_{13}$ (and not ${}^0\$_{11}$). Taking into account that the motion is planar, this vector is uniquely defined. Similarly, $\boldsymbol{\zeta}_2$ is a pure force along $\overrightarrow{A_{22}A_{13}}$ reciprocal to ${}^0\$_{22}$ (and not ${}^0\$_{21}$).

Thus, the matrices **A** and **B** of (7.60) can be written as

$$ \mathbf{A} = \begin{bmatrix}\boldsymbol{\zeta}_1^T \\ \boldsymbol{\zeta}_2^T\end{bmatrix} = \begin{bmatrix}c_{121} & s_{121} & 0 & 0 & 0 & 0 \\ c_{122} & s_{121} & 0 & 0 & 0 & 0\end{bmatrix} \tag{7.103}$$

$$\mathbf{B} = -\begin{bmatrix} \boldsymbol{\zeta}_1^T \, {}^0\$_{1a} & 0 \\ 0 & \boldsymbol{\zeta}_2^T \, {}^0\$_{2a} \end{bmatrix} \tag{7.104}$$

with, for $i = 1, 2$ (after simplifications),

$$\boldsymbol{\zeta}_i^T \, {}^0\$_{ia} = d_{i2} \sin q_{i2}. \tag{7.105}$$

As already mentioned, the five-bar mechanism has only two *DOF*, which are parameterized by the coordinates $\mathbf{x} = [x \ y]^T$. As a result, we choose the two first components of the twist ${}^0\mathbf{t}_p$ as the independent velocities. Thus, ${}^0\mathbf{t}_r = [\dot{x} \ \dot{y}]^T$ and the matrix $\boldsymbol{\Psi}_t^{inv}$ defined in the relation (7.61) as the matrix relating ${}^0\mathbf{t}_r$ to ${}^0\mathbf{t}_p$ by ${}^0\mathbf{t}_r = \boldsymbol{\Psi}_t^{inv} \, {}^0\mathbf{t}_p$ is equal to

$$\boldsymbol{\Psi}_t^{inv} = \begin{bmatrix} 1 \, 0 \, 0 \, 0 \, 0 \, 0 \\ 0 \, 1 \, 0 \, 0 \, 0 \, 0 \end{bmatrix}. \tag{7.106}$$

The matrix $\boldsymbol{\Psi}_t$ defined in the relation (7.61) as the matrix relating ${}^0\mathbf{t}_r$ to ${}^0\mathbf{t}_p$ by ${}^0\mathbf{t}_p = \boldsymbol{\Psi}_t \, {}^0\mathbf{t}_r$ can be found from the following proof. By differentiating w.r.t. time the constraint relationship on the platform coordinates (7.6) defined in Sect. 7.1.2.1, we obtain:

$$\begin{aligned} \frac{\dot{\phi}}{\cos^2 \phi} = & \frac{(\dot{y} - d_{22}\dot{q}_{21} \cos q_{21})(x - d_{22} \cos q_{21} - d_{21})}{(x - d_{22} \cos q_{21} - d_{21})^2} \\ & - \frac{(\dot{x} + d_{22}\dot{q}_{21} \sin q_{21})(y - d_{22} \sin q_{21})}{(x - d_{22} \cos q_{21} - d_{21})^2} \end{aligned} \tag{7.107}$$

or also, after grouping the terms

$$\dot{\phi} = a_x \dot{x} + a_y \dot{y} + a_q \dot{q}_{21} \tag{7.108}$$

with

$$a_y = \frac{\cos^2 \phi}{x - d_{22} \cos q_{21} - d_{21}},$$

$$a_x = -\frac{(y - d_{22} \sin q_{21}) \cos^2 \phi}{(x - d_{22} \cos q_{21} - d_{21})^2},$$

$$a_q = -d_{22} \frac{(x - d_{21}) \cos q_{21} + y \sin q_{21} - d_{22}}{(x - d_{22} \cos q_{21} - d_{21})^2} \cos^2 \phi.$$

Finally, introducing (7.112) into (7.108), we have:

$$\dot{\phi} = (a_x + a_q \, j_{inv}^{21})\dot{x} + (a_y + a_q \, j_{inv}^{22})\dot{y} \tag{7.109}$$

where j_{inv}^{ij} is the term in the ith row and jth column of the matrix $\mathbf{J}_{inv}$ that will be defined later.

At the end, the matrix $\mathbf{\Psi}_t$ is defined by

$$\mathbf{\Psi}_t = \begin{bmatrix} 1 & 0 \\ 0 & 1 \\ 0 & 0 \\ 0 & 0 \\ 0 & 0 \\ (a_x + a_q j_{inv}^{21}) & (a_y + a_q j_{inv}^{22}) \end{bmatrix}. \tag{7.110}$$

As a result, the first-order *IKM* and *FKM* are given by:

$$^0\mathbf{t}_r = -\mathbf{A}_r^{-1}\mathbf{B}\dot{\mathbf{q}}_a = \mathbf{J}\dot{\mathbf{q}}_a \tag{7.111}$$

and

$$\dot{\mathbf{q}}_a = -\mathbf{B}^{-1}\mathbf{A}_r{}^0\mathbf{t}_r = \mathbf{J}_{inv}{}^0\mathbf{t}_r \tag{7.112}$$

where

$$\mathbf{A}_r = \mathbf{A}\,\mathbf{\Psi}_t = \begin{bmatrix} c_{121} & s_{121} \\ c_{122} & s_{122} \end{bmatrix}. \tag{7.113}$$

Now that the first-order *IKM* and *FKM* of the five-bar mechanism are computed, let us find the expression of the passive joint velocities.

By identification between the relations (7.96), (7.97) and (7.85), we have:

$$^0\mathbf{J}_{d1} = {}^0\$_{d1},\ {}^0\mathbf{J}_{d2} = {}^0\$_{d2} \tag{7.114}$$

$$^0\mathbf{t}_p^{c\,1} = {}^0\mathbf{t}_p - {}^0\$_{a1}\dot{q}_{11},\ {}^0\mathbf{t}_p^{c\,2} = {}^0\mathbf{t}_p - {}^0\$_{a2}\dot{q}_{21}. \tag{7.115}$$

As the matrix $^0\mathbf{J}_{d1}$ ($^0\mathbf{J}_{d2}$, resp.) is of rank 2 (rank 1, resp.), only two components (one component, resp.) of $^0\mathbf{t}_p^{c\,1}$ ($^0\mathbf{t}_p^{c\,2}$, resp.) are independent.

Projecting the matrices $^0\mathbf{J}_{di}$ in the frame of the link $\mathcal{B}_{i2}$ ($i = 1, 2$), their expression becomes

$$^{12}\mathbf{J}_{d1} = {}^{12}\mathbf{R}_0{}^0\mathbf{J}_{d1} = \begin{bmatrix} 0 & d_{13} & 0 & 0 & 0 & 1 \\ 0 & 0 & 0 & 0 & 0 & 1 \end{bmatrix}^T \tag{7.116}$$

and

$$^{22}\mathbf{J}_{d2} = {}^{22}\mathbf{R}_0{}^0\mathbf{J}_{d22} = \begin{bmatrix} 0 & d_{23} & 0 & 0 & 0 & 1 \end{bmatrix}^T \tag{7.117}$$

where

$$ {}^{i2}\mathbf{R}_0 = \begin{bmatrix} c_{12i} & s_{12i} & 0 \\ -s_{12i} & c_{12i} & 0 \\ 0 & 0 & 1 \end{bmatrix}. \quad (7.118) $$

In (7.116), we can see directly that the second and last rows of the matrix ${}^{12}\mathbf{J}_{d1}$ are independent. From (7.116), the second row of ${}^{22}\mathbf{J}_{d2}$ is independent. Thus, by identification between (7.116), (7.117) and (7.86), the matrices $\boldsymbol{\Psi}_{t\,1}$ and $\boldsymbol{\Psi}_{t\,2}$ are defined by

$$ \boldsymbol{\Psi}_{t\,1} = \begin{bmatrix} 0\;1\;0\;0\;0\;0 \\ 0\;0\;0\;0\;0\;1 \end{bmatrix} {}^{12}\overline{\mathbf{R}}_0 = \begin{bmatrix} -s_{121} & c_{121} & 0\;0\;0\;0 \\ 0 & 0 & 0\;0\;0\;1 \end{bmatrix} \quad (7.119) $$

$$ \boldsymbol{\Psi}_{t\,2} = \begin{bmatrix} 0\;1\;0\;0\;0\;0 \end{bmatrix} {}^{22}\overline{\mathbf{R}}_0 = \begin{bmatrix} -s_{122} & c_{122} & 0\;0\;0\;0 \end{bmatrix}. \quad (7.120) $$

Then, by straightforwardly using the Eqs. (7.87)–(7.90), we obtain the kinematic relationship:

$$ \mathbf{J}_t {}^0\mathbf{t}_r - \mathbf{J}_{ta}\dot{\mathbf{q}}_a - \mathbf{J}_{td}\dot{\mathbf{q}}_d = \mathbf{0} \quad (7.121) $$

with

$$ \mathbf{J}_t = \begin{bmatrix} -s_{121} & c_{121} \\ (a_x + a_q\, j_{inv}^{21}) & (a_y + a_q\, j_{inv}^{22}) \\ -s_{122} & c_{122} \end{bmatrix} \quad (7.122) $$

$$ \mathbf{J}_{ta} = \begin{bmatrix} d_{21}\cos q_{21} + d_{31} & 0 \\ 1 & 0 \\ 0 & d_{22}\cos q_{22} + d_{32} \end{bmatrix} \quad (7.123) $$

$$ \mathbf{J}_{td} = \begin{bmatrix} d_{13} & 0 & 0 \\ 1 & 1 & 0 \\ 0 & 0 & d_{23} \end{bmatrix} \quad (7.124) $$

with $\dot{\mathbf{q}}_a^T = [\dot{q}_{11}\ \dot{q}_{12}]$ and $\dot{\mathbf{q}}_d^T = [\dot{q}_{21}\ \dot{q}_{31}\ \dot{q}_{22}]$, from which we deduce:

$$ \dot{\mathbf{q}}_d = \mathbf{J}_{td}^{-1}(\mathbf{J}_t {}^0\mathbf{t}_r - \mathbf{J}_{ta}\dot{\mathbf{q}}_a). \quad (7.125) $$

7.3.4.2 First-order Kinematics of the 3–*R<u>P</u>R* Planar Parallel Manipulator

In this section, we study only the input/output kinematic relations of the 3–*R<u>P</u>R* planar parallel manipulator introduced in Sect. 7.1.2.2. The computation of the passive joint velocities is not introduced and left as an exercise for the reader.

Following the method of Sect. 7.3.1, and by using the results presented in the Appendix C.3.1, we have a matrix $\mathbf{A}$ equal to:

$$\mathbf{A}^T = \left[\boldsymbol{\zeta}_1\ \boldsymbol{\zeta}_2\ \boldsymbol{\zeta}_3\right] \tag{7.126}$$

with $\boldsymbol{\zeta}_i$ a unit wrench reciprocal to the unit twists of joints $i1$ and $i3$ and not to the unit twist of joint $i2$. It is a force along $\overrightarrow{A_{i1}A_{i3}}$ whose expression is given by:

$$\begin{aligned}\boldsymbol{\zeta}_i^T &= \frac{1}{q_{i2}}\left[{}^0\mathbf{r}_{A_{i1}A_{i3}}^T\ \left({}^0\mathbf{r}_{PA_{i3}} \times {}^0\mathbf{r}_{A_{i1}A_{i3}}\right)^T\right] \\ &= \left[\cos q_{i1}\ \sin q_{i1}\ 0\ 0\ 0\ y_{A_{i3}P}\cos q_{i1} - x_{A_{i3}P}\sin q_{i1}\right]\end{aligned} \tag{7.127}$$

where the points A_{ij} are described at Fig. 7.5, $q_{i2} = \left\|\mathbf{r}_{A_{i1}A_{i3}}\right\|$ is the active joint variable for leg i, whose expression is given at (7.20), $x_{A_{i3}P} = x - d_{i4}\cos(\phi + \varepsilon_i)$ and $y_{A_{i3}P} = y - d_{i4}\cos(\phi + \varepsilon_i)$, d_{i4} and ε_i being defined in Sect. 7.1.2.2.

Moreover, as shown in the Appendix C.4.1, for any leg of the robot, $\$_{i2} = \$_{ia}$ is a twist representing a pure translation along the $\underline{P}$ joint direction. As a result,

$$\begin{aligned}{}^0\$_{ia}^T &= \frac{1}{q_{i2}}\left[{}^0\mathbf{r}_{A_{i1}A_{i3}}^T\ \mathbf{0}_{1\times 3}\right] \\ &= \left[\cos q_{i1}\ \sin q_{i1}\ 0\ 0\ 0\ 0\right].\end{aligned} \tag{7.128}$$

Thus, the matrix $\mathbf{B}$ is equal to:

$$\mathbf{B} = -\begin{bmatrix} \boldsymbol{\zeta}_1^T\ {}^0\$_{1a} & 0 & 0 \\ 0 & \boldsymbol{\zeta}_2^T\ {}^0\$_{2a} & 0 \\ 0 & 0 & \boldsymbol{\zeta}_3^T\ {}^0\$_{3a} \end{bmatrix} \tag{7.129}$$

with

$$\boldsymbol{\zeta}_i^T\ {}^0\$_{ia} = 1. \tag{7.130}$$

The 3–$R\underline{P}R$ planar parallel manipulator having 3 DOF in the plane $(O, \boldsymbol{x}_0, \boldsymbol{y}_0)$, ${}^0\mathbf{t}_p = \boldsymbol{\Psi}_t\,{}^0\mathbf{t}_r$ with

$$\boldsymbol{\Psi}_t = \begin{bmatrix} 1\,0\,0 \\ 0\,1\,0 \\ 0\,0\,0 \\ 0\,0\,0 \\ 0\,0\,0 \\ 0\,0\,1 \end{bmatrix}. \tag{7.131}$$

As a result,

$$\mathbf{A}_r^T = \boldsymbol{\Psi}_t^T \mathbf{A}^T = \left[\boldsymbol{\zeta}_1^r\ \boldsymbol{\zeta}_2^r\ \boldsymbol{\zeta}_3^r\right] \tag{7.132}$$

with

$$\boldsymbol{\zeta}_i^{rT} = \boldsymbol{\zeta}_i^T \boldsymbol{\Psi}_t = \left[\cos q_{i1} \ \sin q_{i1} \ y_{A_{i3}P} \cos q_{i1} - x_{A_{i3}P} \sin q_{i1}\right]. \tag{7.133}$$

Finally, we have

$$\mathbf{A}_r {}^0\mathbf{t}_r + \mathbf{B}\dot{\mathbf{q}}_a = \mathbf{0}. \tag{7.134}$$

7.3.4.3 First-Order Kinematics of the Orthoglide

In this section, we study only the input/output kinematic relations of the Orthoglide introduced in Sect. 7.1.2.4. The computation of the passive joint velocities is tedious, this is the reason why it is not detailed here but is given in: http://www.irccyn.ec-nantes.fr/~briot/Books.html.

Following the method of the Sect. 7.3.1, and by using the results presented in the Appendix C.4 and using the fact that, when the actuator i is blocked, the leg shown in Fig. 7.8b is a *UU* passive system, we have a matrix **A** equal to:

$$\mathbf{A}^T = \left[\boldsymbol{\zeta}_1 \ \boldsymbol{\zeta}_2 \ \boldsymbol{\zeta}_3\right] \tag{7.135}$$

with

$$\boldsymbol{\zeta}_i^T = \frac{1}{d_4}\left[{}^0\mathbf{r}_{A_{i3}A_{i4}}^T \ \left({}^0\mathbf{r}_{PA_{i4}} \times {}^0\mathbf{r}_{A_{i3}A_{i4}}\right)^T\right] \tag{7.136}$$

where the points A_{ij} are described at Fig. 7.8, and $d_4 = \left\|\mathbf{r}_{A_{i3}A_{i4}}\right\|$ is a constant length defined in Table 7.3. $\boldsymbol{\zeta}_i$ is a force directed along $\overrightarrow{A_{i3}A_{i4}}$.

Moreover, for any leg of the robot, $\$_{i1} = \$_{ia}$ is a twist representing a pure translation along the $\underline{P}$ joint direction. As a result,

$${}^0\$_{1a}^T = \left[0 \ 0 \ 1 \ \mathbf{0}_{1\times 3}\right] \tag{7.137}$$

$${}^0\$_{2a}^T = \left[1 \ 0 \ 0 \ \mathbf{0}_{1\times 3}\right] \tag{7.138}$$

$${}^0\$_{3a}^T = \left[0 \ 1 \ 0 \ \mathbf{0}_{1\times 3}\right]. \tag{7.139}$$

Thus, the matrix **B** is equal to:

$$\mathbf{B} = -\frac{1}{d_4}\begin{bmatrix} \boldsymbol{\zeta}_1^T \ {}^0\$_{1a} & 0 & 0 \\ 0 & \boldsymbol{\zeta}_2^T \ {}^0\$_{2a} & 0 \\ 0 & 0 & \boldsymbol{\zeta}_3^T \ {}^0\$_{3a} \end{bmatrix} \tag{7.140}$$

with

$$\boldsymbol{\zeta}_1^T \ {}^0\$_{1a} = z_{A_{13}A_{14}} \tag{7.141}$$

$$\boldsymbol{\zeta}_2^{T\ 0}\$_{2a} = x_{A_{23}A_{24}} \tag{7.142}$$

and

$$\boldsymbol{\zeta}_3^{T\ 0}\$_{3a} = y_{A_{33}A_{34}} \tag{7.143}$$

in which $x_{A_{i3}A_{i4}}$, $y_{A_{i3}A_{i4}}$ and $z_{A_{i3}A_{i4}}$ are the coordinates of the vector ${}^0\mathbf{r}_{A_{i3}A_{i4}}$ expressed in the base frame.

The Orthoglide having 3 translational *DOF*, ${}^0\mathbf{t}_p = \boldsymbol{\Psi}_t\,{}^0\mathbf{t}_r$ with

$$\boldsymbol{\Psi}_t = \begin{bmatrix} \mathbf{1}_3 \\ \mathbf{0}_{3\times3} \end{bmatrix}. \tag{7.144}$$

As a result,

$$\mathbf{A}_r^T = \left[\boldsymbol{\zeta}_1^r\ \boldsymbol{\zeta}_2^r\ \boldsymbol{\zeta}_3^r\right] \tag{7.145}$$

with

$$\boldsymbol{\zeta}_i^{r\,T} = {}^0\mathbf{r}_{A_{i3}A_{i4}}^T. \tag{7.146}$$

Finally, we have

$$\mathbf{A}_r\,{}^0\mathbf{t}_r + \mathbf{B}\dot{\mathbf{q}}_a = \mathbf{0}. \tag{7.147}$$

7.3.4.4 First-Order Kinematics of the Gough-Stewart Platform

In this section, we study only the input/output kinematic relations of the Gough-Stewart platform introduced in Sect. 7.1.2.5.

Following the method of Sect. 7.3.1, and by using the results presented in the Appendix C.4.1, we have a matrix $\mathbf{A}$ equal to:

$$\mathbf{A}^T = \left[\boldsymbol{\zeta}_1 \dots \boldsymbol{\zeta}_6\right] \tag{7.148}$$

with

$$\boldsymbol{\zeta}_i^T = \frac{1}{q_{i3}}\left[{}^0\mathbf{r}_{A_{i1}A_{i6}}^T\ \left({}^0\mathbf{r}_{PA_{i6}} \times {}^0\mathbf{r}_{A_{i1}A_{i6}}\right)^T\right] \tag{7.149}$$

where the points A_{ij} are described at Fig. 7.10 and $q_{i3} = \left\|\mathbf{r}_{A_{i1}A_{i6}}\right\|$ is the active joint variable for leg i, whose expression is given at (7.37). $\boldsymbol{\zeta}_i$ is a pure force along $\overrightarrow{A_{i1}A_{i6}}$ which is reciprocal to all passive joint twists of the leg i (and not the active joint twists).

Moreover, as shown in the Appendix C.4.1, for any leg of the robot, $\$_{i3} = \$_{ia}$ is a twist representing a pure translation along the $\underline{P}$ joint direction. As a result,

$$ {}^{0}\$_{ia}^{T} = \frac{1}{q_{i3}} \begin{bmatrix} {}^{0}\mathbf{r}_{A_{i1}A_{i6}}^{T} & \mathbf{0}_{1\times 3} \end{bmatrix}. \tag{7.150} $$

Finally, the matrix $\mathbf{B}$ is equal to:

$$ \mathbf{B} = -\begin{bmatrix} \boldsymbol{\zeta}_1^{T}\, {}^{0}\$_{1a} & 0 & \dots & 0 \\ 0 & \boldsymbol{\zeta}_2^{T}\, {}^{0}\$_{2a} & \dots & 0 \\ \dots & \dots & \dots & \dots \\ 0 & 0 & \dots & \boldsymbol{\zeta}_6^{T}\, {}^{0}\$_{6a} \end{bmatrix} \tag{7.151} $$

with

$$ \boldsymbol{\zeta}_i^{T}\, {}^{0}\$_{ia} = 1. \tag{7.152} $$

The Gough-Stewart platform having 6 *DOF*, $\mathbf{A} = \mathbf{A}_r$, ${}^{0}\mathbf{t}_p = {}^{0}\mathbf{t}_r$ and we finally have

$$ \mathbf{A}\,{}^{0}\mathbf{t}_p + \mathbf{B}\dot{\mathbf{q}}_a = \mathbf{0}. \tag{7.153} $$

7.4 Acceleration Analysis

7.4.1 Kinematic Constraint Relations of the Second Order

The second-order kinematic relations could be obtained by differentiating (7.62) w.r.t. time. However, this solution may not be computationally efficient. Therefore, we propose to use the following methodology which can take advantage of the recursive algorithms defined in Sect. 5.3. From (5.30), we can express for the chain i the acceleration of the platform as a function of all joint accelerations of the considered leg as:

$$ {}^{0}\dot{\mathbf{t}}_p = \begin{bmatrix} {}^{0}\dot{\mathbf{v}}_p \\ {}^{0}\dot{\boldsymbol{\omega}}_p \end{bmatrix} = {}^{0}\mathbf{J}_{p_i}\ddot{\mathbf{q}}_i + {}^{0}\overline{\mathbf{b}}_{p_i} = \begin{bmatrix} {}^{0}\$_{i1} & \dots & {}^{0}\$_{i\,m_i} \end{bmatrix} \ddot{\mathbf{q}}_i + {}^{0}\overline{\mathbf{b}}_{p_i} \tag{7.154} $$

where ${}^{0}\mathbf{J}_{p_i}$ is defined at (7.53) and ${}^{0}\overline{\mathbf{b}}_{p_i} = {}^{0}\mathbf{R}_p\,{}^{p}\overline{\mathbf{b}}_{p_i}$ with ${}^{p}\overline{\mathbf{b}}_{p_i}$ defined at (5.31).

As previously, let us rewrite (7.154) by reorganizing matrix ${}^{0}\mathbf{J}_{p_i}$ so that we can group:

- in a sub-matrix ${}^{0}\$_{ia}$ the unit twists ${}^{0}\$_{ik}$ corresponding to the active joints of velocities $\dot{\mathbf{q}}_{ai}$, and
- in a sub-matrix ${}^{0}\$_{id}$ the unit twists ${}^{0}\$_{ik}$ corresponding to the passive joints of velocities $\dot{\mathbf{q}}_{di}$.

Thus (7.154) becomes

$$^0\dot{\mathbf{t}}_p = {}^0\mathbf{\$}_{ia}\,\ddot{\mathbf{q}}_{ai} + {}^0\mathbf{\$}_{id}\,\ddot{\mathbf{q}}_{di} + {}^0\overline{\mathbf{b}}_{p_i}. \tag{7.155}$$

Similarly as in Sect. 7.3.1, to eliminate the passive joint accelerations $\ddot{\mathbf{q}}_{di}$ from (7.155), $^0\dot{\mathbf{t}}_p$ can be multiplied by the wrench $\boldsymbol{\zeta}_i$ which is reciprocal to all passive joint twists $^0\mathbf{\$}_{id}$ but NOT to the active joint twists $^0\mathbf{\$}_{ia}$. As a result, the following scalar equation can be obtained:

$$\boldsymbol{\zeta}_i^T\,{}^0\dot{\mathbf{t}}_p = (\boldsymbol{\zeta}_i^T\,{}^0\mathbf{\$}_{ia})\,\ddot{\mathbf{q}}_{ai} + \boldsymbol{\zeta}_i^T\,{}^0\overline{\mathbf{b}}_{p_i}. \tag{7.156}$$

Finally, considering all legs, Eq. (7.156) can be rewritten under the matrix form:

$$\begin{bmatrix} \boldsymbol{\zeta}_1^T \\ \boldsymbol{\zeta}_2^T \\ \vdots \\ \boldsymbol{\zeta}_n^T \end{bmatrix} {}^0\dot{\mathbf{t}}_p = \begin{bmatrix} \boldsymbol{\zeta}_1^T\,{}^0\mathbf{\$}_{1a} & \mathbf{0} & \dots & \mathbf{0} \\ \mathbf{0} & \boldsymbol{\zeta}_2^T\,{}^0\mathbf{\$}_{2a} & \dots & \mathbf{0} \\ \dots & \dots & \dots & \dots \\ \mathbf{0} & \mathbf{0} & \dots & \boldsymbol{\zeta}_n^T\,{}^0\mathbf{\$}_{na} \end{bmatrix} \ddot{\mathbf{q}}_a + \begin{bmatrix} \boldsymbol{\zeta}_1^T\,{}^0\overline{\mathbf{b}}_{p1} \\ \boldsymbol{\zeta}_2^T\,{}^0\overline{\mathbf{b}}_{p2} \\ \vdots \\ \boldsymbol{\zeta}_n^T\,{}^0\overline{\mathbf{b}}_{pn} \end{bmatrix}. \tag{7.157}$$

Thus, by introducing (7.58) and (7.59) into (7.157), we obtain

$$\mathbf{A}\,{}^0\dot{\mathbf{t}}_p + \mathbf{B}\ddot{\mathbf{q}}_a = {}^0\overline{\mathbf{b}}'_p \tag{7.158}$$

with

$$^0\overline{\mathbf{b}}'_p = \begin{bmatrix} \boldsymbol{\zeta}_1^T\,{}^0\overline{\mathbf{b}}_{p1} \\ \boldsymbol{\zeta}_2^T\,{}^0\overline{\mathbf{b}}_{p2} \\ \vdots \\ \boldsymbol{\zeta}_n^T\,{}^0\overline{\mathbf{b}}_{pn} \end{bmatrix}. \tag{7.159}$$

7.4.2 Forward and Inverse Second-Order Kinematic Models

7.4.2.1 Condensed Form of the Second-Order Kinematic Constraint Relations

Let us define the vector $^0\dot{\mathbf{t}}_r$ as the derivative w.r.t. time of the twist $^0\mathbf{t}_r$. Thus, from (7.61), the following relations hold :

$$^0\dot{\mathbf{t}}_p = \boldsymbol{\Psi}_t\,{}^0\dot{\mathbf{t}}_r + \dot{\boldsymbol{\Psi}}_t\,{}^0\mathbf{t}_r \Leftrightarrow {}^0\dot{\mathbf{t}}_r = \boldsymbol{\Psi}_t^{inv}\,{}^0\dot{\mathbf{t}}_p. \tag{7.160}$$

Using (7.160), we can rewrite (7.158) as:

$$\mathbf{A}_r\,{}^0\dot{\mathbf{t}}_r + \mathbf{B}\ddot{\mathbf{q}}_a = {}^0\overline{\mathbf{b}}_p \tag{7.161}$$

where $\mathbf{A}_r = \mathbf{A}\,\boldsymbol{\Psi}_t$ is defined in (7.62) and

$$^{0}\overline{\mathbf{b}}_p = {}^{0}\overline{\mathbf{b}}'_p - \mathbf{A}\dot{\boldsymbol{\Psi}}_t\,{}^{0}\mathbf{t}_r. \tag{7.162}$$

7.4.2.2 Second-Order Kinematic Models of *PKM* Without Redundancy

Matrices $\mathbf{A}_r$ and $\mathbf{B}$ being square for a *PKM* without actuation redundancy, the second-order *FKM* is given by:

$$^{0}\dot{\mathbf{t}}_r = -\mathbf{A}_r^{-1}(\mathbf{B}\ddot{\mathbf{q}}_a + {}^{0}\overline{\mathbf{b}}_p) = \mathbf{J}\ddot{\mathbf{q}}_a + \mathbf{a}_t \tag{7.163}$$

with $\mathbf{J} = -\mathbf{A}_r^{-1}\mathbf{B}$ is defined in (7.63) and $\mathbf{a}_t = -\mathbf{A}_r^{-1}\,{}^{0}\overline{\mathbf{b}}_p$. This expression is valid as long as matrix $\mathbf{A}_r$ is not singular. The singularity conditions of matrix $\mathbf{A}_r$ are disclosed in Sect. 7.5.

The second-order *IKM* is given by:

$$\ddot{\mathbf{q}}_a = -\mathbf{B}^{-1}(\mathbf{A}_r\,{}^{0}\dot{\mathbf{t}}_r + {}^{0}\overline{\mathbf{b}}_p) = \mathbf{J}_{inv}\,{}^{0}\dot{\mathbf{t}}_r + \mathbf{a}_q \tag{7.164}$$

with $\mathbf{J}_{inv} = -\mathbf{B}^{-1}\mathbf{A}_r$ is defined in (7.64) and $\mathbf{a}_q = -\mathbf{B}^{-1}\,{}^{0}\overline{\mathbf{b}}_p$. This expression is valid as long as matrix $\mathbf{B}$ is not singular. The singularity conditions of matrix $\mathbf{B}$ are disclosed in Sect. 7.5.

7.4.2.3 Second-Order Kinematic Models of *PKM* with Actuation Redundancy

In that case, matrix $\mathbf{A}_r$ being rectangular while matrix $\mathbf{B}$ is square, the second-order *IKM* can be defined without any problem using (7.164)—as long as matrix $\mathbf{B}$ is not singular—while the second-order *FKM* becomes:

$$^{0}\dot{\mathbf{t}}_r = -\mathbf{A}_r^{+}(\mathbf{B}\ddot{\mathbf{q}}_a + {}^{0}\overline{\mathbf{b}}_p) = \mathbf{J}_{inv}^{+}\ddot{\mathbf{q}}_a + \mathbf{a}_t^{r} \tag{7.165}$$

where $\mathbf{A}_r^{+} = \left(\mathbf{A}_r^T\mathbf{A}_r\right)^{-1}\mathbf{A}_r^T$ is the pseudo-inverse of the inverse Jacobian matrix $\mathbf{A}_r$, $\mathbf{J}_{inv}^{+} = \left(\mathbf{J}_{inv}^T\mathbf{J}_{inv}\right)^{-1}\mathbf{J}_{inv}^T$ is the pseudo-inverse of the inverse kinematic Jacobian matrix $\mathbf{J}_{inv}$ defined at (7.64) and $\mathbf{a}_t^{r} = -\mathbf{A}_r^{+}\,{}^{0}\overline{\mathbf{b}}_p$.

7.4.2.4 Second-Order Kinematic Models of *PKM* with Kinematic Redundancy

In that case, matrix $\mathbf{B}$ being rectangular while matrix $\mathbf{A}_r$ is square, the second-order *FKM* can be defined without any problem using (7.163)—as long as matrix $\mathbf{A}_r$ is not singular—while the second-order *IKM* becomes:

$$\ddot{\mathbf{q}}_a = -\mathbf{B}^{+}(\mathbf{A}_r\,{}^{0}\dot{\mathbf{t}}_r + {}^{0}\overline{\mathbf{b}}_p) + \left(\mathbf{1}_{n_a} - \mathbf{B}^{+}\mathbf{B}\right)\boldsymbol{\xi}_d = \mathbf{J}^{+0}\dot{\mathbf{t}}_r + \mathbf{a}_q^{r} \tag{7.166}$$

where $\mathbf{B}^{+}=\mathbf{B}^{T}\left(\mathbf{B}\mathbf{B}^{T}\right)^{-1}$ is the pseudo-inverse of the Jacobian matrix $\mathbf{B}$, $\boldsymbol{\xi}_d$ is an arbitrary vector in the $\ddot{\mathbf{q}}_a$ space (Khalil and Dombre 2002), $\mathbf{J}^{+}=\mathbf{J}^{T}\left(\mathbf{J}\mathbf{J}^{T}\right)^{-1}$ is the pseudo-inverse of the kinematic Jacobian matrix $\mathbf{J}$ defined at (7.63) and $\mathbf{a}_q^r = -\mathbf{B}^{+}\,{}^{0}\overline{\mathbf{b}}_p+\left(\mathbf{1}_{n_a}-\mathbf{B}^{+}\mathbf{B}\right)\boldsymbol{\xi}_d$.

7.4.2.5 Relations Between the Derivative of the Platform Twist and the Derivatives of the Platform Coordinates w.r.t. Time

Considering as in Sect. 7.3.2.5 that $\dot{\mathbf{x}}_p^T=[\dot{x}\ \dot{y}\ \dot{z}\ \dot{\phi}\ \dot{\theta}\ \dot{\psi}]$, and thus that $\ddot{\mathbf{x}}_p^T=[\ddot{x}\ \ddot{y}\ \ddot{z}\ \ddot{\phi}\ \ddot{\theta}\ \ddot{\psi}]$, the rotational acceleration of the platform is given by:

$$
{}^{0}\dot{\boldsymbol{\omega}}_p=\begin{bmatrix}\ddot{\psi}\mathrm{s}_\theta\mathrm{c}_\phi-\ddot{\theta}\mathrm{s}_\phi+\dot{\psi}(\dot{\theta}\mathrm{c}_\theta\mathrm{c}_\phi-\dot{\phi}\mathrm{s}_\theta\mathrm{s}_\phi)-\dot{\phi}\dot{\theta}\mathrm{c}_\phi\\ \ddot{\psi}\mathrm{s}_\theta\mathrm{s}_\phi+\ddot{\theta}\mathrm{c}_\phi+\dot{\psi}(\dot{\theta}\mathrm{c}_\theta\mathrm{s}_\phi+\dot{\phi}\mathrm{s}_\theta\mathrm{c}_\phi)-\dot{\phi}\dot{\theta}\mathrm{s}_\phi\\ \ddot{\phi}+\ddot{\psi}\mathrm{c}_\theta-\dot{\theta}\dot{\psi}\mathrm{s}_\theta\end{bmatrix} \tag{7.167}
$$

which can be rewritten in the matrix form

$$
{}^{0}\dot{\boldsymbol{\omega}}_p=\mathbf{D}\begin{bmatrix}\ddot{\phi}\\ \ddot{\theta}\\ \ddot{\psi}\end{bmatrix}+\begin{bmatrix}\dot{\psi}(\dot{\theta}\mathrm{c}_\theta\mathrm{c}_\phi-\dot{\phi}\mathrm{s}_\theta\mathrm{s}_\phi)-\dot{\phi}\dot{\theta}\mathrm{c}_\phi\\ \dot{\psi}(\dot{\theta}\mathrm{c}_\theta\mathrm{s}_\phi+\dot{\phi}\mathrm{s}_\theta\mathrm{c}_\phi)-\dot{\phi}\dot{\theta}\mathrm{s}_\phi\\ -\dot{\theta}\dot{\psi}\mathrm{s}_\theta\end{bmatrix}=\mathbf{D}\begin{bmatrix}\ddot{\phi}\\ \ddot{\theta}\\ \ddot{\psi}\end{bmatrix}+\mathbf{e} \tag{7.168}
$$

where $\mathbf{D}$ is defined in (7.69).

Thus, we can deduce that the derivative of the platform twist $\dot{\mathbf{t}}_p$ is equal to:

$$
{}^{0}\dot{\mathbf{t}}_p=\overline{\mathbf{D}}\ddot{\mathbf{x}}_p+\begin{bmatrix}\mathbf{0}\\ \mathbf{e}\end{bmatrix}=\overline{\mathbf{D}}\ddot{\mathbf{x}}_p+\overline{\mathbf{e}} \tag{7.169}
$$

where $\overline{\mathbf{D}}$ is defined in (7.70).

Similar equations can be found whatever are the types of parameters.

7.4.2.6 Obtaining the Second-Order Kinematic Models by Differentiating the First-Order Kinematic Constraint Equations

Usually, the second-order kinematic relationships are obtained through the differentiation of (7.72) w.r.t. time, which leads to:

$$
\mathbf{A}_d\ddot{\mathbf{x}}+\dot{\mathbf{A}}_d\dot{\mathbf{x}}+\mathbf{B}_d\ddot{\mathbf{q}}_a+\dot{\mathbf{B}}_d\dot{\mathbf{q}}_a=\mathbf{0} \tag{7.170}
$$

from which we can get (for a non-redundant *PKM*)

$$
\ddot{\mathbf{x}}=\mathbf{J}_x\ddot{\mathbf{q}}_a-\mathbf{A}_d^{-1}\left(\dot{\mathbf{A}}_d\dot{\mathbf{x}}+\dot{\mathbf{B}}_d\dot{\mathbf{q}}_a\right) \tag{7.171}
$$

or also

$$\ddot{\mathbf{q}}_a = \mathbf{J}_x^{-1}\ddot{\mathbf{x}} - \mathbf{B}_d^{-1}\left(\dot{\mathbf{A}}_d\dot{\mathbf{x}} + \dot{\mathbf{B}}_d\dot{\mathbf{q}}_a\right). \tag{7.172}$$

7.4.3 Computation of the Passive Joint Accelerations

Computation of the passive joint accelerations is necessary for the computation of the dynamic model. Therefore, the way to compute them is defined in this section.

Once again, all joint accelerations could be obtained by differentiating (7.3) w.r.t. time twice. However, this solution may not be computationally efficient. Therefore, we propose to use the following methodology which can take advantage of the recursive algorithms defined in Sect. 5.3.

Let us consider the chain i of the *PKM*, which is composed of m_i joints (Fig. 7.1). From (3.20), we can compute the acceleration of the platform at point $A_{i\,m_i}$ (that will be denoted as $\dot{\mathbf{t}}_p^i$) as:

$$ {}^0\dot{\mathbf{t}}_p^i = \begin{bmatrix} \mathbf{1}_3 & -{}^0\hat{\mathbf{r}}_{PA_{i\,m_i}} \\ \mathbf{0}_3 & \mathbf{1}_3 \end{bmatrix} {}^0\dot{\mathbf{t}}_p + \begin{bmatrix} {}^0\boldsymbol{\omega}_p \times ({}^0\boldsymbol{\omega}_p \times {}^0\mathbf{r}_{PA_{i\,m_i}}) \\ \mathbf{0} \end{bmatrix} = \mathbf{J}_{t_i}{}^0\dot{\mathbf{t}}_p + \mathbf{d}_i. \tag{7.173}$$

From (5.30), and as the joint located at $A_{i\,m_i}$ also belongs to the chain i, we also have

$$ {}^0\dot{\mathbf{t}}_p^i = {}^0\mathbf{J}_{i\,m_i}\ddot{\mathbf{q}}_i + {}^0\overline{\mathbf{b}}_{i\,m_i} \tag{7.174}$$

where $\ddot{\mathbf{q}}_i$ represents the vector of all joint accelerations of the chain i, ${}^0\mathbf{J}_{i\,m_i} = {}^0\overline{\mathbf{R}}_{i\,m_i}{}^{i\,m_i}\mathbf{J}_{i\,m_i}$ is the chain i Jacobian matrix also found in (7.83), and ${}^0\overline{\mathbf{b}}_{i\,m_i} = {}^0\overline{\mathbf{R}}_{i\,m_i}{}^{i\,m_i}\overline{\mathbf{b}}_{i\,m_i}$, where ${}^{i\,m_i}\overline{\mathbf{b}}_{i\,m_i}$ can be obtained by the recursive algorithm (5.31).

As in the Sect. 7.3.3, Eq. (7.174) can be rewritten by reorganizing matrix ${}^0\mathbf{J}_{i\,m_i}$ so that we can regroup:

- in a sub-matrix ${}^0\mathbf{J}_{ai}$ the unit twists ${}^0\$_{i\,m_i}^{ik}$ corresponding to the active joints of velocities $\dot{\mathbf{q}}_{ai}$, and
- in a sub-matrix ${}^0\mathbf{J}_{di}$ the unit twists ${}^0\$_{i\,m_i}^{ik}$ corresponding to the passive joints of velocities $\dot{\mathbf{q}}_{di}$.

Thus (7.174) becomes

$$ {}^0\dot{\mathbf{t}}_p^i = {}^0\mathbf{J}_{i\,m_i}{}^i\ddot{\mathbf{q}}_i + {}^0\overline{\mathbf{b}}_{i\,m_i} = \begin{bmatrix} {}^0\mathbf{J}_{ai} & {}^0\mathbf{J}_{di} \end{bmatrix} \begin{bmatrix} \ddot{\mathbf{q}}_{ai} \\ \ddot{\mathbf{q}}_{di} \end{bmatrix} + {}^0\overline{\mathbf{b}}_{i\,m_i} = {}^0\mathbf{J}_{ai}\ddot{\mathbf{q}}_{ai} + {}^0\mathbf{J}_{di}\ddot{\mathbf{q}}_{di} + {}^0\overline{\mathbf{b}}_{i\,m_i} \tag{7.175}$$

or also

$$
{}^{0}\mathbf{a}_{p}^{i} = {}^{0}\mathbf{J}_{di}\ddot{\mathbf{q}}_{di} \tag{7.176}
$$

where ${}^{0}\mathbf{a}_{p}^{i} = {}^{0}\dot{\mathbf{t}}_{p}^{i} - {}^{0}\mathbf{J}_{ai}\ddot{\mathbf{q}}_{ai} - {}^{0}\overline{\mathbf{b}}_{i\,m_i}$.

As $\ddot{\mathbf{q}}_{di}$ is a vector of dimension $n_{di} < 6$, only n_{di} components of ${}^{0}\mathbf{a}_{p}^{i}$ are independent. Let us define a vector $\mathbf{a}_{p}^{c\,i}$ of n_{di} independent components such that

$$
\mathbf{a}_{p}^{c\,i} = \boldsymbol{\Psi}_{ti}{}^{0}\mathbf{a}_{p}^{i} = \boldsymbol{\Psi}_{ti}({}^{0}\dot{\mathbf{t}}_{p}^{i} - {}^{0}\mathbf{J}_{ai}\ddot{\mathbf{q}}_{ai} - {}^{0}\overline{\mathbf{b}}_{i\,m_i}) \tag{7.177}
$$

with $\boldsymbol{\Psi}_{ti}$ a $(n_{di} \times 6)$ matrix defined in Sect. 7.3.3. Introducing (7.160), (7.173) into (7.177) leads to

$$
\mathbf{a}_{p}^{c\,i} = \mathbf{J}_{t_i}^{c}{}^{0}\dot{\mathbf{t}}_{r} - \mathbf{J}_{tai}\ddot{\mathbf{q}}_{ai} + \mathbf{d}_{i}^{c} \tag{7.178}
$$

where $\mathbf{J}_{t_i}^{c}$ and $\mathbf{J}_{tai}$ are defined in (7.88) and (7.89), and

$$
\mathbf{d}_{i}^{c} = \boldsymbol{\Psi}_{ti}\left(\mathbf{J}_{t_i}\dot{\boldsymbol{\Psi}}_{t}{}^{0}\mathbf{t}_{r} + \mathbf{d}_{i} - {}^{0}\overline{\mathbf{b}}_{i\,m_i}\right). \tag{7.179}
$$

Introducing (7.90) and (7.178) into (7.176), then

$$
\mathbf{J}_{t_i}^{c}{}^{0}\dot{\mathbf{t}}_{r} - \mathbf{J}_{tai}\ddot{\mathbf{q}}_{ai} + \mathbf{d}_{i}^{c} = \mathbf{J}_{tdi}\ddot{\mathbf{q}}_{di}. \tag{7.180}
$$

Thus, as the platform twist/acceleration and active joint velocities/accelerations are considered known by using the equations presented in the previous Sections, the vector $\mathbf{a}_{p}^{c\,i}$ is known and the joint accelerations $\ddot{\mathbf{q}}_{di}$ of the chain i can be found by inverting the matrix $\mathbf{J}_{tdi}$ in (7.180). The conditions of singularity of matrix $\mathbf{J}_{tdi}$ are disclosed in Sect. 7.5.2.

Now, considering all legs, we have

$$
\mathbf{J}_{t}{}^{0}\dot{\mathbf{t}}_{r} - \mathbf{J}_{ta}\ddot{\mathbf{q}}_{a} + \mathbf{d}_{c} = \mathbf{J}_{td}\ddot{\mathbf{q}}_{d} \tag{7.181}
$$

with $\mathbf{J}_{t}$, $\mathbf{J}_{ta}$ and $\mathbf{J}_{td}$ three matrices defined in (7.93), (7.94) and (7.92), and

$$
\mathbf{d}_{c} = \begin{bmatrix} \mathbf{d}_{1}^{c} \\ \vdots \\ \mathbf{d}_{n}^{c} \end{bmatrix} \tag{7.182}
$$

or also

$$
\ddot{\mathbf{q}}_{d} = \mathbf{J}_{td}^{-1}(\mathbf{J}_{t}{}^{0}\dot{\mathbf{t}}_{r} - \mathbf{J}_{ta}\ddot{\mathbf{q}}_{a} + \mathbf{d}_{c}). \tag{7.183}
$$

7.4.4 Examples

In this section, we present the second-order kinematic models of some *PKM* described in previous sections.

7.4.4.1 Second-Order Kinematics of the Five-Bar Mechanism

Let us consider again the five-bar mechanism presented in Sect. 7.1.2.1. For this mechanism, one can show that the relation (7.154) can be expressed as:

$$ {}^0\dot{\mathbf{t}}_p = \left[{}^0\mathbf{\$}_{11}\ {}^0\mathbf{\$}_{12}\ {}^0\mathbf{\$}_{13}\right]\ddot{\mathbf{q}}_1 + {}^0\overline{\mathbf{b}}_{p1}, \tag{7.184} $$

$$ {}^0\dot{\mathbf{t}}_p = \left[{}^0\mathbf{\$}_{21}\ {}^0\mathbf{\$}_{22}\right]\ddot{\mathbf{q}}_2 + {}^0\overline{\mathbf{b}}_{p2}, \tag{7.185} $$

where the unit twist ${}^0\mathbf{\$}_{ij}$ are defined at (7.96) and (7.97), ${}^0\dot{\mathbf{t}}_p = [\ddot{x}\ \ddot{y}\ 0\ 0\ 0\ \ddot{\phi}]^T$, $\ddot{\mathbf{q}}_1 = [\ddot{q}_{11}\ \ddot{q}_{12}\ \ddot{q}_{13}]^T$, and $\ddot{\mathbf{q}}_2 = [\ddot{q}_{21}\ \ddot{q}_{22}]^T$, and

$$ {}^0\overline{\mathbf{b}}_{pi} = \begin{bmatrix} -d_{i2}\dot{q}_{i1}\cos q_{i1} - d_{i3}c_{12i}(\dot{q}_{i1}+\dot{q}_{i2}) \\ -d_{i2}\dot{q}_{i1}\sin q_{i1} - d_{i3}s_{12i}(\dot{q}_{i1}+\dot{q}_{i2}) \\ 0 \\ 0 \\ 0 \\ 0 \end{bmatrix} \dot{q}_{i1} + \begin{bmatrix} -d_{i3}c_{12i} \\ -d_{i3}s_{12i} \\ 0 \\ 0 \\ 0 \\ 0 \end{bmatrix} \dot{q}_{i2}(\dot{q}_{i1}+\dot{q}_{i2}). \tag{7.186} $$

Then, using the fact that the wrenches $\boldsymbol{\zeta}_i$ are already defined in Sect. 7.3.4.1, as well as the matrix $\mathbf{\Psi}_t^{inv}$, the relation (7.161) is given by

$$ \mathbf{A}_r\,{}^0\dot{\mathbf{t}}_r + \mathbf{B}\ddot{\mathbf{q}}_a = {}^0\overline{\mathbf{b}}_p \tag{7.187} $$

where the matrices $\mathbf{A}_r$ and $\mathbf{B}$ are defined in (7.104) and (7.113) and

$$ {}^0\overline{\mathbf{b}}_p = \begin{bmatrix} \boldsymbol{\zeta}_1^T\,{}^0\overline{\mathbf{b}}_{p1} \\ \boldsymbol{\zeta}_2^T\,{}^0\overline{\mathbf{b}}_{p2} \end{bmatrix}. \tag{7.188} $$

From (7.187), the second-order *FKM* and *IKM* can be straightforwardly defined.

Then, starting from the fact that the relation (7.174) for the five-bar mechanism is already defined in (7.184) and (7.185) by assimilating the fact that, for $i = 1, 2$,

- ${}^0\dot{\mathbf{t}}_p^i = {}^0\dot{\mathbf{t}}_p$,
- ${}^0\mathbf{J}_{1m_1} = \left[{}^0\mathbf{\$}_{11}\ {}^0\mathbf{\$}_{12}\ {}^0\mathbf{\$}_{13}\right]$, ${}^0\mathbf{J}_{2m_2} = \left[{}^0\mathbf{\$}_{21}\ {}^0\mathbf{\$}_{22}\right]$,
- ${}^0\overline{\mathbf{b}}_{im_i} = {}^0\overline{\mathbf{b}}_{pi}$,

the passive joint acceleration given by the relation (7.183) can be straightforwardly computed as

$$\ddot{\mathbf{q}}_d = \mathbf{J}_{td}^{-1}(\mathbf{J}_t\,{}^0\dot{\mathbf{t}}_r - \mathbf{J}_{ta}\ddot{\mathbf{q}}_a + \mathbf{d}_c) \tag{7.189}$$

where the matrices $\mathbf{J}_t$, $\mathbf{J}_{ta}$ and $\mathbf{J}_{td}$ are given in (7.122)–(7.124) and

$$\mathbf{d}_c = \begin{bmatrix} \mathbf{d}_1^c \\ \mathbf{d}_2^c \end{bmatrix} \tag{7.190}$$

with $\mathbf{d}_i^c = \mathbf{\Psi}_{ti}\left(\mathbf{J}_{t_i}\dot{\mathbf{\Psi}}_t\,{}^0\mathbf{t}_r + \mathbf{d}_i - {}^0\overline{\mathbf{b}}_{i\,m_i}\right)$ in which the matrices $\mathbf{\Psi}_{ti}$ defined in (7.119) and (7.120).

7.4.4.2 Second-Order Kinematics of the Orthoglide

The second-order kinematics are tedious, this is the reason why they are not detailed here but are given in:
http://www.irccyn.ec-nantes.fr/~briot/Books.html.

7.5 Singularity Analysis

In the present Section, we will deal with the problem of singular configurations of *PKM*. The aim of the information given below is not to present a straightforward way to analyze and find the singularity of a *PKM*, but to show the most current singular configurations that we can meet, and above all, to disclose information about the configurations that impact the robot dynamic model.

For a complete and comprehensive discussion about the singularity problem, the reader is referred to Conconi and Carricato (2009) and Zlatanov et al. (1994a, b).

7.5.1 Input-Output Singularities

The first kind of singular configurations we will analyze are those that we call *input-output* singularities. They can be defined through analysis of the input-output kinematic relationship described in (7.62), from which three main types of singularity can be defined (Gosselin and Angeles 1990):

- when matrix **B** is rank-deficient: such kind of singularity is called a *Type 1* singularity.[3] In such singularities, the *PKM* loses the ability to move along one (or more) direction of the workspace, i.e. a motion of the actuators does not lead to the displacement of the robot platform (Fig. 7.18).

[3]They are also called *serial* singularities in some works because it is similar to the singularities of serial robots. However, in the present book, a *serial* singularity has another meaning.

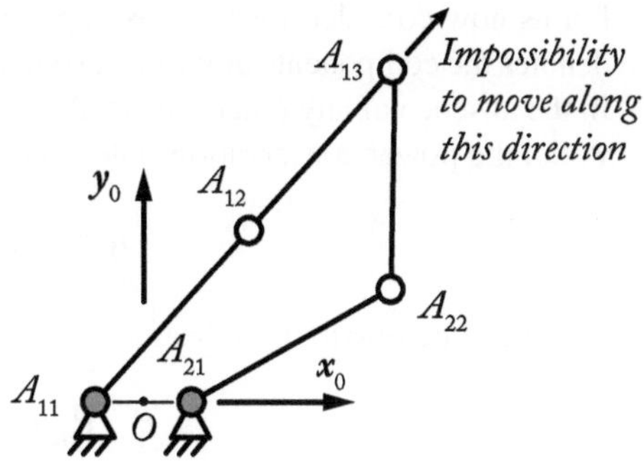

Fig. 7.18 Example of a parallel robot (here, a five-bar mechanism) in a Type 1 singularity

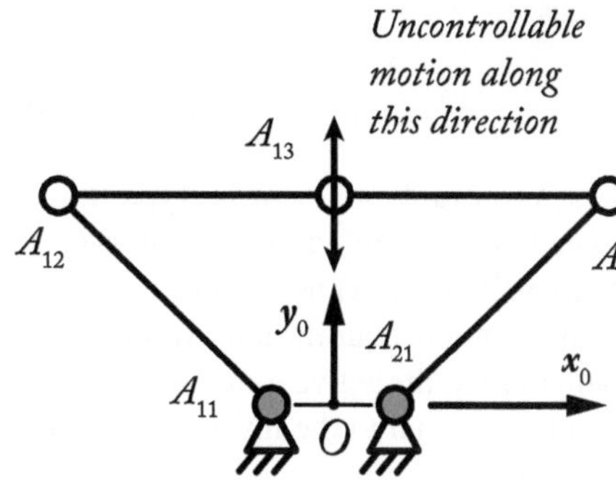

Fig. 7.19 Example of parallel robot (here, a five-bar mechanism) in a Type 2 singularity

- when matrix $\mathbf{A}_r$ (as well as matrices $\mathbf{A}$ and $\mathbf{A}_d$) is rank-deficient: such kind of singularity is called *Type 2* singularity.[4] In such singularities, the *PKM* gains one (or more) uncontrollable motion, i.e. it becomes shaky. Kinematically speaking, there can exist a non-null vector $\mathbf{t}_s$ defined such that $\mathbf{A}_r\mathbf{t}_s = \mathbf{0}$ (which means that $\mathbf{t}_s$ is in the null space of $\mathbf{A}_r$) while $\dot{\mathbf{q}}_a = \mathbf{0}$, i.e. the actuators are fixed (Fig. 7.19). Statically speaking, the robot cannot resist an external wrench applied on the platform. This can be easily proven as follows.

Let us consider a virtual platform twist $\mathbf{t}_r^*$ linked from (7.62) to the virtual joint velocity $\dot{\mathbf{q}}_a^*$ by

$$\mathbf{A}_r\mathbf{t}_r^* + \mathbf{B}\dot{\mathbf{q}}_a^* = \mathbf{0} \tag{7.191}$$

which, in the case of *PKM* without kinematic redundancy can be rewritten as

$$\dot{\mathbf{q}}_a^* = -\mathbf{B}^{-1}\mathbf{A}_r\mathbf{t}_r^* = \mathbf{J}_{inv}\mathbf{t}_r^*. \tag{7.192}$$

[4]They are also called *parallel* singularities in some works. However, we do not want to use this term because it may concern also the *constraint singularities* and thus may be confusing.

Let us now consider a wrench $\mathbf{w}_p$ applied on the *PKM* platform. The vector $\mathbf{w}_r$ denotes the components of $\mathbf{w}_p$ corresponding to the independent coordinates of $\mathbf{t}_r^*$. In the absence of any other effects, the wrench $\mathbf{w}_p$ leads to the robot input efforts $\boldsymbol{\tau}$. So, the power conservation states that:

$$\dot{\mathbf{q}}_a^{*\,T}\boldsymbol{\tau} = \mathbf{t}_r^{*\,T}\mathbf{w}_r \tag{7.193}$$

which can be rewritten as, from (7.64)

$$\mathbf{t}_r^{*\,T}\mathbf{J}_{inv}^T\boldsymbol{\tau} = \mathbf{t}_r^{*\,T}\mathbf{w}_r. \tag{7.194}$$

Thus,

$$\mathbf{J}_{inv}^T\boldsymbol{\tau} = -\mathbf{A}_r^T\mathbf{B}^{-T}\boldsymbol{\tau} = \mathbf{w}_r. \tag{7.195}$$

As $\mathbf{A}_r$ is rank deficient in Type 2 singularities, $\mathbf{J}_{inv}$ is also singular. Thus, this means that there can exist a non-null input effort vector $\boldsymbol{\tau}$ corresponding to a null wrench $\mathbf{w}_r$. Thus, the linear system (7.195) has an infinity of solutions and the robot is not in equilibrium. Another consequence is that in the neighborhood of the singularity, the active joint efforts $\boldsymbol{\tau}$ may increase considerably as their expression is proportional to the inverse of the determinant of $\mathbf{A}_r\mathbf{A}_r^T$, which is close to zero in that area.

Note that a similar proof can be given for the kinematically redundant *PKM*.

- when matrices $\mathbf{A}_r$ and $\mathbf{B}$ are simultaneously rank-deficient: such kind of singularity is called *Type 3* singularity and are a combination of both Type 1 and Type 2 singularities. For such configurations, the mechanism loses locally the ability to move along one (or more) direction of the workspace and gains one (or more) uncontrollable motion along another direction.

7.5.2 Serial Singularities

The second kind of singular configurations we will analyze are those that we call the *serial* singularities, i.e. the singularities that are due to the degeneracy of the leg i configuration only. Type 1 singularities are a particular case of serial singularities.

If the leg i meets such a configuration, the matrix ${}^0\mathbf{J}_{i\,m_i}$ of (7.83) becomes rank-deficient.

From (5.7) and (7.83), the matrix ${}^0\mathbf{J}_{i\,m_i}$ has the following form:

$${}^0\mathbf{J}_{i\,m_i} = \begin{bmatrix} {}^0\$_{i\,m_i}^{i1} & {}^0\$_{i\,m_i}^{i2} & \cdots & {}^0\$_{i\,m_i}^{i\,m_i} \end{bmatrix} \tag{7.196}$$

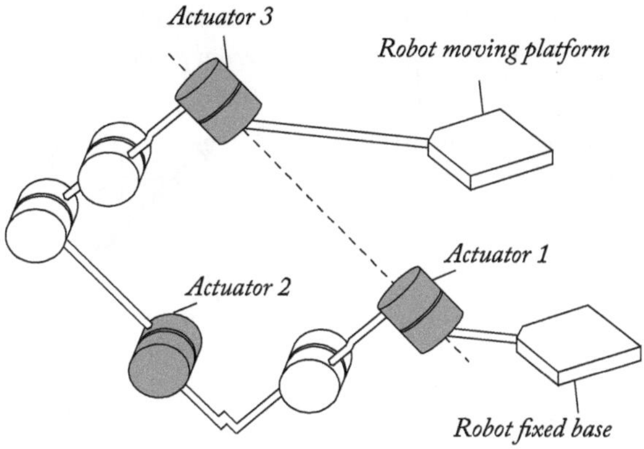

Fig. 7.20 Case 1 of serial singularity: in that case, the axes of actuators 1 and 3 are aligned and if $\dot{q}_1 = -\dot{q}_3$ ($\dot{q}_i$ being the velocity of the actuator i), actuator 1 and 3 motions do not lead to a platform motion

where ${}^0\$^{ik}_{i\,m_i}$ is a unit twist parameterizing the displacement of the leg's tip (expressed in the reference frame $\mathscr{F}_0$) when the joint located at $A_{i\,k}$ is moving only. Thus, singularity conditions appear when the basis B_i defined by the unit twists $\{{}^0\$^{i1}_{i\,m_i}\ {}^0\$^{i2}_{i\,m_i} \ldots {}^0\$^{i\,m_i}_{i\,m_i}\}$ is degenerated.

Let us rewrite this basis into two sub-bases written under a matrix form as: $B_i = \{B_{ia}\ B_{id}\} : \left[{}^0\mathbf{J}_{ai}\ {}^0\mathbf{J}_{di}\right]$, where $B_{ia} :\ {}^0\mathbf{J}_{ai}$ groups the columns of ${}^0\mathbf{J}_{i\,m_i}$ corresponding to the active joints and $B_{id} :\ {}^0\mathbf{J}_{di}$ the columns corresponding to the passive joints. Three cases can then be met:

- **Case 1**: the sub-system ${}^0\mathbf{J}_{ai}$ is rank-deficient (this also corresponds to the degeneracy of the matrix $\mathbf{J}_{tai}$ in (7.87) and, as a result, to the degeneracy of the matrix $\mathbf{J}_{ta}$ in (7.91))—in that case, a displacement of the active joints of the leg does not necessarily bring a displacement of the end-effector along one given direction (Fig. 7.20). We will call them *Leg Active Joint Twist System* (*LAJTS*) singularities.
- **Case 2**: the sub-system ${}^0\mathbf{J}_{di}$ is rank-deficient (this also corresponds to the degeneracy of the matrix $\mathbf{J}_{tdi}$ in (7.87) and, as a result, to the degeneracy of the matrix $\mathbf{J}_{td}$ in (7.91))—in that case, a displacement of the passive joints of the leg does not necessarily bring a displacement of the end-effector along one given direction. Moreover, in such configuration, the leg instantaneously gains an uncontrolled motion (Fig. 7.21). Later in the book, we will call them *Leg Passive Joint Twist System* (*LPJTS*) singularities.
- **Case 3**: the system $\left[{}^0\mathbf{J}_{ai}\ {}^0\mathbf{J}_{di}\right]$ is rank-deficient, while the systems ${}^0\mathbf{J}_{ai}$ and ${}^0\mathbf{J}_{di}$ are not—in that case, a displacement of any types of the leg joint does not necessarily bring a displacement of the end-effector along one given direction (Fig. 7.22). Such singularities are usually similar to the Type 1 singularities (see Sect. 7.5.1).

It will be shown later that the degeneracy of the system ${}^0\mathbf{J}_{di}$ lead to the degeneracy of the *PKM* dynamic model.

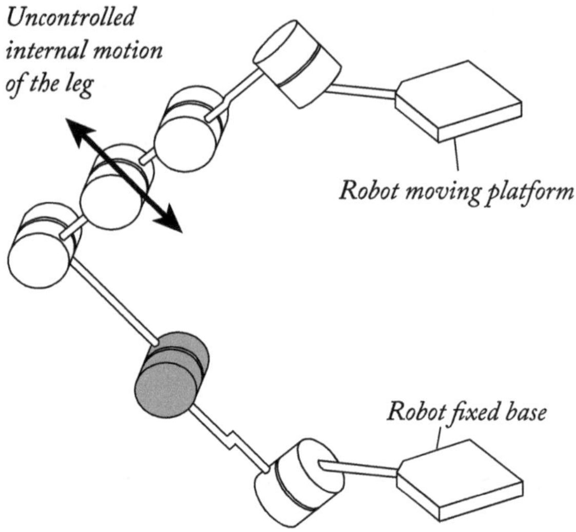

Fig. 7.21 Case 2 of serial singularity: in that case, the actuator being fixed, the leg gains one internal mobility and the motion of the passive joints does not lead to a platform motion

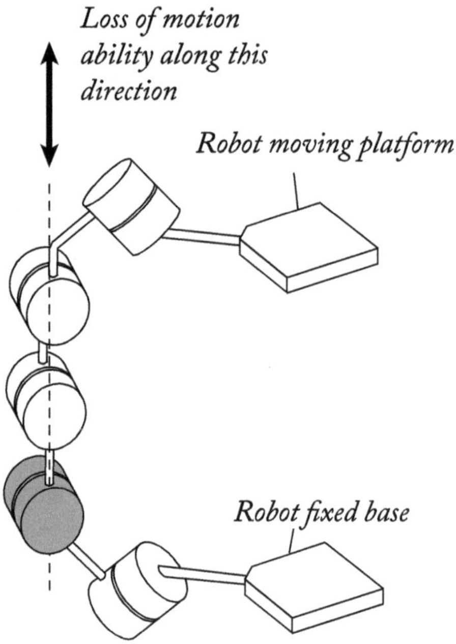

Fig. 7.22 Case 3 of serial singularity: in that case, the motion of all joints does not lead to a platform motion in the vertical direction

7.5.3 Other Types of Singularities

Other types of singularities appearing for lower-mobility *PKM* (for which $n_{dof} < 6$), such as the constraint singularities (Zlatanov et al. 2002), can also appear. They are due to the degeneracy of the constraint wrenches applied on the platform by the legs and lead to the apparition of uncontrollable motions of the platform. They are quite analogous to Type 2 singularities. However, their main specificity is that, in constraint singularities, the mobility of the platform is changed. For example, if a robot with three translational *DOF* encounters a constraint singularity, the uncontrollable motion will be a rotation of the platform.

They have been discovered for the first time on a 3–$U\underline{P}U$ mechanism with special arrangement of the *U* joints.

Constraints singularities are not the last case of singularities that a *PKM* can meet (see for example Amine et al. 2012a) but all such types of singularities are not crucial for the dynamic model computation and are therefore out of the scope of our book. As mentioned above, for a complete and comprehensive discussion about the singularity problem, the reader is referred to Conconi and Carricato (2009) and Zlatanov et al. (1994a, b).

7.5.4 Finding Robot Singular Configurations

For simple robots (see Sect. 7.5.6), the singular configurations can be found by analyzing the conditions of vanishing the determinant of the corresponding Jacobian matrices (in the case where they are square). However, such kind of analysis can be rapidly complicated and other methods should be preferred. Note that the singularity conditions of matrix **B** in (7.62) are usually simple and do not require more complicated analysis.

There exists two main methods in the literature that allows one to find robot singular configurations: the Grassmann geometry and the Grassmann-Cayley algebra. The next Sections present brief summaries of both methods. However, first of all, we define a methodology for finding the unit wrenches $\boldsymbol{\zeta}_k$ which are the rows of the matrix **A** in (7.58) and that represent the wrenches applied by the legs on the platform when the actuators are developing an input effort in a static mode of operation and in absence of any other type of external effects (Bonev 2002).

7.5.4.1 Computation of the Static Wrenches Applied by the Legs on the Platform

Let us consider the robot leg i whose kinematic Jacobian matrix is given at expression (7.196). As mentioned above, the columns of ${}^0\mathbf{J}_{i\,m_i} = \begin{bmatrix} {}^0\$^{i1}_{i\,m_i} & {}^0\$^{i2}_{i\,m_i} & \cdots & {}^0\$^{i\,m_i}_{i\,m_i} \end{bmatrix}$ are unit twists which compose a basis B_i and that can be divided into two sub-systems $B_{i\,a}$ (which regroups the twists corresponding to the active joints) and $B_{i\,d}$ (which groups the twists corresponding to the passive joints).

Let us denote as $\boldsymbol{\zeta}_{c\,i}$ the system of wrenches defined such as:

$$\boldsymbol{\zeta}_{c\,i}^T\ {}^0\$^{i\,k}_{i\,m_i} = \mathbf{0} \tag{7.197}$$

for $k = 1, \ldots, m_i$. Wrenches $\boldsymbol{\zeta}_{ci}$ are reciprocal to all unit twists of the leg i and are defined as the *constraint wrenches* applied by the legs on the platform.

Let us now denote as $\boldsymbol{\zeta}_{ti}$ the system of wrenches defined such as:

$$\boldsymbol{\zeta}_{ti}^T \; {}^0\mathbf{J}_{di} = \mathbf{0} \tag{7.198}$$

where ${}^0\mathbf{J}_{di}$ is defined in Sect. 7.5.2 and groups the columns of ${}^0\mathbf{J}_{i\,m_i}$ corresponding to the passive joints. Vectors $\boldsymbol{\zeta}_{ti}$ are reciprocal to the unit twists of the leg i due to the passive joints. By definition, $\boldsymbol{\zeta}_{ti} = \{\boldsymbol{\zeta}_{ci} \; \boldsymbol{\zeta}_{ai}\}$, where $\boldsymbol{\zeta}_{ai}$ are defined as the actuation wrenches, i.e. the wrenches applied by the leg i on the platform when its actuators are developing an input effort in a static mode of operation and in absence of any other type of external effects (Bonev 2002).

$\boldsymbol{\zeta}_{ai}$, $i = 1, \ldots, n$, are the vectors representing the rows of matrix $\mathbf{A}$ and is reciprocal to all unit twists of the leg i, except those corresponding to the active joints. Examples of computations of vectors $\boldsymbol{\zeta}_{ai}$ and $\boldsymbol{\zeta}_{ci}$ are given in Appendix C.

7.5.4.2 Basics of Grassmann Geometry

The use of the Grassmann geometry is based on the fact that the rows of matrix $\mathbf{A}$ in (7.60) (or the columns of matrix ${}^0\mathbf{J}_{i\,m_i}$ in (7.196)) are unit screws which can be seen as a Plücker representation of lines.

Using the Plücker representation, a line $\mathscr{L}$ can be represented by a direction vector $\mathbf{u}$ and a moment (given w.r.t. any frame, e.g. here $\mathscr{F}_0$) $\mathbf{r}_{O_0M} \times \mathbf{u}$, where M is any point belonging to the line $\mathscr{L}$. Thus:

$$\mathscr{L} : \begin{bmatrix} \mathbf{u} \\ \mathbf{r}_{O_0M} \times \mathbf{u} \end{bmatrix} \tag{7.199}$$

which corresponds typically to the expression of a screw (see Sect. 3.1) and Appendix C.

If k Plücker vectors are linearly independent (which may correspond to the linear dependency of k rows of matrix $\mathbf{A}$ or also k columns of matrix ${}^0\mathbf{J}_{i\,m_i}$), they will span a *variety* with dimension $k \leq 6$. The idea of Grassmann geometry is to show that the linear dependency of Plücker vectors induced geometric relations between the corresponding lines.

The varieties of dimension 1 and 2, which are quite common and easy to use for *PPM*, are defined as follows:

- *Variety of dimension 1*: it is represented by a single Plücker vector.
- *Variety of dimension* 2: it is spanned by:
  - (2a) either two lines that are skew, i.e. that do not intersect and are not parallel,
  - (2b) or by more than two lines that are (i) all parallel or (ii) are lying in the same plane and are intersecting in a single point. Such conditions are called *flat pencil of lines*.

- *Variety of dimension 3*: it is spanned by:
  - (3a) a *regulus*, i.e. a set of lines able to intersect three other skew lines,
  - (3b) the union of two *flat pencils* having a line in common but lying in distinct planes and with distinct centers,
  - (3c) all lines through a point,
  - (3d) all lines in a plane.
- *Variety of dimension 4*: it is spanned by:
  - (4a) an *elliptic congruence*: the variety is spanned by four skew lines such that none of these lines intersect the *regulus* generated by the other three,
  - (4b) a *hyperbolic congruence*: all the lines intersect two given skew lines,
  - (4c) a *parabolic congruence*: all the lines belong to the union of three *flat pencils of lines*, in different planes but with a common line.
  - (4d) a *degenerate congruence*: all lines lie in a plane or meet a a common point that lies within that plane.
- *Variety of dimension 5*: it is spanned by:
  - (5a) a *non-singular complex*: generated by five skew lines,
  - (5b) a *singular complex*: all the lines meeting one given line.

For more explanations, the reader is invited to read (Merlet 1989, 2006b).

7.5.4.3 Basics of Grassmann-Cayley Algebra

The Grassmann-Cayley algebra was developed by H. Grassmann as a calculus for linear varieties operating on extensors with the join "$\vee$" and meet "$\wedge$" operators. The latter are associated with the span and intersection of vector spaces. Grassmann-Cayley algebra makes it possible to work at the symbolic level, and therefore, to produce coordinate-free algebraic expressions for the singularity conditions of *SPM*. For further details on Grassmann-Cayley algebra, the reader is referred to Ben-Horin and Shoham (2006, 2009), Kanaan et al. (2009) and White (2008).

In order to use the Grassmann-Cayley algebra, we need to build a system of screws composed of at least 6 screws. However, for lower mobility *PKM*, i.e. *PKM* with $n_{dof} < 6$, the system of wrenches $\boldsymbol{\zeta}_{ai}$ $(i = 1, \ldots, n)$ composing the matrix $\mathbf{A}$ is not of dimension 6. Thus, in order to be able to use that tool, we need to complete the system of wrenches by adding the constraint wrenches $\boldsymbol{\zeta}_{ci}$ $(i = 1, \ldots, n)$ defined at (7.197).

Let us denote as $\mathbf{W} = [\boldsymbol{\zeta}_a \; \boldsymbol{\zeta}_c]$ the matrix composed of $\boldsymbol{\zeta}_a = [\boldsymbol{\zeta}_{a1} \cdots \boldsymbol{\zeta}_{an}]$ and $\boldsymbol{\zeta}_c = [\boldsymbol{\zeta}_{c1} \cdots \boldsymbol{\zeta}_{cb}]$. Let us consider first that $\mathbf{W}$ is a (6×6) matrix. The superjoin of the six wrenches composing $\mathbf{W}$ corresponds to the determinant of $\mathbf{W}$ up to a scalar multiple, which is the superbracket in Grassmann-Cayley algebra (White 2008). Thus, a singularity occurs when this superbracket vanishes. The superbracket is an expression involving 12 points selected on the six lines (for which the six wrenches represent the Plücker coordinates) and can be developed into a linear combination of

24 bracket monomials (Ben-Horin and Shoham 2006), each one being the product of three brackets of four projective points:

$$[\mathbf{ab},\ \mathbf{cd},\ \mathbf{ef},\ \mathbf{gh},\ \mathbf{ij},\ \mathbf{kl}] = \sum_{i=1}^{24} y_i \tag{7.200}$$

where

$$\begin{array}{lll}
y_1 = -[\mathbf{abcd}][\mathbf{efgi}][\mathbf{hjkl}] & y_2 = [\mathbf{abcd}][\mathbf{efhi}][\mathbf{gjkl}] & y_3 = [\mathbf{abcd}][\mathbf{efgj}][\mathbf{hikl}] \\
y_4 = -[\mathbf{abcd}][\mathbf{efhj}][\mathbf{gikl}] & y_5 = [\mathbf{abce}][\mathbf{dfgh}][\mathbf{ijkl}] & y_6 = -[\mathbf{abde}][\mathbf{cfgh}][\mathbf{ijkl}] \\
y_7 = -[\mathbf{abcf}][\mathbf{degh}][\mathbf{ijkl}] & y_8 = [\mathbf{abdf}][\mathbf{cegh}][\mathbf{ijkl}] & y_9 = -[\mathbf{abce}][\mathbf{dghi}][\mathbf{fjkl}] \\
y_{10} = [\mathbf{abde}][\mathbf{cghi}][\mathbf{fjkl}] & y_{11} = [\mathbf{abcf}][\mathbf{dghi}][\mathbf{ejkl}] & y_{12} = [\mathbf{abce}][\mathbf{dghj}][\mathbf{fikl}] \\
y_{13} = -[\mathbf{abdf}][\mathbf{cghi}][\mathbf{ejkl}] & y_{14} = -[\mathbf{abde}][\mathbf{cghj}][\mathbf{fikl}] & y_{15} = -[\mathbf{abcf}][\mathbf{dghj}][\mathbf{eikl}] \\
y_{16} = [\mathbf{abdf}][\mathbf{cghj}][\mathbf{eikl}] & y_{17} = [\mathbf{abcg}][\mathbf{defi}][\mathbf{hjkl}] & y_{18} = -[\mathbf{abdg}][\mathbf{cefi}][\mathbf{hjkl}] \\
y_{19} = -[\mathbf{abch}][\mathbf{defi}][\mathbf{gjkl}] & y_{20} = -[\mathbf{abcg}][\mathbf{defj}][\mathbf{hikl}] & y_{21} = [\mathbf{abdh}][\mathbf{cefi}][\mathbf{gjkl}] \\
y_{22} = [\mathbf{abdg}][\mathbf{cefj}][\mathbf{hikl}] & y_{23} = [\mathbf{abch}][\mathbf{defj}][\mathbf{gikl}] & y_{24} = -[\mathbf{abdh}][\mathbf{cefj}][\mathbf{gikl}]
\end{array} \tag{7.201}$$

In (7.200), the notation **ab** denotes the line passing through the points of homogeneous coordinates **a** and **b**, while in (7.200), the notation [**abcd**] (called a bracket) is for the determinant of the (4×4) matrix whose columns are composed of the homogeneous coordinates **a**, **b**, **c** and **d**. As a result, [**abcd**] represents the volume of a tetrahedron passing through the points of homogeneous coordinates **a**, **b**, **c** and **d** and it vanishes if and only if those points are coplanar.

The used points for computing the bracket [**abcd**] can be in real space or on a plane at infinity. In that case, the last component of the homogeneous coordinate vector is equal to zero and the representation of that point is similar to the representation of a vector in the homogeneous representation (see Sect. 2.1).

In the case where the matrix $\mathbf{W}$ is not a square matrix, the problem must be divided into k sub-problems, by defining the k possible (6×6) matrices $\mathbf{W}'$ that regroup subsets of columns of matrix $\mathbf{W}$ and by making the intersection of the found solutions.

The smart definition of the points whose coordinates are represented in the super-brack*et al.*lows the simplification of its expression (Amine et al. 2012a; Ben-Horin and Shoham 2006, 2009; Kanaan et al. 2009). Note that, to simplify the use of the Grassmann-Cayley algebra, a *Matlab* interface has been proposed in Ben-Horin et al. (2008).

7.5.5 Finding Robot Serial Singular Configurations

As the columns of the matrix ${}^0\mathbf{J}_{i\,m_i}$ are unit screws which can be seen as a Plücker representation of lines, the methods proposed in Sect. 7.5.4 can be applied to find serial singularities.

7.5.6 Examples

7.5.6.1 Singular Configurations of the Five-Bar Mechanism

Let us consider again the five-bar mechanism presented in Sect. 7.1.2.1. From the analysis of its input-output kinematic relations defined in Sect. 7.3.4.1, we can see that:

- Type 1 singularities (when matrix **B** defined at (7.104) is rank-deficient) appear when $q_{i2} = 0$ or π, i.e. when the leg is full stretched or folded. An example of such type of singularity for the five-bar mechanism was provided in Fig. 7.18.

Table 7.8 List of references dealing with the singularity problem of some interesting *SPM*

Type of *DOF*[a]	Robot name	References
2T1R	Zero-torsion *PKM*	Briot and Bonev (2008)
3T	Delta	Di Gregorio (2004b)
	Orthoglide-like family[b]	Pashkevich et al. (2006)
	3-$U\underline{P}U$	Zlatanov et al. (2002)
	Pantopteron	Briot and Bonev (2009b)
3R	Agile Eye	Bonev et al. (2002b)
	3–$\underline{R}SS$ wrist	Di Gregorio (2004a)
	3–$U\underline{P}U$ wrist	Di Gregorio (2003)
	other types of spherical *SPM*	Bonev and Gosselin (2005b)
		Gosselin and Sefrioui (1992)
3T1R	Quattro/Par4	Nabat (2007)
	H4, I4, I4L	Company and Pierrot (1999)
	PAMINSA	Briot et al. (2008)
	MacGill SMG	Alizadeh (2009)
	Quadrupteron	Kong and Gosselin (2011a)
	Pantopteron–4	Briot and Bonev (2010)
	4–$\underline{R}UU$	Amine et al. (2011)
3T2R	5–$R\underline{P}UR$	Amine et al. (2012b)
	Verne Machine	Kanaan (2008)
3T3R	Gough-Stewart platform family	Ben-Horin and Shoham (2006, 2009)
		Husty and Karger (2000), Merlet (1989)
		Innocenti and Parenti-Castelli (1992)
	Hexa family (6–$\underline{R}US$/ 6–$\underline{P}US$)	Same approach as for Gough-Stewart platforms
	Hexapteron	Seward and Bonev (2014)

[a]In that column, the symbol "iTjR" denotes that the considered mechanism has i translational *DOF* and j rotational *DOF*

[b]Orthoglide-like family regroups all 3 T robots with linear actuators mounted onto the base (whatever is their orientation) followed by passive legs of the Orthoglide type (Fig. 7.8)

- Type 2 singularities (when matrix $\mathbf{A}_r$ defined at (7.113) is rank-deficient) appear when $q_{11} + q_{12} - q_{21} - q_{22} = 0$ or π, i.e. when the points A_{12}, A_{13} and A_{22} are aligned. An example of such type of singularity for the five-bar mechanism was provided in Fig. 7.19.
  It can be also shown that, when $q_{11} + q_{12} - q_{21} - q_{22} = 0$ or π, there exists a non zero vector $\mathbf{t}_s = [-\sin(q_{11} + q_{12}) \;\; \cos(q_{11} + q_{12})]^T$ such that: $\mathbf{At}_s = \mathbf{0}$. $\mathbf{t}_s$ is orthogonal to the direction defined by the line passing through the points A_{12}, A_{13} and A_{22} and represents the direction of the uncontrollable motion inside the singularity (Fig. 7.19).

For the five-bar mechanism, there are no other types of singular configurations.

7.5.6.2 Further Readings

The singularities of all *PPM* have been deeply studied in Bonev (2002), Bonev et al. (2003) and Briot et al. (2008).

Moreover, the Table 7.8 presents a list of papers dealing with the singularity problem of some interesting *SPM*. It should be mentioned that this list is far from being exhaustive but only presents *SPM* quite often met in the literature. For a longer list of references, please visit the website of Merlet (2014).

Chapter 8
Dynamic Modeling of Parallel Robots

Abstract In this chapter, we present the computation of the dynamic models of redundant and non-redundant parallel robots. In order to obtain the inverse and direct dynamic models, first, all closed loops must be virtually opened to make the platform virtually disassembled from the rest of the structure which becomes a tree structure with all joints actuated. The dynamic model of the tree structure and of the free platform is then computed using a systematic procedure based on the Newton-Euler principle, that makes it possible to reduce the computational complexity of the model. Then, the loops are closed using the loop-closure equations and the principle of virtual powers. As a matter of fact, after an introductory section, this chapter will present an effective way to compute the dynamic models of the tree structure robots. Then, the computation of the dynamic models of redundant and non-redundant parallel robots is investigated. Other types of models are also detailed, such that the energy models and the ground reaction models. The chapter ends with a section on the computation of the base dynamic parameters of parallel robots. The dynamic models of some examples of parallel robots are detailed and compared with experiments.

8.1 Introduction

The work on the dynamics of parallel manipulator started with the dynamic analysis of Stewart platforms (Fichter 1986; Hoffman and Hoffman 1979). Those studies mostly dealt with either the oscillation or the inverse dynamics problem under very simple frameworks. Later, other works presented more elaborated analysis to solve the dynamic modeling of parallel manipulators using different mechanical formalisms. For example Lee and Shah (1988), Geng et al. (1992), Lebret et al. (1993), Ait-Ahmed (1993), Bhattacharya et al. (1997, 1998), Liu et al. (2000), Abdellatif and Heimann (2009), Gugliemetti and Longchamp (1994) and Miller (2004) used Lagrange-Euler formalism. The principle of virtual work has been used by Codourey and Burdet (1997) and Tsai (2000). On the other hand, Newton-Euler equations have been used by Sugimoto (1989), Reboulet and Berthomieu (1991), Ji (1993), Gosselin (1993), Dasgupta and Choudhury (1999), Dasgupta and Mruthyunjaya (1998a, b). However, recently, Fu et al. (2007), Vakil et al. (2008), Carricato and

S. Briot and W. Khalil, *Dynamics of Parallel Robots*,
Mechanisms and Machine Science 35, DOI 10.1007/978-3-319-19788-3_8

Gosselin (2009) and Afroun et al. (2012) have pointed out common errors in many methods related to parametrization and instantaneous kinematic behavior of the legs. These errors may cause kinematic and dynamic miscalculation. The aim of this chapter is to present (i) a systematic procedure that provide the full dynamics of rigid bodies, taking into account all the dynamics of the legs and the platform and (ii) to highlight the effect of singularities.

This chapter is based on the works (Briot and Arakelian 2008; Briot and Gautier 2014; Khalil and Guégan 2004; Khalil and Ibrahim 2007). It gives a simple and general closed form solution for the inverse and direct dynamic models of parallel robots. Four types of models will be investigated:

- the *Inverse Dynamic Model* (*IDM*) that provides actuator torques and forces in terms of active joint positions, velocities and accelerations. It is described by:

$$\boldsymbol{\tau} = \mathbf{idm}(\ddot{\mathbf{q}}_a, \dot{\mathbf{q}}_a, \mathbf{q}_a, \mathbf{w}_e) \tag{8.1}$$

  where

  - $\boldsymbol{\tau}$ is the vector of the input efforts (torque in the case of a revolute actuator, force in the case of a prismatic actuator), i.e. the efforts produced by the actuators for moving the mechanism along the trajectory defined by $(\ddot{\mathbf{q}}_a, \dot{\mathbf{q}}_a, \mathbf{q}_a)$,
  - $\mathbf{w}_e$ is the system of wrenches applied by the robot on the environment.

  We call Eq. (8.1) the inverse dynamic model because it defines the system input $\boldsymbol{\tau}$ as a function of the output variables $(\ddot{\mathbf{q}}_a, \dot{\mathbf{q}}_a, \mathbf{q}_a)$. This form of model which is expressed in terms of *Lagrangian* variables (joint variables and their derivatives) is called "Lagrangian Model". The Euler model makes use of the *Eulerian* variables (linear and rotational Cartesian velocities and accelerations).
- the *Direct Dynamic Model* (*DDM*) that provides active joint accelerations as a function of the input effort and the active joint positions and velocities. It is described by:

$$\ddot{\mathbf{q}}_a = \mathbf{ddm}(\dot{\mathbf{q}}_a, \mathbf{q}_a, \boldsymbol{\tau}, \mathbf{w}_e) \tag{8.2}$$

- the *Ground Reaction Model* (*GRM*) that computes the wrench $\mathbf{w}_g$ transmitted on the ground by a moving robot as a function of its active joint positions, velocities and accelerations. It is described by:

$$\mathbf{w}_g = \mathbf{grm}(\ddot{\mathbf{q}}_a, \dot{\mathbf{q}}_a, \mathbf{q}_a, \mathbf{w}_e) \tag{8.3}$$

- the *Energy Model* that computes the robot's total energy as a function of its active joint positions and velocities. It is described by:

$$H = E + U, \;\; E = E(\dot{\mathbf{q}}_a, \mathbf{q}_a), \;\; U = U(\mathbf{q}_a) \tag{8.4}$$

  where H is the total energy, E is the kinetic energy and U the potential energy (see Sect. 6.1).

The *IDM* and *DDM* of robots play an important role in design and control. For robot design, the *IDM* can be used to select actuators (Chedmail and Gautier 1990; Germain et al. 2013; Potkonjak 1986), while the *DDM* is employed to carry out simulations that test the performance of the robot and to study the relative merits of possible control schemes. Regarding robot control, the *IDM* is used to compute the actuator efforts, which are needed to achieve a desired motion (Khalil and Dombre 2002). It is also used to identify the dynamic parameters that are necessary for both control and simulation applications (Gautier 1986; Gautier and Briot 2011a, b, 2012; Hollerbach et al. 2008; Khalil et al. 2007b).

The *GRM* is less known but can be used to identify the robot dynamic parameters (Ayusawa et al. 2008; Raucent et al. 1992) or for design purpose in shaking force and shaking moment balancing (Briot and Arakelian 2009; Briot et al. 2012a; Foucault and Gosselin 2004).

Finally, energy models can also be used to identify the robot's dynamic parameters (Gautier 1997; Gautier and Briot 2013) or for design purpose or trajectory planning to reduce the robot's energy consumption (Ur-Rehman et al. 2009).

To obtain the desired equations for the *IDM* and *DDM* of *PKM*, we will use a method proposed in (Ibrahim and Khalil 2010). The main idea for the *IDM* computation is to decompose the problem into two steps:

1. first, all closed loops are virtually opened to make the platform virtually disassembled from the rest of the structure (Fig. 8.1b); each leg joint is virtually considered actuated (even for unactuated actual joints) so that the robot becomes a tree structure and a free body: the platform. The dynamic model of the tree structure and of the free platform is then computed using a systematic procedure based on the Newton-Euler principle, that makes it possible to obtain

$$\boldsymbol{\tau}_t = \mathbf{idm}_t(\ddot{\mathbf{q}}_t, \dot{\mathbf{q}}_t, \mathbf{q}_t, \mathbf{w}_t) \tag{8.5}$$

$$\mathbf{w}_p = \mathbf{idm}_p(\dot{\mathbf{t}}_p, \mathbf{t}_p, \mathbf{x}_p, \mathbf{w}_e) \tag{8.6}$$

   where $\mathbf{idm}_t$ represents the *IDM* of the tree structure, $\mathbf{idm}_p$ the *IDM* of the platform, $\mathbf{q}_t$ are the joint coordinates of the tree structure, and $\mathbf{t}_p$, $\mathbf{x}_p$ are the platform twist and pose, $\mathbf{w}_t$ is the system of wrenches applied by the tree-structure robot on the environment and $\mathbf{w}_e$ is the system of wrenches applied by the free platform on the environment.
2. then, the loops are closed using the loop-closure equations presented in Sect. 7.3—that relate $\mathbf{t}_p$ and $\dot{\mathbf{q}}_t$ to $\dot{\mathbf{q}}_a$—and the Lagrange multipliers presented in Sect. 6.1.4.[1]

Computation of the *DDM* is a bit different, even if obtained in a similar manner, and is based on calculation of the inertia matrix and vector of Coriolis, centrifugal and gravity effects of its virtual structure. The way to compute it will be detailed later.

[1]For closing the loops, the *PVP* could also be used.

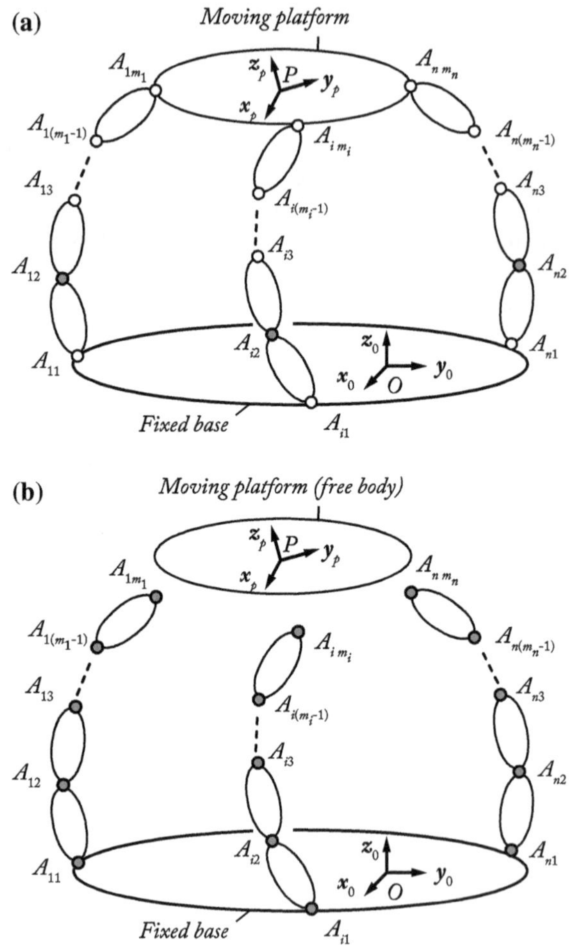

Fig. 8.1 A general parallel robot (the gray circles denote the actuated joints). **a** Kinematic chain (A_{jk} k is the joint j of the leg k, and m_k is the total number of joints for the leg k). **b** Virtual tree structure and free moving platform

As a result, the present chapter is organized as follows:

- Section 8.2 introduces a systematic formulation based on the *NE* principle for the computation of the dynamics of tree-structure robots; this formulation is optimized so that the number of operators "+", "−", "×" and "/" used for computing the models is minimized.
- Section 8.3 presents the dynamic model of the free moving platform;
- Sections 8.4 and 8.5 present the computation of the dynamic models of:

1. non-redundant *PKM*,
2. *PKM* with actuation redundancy.

Note that the computation of the dynamic models of *PKM* with kinematic redundancy is obtained in the same way as for non-redundant *PKM*.

- Section 8.6 presents the computation of the robot energy and ground reaction models.
- Section 8.7 presents the computation of the robot base dynamic parameters which constitute the minimal number of parameters needed to compute the dynamic models.

8.2 Dynamics of Tree-Structure Robots

It is straightforward to get expressions of the *IDM*, inertia matrix and vector of Coriolis/centrifugal/gravity/friction effects of any tree-structure robot "by hand". This will be of course tedious, but in such a case, the simplest method is probably to use the Lagrange equations which state that:

$$\boldsymbol{\tau}_t = \frac{\mathrm{d}}{\mathrm{d}t}\left(\frac{\partial L_t}{\partial \dot{\mathbf{q}}_t}\right)^T - \left(\frac{\partial L_t}{\partial \mathbf{q}_t}\right)^T \tag{8.7}$$

where L_t is the Lagrangian of the virtual tree-structured robot.

In the present section, we prefer to give an efficient Newton-Euler formulation based on a recursive algorithm which allows for decreasing the computational complexity of the model.

8.2.1 Newton-Euler Formulation for Computation of the Inverse Dynamic Model

Let us consider the tree-structure of the general *PKM* presented in Sect. 7.1.1 obtained when disassembling the platform from the rest of the robot and by virtually actuating all passive joints (Fig. 8.1b). This tree-structure is made of n legs (see Fig. 8.1a), each leg being a kinematic chain (which is serial or tree-structure type most of the time, but can also be composed of closed-loop sub-chains) composed of $m_i - 1$ links connected by m_i joints located at points A_{ij}. The jth link of the leg i will be denoted in what follows as the link $\mathcal{B}_{ij}$. Moreover, the joint located at point A_{ij} will be parameterized by the variable q_{ij}.

Let us recall from Sect. 6.2 the *NE* equations giving the total forces $\Sigma\mathbf{f}_{ij}$ and moments $\Sigma\mathbf{m}_{ij}$ on link $\mathcal{B}_{ij}$ at the origin O_{ij} of frame $\mathcal{F}_{ij}$:

$$\begin{aligned}\Sigma\mathbf{f}_{ij} &= m_{ij}\,\dot{\mathbf{v}}_{ij} + \dot{\boldsymbol{\omega}}_{ij} \times \mathbf{ms}_{ij} + \boldsymbol{\omega}_{ij} \times (\boldsymbol{\omega}_{ij} \times \mathbf{ms}_{ij})\\ \Sigma\mathbf{m}_{ij} &= \mathbf{I}_{O_{ij}}\dot{\boldsymbol{\omega}}_{ij} + \mathbf{ms}_{ij} \times \dot{\mathbf{v}}_{ij} + \boldsymbol{\omega}_{ij} \times (\mathbf{I}_{O_{ij}}\boldsymbol{\omega}_{ij})\end{aligned} \tag{8.8}$$

in which:

- $\dot{\mathbf{v}}_{ij}$ is the acceleration of the origin of the frame $\mathcal{F}_{ij}$,
- $\boldsymbol{\omega}_{ij}$ is the rotational velocity of body $\mathcal{B}_{ij}$,

- $\dot{\boldsymbol{\omega}}_{ij}$ is the rotational acceleration of body $\mathscr{B}_{ij}$,
- $\mathbf{ms}_{ij}$ is the vector of the first moment of inertia which represents the mass of the body $\mathscr{B}_{ij}$ multiplied by the position of its , w.r.t. frame $\mathscr{F}_{ij}$ (see Sect. 6.1.2),
- $\mathbf{I}_{O_{ij}}$ is the inertia matrix expressed at O_{ij} defined in Sect. 6.1.2.

The inverse dynamic model of a tree-structure can be obtained by using two recursive algorithms sequentially (Khalil and Kleinfinger 1987; Luh et al. 1980).

Forward recursive computation: To compute $\Sigma\mathbf{f}_{ij}$ and $\Sigma\mathbf{m}_{ij}$ for $i = 1, \ldots, n$ and $j = 1, \ldots, m_i$, using (8.8), we need $\dot{\mathbf{v}}_{ij}$, $\boldsymbol{\omega}_{ij}$ and $\dot{\boldsymbol{\omega}}_{ij}$. Let us denote as link $\mathscr{B}_{il}$ the antecedent of link $\mathscr{B}_{ij}$. The velocities are given by the recursive Eqs. (5.18) and (5.19) rewritten hereafter as:

$$^{ij}\boldsymbol{\omega}_{ij} = {}^{ij}\boldsymbol{\omega}_{il} + \bar{\sigma}_{ij}\ {}^{ij}\mathbf{a}_{ij}\ \dot{q}_{ij} \tag{8.9}$$

$$^{ij}\mathbf{v}_{ij} = {}^{ij}\mathbf{v}_{il} + {}^{ij}\mathbf{R}_{il}({}^{il}\boldsymbol{\omega}_{il} \times {}^{il}\mathbf{r}_{ij}) + \sigma_{ij}\ {}^{ij}\mathbf{a}_{ij}\ \dot{q}_{ij} \tag{8.10}$$

where $il = a(ij)$.

Moreover, the accelerations are given by the recursive Eqs. (5.27)–(5.29) rewritten hereafter as:

$$^{ij}\dot{\boldsymbol{\omega}}_{ij} = {}^{ij}\mathbf{R}_{il}\ {}^{il}\dot{\boldsymbol{\omega}}_{il} + \bar{\sigma}_{ij}({}^{ij}\mathbf{a}_{ij}\ \ddot{q}_{ij} + {}^{ij}\boldsymbol{\omega}_{il} \times {}^{ij}\mathbf{a}_{ij}\ \dot{q}_{ij}) \tag{8.11}$$

$$^{ij}\dot{\mathbf{v}}_{ij} = {}^{ij}\mathbf{R}_{il}\,({}^{il}\dot{\mathbf{v}}_{il} + {}^{il}\mathbf{U}_{il}\ {}^{il}\mathbf{r}_{ij}) + \sigma_{ij}({}^{ij}\mathbf{a}_{ij}\ \ddot{q}_{ij} + 2\,{}^{ij}\boldsymbol{\omega}_{il} \times {}^{ij}\mathbf{a}_{ij}\ \dot{q}_{ij}) \tag{8.12}$$

with

$$^{il}\mathbf{U}_{il} = {}^{il}\hat{\dot{\boldsymbol{\omega}}}_{il} + {}^{il}\hat{\boldsymbol{\omega}}_{il}\ {}^{il}\hat{\boldsymbol{\omega}}_{il}. \tag{8.13}$$

The initial conditions for a robot with a fixed base are ${}^0\boldsymbol{\omega}_0 = \mathbf{0}$, ${}^0\dot{\boldsymbol{\omega}}_0 = \mathbf{0}$, ${}^0\mathbf{v}_0 = \mathbf{0}$ and ${}^0\dot{\mathbf{v}}_0 = \mathbf{0}$.

Finally, we compute the expressions of $\Sigma\mathbf{f}_{ij}$ and $\Sigma\mathbf{m}_{ij}$ given in (8.8) and we project them into the frame $\mathscr{F}_{ij}$.

Backward recursive computation: Let us suppose that the notation ik denotes all the links such that $a(ik) = ij$ (Fig. 8.2). An algorithm for computation of the dynamic model is based on writing for each link $\mathscr{B}_{ij}$, for $j = m_i, \ldots, 1$, for all i, the *NE* equations at the origin O_{ij}, as follows (Fig. 8.2):

$$\begin{aligned}
^{ij}\Sigma\mathbf{f}_{ij} &= {}^{ij}\mathbf{f}_{ij} - \sum_k {}^{ij}\mathbf{f}_{ik} + m_{ij}\ {}^{ij}\mathbf{g} - {}^{ij}\mathbf{f}_{e_{ij}} \\
^{ij}\Sigma\mathbf{m}_{ij} &= {}^{ij}\mathbf{m}_{ij} - \sum_k \left({}^{ij}\mathbf{R}_{ik}\ {}^{ik}\mathbf{m}_{ik} + {}^{ij}\mathbf{r}_{ik} \times {}^{ij}\mathbf{f}_{ik}\right) + \mathbf{ms}_{ij} \times {}^{ij}\mathbf{g} - {}^{ij}\mathbf{m}_{e_{ij}}
\end{aligned} \tag{8.14}$$

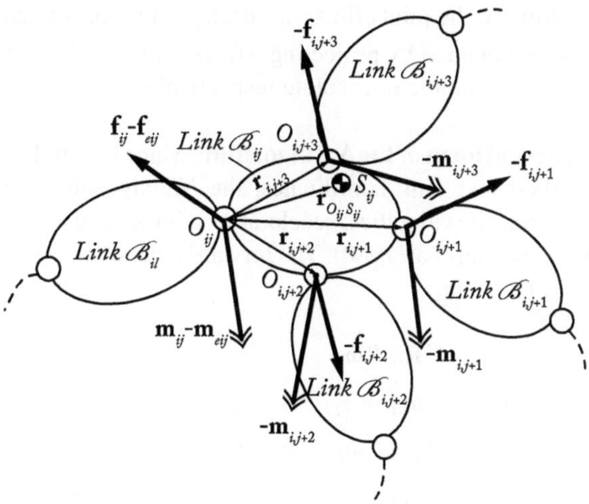

Fig. 8.2 Forces and moments on link $\mathcal{B}_{ij}$

where

- ${}^{ij}\mathbf{r}_{ik} = {}^{ij}\mathbf{r}_{O_{ij}O_{ik}}$,
- $\mathbf{ms}_{ij} \times {}^{ij}\mathbf{g} = \mathbf{r}_{O_{ij}S_{ij}} \times m_{ij}{}^{ij}\mathbf{g}$,
- $\mathbf{f}_{ij}$ and $\mathbf{m}_{ij}$ are the reaction force and moment, respectively, exerted by the body $\mathcal{B}_{il}$ on the body $\mathcal{B}_{ij}$ at point O_{ij},
- $\mathbf{f}_{ik}$ and $\mathbf{m}_{ik}$ are the reaction force and moment, respectively, exerted by the body $\mathcal{B}_{ij}$ on the body $\mathcal{B}_{ik}$ at point O_{ik},
- $\mathbf{f}_{e_{ij}}$ and $\mathbf{m}_{e_{ij}}$ represent the force and moment exerted by link $\mathcal{B}_{ij}$ on the environment. Their values are assumed to be known, or at least to be calculated from known quantities.

We can cancel the gravity terms from Eq. (8.14) and take into account their effects by setting up the initial linear acceleration such that:

$$ {}^{0}\dot{\mathbf{v}}_0 = -{}^{0}\mathbf{g}. \tag{8.15}$$

Thus, using Eq. (8.14), we obtain:

$$\begin{aligned} {}^{ij}\mathbf{f}_{ij} &= {}^{ij}\Sigma\mathbf{f}_{ij} + \sum_k {}^{ij}\mathbf{f}_{ik} + {}^{ij}\mathbf{f}_{e_{ij}} \\ {}^{ij}\mathbf{m}_{ij} &= {}^{ij}\Sigma\mathbf{m}_{ij} + \sum_k \left({}^{ij}\mathbf{R}_{ik}\,{}^{ik}\mathbf{m}_{ik} + {}^{ij}\mathbf{r}_{ik} \times {}^{ij}\mathbf{f}_{ik}\right) + {}^{ij}\mathbf{m}_{e_{ij}}. \end{aligned} \tag{8.16}$$

This backward recursive algorithm is initialized by ${}^{i\,m_i}\mathbf{f}_{i\,(m_i+1)} = \mathbf{0}$ and ${}^{i\,m_i}\mathbf{m}_{i\,(m_i+1)} = \mathbf{0}$, the body $\mathcal{B}_{i\,m_i}$ being a terminal body of the tree-structure.

Finally, the joint effort $\tau_{t_{ij}}$ (each joint is considered virtually active, see Sect. 8.1) can be obtained by projecting ${}^{ij}\mathbf{f}_{ij}$ or ${}^{ij}\mathbf{m}_{ij}$ on the joint axis, depending whether the joint is prismatic or revolute respectively.

Practical form of the *NE* algorithm: Since $\mathbf{I}_{O_{ij}}$ and $\mathbf{ms}_{ij}$ are constants when referred to their own link coordinates, the *NE* algorithm can be efficiently computed by referring the velocities, accelerations, forces and moments to the local link coordinate system (Luh et al. 1980). The forward recursive equations becomes, for $i = 1, \ldots, n$, $j = 1, \ldots, m_i$:

$$
\begin{aligned}
{}^{ij}\boldsymbol{\omega}_{il} &= {}^{ij}\mathbf{R}_{il}\ {}^{il}\boldsymbol{\omega}_{il} && (8.17)\\
{}^{ij}\boldsymbol{\omega}_{ij} &= {}^{ij}\boldsymbol{\omega}_{il} + \bar{\sigma}_{ij}\ {}^{ij}\mathbf{a}_{ij}\ \dot{q}_{ij} && (8.18)\\
{}^{ij}\dot{\boldsymbol{\omega}}_{ij} &= {}^{ij}\mathbf{R}_{il}\ {}^{il}\dot{\boldsymbol{\omega}}_{il} + \bar{\sigma}_{ij}({}^{ij}\mathbf{a}_{ij}\ \ddot{q}_{ij} + {}^{ij}\boldsymbol{\omega}_{il} \times {}^{ij}\mathbf{a}_{ij}\ \dot{q}_{ij}) && (8.19)\\
{}^{ij}\dot{\mathbf{v}}_{ij} &= {}^{ij}\mathbf{R}_{il}\,({}^{il}\dot{\mathbf{v}}_{il} + {}^{il}\mathbf{U}_{il}\ {}^{il}\mathbf{r}_{ij}) + \sigma_{ij}({}^{ij}\mathbf{a}_{ij}\ \ddot{q}_{ij} + 2\,{}^{ij}\boldsymbol{\omega}_{il} \times {}^{ij}\mathbf{a}_{ij}\ \dot{q}_{ij}) && (8.20)\\
{}^{ij}\mathbf{U}_{ij} &= {}^{ij}\dot{\hat{\boldsymbol{\omega}}}_{ij} + {}^{ij}\hat{\boldsymbol{\omega}}_{ij}\ {}^{ij}\hat{\boldsymbol{\omega}}_{ij} && (8.21)\\
{}^{ij}\Sigma\mathbf{f}_{ij} &= m_{ij}\,{}^{ij}\dot{\mathbf{v}}_{ij} + {}^{ij}\mathbf{U}_{ij}\ {}^{ij}\mathbf{ms}_{ij} && (8.22)\\
{}^{ij}\Sigma\mathbf{m}_{ij} &= {}^{ij}\mathbf{I}_{O_{ij}}\,{}^{ij}\dot{\boldsymbol{\omega}}_{ij} + {}^{ij}\mathbf{ms}_{ij} \times {}^{ij}\dot{\mathbf{v}}_{ij} + {}^{ij}\boldsymbol{\omega}_{ij} \times \left({}^{ij}\mathbf{I}_{O_{ij}}\ {}^{ij}\boldsymbol{\omega}_{ij}\right). && (8.23)
\end{aligned}
$$

For a stationary base, the initial conditions are ${}^0\boldsymbol{\omega}_0 = \mathbf{0}$, ${}^0\dot{\boldsymbol{\omega}}_0 = \mathbf{0}$ and ${}^0\dot{\mathbf{v}}_0 = -{}^0\mathbf{g}$.

The use of ${}^{ij}\mathbf{U}_{ij}$ saves $21n_t$ multiplications and $6n_t$ additions ($n_t = \sum_{i=1}^{n} m_i$) in the computation of the inverse dynamic model of a general robot (Khalil and Kleinfinger 1986).

The backward recursive equations, for $j = m_i, \ldots, 1$, for any i are:

$$
\begin{aligned}
{}^{ij}\mathbf{f}_{ij} &= {}^{ij}\Sigma\mathbf{f}_{ij} + \sum_k {}^{ij}\mathbf{f}_{ik} + {}^{ij}\mathbf{f}_{e_{ij}} && (8.24)\\
{}^{il}\mathbf{f}_{ij} &= {}^{il}\mathbf{R}_{ij}\ {}^{ij}\mathbf{f}_{ij} && (8.25)\\
{}^{ij}\mathbf{m}_{ij} &= {}^{ij}\Sigma\mathbf{m}_{ij} + \sum_k \left({}^{ij}\mathbf{R}_{ik}\ {}^{ik}\mathbf{m}_{ik} + {}^{ij}\mathbf{r}_{ik} \times {}^{ij}\mathbf{f}_{ik}\right) + {}^{ij}\mathbf{m}_{e_{ij}} && (8.26)\\
\tau_{t_{ij}} &= (\sigma_{ij}\ {}^{ij}\mathbf{f}_{ij} + \bar{\sigma}_{ij}\ {}^{ij}\mathbf{m}_{ij})^T\ {}^{ij}\mathbf{a}_{ij}. && (8.27)
\end{aligned}
$$

The previous algorithm can be numerically programmed for a general tree structure robot. Its computational complexity is $O(n_t)$, which means that the number of operations is linear in the number of degrees of freedom. However, as we will see in Sect. 8.7, the use of the base inertial parameters in a customized symbolic algorithm considerably reduces the number of operations of the dynamic model.

Note that:

- The symbol ${}^{ij}\dot{\mathbf{v}}_{ij}$ means ${}^{ij}\mathbf{R}_0{}^0\dot{\mathbf{v}}_{ij}$, and not the time derivative of ${}^{ij}\mathbf{v}_{ij}$.
- As mentioned in Sect. 4.1, we use $\sigma_{ij} = 2$ to define a fixed frame $\mathscr{F}_{ij}$ with respect to $\mathscr{F}_{il}$. In that case, $\dot{q}_{ij}$ and $\ddot{q}_{ij}$ are set to 0 and $\bar{\sigma}_{ij}$ is not defined. Moreover, $\tau_{t_{ij}}$ defined at (8.27) has no physical meaning and should not be calculated, whereas the velocity and acceleration equations can be used after eliminating the terms multiplied by σ_{ij} and $\bar{\sigma}_{ij}$.

8.2.2 Considering the Inertia of Actuators

The kinetic energy of the rotor (and transmission system) of actuator ij is given by the expression $\frac{1}{2} Ia_{ij}\dot{q}_{ij}^2$. The inertial parameter Ia_{ij} denotes the equivalent inertia referred to the joint velocity. It is given by:

$$Ia_{ij} = N_{ij}^2 \; Im_{ij} \tag{8.28}$$

where Im_{ij} is the moment of inertia of the rotor and transmissions of actuator ij, N_{ij} is the gear transmission ratio of joint ij axis, equal to $N_{ij} = \dot{q}_{m_{ij}} / \dot{q}_{ij}$ where $\dot{q}_{m_{ij}}$ denotes the rotor velocity of actuator ij while $\dot{q}_{ij}$ denotes the joint ij velocity. In the case of a prismatic joint, Ia_{ij} is an equivalent mass.

In order to consider the rotor inertia in the dynamic model of the robot, (8.27) becomes

$$\tau_{t_{ij}} = (\sigma_{ij} \; {}^{ij}\mathbf{f}_{ij} + \bar{\sigma}_{ij} \; {}^{ij}\mathbf{m}_{ij})^T \; {}^{ij}\mathbf{a}_{ij} + Ia_{ij}\ddot{q}_{ij} \tag{8.29}$$

Note that such modeling neglects the gyroscopic effects of the rotors, which take place when the actuator is fixed on a moving link. However, this approximation is justified for high gear transmission ratios. For more accurate modeling of the rotors the reader is referred to (Chedmail et al. 1986; Llibre et al. 1983; Murphy and Ting-Yung Wen 1993; Sciavicco et al. 1994) where the rotor is considered as a body articulated on the link on which it is fixed.

8.2.3 Considering Friction

Friction plays a dominant role in limiting the quality of robot performance. Non-compensated friction produces static error, delay, and limit cycle behavior (Canudas de Wit and Aubin 1990). Many works have been devoted to studying friction torque in joint and transmission systems. Various friction models have been proposed in the literature (Armstrong 1988, 1991; Canudas de Wit et al. 1989; Dahl 1977). In general, three kinds of frictions are noted: Coulomb friction, static friction, and viscous friction.

Fig. 8.3 Friction model

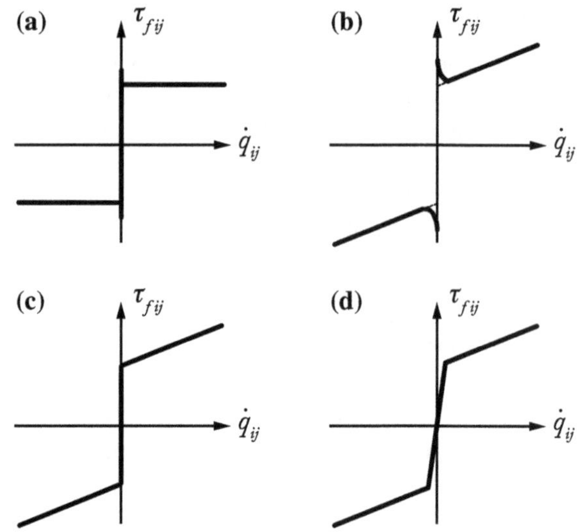

The model based on Coulomb friction assumes a constant friction component that is independent of the magnitude of the velocity. The static friction is the torque necessary to initiate motion from rest. It is often greater than the Coulomb friction (Fig. 8.3a). The viscous friction is generally represented as being proportional to the velocity, but experimental studies (Armstrong 1988) have pointed out the Stribeck phenomenon that arises from the use of fluid lubrication. It results in decreasing friction with increasing velocity at low velocity, then the friction becomes proportional to velocity (Fig. 8.3b). A general friction model describing these components is given by:

$$\tau_{fij} = fs_{ij}\text{sign}(\dot{q}_{ij}) + fv_{ij}\dot{q}_{ij} + (fs_{tij} - fs_{ij})\text{sign}(\dot{q}_{ij})e^{-|\dot{q}_{ij}|B_{ij}}. \tag{8.30}$$

In this expression, τ_{fij} denotes the friction torque of joint ij, fs_{ij} and fv_{ij} indicate the Coulomb and viscous friction parameters respectively. The static torque is equal to $fs_{tij}\text{sign}(\dot{q}_{ij})$.

The most often employed model is composed of Coulomb friction together with viscous friction (Fig. 8.3c). Therefore, the friction torque at joint ij is written as:

$$\tau_{fij} = fs_{ij}\text{sign}(\dot{q}_{ij}) + fv_{ij}\dot{q}_{ij}. \tag{8.31}$$

This friction model can be approximated by a piecewise linear model as shown in Fig. 8.3d.

In order to consider the friction effects in the dynamic model of the robot, (8.29) becomes

$$\tau_{t_{ij}} = (\sigma_{ij}\ {}^{ij}\mathbf{f}_{ij} + \bar{\sigma}_{ij}\ {}^{ij}\mathbf{m}_{ij})^T\ {}^{ij}\mathbf{a}_{ij} + Ia_{ij}\ddot{q}_{ij} + \tau_{fij}. \tag{8.32}$$

Remark on the Friction Models

In Robotics, we usually made the assumption that the terms fs_{ij} and fv_{ij} in the model (8.31) are constant. Of course, in reality, this is not the case as they should vary as a function of the load in the joints (Hamon et al. 2010; Leonesio and Bianchi 2009).

However, this assumption is made usually for two main reasons:

- the obtained model is linear and thus simpler to use,
- for robots with accelerations lower than the acceleration of the gravity field, the load in joints is mainly due to the gravity effects and its variation is quite small.

8.2.4 Computing the Vector of Coriolis, Centrifugal, Gravity Effects, Friction and External Wrenches

As mentioned in Sect. 6.1.1, for any rigid robot, the *IDM* can be written in the form (Khalil and Dombre 2002):

$$\boldsymbol{\tau}_t = \begin{bmatrix} \tau_{t_{11}} \\ \vdots \\ \tau_{t_{n\,m_n}} \end{bmatrix} = \mathbf{idm}_t(\ddot{\mathbf{q}}_t, \dot{\mathbf{q}}_t, \mathbf{q}_t, \mathbf{w}_e) = \mathbf{M}_t(\mathbf{q}_t)\ddot{\mathbf{q}}_t + \mathbf{c}_t(\mathbf{q}_t, \dot{\mathbf{q}}_t) \tag{8.33}$$

where $\mathbf{q}_t$ groups all joint variables, $\mathbf{M}_t(\mathbf{q}_t)$ is the inertia matrix of the robot tree-structure and $\mathbf{c}_t(\mathbf{q}_t, \dot{\mathbf{q}}_t)$ is the vector of Coriolis, centrifugal, gravity effects, friction and external wrenches.

Computation of the inertia matrix $\mathbf{M}_t(\mathbf{q}_t)$ and vector $\mathbf{c}_t(\mathbf{q}_t, \dot{\mathbf{q}}_t)$ of the tree structure is important for obtaining the *DDM* of the *PKM*. Therefore, in the present section and the following one, procedures for computing them are given.

By analyzing the expression (8.33), we can deduce that $\boldsymbol{\tau}_t = \mathbf{c}_t(\mathbf{q}_t, \dot{\mathbf{q}}_t)$ if $\ddot{\mathbf{q}}_t = \mathbf{0}$. As a result, the simplest and most effective way to obtain the vector $\mathbf{c}_t(\mathbf{q}_t, \dot{\mathbf{q}}_t)$ is to run Eqs. (8.17)–(8.27) by imposing $\ddot{\mathbf{q}}_t = \mathbf{0}$.

In the next section, the computation of the inertia matrix is investigated.

8.2.5 Computing the Inertia Matrix

The inertia matrix can be calculated one column at a time, using *NE IDM* Eqs. (8.17)–(8.27) (Walker and Orin 1982). From relation (8.33), we deduce that the rth column of $\mathbf{M}_t$ is equal to $\boldsymbol{\tau}_t$ if, for $s = 1, \ldots, n_t$.:

$$\ddot{\mathbf{q}}_t = \mathbf{u}_r,\ \ \dot{\mathbf{q}}_t = \mathbf{0},\ \ \mathbf{g} = \mathbf{0},\ \ \mathbf{f}_{e_s} = \mathbf{0},\ \ \mathbf{m}_{e_s} = \mathbf{0} \tag{8.34}$$

where $\mathbf{u}_r$ is an $(n_t \times 1)$ unit vector with 1 in the rth row and zeros elsewhere. Iterating the procedure for $r = 1, \ldots, n_t$ leads to construction of the entire inertia matrix.

To reduce the computational complexity of this algorithm, we can make use of its base inertial parameters and its customized symbolic techniques. Moreover, we can take advantage of the fact that the inertia matrix $\mathbf{M}_t$ is symmetric. A more efficient procedure for computing the inertia matrix using the concept of composite links is described below (Khalil and Dombre 2002; Walker and Orin 1982).

8.2.5.1 Inertial Parameters of a Composite Link

The composite link $\mathcal{B}_{ij}^{+}$ is composed of link $\mathcal{B}_{ij}$ and of the links supported by link $\mathcal{B}_{ij}$ (Fig. 8.4). The idea is to compute, using the well-known Huygens-Steiner theorem, the first and second moment of inertia of this link $\mathcal{B}_{ij}^{+}$ as a function of the inertial parameters of all the links composing it and of the joint variables.

The inertial parameters of the composite link $\mathcal{B}_{ij}^{+}$ can be calculated using the following recursive algorithm:

1. *Initialization*: for $j = 1, \ldots, m_i$, for any i:

$$^{ij}\mathbf{I}_{O_{ij}}^{+} = {}^{ij}\mathbf{I}_{O_{ij}},\quad {}^{ij}\mathbf{ms}_{ij}^{+} = {}^{ij}\mathbf{ms}_{ij},\quad m_{ij}^{+} = m_{ij}.$$

2. *Initialization*: for $j = m_i, \ldots, 2$, for any i: We recall that il is the number of the link that is antecedent to link $\mathcal{B}_{ij}$;

$$^{il}\mathbf{I}_{O_{il}}^{+} = {}^{il}\mathbf{I}_{O_{il}}^{+} + {}^{il}\mathbf{R}_{ij}\,{}^{ij}\mathbf{I}_{O_{ij}}^{+}\,{}^{ij}\mathbf{R}_{il} - \left[{}^{il}\hat{\mathbf{r}}_{ij}\,{}^{il}\widehat{\mathbf{ms}}_{ij}^{+} + \left({}^{il}\hat{\mathbf{r}}_{ij}\,{}^{il}\widehat{\mathbf{ms}}_{ij}^{+}\right)^{T}\right] + {}^{il}\hat{\mathbf{r}}_{ij}\,{}^{il}\hat{\mathbf{r}}_{ij}^{T} m_{ij}^{+} \tag{8.35}$$

$$^{il}\mathbf{ms}_{il}^{+} = {}^{il}\mathbf{ms}_{il}^{+} + {}^{il}\mathbf{R}_{ij}\,{}^{ij}\mathbf{ms}_{ij}^{+} + {}^{il}\mathbf{r}_{ij}\, m_{ij}^{+} \tag{8.36}$$

$$m_{il}^{+} = m_{il}^{+} + m_{ij}^{+}. \tag{8.37}$$

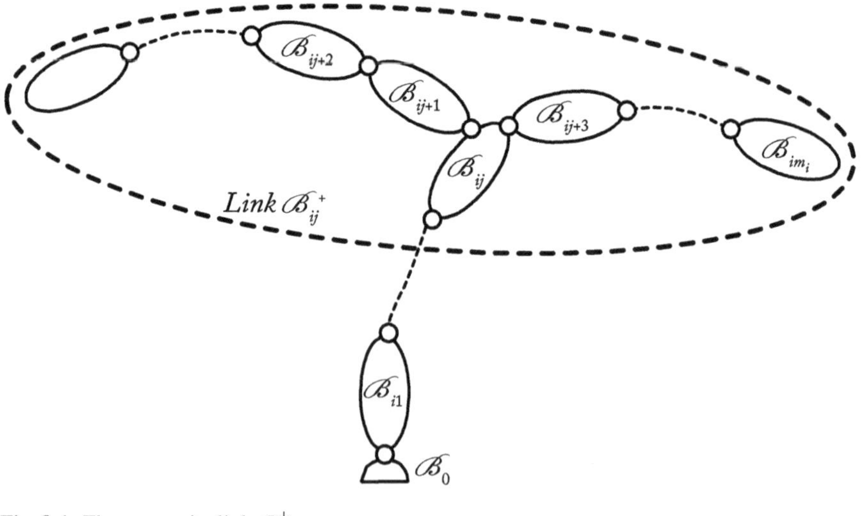

Fig. 8.4 The composite link $\mathcal{B}_{ij}^{+}$

8.2.5.2 Computation of the Inertia Matrix

We have seen above that the rth column of the inertia matrix $\mathbf{M}_t$ can be computed by the Newton-Euler inverse dynamic algorithm by setting $\ddot{\mathbf{q}}_t = \mathbf{u}_r$, $\dot{\mathbf{q}}_t = \mathbf{0}$, $\mathbf{g} = \mathbf{0}$, $\mathbf{f}_{e_s} = \mathbf{0}$ and $\mathbf{m}_{e_s} = \mathbf{0}$ for $s = 1, \ldots, n_t$, where $\mathbf{u}_r$ is an $(n_r \times 1)$ vector with a 1 in the rth row and zeros elsewhere.

Under these conditions, the forward recursive equations of the *NE* inverse dynamic (Sect. 8.2.1) are only applied to link $\mathscr{B}_{ij}^{+}$:

$$ {}^{il}\boldsymbol{\omega}_{il} = \mathbf{0},\ {}^{il}\dot{\boldsymbol{\omega}}_{il} = \mathbf{0},\ {}^{il}\dot{\mathbf{v}}_{il} = \mathbf{0},\ {}^{il}\Sigma\mathbf{f}_{il} = \mathbf{0},\ {}^{il}\Sigma\mathbf{m}_{il} = \mathbf{0},\ \text{for } l < j \tag{8.38} $$

$$ {}^{ij}\boldsymbol{\omega}_{ij} = \mathbf{0} \tag{8.39} $$

$$ {}^{ij}\dot{\boldsymbol{\omega}}_{ij} = \bar{\sigma}_{ij}{}^{ij}\mathbf{a}_{ij} \tag{8.40} $$

$$ {}^{ij}\dot{\mathbf{v}}_{ij} = \sigma_{ij}{}^{ij}\mathbf{a}_{ij} \tag{8.41} $$

$$ {}^{ij}\Sigma\mathbf{f}_{ij} = m_{ij}^{+}{}^{ij}\dot{\mathbf{v}}_{ij} + {}^{ij}\dot{\boldsymbol{\omega}}_{ij} \times {}^{ij}\mathbf{ms}_{ij}^{+} \tag{8.42} $$

$$ {}^{ij}\Sigma\mathbf{m}_{ij} = {}^{ij}\mathbf{I}_{O_{ij}}^{+}\,{}^{ij}\dot{\boldsymbol{\omega}}_{ij} + {}^{ij}\mathbf{ms}_{ij} \times {}^{ij}\dot{\mathbf{v}}_{ij}. \tag{8.43} $$

We deduce that:

- if joint ij is prismatic (${}^{ij}\dot{\boldsymbol{\omega}}_{ij} = \mathbf{0}$ and ${}^{ij}\dot{\mathbf{v}}_{ij} = [0\ 0\ 1]^T$), then:

$$ {}^{ij}\Sigma\mathbf{f}_{ij} = \begin{bmatrix} 0 & 0 & m_{ij}^{+} \end{bmatrix}^T \tag{8.44} $$

$$ {}^{ij}\Sigma\mathbf{m}_{ij} = \begin{bmatrix} my_{ij}^{+} & -mx_{ij}^{+} & 0 \end{bmatrix}^T \tag{8.45} $$

- if joint ij is revolute (${}^{ij}\dot{\mathbf{v}}_{ij} = \mathbf{0}$ and ${}^{ij}\dot{\boldsymbol{\omega}}_{ij} = [0\ 0\ 1]^T$), then:

$$ {}^{ij}\Sigma\mathbf{f}_{ij} = \begin{bmatrix} -my_{ij}^{+} & mx_{ij}^{+} & 0 \end{bmatrix}^T \tag{8.46} $$

$$ {}^{ij}\Sigma\mathbf{m}_{ij} = \begin{bmatrix} xz_{ij}^{+} & yz_{ij}^{+} & zz_{ij}^{+} \end{bmatrix}^T. \tag{8.47} $$

The recursive backward computation starts by link $\mathscr{B}_{ij}$ and ends with link $\mathscr{B}_{i1}$, where $a(i1) = 0$. The algorithm is given by the following equations:

- if joint ij is prismatic, then:

$$ {}^{ij}\mathbf{f}_{ij} = {}^{ij}\Sigma\mathbf{f}_{ij} = \begin{bmatrix} 0 & 0 & m_{ij}^{+} \end{bmatrix}^T \tag{8.48} $$

$$ {}^{ij}\mathbf{m}_{ij} = {}^{ij}\Sigma\mathbf{m}_{ij} = \begin{bmatrix} my_{ij}^{+} & -mx_{ij}^{+} & 0 \end{bmatrix}^T \tag{8.49} $$

$$ \mathbf{M}_t(n_{ij}, n_{ij}) = m_{ij}^{+} + Ia_{ij} \tag{8.50} $$

where $\mathbf{M}_t(n_{ij}, n_{ij})$ denotes the elements of the n_{ij}th row and n_{ij}th column of the matrix $\mathbf{M}_t$, n_{ij} being the position of the variable q_{ij} in the vector $\mathbf{q}_t$.

- if joint ij is revolute, then:

$$^{ij}\mathbf{f}_{ij} = {}^{ij}\Sigma\mathbf{f}_{ij} = \begin{bmatrix} -my_{ij}^{+} & mx_{ij}^{+} & 0 \end{bmatrix}^T \tag{8.51}$$

$$^{ij}\mathbf{m}_{ij} = {}^{ij}\Sigma\mathbf{m}_{ij} = \begin{bmatrix} xz_{ij}^{+} & yz_{ij}^{+} & zz_{ij}^{+} \end{bmatrix}^T \tag{8.52}$$

$$\mathbf{M}_t(n_{ij}, n_{ij}) = zz_{ij}^{+} + Ia_{ij}. \tag{8.53}$$

Then, the following equations are computed for $u = ij, a(ij), a(a(ij)), \ldots, i1$, where $a(i1) = 0$:

$$^{a(u)}\mathbf{f}_{a(u)} = {}^{a(u)}\mathbf{R}_u \; {}^u\mathbf{f}_u \tag{8.54}$$

$$^{a(u)}\mathbf{m}_{a(u)} = {}^{a(u)}\mathbf{R}_u \; {}^u\mathbf{m}_u + {}^{a(u)}\mathbf{r}_u \times {}^{a(u)}\mathbf{f}_{a(u)} \tag{8.55}$$

$$\mathbf{M}_t(n_{a(u)}, n_{ij}) = (\sigma_{a(u)} \; {}^{a(u)}\mathbf{f}_{a(u)} + \bar{\sigma}_{a(u)} \; {}^{a(u)}\mathbf{m}_{a(u)})^T \; {}^{a(u)}\mathbf{a}_{a(u)}, \tag{8.56}$$

$n_{a(u)}$ being the position of the variable $q_{a(u)}$ in the vector $\mathbf{q}_t$.

Note that:

- the element $\mathbf{M}_t(n_{il}, n_{ij})$ of the inertia matrix is set to zero if link $\mathscr{B}_{il}$ does not belong to the path between the base and link $\mathscr{B}_{ij}$;
- this algorithm provides the elements of the lower part of the inertia matrix. The other elements are deduced using the fact that the inertia matrix $\mathbf{M}_t$ is symmetric.

8.2.6 Automatic Computation of the IDM, *Inertia Matrix and Vector of Coriolis, Centrifugal/Gravity/Friction Effects*

In order to finally obtain symbolic equations for the model with a minimum number of operations, the following method is used. First, the rigid kinematics of each element are modeled using the modified Denavit-Hartenberg notations (Sect. 4.2). Then, the customized algorithms defined in Sects. 8.2.1, 8.2.4 and 8.2.5 are run.

For each computation, the elements of a vector or a matrix containing at least one mathematical operation are replaced by an intermediate variable. This variable is written in an output file which contains the model. The elements that do not contain any operations are not modified. The obtained vectors and matrices are propagated in the subsequent equations. Consequently, at the end, the models are obtained as a set of intermediate variables. Those that have no effect on the desired output can be eliminated by scanning the intermediate variables from the end to the beginning. With this procedure, it is also possible to know the exact number of operators necessary for computation of the model. These algorithms have been successfully implemented with Mathematica (Khalil and Creusot 1997) and are now freely available in Python (Khalil et al. 2014).

8.3 Dynamic Model of the Free Moving Platform

The dynamic equations of a free moving platform are given by the *NE* Eq. (6.35) rewritten here as

$$\begin{aligned}\mathbf{w}_p = \begin{bmatrix}\mathbf{f}_p \\ \mathbf{m}_p\end{bmatrix} = \begin{bmatrix} m_p\,\mathbf{1}_3 & \widehat{\mathbf{ms}}_p^T \\ \widehat{\mathbf{ms}}_p & \mathbf{I}_p\end{bmatrix}\begin{bmatrix}\dot{\mathbf{v}}_p - \mathbf{g} \\ \dot{\boldsymbol{\omega}}_p\end{bmatrix} + \begin{bmatrix}\boldsymbol{\omega}_p \times (\boldsymbol{\omega}_p \times \mathbf{ms}_p) \\ \boldsymbol{\omega}_p \times (\mathbf{I}_p\boldsymbol{\omega}_p)\end{bmatrix} - \begin{bmatrix}\mathbf{f}_p^{ext} \\ \mathbf{m}_p^{ext}\end{bmatrix} \\ = \mathbf{M}_p\dot{\mathbf{t}}_p + \mathbf{c}_p \end{aligned} \tag{8.57}$$

where

- m_p is the platform mass,
- $\mathbf{ms}_p$ is the vector of the fist moment of inertia of the platform,
- $\mathbf{I}_p$ is the matrix of inertia for the platform,
- $\mathbf{t}_p^T = [\mathbf{v}_p^T\ \boldsymbol{\omega}_p^T]$ is the platform twist, while $\dot{\mathbf{t}}_p^T = [\dot{\mathbf{v}}_p^T\ \dot{\boldsymbol{\omega}}_p^T]$ represents the platform acceleration,
- $\mathbf{w}_p^T = [\mathbf{f}_p^T\ \mathbf{m}_p^T]$ is the total wrench applied by the platform,
- $\mathbf{w}_p^{ext\ T} = [\mathbf{f}_p^{ext\ T}\ \mathbf{m}_p^{ext\ T}]$ is the interaction wrench applied by the platform on the environment,
- $\mathbf{M}_p$ is the platform inertia matrix,
- $\mathbf{c}_p$ is the vector of Coriolis, centrifugal, gravity and external effects for the platform (the gravity forces can be eliminated from $\mathbf{c}_p$ by adding $-\mathbf{g}$ to $\dot{\mathbf{t}}_p$).

8.4 Inverse and Direct Dynamic Models of Non-redundant Parallel Robots

The dynamic equations of the virtual tree structure (Sect. 8.2) and of the free moving platform (8.57) do not take into account the closed-loop characteristics of parallel robots: among all joint and platform coordinates $\mathbf{q}_t$ and $\mathbf{x}$ of the virtual robot (Fig. 8.1b), resp., only a subset denoted as $\mathbf{q}_a$ is independent in the real robot (the actual actuated joints positions, that are indeed a subset of $\mathbf{q}_t$). All these variables are linked through the loop-closure equations of the real robot defined in Sect. 7.3 that will be used to define a set of Lagrange equations based on the use of Lagrange multipliers (Sect. 6.1.4).

8.4.1 Inverse Dynamic Model

8.4.1.1 Prerequisite Notions

Before computing the *PKM IDM*, it is necessary to note the two following things:

- By identification between (6.1), which gives the general form of the Lagrange equations, and (8.33), which gives the general form of the *IDM* of the virtual tree structure, we have:

$$\boldsymbol{\tau}_t = \mathbf{M}_t(\mathbf{q}_t)\ddot{\mathbf{q}}_t + \mathbf{c}_t(\mathbf{q}_t, \dot{\mathbf{q}}_t) = \frac{\mathrm{d}}{\mathrm{d}t}\left(\frac{\partial L_t}{\partial \dot{\mathbf{q}}_t}\right)^T - \left(\frac{\partial L_t}{\partial \mathbf{q}_t}\right)^T \tag{8.58}$$

  where L_t is the Lagrangian of the virtual tree-structured robot. We decompose the vector $\boldsymbol{\tau}_t$ into two sub-parts $\boldsymbol{\tau}_{ta}$ and $\boldsymbol{\tau}_{td}$ defined such as $\boldsymbol{\tau}_t = \mathbf{E}_\tau \begin{bmatrix} \boldsymbol{\tau}_{ta} \\ \boldsymbol{\tau}_{td} \end{bmatrix}$, with $\mathbf{E}_\tau$ a square matrix of dimension n_t which is used to sort the vector $\boldsymbol{\tau}_t$ so that:

$$\boldsymbol{\tau}_{ta} = \mathbf{M}_{ta}(\mathbf{q}_t)\begin{bmatrix} \ddot{\mathbf{q}}_a \\ \ddot{\mathbf{q}}_d \end{bmatrix} + \mathbf{c}_{ta}(\mathbf{q}_t, \dot{\mathbf{q}}_t) = \frac{\mathrm{d}}{\mathrm{d}t}\left(\frac{\partial L_t}{\partial \dot{\mathbf{q}}_a}\right)^T - \left(\frac{\partial L_t}{\partial \mathbf{q}_a}\right)^T = \mathbf{idm}_a(\ddot{\mathbf{q}}_t, \dot{\mathbf{q}}_t, \mathbf{q}_t, \mathbf{w}_t) \tag{8.59}$$

$$\boldsymbol{\tau}_{td} = \mathbf{M}_{td}(\mathbf{q}_t)\begin{bmatrix} \ddot{\mathbf{q}}_a \\ \ddot{\mathbf{q}}_d \end{bmatrix} + \mathbf{c}_{td}(\mathbf{q}_t, \dot{\mathbf{q}}_t) = \frac{\mathrm{d}}{\mathrm{d}t}\left(\frac{\partial L_t}{\partial \dot{\mathbf{q}}_d}\right)^T - \left(\frac{\partial L_t}{\partial \mathbf{q}_d}\right)^T = \mathbf{idm}_d(\ddot{\mathbf{q}}_t, \dot{\mathbf{q}}_t, \mathbf{q}_t, \mathbf{w}_t) \tag{8.60}$$

  where

  - $\boldsymbol{\tau}_{ta}$ is a $(n_a \times 1)$ which corresponds to the virtual input efforts of the virtual structure in the joints corresponding to the actuated joints of the real parallel robot,
  - $\boldsymbol{\tau}_{td}$ is a $(n_d \times 1)$ which corresponds to the virtual input efforts of the virtual structure in the joints corresponding to the passive joints of the real parallel robot,
  - $\mathbf{M}_{ta}$ is a $(n_a \times n_t)$ matrix and $\mathbf{M}_{td}$ is a $(n_d \times n_t)$ matrix,
  - $\mathbf{c}_{ta}$ is a vector of dimension n_a and $\mathbf{c}_{td}$ is a vector of dimension n_d.

  $\mathbf{M}_{ta}$ and $\mathbf{M}_{td}$ are defined such that

$$\mathbf{M}_t = \mathbf{E}_\tau \begin{bmatrix} \mathbf{M}_{ta} \\ \mathbf{M}_{td} \end{bmatrix} \mathbf{E}_\tau^T \Leftrightarrow \mathbf{E}_\tau^T \mathbf{M}_t \mathbf{E}_\tau = \begin{bmatrix} \mathbf{M}_{ta} \\ \mathbf{M}_{td} \end{bmatrix} \tag{8.61}$$

  $\mathbf{c}_{ta}$ and $\mathbf{c}_{td}$ are defined such that

$$\mathbf{c}_t = \mathbf{E}_\tau \begin{bmatrix} \mathbf{c}_{ta} \\ \mathbf{c}_{td} \end{bmatrix} \Leftrightarrow \mathbf{E}_\tau^T \mathbf{c}_t = \begin{bmatrix} \mathbf{c}_{ta} \\ \mathbf{c}_{td} \end{bmatrix}. \tag{8.62}$$

Please note that $\mathbf{E}_\tau$ can also be used to sort the vector $\ddot{\mathbf{q}}_t$ so that

$$\mathbf{E}_\tau^T \ddot{\mathbf{q}}_t = \begin{bmatrix} \ddot{\mathbf{q}}_a \\ \ddot{\mathbf{q}}_d \end{bmatrix} \Leftrightarrow \ddot{\mathbf{q}}_t = \mathbf{E}_\tau \begin{bmatrix} \ddot{\mathbf{q}}_a \\ \ddot{\mathbf{q}}_d \end{bmatrix}. \tag{8.63}$$

- By using the *PVP* and the Eq. (7.61) that links, for a *PKM* with a lower mobility (i.e. a spatial robot with less than 6 *DOF*), the independent platform twist coordinates ${}^0\mathbf{t}_r$ to all the twist components ${}^0\mathbf{t}_p$ by ${}^0\mathbf{t}_p = \boldsymbol{\Psi}_t\,{}^0\mathbf{t}_r$, we can prove that:

$${}^0\mathbf{t}_p^{*\,T}\,{}^0\mathbf{w}_p = {}^0\mathbf{t}_r^{*\,T}\,{}^0\mathbf{w}_r \Leftrightarrow {}^0\mathbf{t}_r^{*\,T}\,\boldsymbol{\Psi}_t^T\,{}^0\mathbf{w}_p = {}^0\mathbf{t}_r^{*\,T}\,{}^0\mathbf{w}_r \tag{8.64}$$

or also, by identifying the terms of the right-hand side of the equation,

$${}^0\mathbf{w}_r = \mathbf{idm}_p(\dot{\mathbf{t}}_p, \mathbf{t}_p, \mathbf{x}_p, \mathbf{w}_e) = \boldsymbol{\Psi}_t^T\,{}^0\mathbf{w}_p \tag{8.65}$$

where ${}^0\mathbf{w}_r$ represents the independent components of the platform wrench ${}^0\mathbf{w}_p$. Moreover, by using (7.71) which states that ${}^0\mathbf{t}_r = \overline{\mathbf{D}}_r\dot{\mathbf{x}}$, where $\mathbf{x}$ is the vector of the independent platform coordinates, we can prove that:

$${}^0\mathbf{t}_r^{*\,T}\,{}^0\mathbf{w}_r = \dot{\mathbf{x}}^{*\,T}\,\boldsymbol{\tau}_p \Leftrightarrow \dot{\mathbf{x}}^{*\,T}\,\overline{\mathbf{D}}_r^T\,{}^0\mathbf{w}_r = \dot{\mathbf{x}}^{*\,T}\,\boldsymbol{\tau}_p \tag{8.66}$$

or also, by identifying the terms of the right-hand side of the equation,

$$\boldsymbol{\tau}_p = \overline{\mathbf{D}}_r^T\,{}^0\mathbf{w}_r \Leftrightarrow {}^0\mathbf{w}_r = \overline{\mathbf{D}}_r^{-T}\,\boldsymbol{\tau}_p \tag{8.67}$$

where $\overline{\mathbf{D}}_r^T$ is a square and invertible matrix and,

$$\boldsymbol{\tau}_p = \frac{\mathrm{d}}{\mathrm{d}t}\left(\frac{\partial L_p}{\partial \dot{\mathbf{x}}}\right)^T - \left(\frac{\partial L_p}{\partial \mathbf{x}}\right)^T \tag{8.68}$$

in which L_p is the Lagrangian of the free moving platform.

8.4.1.2 Computation of the Inverse Dynamic Model

By considering the two constraints relations (7.62) and (7.91) which state that:

$$\mathbf{A}_r\,{}^0\mathbf{t}_r + \mathbf{B}\dot{\mathbf{q}}_a = \mathbf{0} \tag{8.69}$$

$$\mathbf{J}_t\,{}^0\mathbf{t}_r - \mathbf{J}_{ta}\dot{\mathbf{q}}_a = \mathbf{J}_{td}\dot{\mathbf{q}}_d, \tag{8.70}$$

we can now use the Lagrange equations with multipliers (see Sect. 6.1.4) (or directly the *PVP*) to obtain the *IDM* of the *PKM* as:

$$\boxed{\boldsymbol{\tau} = \boldsymbol{\tau}_{ta} + \mathbf{J}^{T\,0}\mathbf{w}_r + \mathbf{J}_d^T \boldsymbol{\tau}_{td} = \begin{bmatrix}\mathbf{1}_{n_a} & \mathbf{J}_d^T\end{bmatrix} \mathbf{E}_\tau^T \boldsymbol{\tau}_t + \mathbf{J}^{T\,0}\mathbf{w}_r} \tag{8.71}$$

where $\boldsymbol{\tau}$ is the $(n_a \times 1)$ vector of the real robot input efforts, the matrix $\mathbf{J}$ is defined in Eq. (7.63) as

$$\mathbf{J} = -\mathbf{A}_r^{-1}\mathbf{B} \tag{8.72}$$

and, by using (7.95), we get

$$\mathbf{J}_d = \mathbf{J}_{td}^{-1}(\mathbf{J}_t\mathbf{J} - \mathbf{J}_{ta}). \tag{8.73}$$

In these expressions,

- $\boldsymbol{\tau}_t$ is the vector of the virtual tree structure input efforts provided in Sect. 8.2.
- ${}^0\mathbf{w}_r$ is the wrench of the free platform expressed in the base frame $\mathscr{F}_0$ and given at (8.65).
- $\mathbf{E}_\tau$ is a matrix ordering the vector $\boldsymbol{\tau}_t$ into two sub-parts $\boldsymbol{\tau}_{ta}$ and $\boldsymbol{\tau}_{td}$ and is defined at (8.61).
- $\mathbf{J}$ is the *PKM* kinematic Jacobian matrix defined in (7.63), $\mathbf{J} = [\partial^0\mathbf{t}_r / \partial\dot{\mathbf{q}}_a]$.
- $\mathbf{J}_t$, $\mathbf{J}_{ta}$ and $\mathbf{J}_{td}$ are defined in (7.93), (7.94) and (7.92), respectively.

Equation (8.71) shows that the torque of the actuated joints of the closed loop is the sum of $\boldsymbol{\tau}_{ta}$ and the projection of $\boldsymbol{\tau}_{td}$ and ${}^0\mathbf{w}_r$ on the joint axes via the rules of the transpose of the Jacobian matrix.

Proof

Let us first rewrite (8.69) as

$$\mathbf{A}_d\dot{\mathbf{x}} + \mathbf{B}\dot{\mathbf{q}}_a = \mathbf{0} \tag{8.74}$$

where, from (7.76),

$$\mathbf{A}_d = \mathbf{A}_r\overline{\mathbf{D}}_r \text{ or also } \mathbf{A}_r = \mathbf{A}_d\overline{\mathbf{D}}_r^{-1}. \tag{8.75}$$

Moreover, let us also express (8.70) as a function of $\dot{\mathbf{x}}$:

$$\mathbf{J}_t\overline{\mathbf{D}}_r\,\dot{\mathbf{x}} - \mathbf{J}_{ta}\dot{\mathbf{q}}_a = \mathbf{J}_{td}\dot{\mathbf{q}}_d. \tag{8.76}$$

Using (8.74) and (8.76) in combination with the Lagrange equations with multipliers, we have

$$\boldsymbol{\tau} = \boldsymbol{\tau}_{ta} - \mathbf{B}^T\boldsymbol{\lambda}_1 - \mathbf{J}_{ta}^T\boldsymbol{\lambda}_2 \text{ with } \boldsymbol{\tau}_{ta} = \mathbf{idm}_a(\ddot{\mathbf{q}}_t, \dot{\mathbf{q}}_t, \mathbf{q}_t, \mathbf{w}_e) = \frac{\mathrm{d}}{\mathrm{d}t}\left(\frac{\partial L_t}{\partial \dot{\mathbf{q}}_a}\right)^T - \left(\frac{\partial L_t}{\partial \mathbf{q}_a}\right)^T \tag{8.77}$$

$$\boldsymbol{\tau}_p = \mathbf{A}_d^T\boldsymbol{\lambda}_1 - \overline{\mathbf{D}}_r^T\mathbf{J}_t^T\boldsymbol{\lambda}_2 \text{ with } \boldsymbol{\tau}_p = \frac{\mathrm{d}}{\mathrm{d}t}\left(\frac{\partial L_p}{\partial \dot{\mathbf{x}}}\right)^T - \left(\frac{\partial L_p}{\partial \mathbf{x}}\right)^T \tag{8.78}$$

$$\boldsymbol{\tau}_{td} = \mathbf{J}_{td}^T \boldsymbol{\lambda}_2 \text{ with } \boldsymbol{\tau}_{td} = \mathbf{idm}_d(\ddot{\mathbf{q}}_t, \dot{\mathbf{q}}_t, \mathbf{q}_t, \mathbf{w}_e) = \frac{\mathrm{d}}{\mathrm{d}t}\left(\frac{\partial L_t}{\partial \dot{\mathbf{q}}_d}\right)^T - \left(\frac{\partial L_t}{\partial \mathbf{q}_d}\right)^T \tag{8.79}$$

where $\boldsymbol{\lambda}_1$ is a first vector of Lagrange multipliers of dimension n_a and $\boldsymbol{\lambda}_2$ is a second vector of Lagrange multipliers of dimension n_d. Their physical meaning is disclosed below.

Left multiplying Eq. (8.78) by $\overline{\mathbf{D}}_r^{-T}$, we get:

$$\overline{\mathbf{D}}_r^{-T} \mathbf{A}_d^T \boldsymbol{\lambda}_1 - \overline{\mathbf{D}}_r^{-T} \overline{\mathbf{D}}_r^T \mathbf{J}_t^T \boldsymbol{\lambda}_2 = \overline{\mathbf{D}}_r^{-T} \boldsymbol{\tau}_p. \tag{8.80}$$

Now introducing (8.67) and (8.75) into (8.80), Eq. (8.78) can be rewritten as

$$\mathbf{A}_r^T \boldsymbol{\lambda}_1 - \mathbf{J}_t^T \boldsymbol{\lambda}_2 = {}^0\mathbf{w}_r. \tag{8.81}$$

Thus, we have the new set of equations

$$\boldsymbol{\tau} = \boldsymbol{\tau}_{ta} - \mathbf{B}^T \boldsymbol{\lambda}_1 - \mathbf{J}_{ta}^T \boldsymbol{\lambda}_2 \tag{8.82}$$

$${}^0\mathbf{w}_r = \mathbf{A}_r^T \boldsymbol{\lambda}_1 - \mathbf{J}_t^T \boldsymbol{\lambda}_2 \tag{8.83}$$

$$\boldsymbol{\tau}_{td} = \mathbf{J}_{td}^T \boldsymbol{\lambda}_2 \tag{8.84}$$

which is simpler to explain in terms of physics than the previous set (8.77) and (8.78). Please note that:

- In the case of a *PKM* without redundancy, matrix $\mathbf{A}_r^T$ is square and can thus be inverted as long as there are no Type 2 singularities (see Sect. 7.5.1),
- $\mathbf{J}_{td}^T$ is always square and can thus be inverted as long as it is not singular (see Sect. 7.5.2).

Indeed, the physical meaning of the Eqs. (8.82)–(8.84) is the following:

- For Eq. (8.84): first, we must recall from Sect. 7.3.3 that the matrix $\mathbf{J}_{td}$ is the kinematic Jacobian matrix which links the displacement of the leg extremities to the displacement of the passive joints. Thus, from Sect. 6.4, we can see that the vector $\boldsymbol{\lambda}_2$ is composed of all wrenches applied by the virtual tree-structure by its terminal links on the platform in points $A_{i\,m_i}$ when the joints corresponding to the passive joints of the real robot are moving (Fig. 8.5),
- For Eq. (8.83): this equation represents the platform equilibrium equation. The wrenches $\boldsymbol{\lambda}_2$ are projected towards the platform controlled point through the use of the matrix $\mathbf{J}_t^T$. Moreover, as explained in Sect. 7.3.1, the column i of the matrix $\mathbf{A}_r^T$ is a unit wrench which is proportional to the wrenches applied by the leg i on the platform when its actuator is developing an input effort in a static mode of operation and in absence of any other type of external effects. Thus, $\boldsymbol{\lambda}_1$ are the norm of those wrenches. So, Eq. (8.83) represents the platform equilibrium so that the virtual structure can have the same motion as the real parallel robot.

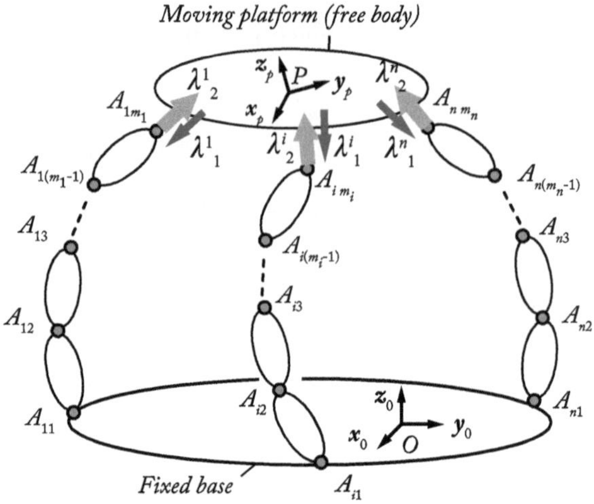

Fig. 8.5 Platform equilibrium

From (8.84), the values of $\boldsymbol{\lambda}_2$ can be found:

$$\boldsymbol{\lambda}_2 = \mathbf{J}_{td}^{-T}\boldsymbol{\tau}_{td} = \mathbf{J}_{td}^{-T}\mathbf{idm}_d(\ddot{\mathbf{q}}_t, \dot{\mathbf{q}}_t, \mathbf{q}_t, \mathbf{w}_t). \tag{8.85}$$

Then, introducing (8.85) into (8.83), we have

$$\begin{aligned} &\mathbf{A}_r^T\boldsymbol{\lambda}_1 = {}^0\mathbf{w}_r + \mathbf{J}_t^T\boldsymbol{\lambda}_2 = {}^0\mathbf{w}_r + \mathbf{J}_t^T\mathbf{J}_{td}^{-T}\boldsymbol{\tau}_{td} \\ &\Leftrightarrow \boldsymbol{\lambda}_1 = \mathbf{A}_r^{-T}\left({}^0\mathbf{w}_r + \mathbf{J}_t^T\mathbf{J}_{td}^{-T}\boldsymbol{\tau}_{td}\right). \end{aligned} \tag{8.86}$$

Finally, the *IDM* of the *PKM* is given from (8.82) and (8.86) by:

$$\begin{aligned} &\boldsymbol{\tau} = \boldsymbol{\tau}_{ta} - \mathbf{B}^T\boldsymbol{\lambda}_1 - \mathbf{J}_{ta}^T\boldsymbol{\lambda}_2 \\ \Leftrightarrow\ &\boldsymbol{\tau} = \boldsymbol{\tau}_{ta} - \mathbf{B}^T\mathbf{A}_r^{-T}\left({}^0\mathbf{w}_r + \mathbf{J}_t^T\mathbf{J}_{td}^{-T}\boldsymbol{\tau}_{td}\right) - \mathbf{J}_{ta}^T\mathbf{J}_{td}^{-T}\boldsymbol{\tau}_{td} \\ \Leftrightarrow\ &\boldsymbol{\tau} = \boldsymbol{\tau}_{ta} + \mathbf{J}^{T\,0}\mathbf{w}_r + (\mathbf{J}^T\mathbf{J}_t^T - \mathbf{J}_{ta}^T)\mathbf{J}_{td}^{-T}\boldsymbol{\tau}_{td} \\ \Leftrightarrow\ &\boldsymbol{\tau} = \boldsymbol{\tau}_{ta} + \mathbf{J}^{T\,0}\mathbf{w}_r + \mathbf{J}_d^T\boldsymbol{\tau}_{td}. \end{aligned} \tag{8.87}$$

Note that the matrix $\mathbf{J}_d$ defined in (8.73) is the matrix that allows us to express the passive joint velocities $\dot{\mathbf{q}}_d$ as a function of the active joint velocities $\dot{\mathbf{q}}_a$ only. This can be proven as follows.

Let us recall that, from (7.91), we have

$$\mathbf{J}_t\,{}^0\mathbf{t}_r - \mathbf{J}_{ta}\dot{\mathbf{q}}_a = \mathbf{J}_{td}\dot{\mathbf{q}}_d. \tag{8.88}$$

From (7.63), we also have

$${}^0\mathbf{t}_r = \mathbf{J}\dot{\mathbf{q}}_a. \tag{8.89}$$

Introducing (8.89) into (8.88), we obtain

$$\mathbf{J}_{td}\dot{\mathbf{q}}_d = \mathbf{J}_t\,\mathbf{J}\dot{\mathbf{q}}_a - \mathbf{J}_{ta}\dot{\mathbf{q}}_a = (\mathbf{J}_t\,\mathbf{J} - \mathbf{J}_{ta})\dot{\mathbf{q}}_a \tag{8.90}$$

which leads to

$$\dot{\mathbf{q}}_d = \mathbf{J}_{td}^{-1}(\mathbf{J}_t\,\mathbf{J} - \mathbf{J}_{ta})\dot{\mathbf{q}}_a = \mathbf{J}_d\dot{\mathbf{q}}_a. \tag{8.91}$$

8.4.2 Direct Dynamic Model

The *DDM* of the *PKM* which expresses the input joint accelerations $\ddot{\mathbf{q}}_a$ as a function of the input efforts $\boldsymbol{\tau}$ is given by

$$\boxed{\ddot{\mathbf{q}}_a = \mathbf{M}^{-1}(\mathbf{q}_a)\,(\boldsymbol{\tau} - \mathbf{c}(\mathbf{q}_a, \dot{\mathbf{q}}_a))} \tag{8.92}$$

where

$$\boxed{\mathbf{M}(\mathbf{q}_a) = \left[\mathbf{1}_{na}\ \mathbf{J}_d^T\right]\mathbf{E}_\tau^T\mathbf{M}_t\mathbf{E}_\tau\begin{bmatrix}\mathbf{1}_{na}\\ \mathbf{J}_d\end{bmatrix} + \mathbf{J}^T\,\boldsymbol{\Psi}_t^T\,{}^0\mathbf{M}_p\,\boldsymbol{\Psi}_t\,\mathbf{J}} \tag{8.93}$$

is the inertia matrix of the *PKM*, and

$$\boxed{\mathbf{c}(\mathbf{q}_a, \dot{\mathbf{q}}_a) = \left[\mathbf{1}_{na}\ \mathbf{J}_d^T\right]\mathbf{E}_\tau^T\left(\mathbf{c}_t + \mathbf{M}_t\mathbf{E}_\tau\begin{bmatrix}\mathbf{0}_{n_a\times 1}\\ \mathbf{a}_d\end{bmatrix}\right) + \mathbf{J}^T\,\boldsymbol{\Psi}_t^T\,({}^0\mathbf{M}_p\,\mathbf{a}_t' + {}^0\mathbf{c}_p)} \tag{8.94}$$

is the vector of Coriolis, centrifugal, gravity and external effects of the *PKM*, in which

$$\mathbf{a}_d = \mathbf{J}_{td}^{-1}(\mathbf{d}_c + \mathbf{J}_t\mathbf{a}_t) \tag{8.95}$$

and

$$\mathbf{a}_t' = \boldsymbol{\Psi}_t\,\mathbf{a}_t + \dot{\boldsymbol{\Psi}}_t\,{}^0\mathbf{t}_r. \tag{8.96}$$

Moreover, in these expressions,

- ${}^0\mathbf{M}_p$ is the inertia matrix of the free platform and ${}^0\mathbf{c}_p$ is its vector of Coriolis, centrifugal, gravity and external effects expressed in the base frame $\mathscr{F}_0$. They are defined from (8.57) by

$$ {}^0\mathbf{M}_p = {}^0\overline{\mathbf{R}}_p{}^p\mathbf{M}_p{}^0\overline{\mathbf{R}}_p^T \tag{8.97}$$

(${}^{p}\mathbf{M}_p$ being a constant matrix) and

$$ {}^{0}\mathbf{c}_p = {}^{0}\overline{\mathbf{R}}_p {}^{p}\mathbf{c}_p \tag{8.98}$$

in which ${}^{0}\overline{\mathbf{R}}_p$ is the (6×6) rotation matrix between the frames $\mathscr{F}_p$ and $\mathscr{F}_0$.

- $\mathbf{M}_t$ is the inertia matrix of the virtual tree structure and $\mathbf{c}_t$ is its vector of Coriolis, centrifugal, gravity and external effects. Their expression is provided in Sect. 8.2.
- $\mathbf{E}_\tau$ is an ordering matrix defined at (8.61).
- $\mathbf{J}_d$ is defined in (8.73) and is the matrix that allows expression of the passive joint velocities $\dot{\mathbf{q}}_d$ as a function of the active joint velocities $\dot{\mathbf{q}}_a$ only.
- $\boldsymbol{\Psi}_t$ is a $(6 \times n_{dof})$ matrix defining the independent coordinates of the platform twist (see Eq. (7.61)).
- $\mathbf{J}$ is the *PKM* kinematic Jacobian matrix defined at (7.63).
- $\mathbf{a}_t$, $\mathbf{J}_t$, $\mathbf{J}_{td}$ and $\mathbf{d}_c$ are defined in (7.163), (7.93), (7.92) and (7.182), respectively.

Proof

To obtain the *DDM* of the *PKM*, we first need to develop the *IDM* equations in order to obtain an expression in the form:

$$\boldsymbol{\tau} = \mathbf{M}(\mathbf{q}_a)\ddot{\mathbf{q}}_a + \mathbf{c}(\mathbf{q}_a, \dot{\mathbf{q}}_a). \tag{8.99}$$

So, starting from (8.71), let us decompose the expression into two sub-terms $\boldsymbol{\tau}_1$ and $\boldsymbol{\tau}_2$, such as

$$\boldsymbol{\tau} = \boldsymbol{\tau}_1 + \boldsymbol{\tau}_2 \tag{8.100}$$

with

$$\boldsymbol{\tau}_1 = \left[\mathbf{1}_{n_a}\ \mathbf{J}_d^T\right] \mathbf{E}_\tau^T \boldsymbol{\tau}_t \tag{8.101}$$

and

$$\boldsymbol{\tau}_2 = \mathbf{J}^{T\,0}\mathbf{w}_r. \tag{8.102}$$

Let us first concentrate on the term $\boldsymbol{\tau}_1$. Introducing (8.58) into (8.101) leads to

$$\boldsymbol{\tau}_1 = \left[\mathbf{1}_{n_a}\ \mathbf{J}_d^T\right] \mathbf{E}_\tau^T (\mathbf{M}_t \ddot{\mathbf{q}}_t + \mathbf{c}_t). \tag{8.103}$$

Then, from (8.63), we deduce that

$$\boldsymbol{\tau}_1 = \left[\mathbf{1}_{n_a}\ \mathbf{J}_d^T\right] \mathbf{E}_\tau^T \left(\mathbf{M}_t \mathbf{E}_\tau \begin{bmatrix} \ddot{\mathbf{q}}_a \\ \ddot{\mathbf{q}}_d \end{bmatrix} + \mathbf{c}_t\right). \tag{8.104}$$

Equation (7.183) states that

$$\ddot{\mathbf{q}}_d = \mathbf{J}_{td}^{-1} (\mathbf{J}_t {}^{0}\dot{\mathbf{t}}_r - \mathbf{J}_{ta}\ddot{\mathbf{q}}_a + \mathbf{d}_c) \tag{8.105}$$

which can be expanded by introducing (7.163) into it as

$$\ddot{\mathbf{q}}_d = \mathbf{J}_{td}^{-1}(\mathbf{J}_t(\mathbf{J}\ddot{\mathbf{q}}_a + \mathbf{a}_t) - \mathbf{J}_{ta}\ddot{\mathbf{q}}_a + \mathbf{d}_c) = \mathbf{J}_{td}^{-1}(\mathbf{J}_t\mathbf{J} - \mathbf{J}_{ta})\ddot{\mathbf{q}}_a + \mathbf{J}_{td}^{-1}(\mathbf{d}_c + \mathbf{J}_t\mathbf{a}_t). \tag{8.106}$$

Then, from (8.73), we deduce that

$$\ddot{\mathbf{q}}_d = \mathbf{J}_d\ddot{\mathbf{q}}_a + \mathbf{a}_d \tag{8.107}$$

with $\mathbf{a}_d = \mathbf{J}_{td}^{-1}(\mathbf{d}_c + \mathbf{J}_t\mathbf{a}_t)$.

Introducing (8.107) into (8.104), we get

$$\begin{aligned}\boldsymbol{\tau}_1 &= \left[\mathbf{1}_{n_a}\ \mathbf{J}_d^T\right]\mathbf{E}_\tau^T\left(\mathbf{M}_t\mathbf{E}_\tau\left(\begin{bmatrix}\mathbf{1}_{n_a}\\ \mathbf{J}_d\end{bmatrix}\ddot{\mathbf{q}}_a + \begin{bmatrix}\mathbf{0}_{n_a\times 1}\\ \mathbf{a}_d\end{bmatrix}\right) + \mathbf{c}_t\right)\\ &= \left[\mathbf{1}_{n_a}\ \mathbf{J}_d^T\right]\mathbf{E}_\tau^T\mathbf{M}_t\mathbf{E}_\tau\begin{bmatrix}\mathbf{1}_{n_a}\\ \mathbf{J}_d\end{bmatrix}\ddot{\mathbf{q}}_a + \left[\mathbf{1}_{n_a}\ \mathbf{J}_d^T\right]\mathbf{E}_\tau^T\left(\mathbf{M}_t\mathbf{E}_\tau\begin{bmatrix}\mathbf{0}_{n_a\times 1}\\ \mathbf{a}_d\end{bmatrix} + \mathbf{c}_t\right).\end{aligned} \tag{8.108}$$

Let us now consider the term $\boldsymbol{\tau}_2$. Introducing (8.65) first and then (8.57) into (8.102) leads to

$$\boldsymbol{\tau}_2 = \mathbf{J}^T\boldsymbol{\Psi}_t^T\ {}^0\mathbf{w}_p = \mathbf{J}^T\boldsymbol{\Psi}_t^T\ ({}^0\mathbf{M}_p{}^0\dot{\mathbf{t}}_p + {}^0\mathbf{c}_p). \tag{8.109}$$

Then, from (7.160) and (7.163), we have

$${}^0\dot{\mathbf{t}}_p = \boldsymbol{\Psi}_t\,{}^0\dot{\mathbf{t}}_r + \dot{\boldsymbol{\Psi}}_t\,{}^0\mathbf{t}_r = \boldsymbol{\Psi}_t\,\mathbf{J}\ddot{\mathbf{q}}_a + \mathbf{a}_t' \text{ where } \mathbf{a}_t' = \boldsymbol{\Psi}_t\,\mathbf{a}_t + \dot{\boldsymbol{\Psi}}_t\,{}^0\mathbf{t}_r \tag{8.110}$$

which can be introduced into (8.109) in order to obtain

$$\boldsymbol{\tau}_2 = \mathbf{J}^T\boldsymbol{\Psi}_t^T\ ({}^0\mathbf{M}_p(\boldsymbol{\Psi}_t\,\mathbf{J}\ddot{\mathbf{q}}_a + \mathbf{a}_t') + {}^0\mathbf{c}_p) \tag{8.111}$$

or also

$$\boldsymbol{\tau}_2 = \mathbf{J}^T\boldsymbol{\Psi}_t^T\,{}^0\mathbf{M}_p\boldsymbol{\Psi}_t\,\mathbf{J}\ddot{\mathbf{q}}_a + \mathbf{J}^T\boldsymbol{\Psi}_t^T\ ({}^0\mathbf{M}_p\mathbf{a}_t' + {}^0\mathbf{c}_p). \tag{8.112}$$

By summing (8.108) and (8.112), we obtain

$$\boldsymbol{\tau} = \mathbf{M}\ddot{\mathbf{q}}_a + \mathbf{c} \tag{8.113}$$

with

$$\mathbf{M} = \left[\mathbf{1}_{na}\ \mathbf{J}_d^T\right]\mathbf{E}_\tau^T\mathbf{M}_t\mathbf{E}_\tau\begin{bmatrix}\mathbf{1}_{na}\\ \mathbf{J}_d\end{bmatrix} + \mathbf{J}^T\ \boldsymbol{\Psi}_t^T\,{}^0\mathbf{M}_p\ \boldsymbol{\Psi}_t\,\mathbf{J} \tag{8.114}$$

and

$$\mathbf{c} = \begin{bmatrix} \mathbf{1}_{na} & \mathbf{J}_d^T \end{bmatrix} \mathbf{E}_\tau^T \left(\mathbf{c}_t + \mathbf{M}_t \mathbf{E}_\tau \begin{bmatrix} \mathbf{0}_{n_a \times 1} \\ \mathbf{a}_d \end{bmatrix} \right) + \mathbf{J}^T \, \boldsymbol{\Psi}_t^T \, ({}^0\mathbf{M}_p \, \mathbf{a}_t' + {}^0\mathbf{c}_p). \quad (8.115)$$

Finally, the *DDM* in (8.92) can be obtained by solving the Eq. (8.113).

8.4.3 Examples

8.4.3.1 Example 1: The Five-Bar Mechanism

Let us deal here with the example of the five-bar mechanism (Fig. 7.3). Its *MDH* parameters, geometric and kinematic models have been presented in Sects. 7.1.2.1, 7.2.2.1, 7.3.4.1, 7.4.4.1 and will be reused in the present example.

To follow the approach presented in the present Chapter, let us first virtually open the closed loop to obtain a virtual tree-structure robot which is presented in Fig. 8.6. The obtained virtual robot is composed of:

- one 3*R* serial planar robot composed of the links $\mathscr{B}_{11}$, $\mathscr{B}_{12}$ and $\mathscr{B}_{13}$ with all joints active,
- one 2*R* serial planar robot composed of the links $\mathscr{B}_{21}$ and $\mathscr{B}_{22}$ with all joints active.

Both of the chains are fixed on the ground denoted as the body $\mathscr{B}_0$. Moreover, the end-effector is modeled by a punctual mass.

Computation of the *IDM*: Following the approach proposed in Sect. 8.2.1, we can prove, using the developed symbolic form, that the *IDM* of the 3*R* serial planar branch of the first leg is given by:

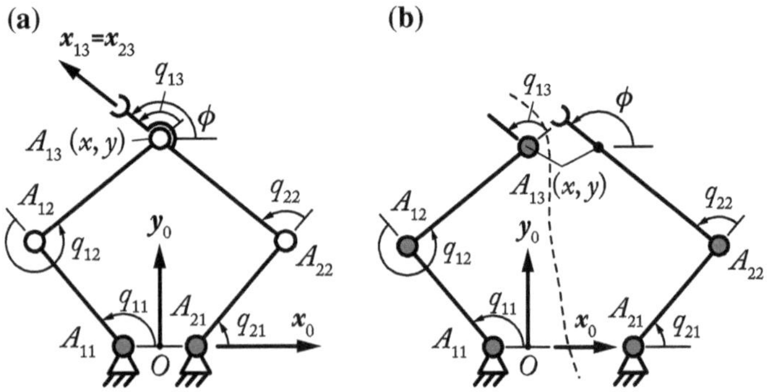

Fig. 8.6 The planar five-bar mechanism (the *gray* pairs denote the actuated joints) and the corresponding virtual tree-structure. **a** Kinematic chain of the real robot. **b** Virtual tree structure (virtually opened)

$$\begin{aligned}\tau_{t_{11}} &= zz_{11R}\ddot{q}_{11} + zz_{12}(\ddot{q}_{11} + \ddot{q}_{12}) \\ &\quad + d_{12}mx_{12}\left((2\ddot{q}_{11} + \ddot{q}_{12})\cos q_{12} - \dot{q}_{12}(2\dot{q}_{11} + \dot{q}_{12})\sin q_{12}\right) \\ &\quad + d_{12}my_{12}\left((2\ddot{q}_{11} + \ddot{q}_{12})\sin q_{12} + \dot{q}_{12}(2\dot{q}_{11} + \dot{q}_{12})\cos q_{12}\right) \\ &\quad + fs_{11}\mathrm{sign}(\dot{q}_{11}) + fv_{11}\dot{q}_{11}\end{aligned} \tag{8.116}$$

$$\begin{aligned}\tau_{t_{12}} &= zz_{12}(\ddot{q}_{11} + \ddot{q}_{12}) + d_{12}mx_{12}\left(\ddot{q}_{11}\cos q_{12} + \dot{q}_{11}^2\sin q_{12}\right) \\ &\quad + d_{12}my_{12}\left(\ddot{q}_{11}\sin q_{12} - \dot{q}_{11}^2\cos q_{12}\right) + fs_{12}\mathrm{sign}(\dot{q}_{12}) + fv_{12}\dot{q}_{12}\end{aligned} \tag{8.117}$$

$$\tau_{t_{13}} = fs_{13}\mathrm{sign}(\dot{q}_{13}) + fv_{13}\dot{q}_{13} \tag{8.118}$$

where

$$zz_{11R} = zz_{11} + Ia_{11} + d_{12}^2 m_{12} \tag{8.119}$$

and d_{12} is given in Table 7.1.

For the 2*R* serial planar branch of the second leg, we can prove that the *IDM* is given by:

$$\begin{aligned}\tau_{t_{21}} &= zz_{21R}\ddot{q}_{21} + zz_{22}(\ddot{q}_{21} + \ddot{q}_{22}) \\ &\quad + d_{22}mx_{22}\left((2\ddot{q}_{21} + \ddot{q}_{22})\cos q_{22} - \dot{q}_{22}(2\dot{q}_{21} + \dot{q}_{22})\sin q_{22}\right) \\ &\quad + d_{22}my_{22}\left((2\ddot{q}_{21} + \ddot{q}_{22})\sin q_{22} + \dot{q}_{22}(2\dot{q}_{21} + \dot{q}_{22})\cos q_{22}\right) \\ &\quad + fs_{21}\mathrm{sign}(\dot{q}_{21}) + fv_{21}\dot{q}_{21}\end{aligned} \tag{8.120}$$

$$\begin{aligned}\tau_{t_{22}} &= zz_{22}(\ddot{q}_{21} + \ddot{q}_{22}) + d_{22}mx_{22}\left(\ddot{q}_{21}\cos q_{22} + \dot{q}_{21}^2\sin q_{22}\right) \\ &\quad + d_{22}my_{22}\left(\ddot{q}_{21}\sin q_{22} - \dot{q}_{21}^2\cos q_{22}\right) + fs_{22}\mathrm{sign}(\dot{q}_{22}) + fv_{22}\dot{q}_{22}\end{aligned} \tag{8.121}$$

where

$$zz_{21R} = zz_{21} + Ia_{21} + d_{22}^2 m_{22} \tag{8.122}$$

and d_{22} is given in Table 7.1.

Note that:

- for the 3*R* serial planar robot, the last body $\mathcal{B}_{13}$ is a virtual body which is considered massless as that body has been added to close the loop,
- for both chains, the term Ia_{ij} is canceled for the passive joints as such terms concern only the real active joints.

Supposing that a desired trajectory $(\mathbf{x}, \dot{\mathbf{x}}, \ddot{\mathbf{x}})$ for the terminal point is given, the corresponding variables q_{i1}, q_{i2}, and q_{i3} and their derivatives are computed using the geometric and kinematic models presented in Sects. 7.1.2.1, 7.2.2.1, 7.3.4.1 and 7.4.4.1.

For this robot, the end-effector is a punctual mass fixed on body $\mathscr{B}_{22}$. The dynamic model of the end-effector (denoted as the body $\mathscr{B}_4$) is given by:

$$ {}^0\mathbf{w}_p^T = m_4 \left[\ddot{x}\ \ddot{y}\ 0\ 0\ 0\ 0\right] \tag{8.123}$$

or in the reduced form, by keeping only the non-null components

$$ {}^0\mathbf{w}_r = \boldsymbol{\Psi}_t^{T\,0}\mathbf{w}_p^T = m_4 \begin{bmatrix} \ddot{x} \\ \ddot{y} \end{bmatrix} \tag{8.124}$$

where $\boldsymbol{\Psi}_t$ is defined at (7.110) and $\ddot{x}$ and $\ddot{y}$ are given by the desired trajectory.

Following the procedure described in Sect. 8.4.1, the *IDM* of the five-bar mechanism is given from Eq. (8.71) by:

$$ \boldsymbol{\tau} = \begin{bmatrix} \tau_1 \\ \tau_2 \end{bmatrix} = \boldsymbol{\tau}_{ta} + \mathbf{J}^{T\,0}\mathbf{w}_r + \mathbf{J}_d^T \boldsymbol{\tau}_{td} \tag{8.125}$$

where

- ${}^0\mathbf{w}_r$ is given in (8.124),
- $\boldsymbol{\tau}_{ta} = [\tau_{t_{11}}\ \tau_{t_{21}}]^T$ is given in the Eqs. (8.116) and (8.117),
- $\boldsymbol{\tau}_{td} = [\tau_{t_{12}}\ \tau_{t_{13}}\ \tau_{t_{22}}]^T$ is given in the Eqs. (8.116) and (8.117),
- $\mathbf{J} = -\mathbf{A}_r^{-1}\mathbf{B}$ can be obtained from the Eqs. (7.104)–(7.113),
- $\mathbf{J}_d = \mathbf{J}_{td}^{-1}(\mathbf{J}_t\mathbf{J} - \mathbf{J}_{ta})$ can be obtained from the Eqs. (7.122)–(7.124).

Computation of the *DDM*: Following the approach proposed in Sects. 8.2.4 and 8.2.5, or by a direct analysis of the Eqs. (8.116) and (8.117), we can prove that the *IDM* of virtual tree-structure can be written in the following matrix form:

$$ \boldsymbol{\tau}_t = \mathbf{M}_t(\mathbf{q}_t)\ddot{\mathbf{q}}_t + \mathbf{c}_t(\mathbf{q}_t, \dot{\mathbf{q}}_t) \tag{8.126}$$

where $\mathbf{q}_t^T = [q_{11}\ q_{12}\ q_{21}\ q_{22}\ q_{23}]$,

$$ \mathbf{M}_t(\mathbf{q}_t) = \begin{bmatrix} M_{11} & M_{12} & 0 & 0 & 0 \\ M_{21} & zz_{12} & 0 & 0 & 0 \\ 0 & 0 & 0 & 0 & 0 \\ 0 & 0 & 0 & M_{44} & M_{45} \\ 0 & 0 & 0 & M_{54} & zz_{22} \end{bmatrix} \tag{8.127}$$

with

$$ M_{11} = zz_{11R} + zz_{12} + 2d_{12}mx_{12}\cos q_{12} + 2d_{12}my_{12}\sin q_{12} \tag{8.128}$$

$$ M_{12} = M_{21} = zz_{12} + d_{12}mx_{12}\cos q_{12} + d_{12}my_{12}\sin q_{12} \tag{8.129}$$

$$M_{44} = zz_{21R} + zz_{22} + 2d_{22}mx_{12}\cos q_{22} + 2d_{22}my_{22}\sin q_{22} \tag{8.130}$$

$$M_{45} = M_{54} = zz_{22} + d_{22}mx_{22}\cos q_{22} + d_{22}my_{22}\sin q_{22} \tag{8.131}$$

and $\mathbf{c}_t(\mathbf{q}_t, \dot{\mathbf{q}}_t) = [c_1\ c_2\ c_3\ c_4\ c_5]^T$ with

$$\begin{aligned} c_1 = & - d_{12}mx_{12}\dot{q}_{12}(2\dot{q}_{11} + \dot{q}_{12})\sin q_{12} \\ & + d_{12}my_{12}\dot{q}_{12}(2\dot{q}_{11} + \dot{q}_{12})\cos q_{12} \\ & + fs_{11}\text{sign}(\dot{q}_{11}) + fv_{11}\dot{q}_{11} \end{aligned} \tag{8.132}$$

$$c_2 = d_{12}mx_{12}\dot{q}_{11}^2\sin q_{12} - d_{12}my_{12}\dot{q}_{11}^2\cos q_{12} + fs_{12}\text{sign}(\dot{q}_{12}) + fv_{12}\dot{q}_{12} \tag{8.133}$$

$$c_3 = fs_{13}\text{sign}(\dot{q}_{13}) + fv_{13}\dot{q}_{13} \tag{8.134}$$

$$\begin{aligned} c_4 = & - d_{22}mx_{22}\dot{q}_{22}(2\dot{q}_{21} + \dot{q}_{22})\sin q_{22} \\ & + d_{22}my_{22}\dot{q}_{22}(2\dot{q}_{21} + \dot{q}_{22})\cos q_{22} \\ & + fs_{21}\text{sign}(\dot{q}_{21}) + fv_{21}\dot{q}_{21} \end{aligned} \tag{8.135}$$

$$c_5 = d_{22}mx_{22}\dot{q}_{21}^2\sin q_{22} - d_{22}my_{22}\dot{q}_{21}^2\cos q_{22} + fs_{22}\text{sign}(\dot{q}_{22}) + fv_{22}\dot{q}_{22}. \tag{8.136}$$

Then, by noticing that $\boldsymbol{\tau}_{ta} = [\tau_{t_{11}}\ \tau_{t_{21}}]^T$ and $\boldsymbol{\tau}_{td} = [\tau_{t_{12}}\ \tau_{t_{22}}\ \tau_{t_{23}}]^T$ and that $\mathbf{q}_a^T = [q_{11}\ q_{21}]$ and $\mathbf{q}_d^T = [q_{12}\ q_{13}\ q_{22}]$, the sorting matrix $\mathbf{E}_\tau$ can be defined such that

$$\mathbf{E}_\tau = \begin{bmatrix} 1 & 0 & 0 & 0 & 0 \\ 0 & 0 & 0 & 1 & 0 \\ 0 & 1 & 0 & 0 & 0 \\ 0 & 0 & 1 & 0 & 0 \\ 0 & 0 & 0 & 0 & 1 \end{bmatrix} \tag{8.137}$$

The dynamic model of the end-effector which is a punctual mass can be put in the matrix form:

$$^0\mathbf{w}_p = {^0\mathbf{M}_p} \begin{bmatrix} \ddot{x} \\ \ddot{y} \\ \mathbf{0}_{4\times1} \end{bmatrix} + {^0\mathbf{c}_p} \tag{8.138}$$

where

$$^0\mathbf{M}_p = m_4 \begin{bmatrix} \mathbf{1}_2 & \mathbf{0}_{2\times4} \\ \mathbf{0}_{4\times2} & \mathbf{0}_{4\times4} \end{bmatrix} \tag{8.139}$$

and $^0\mathbf{c}_p = \mathbf{0}_{6\times1}$.

Following the procedure described in Sect. 8.4.1, the inertia matrix and Coriolis, centrifugal, gravity effects, friction and external wrenches of the five-bar mechanism is given from Eqs. (8.93) and (8.94) by:

$$\mathbf{M} = \begin{bmatrix}\mathbf{1}_2 & \mathbf{J}_d^T\end{bmatrix} \mathbf{E}_\tau^T \mathbf{M}_t \mathbf{E}_\tau \begin{bmatrix}\mathbf{1}_2 \\ \mathbf{J}_d\end{bmatrix} + \mathbf{J}^T\, {}^0\mathbf{M}_r\, \mathbf{J} \tag{8.140}$$

with ${}^0\mathbf{M}_r = \boldsymbol{\Psi}_t^T\, {}^0\mathbf{M}_p\, \boldsymbol{\Psi}_t = m_4 \mathbf{1}_2$ and

$$\mathbf{c} = \begin{bmatrix}\mathbf{1}_2 & \mathbf{J}_d^T\end{bmatrix} \mathbf{E}_\tau^T \left(\mathbf{c}_t + \mathbf{M}_t \mathbf{E}_\tau \begin{bmatrix}\mathbf{0}_{2\times 1} \\ \mathbf{a}_d\end{bmatrix}\right) + \mathbf{J}^T\, \boldsymbol{\Psi}_t^T \left({}^0\mathbf{c}_r + {}^0\mathbf{c}_p\right) \tag{8.141}$$

with ${}^0\mathbf{c}_r = {}^0\mathbf{M}_p\, \mathbf{a}'_t = \mathbf{0}$.

In these equations,

- ${}^0\mathbf{w}_r$ is given in (8.124),
- $\boldsymbol{\Psi}_t^T$ is given in (7.110)
- $\mathbf{J} = -\mathbf{A}_r^{-1}\mathbf{B}$ can be obtained from the Eqs. (7.104)–(7.113),
- $\mathbf{J}_d = \mathbf{J}_{td}^{-1}(\mathbf{J}_t\mathbf{J} - \mathbf{J}_{ta})$ can be obtained from the Eqs. (7.122)–(7.124),
- $\mathbf{a}'_t$ can be deduced from the Eq. (8.96),
- $\mathbf{a}_d$ from the Eq. (8.95).

Finally:

$$\ddot{\mathbf{q}}_a = \begin{bmatrix}\ddot{q}_{11} \\ \ddot{q}_{21}\end{bmatrix} = \mathbf{M}^{-1}\left(\boldsymbol{\tau} - \mathbf{c}\right) \tag{8.142}$$

Experimental validation of the model: Let us now validate experimentally the *IDM* of the five-bar mechanism presented in Fig. 8.7.

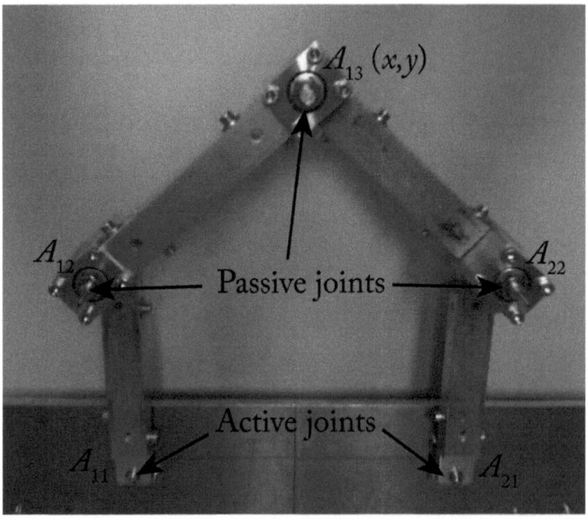

Fig. 8.7 The prototype of planar *five bar* mechanism used for the experimentation, and designed at IFMA (Clermont-Ferrand, France)

For this mechanism, the constant geometric parameters given in Table 7.1 and Eq. (7.7) are:

- $d_{11} = -0.14$ m, $d_{12} = 0.213$ m, $d_{13} = 0.1878$ m,
- $d_{21} = 0.14$ m, $d_{22} = 0.213$ m, $d_{23} = 0.1878$ m.

Its dynamic parameters have been calculated using identification procedures detailed in (Briot and Gautier 2014; Gautier et al. 2013). Their values are:

- $zz_{11R} = 2.11 \cdot 10^{-2}$ kg m^2, $fv_{11} = 6.76$ N m/s, $fs_{11} = 2.94$ N m,
- $zz_{12} = 2.23 \cdot 10^{-5}$ kg m^2, $mx_{12} = 0.012$ kg m, $my_{12} = 0$ kg m, $fv_{12} = 0$ N m/s, $fs_{12} = 0$ N m,
- $fv_{13} = 0$ N m/s, $fs_{13} = 0$ N m,
- $zz_{21R} = 2.24 \cdot 10^{-2}$ kg m^2, $fv_{21} = 6.75$ N m/s, $fs_{21} = 2.95$ N m,
- $zz_{22} = 2.44 \cdot 10^{-5}$ kg m^2, $mx_{22} = 0.012$ kg m, $my_{22} = 0$ kg m, $fv_{22} = 0$ N m/s, $fs_{22} = 0$ N m,
- $m_4 = 0.272$ kg.

So now, let the robot move on the trajectory shown in Figs. 8.8 and 8.9.

To validate the model, we compare the robot input torques with those predicted by the model. The comparison is shown in Fig. 8.10. It can be shown that the error between the prediction and the measure is very small.

8.4.3.2 Example 2: The Orthoglide

Let us treat here the example of the Orthoglide (Figs. 7.7 and 7.8). Its *MDH* parameters, geometric and kinematic models and kinematic Jacobian matrices useful for the dynamic model computation have been presented in Sects. 7.1.2.4 and 7.2.2.4.

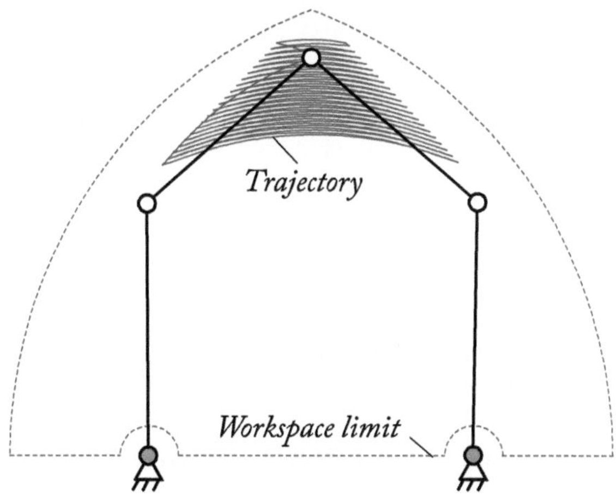

Fig. 8.8 Cartesian displacement (scaled) the *five-bar* mechanism

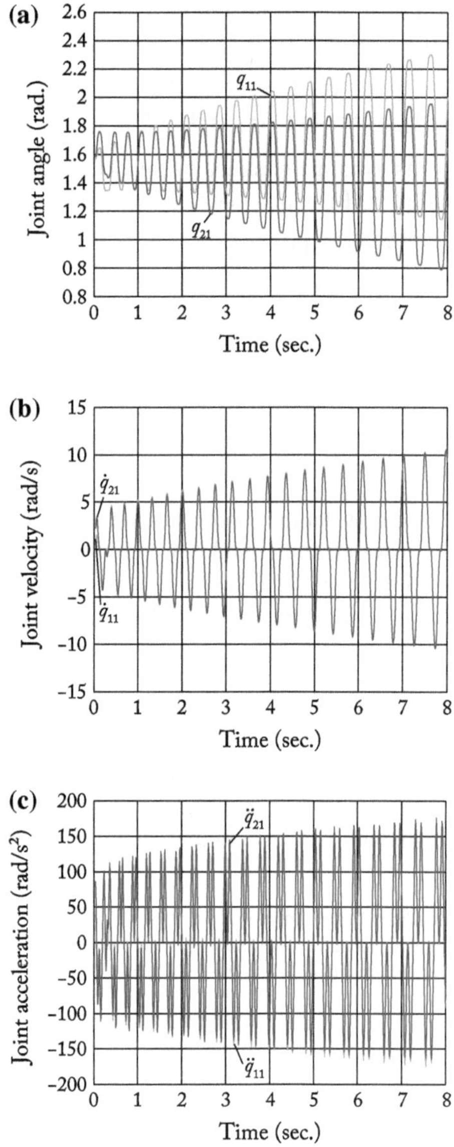

Fig. 8.9 Active joint positions, velocities, accelerations of the five-bar mechanism. **a** Active joint positions. **b** Active joint velocities. **c** Active joiny accclerations

Due to the symbolic complexity of the Orthoglide *IDM* and *DDM*, they are not detailed here, but are given in (using customized symbolic techniques):

http://www.irccyn.ec-nantes.fr/~briot/Books.html

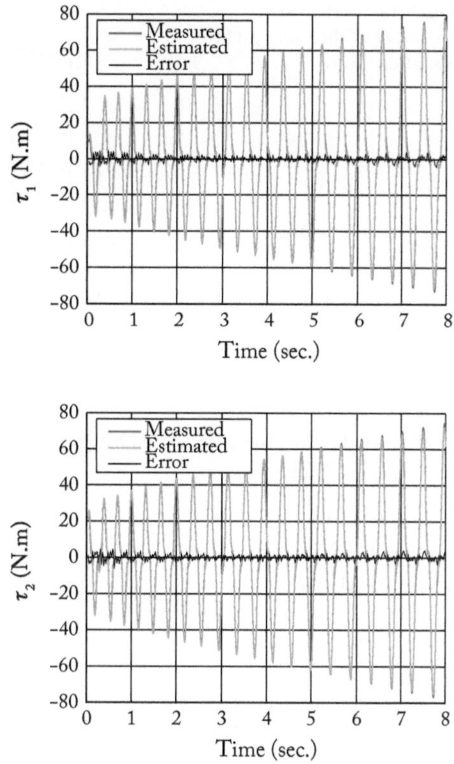

Fig. 8.10 Measured and computed torques of the five-bar mechanism (the torques are expressed in joint side units, not on the motor side)

Table 8.1 Dynamic parameters of the Orthoglide with non-null values and explicitly appearing in the dynamic model (SI units)

Param.	Val.	Param.	Val.	Param.	Val.
m_{11}	1.32	m_{21}	1.25	m_{31}	$6.26 \cdot 10^{-1}$
Ia_{11}	7.13	Ia_{21}	7.13	Ia_{31}	7.82
fv_{11}	79.70	fv_{21}	100.00	fv_{31}	79.40
fs_{11}	3.21	fs_{21}	3.23	fs_{31}	2.21
fs_{12}	0.00	fs_{22}	0.00	fs_{32}	1.28
xx_{13}	$8.69 \cdot 10^{-6}$	xx_{23}	$8.69 \cdot 10^{-6}$	xx_{33}	$8.69 \cdot 10^{-6}$
yy_{13}	$7.91 \cdot 10^{-3}$	yy_{23}	$7.91 \cdot 10^{-3}$	yy_{33}	$7.91 \cdot 10^{-3}$
zz_{13}	$7.91 \cdot 10^{-3}$	zz_{23}	$7.91 \cdot 10^{-3}$	zz_{33}	$7.91 \cdot 10^{-3}$
mx_{13}	$3.81 \cdot 10^{-2}$	mx_{23}	$3.81 \cdot 10^{-2}$	mx_{33}	$3.81 \cdot 10^{-2}$
m_{13}	$2.46 \cdot 10^{-1}$	m_{23}	$2.46 \cdot 10^{-1}$	m_{33}	$2.46 \cdot 10^{-1}$
fs_{14}	$1.22 \cdot 10^{-3}$	fs_{24}	0.00	fs_{34}	0.00
m_6	$5.13 \cdot 10^{-1}$				

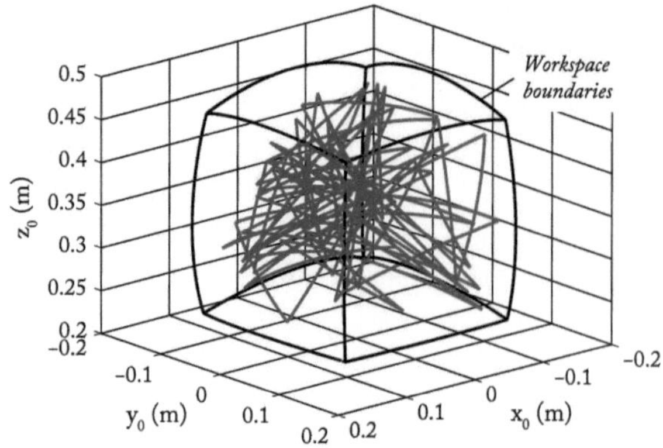

Fig. 8.11 Cartesian trajectory of the Orthoglide (duration: 20 s)

For the Orthoglide, the constant geometric parameters given in Tables 7.3 and 7.4 and in Eq. (7.23) are:

- $a = 0.34$ m,
- $d_4 = 0.31$ m,
- $d_6 = 0.03$ m.

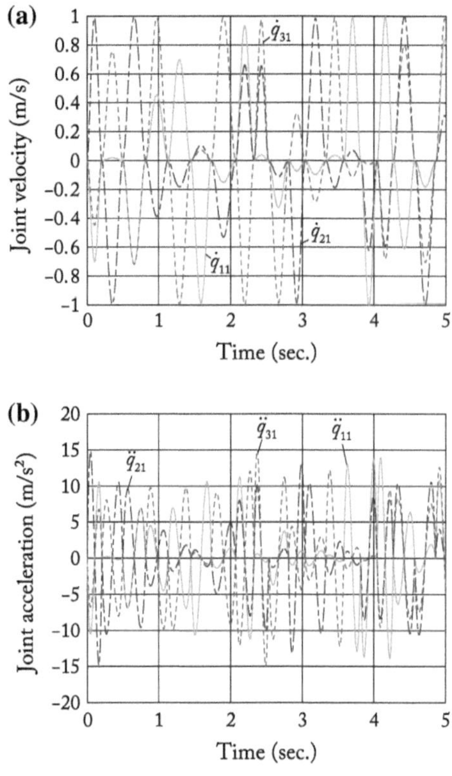

Fig. 8.12 Actuator velocities and accelerations of the Orthoglide. **a** Actuator velocities on the time interval (0 − 5)s. **b** Actuator velocities on the time interval (0 − 5 *s*)s

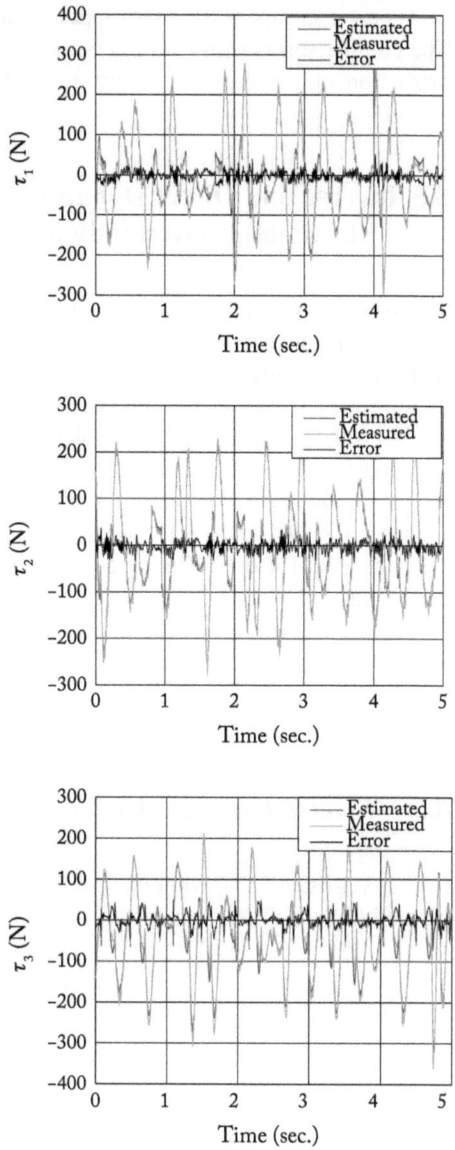

Fig. 8.13 Measured and computed forces of the Orthoglide (the forces are expressed in joint side units, not on the motor side) on the time interval $(0, 5\,s)$

Its standard dynamic parameters have been estimated using identification procedures detailed in (Briot and Gautier 2014; Gautier et al. 2013). The parameters with non-null values and explicitly appearing in the model are given in Table 8.1. In that Table, the body $\mathcal{B}_6$ is the moving platform.

So now, we make the robot move on the trajectory shown in Figs. 8.11 and 8.12.

Then we compare the robot input torques with those predicted by the model. The comparison is shown in Fig. 8.13. It can be shown that the error between the prediction and the measure is very small: thus the model is accurate.

8.5 Inverse and Direct Dynamic Models of Parallel Robots with Actuation Redundancy

The dynamic model of *PKM* with actuation redundancy has a particularity. To better understand it, let us consider the simple example shown in Fig. 8.14. This mechanism with 1 *DOF* is moved using two actuators mounted in parallel that can apply, on the moving body of mass m, two independent forces denoted as f_1 and f_2. For moving the mechanism, there exists an infinity of possible forces to apply, e.g. $[f_1\ f_2] = [m\ddot{x}\ 0]$, or also $[f_1\ f_2] = [0\ m\ddot{x}]$ and even $[f_1\ f_2] = 0.5\,[m\ddot{x}\ m\ddot{x}]$ or many other force combinations $[f_1\ f_2] = m\ddot{x}\,[(1-\alpha)\ \alpha]$ $(\alpha \in \mathbb{R})$ such that $f_1 + f_2 = m\ddot{x}$. The function $\eta = |f_1 - f_2| = |(1-2\alpha)\,m\ddot{x}|$ is called the overconstraint and represents the additional forces in the system that do not impact the motion but add internal constraints in it.

The infinity of possible solutions for *IDM* of *PKM* with actuation redundancy is the particularity of such type of mechanisms. Let us now understand the way to compute it.

8.5.1 Inverse Dynamic Model

Contrary to the *PKM* without actuation redundancy, in the two constraints relations (7.62) and (7.91) which state that:

$$\mathbf{A}_r\,{}^0\mathbf{t}_r + \mathbf{B}\dot{\mathbf{q}}_a = \mathbf{0} \tag{8.143}$$

$$\mathbf{J}_t\,{}^0\mathbf{t}_r - \mathbf{J}_{ta}\dot{\mathbf{q}}_a = \mathbf{J}_{td}\dot{\mathbf{q}}_d. \tag{8.144}$$

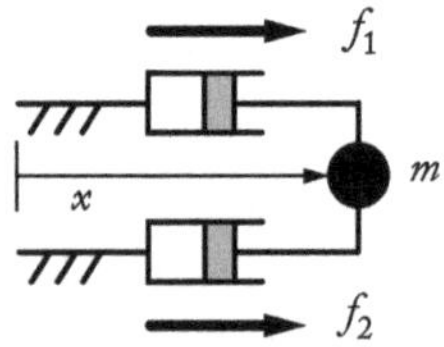

Fig. 8.14 A simple parallel mechanism with actuation redundancy

The matrix $\mathbf{B}$ is square of dimension $(n_a \times n_a)$, and the matrix $\mathbf{A}_r$ is of dimension $(n_a \times n_{dof})$, while the matrices $\mathbf{J}_t$, $\mathbf{J}_{ta}$ and $\mathbf{J}_{td}$ being still of respective dimensions $(n_d \times n_{dof})$, $(n_d \times n_a)$ and $(n_d \times n_d)$. We should remark here that the dimensions of the rectangular matrix $\mathbf{A}_r$ relates to the second case of Lagrange equations with multipliers defined in Sect. 6.1.4.

Thus, the *IDM* of the *PKM* with actuation redundancy can be written as:

$$\boxed{\mathbf{J}_{inv}^T \boldsymbol{\tau} = {}^0\mathbf{w}_r + \mathbf{J}_{inv}^T \boldsymbol{\tau}_{ta} + \mathbf{J}_r^T \boldsymbol{\tau}_{td} = {}^0\mathbf{w}_r + \left[\mathbf{J}_{inv}^T \; \mathbf{J}_r^T\right] \mathbf{E}_\tau^T \boldsymbol{\tau}_t} \tag{8.145}$$

where $\boldsymbol{\tau}$ is the $(n_a \times 1)$ vector of the real robot input efforts, the matrix

$$\mathbf{J}_{inv} = -\mathbf{B}^{-1}\mathbf{A}_r \tag{8.146}$$

is defined in Eq. (7.64) and

$$\mathbf{J}_r = \mathbf{J}_{td}^{-1}(\mathbf{J}_t - \mathbf{J}_{ta}\mathbf{J}_{inv}). \tag{8.147}$$

In these expressions,

- $\boldsymbol{\tau}_t$ is the vector of the virtual tree structure input efforts whose expression is provided in Sect. 8.2.
- ${}^0\mathbf{w}_r$ is the wrench of the free platform expressed in the base frame $\mathscr{F}_0$ and given at (8.65).
- $\mathbf{E}_\tau$ is a matrix ordering the vector $\boldsymbol{\tau}_t$ into two sub-part $\boldsymbol{\tau}_{ta}$ and $\boldsymbol{\tau}_{td}$ and is defined at (8.61).
- $\mathbf{J}_{inv}$ is the *PKM* inverse kinematic Jacobian matrix defined at (7.64).
- $\mathbf{J}_t$, $\mathbf{J}_{ta}$ and $\mathbf{J}_{td}$ are defined in (7.93), (7.94) and (7.92), respectively.

Matrix $\mathbf{J}_{inv}^T$ having more columns than rows, there is an infinity of solutions for $\boldsymbol{\tau}$ that can satisfy (8.145). All those solutions are given by:

$$\boxed{\boldsymbol{\tau} = \boldsymbol{\tau}_{ta} + \mathbf{J}_{inv}^{T\,+} \left({}^0\mathbf{w}_r + \mathbf{J}_r^T \boldsymbol{\tau}_{td}\right) + \left(\mathbf{1}_{n_a} - \mathbf{J}_{inv}^{T\,+}\mathbf{J}_{inv}^T\right)\boldsymbol{\eta}} \tag{8.148}$$

where $\mathbf{J}_{inv}^{T\,+}$ is the pseudo-inverse of $\mathbf{J}_{inv}^T$ and $\boldsymbol{\eta}$ is an arbitrary vector in the $\boldsymbol{\tau}$ space which is called the *overconstraint*. If $\boldsymbol{\eta} = \mathbf{0}$, we get the solution for $\boldsymbol{\tau}$ with the minimal norm. This solution is usually kept as it leads to a minimal power energy consumption for a given set of active joint velocity $\dot{\mathbf{q}}_a$. However this is not always the case as a proper use of the overconstraint $\boldsymbol{\eta}$ can help, for instance, increasing the acceleration or payload capacities (Nahon and Angeles 1989) or reducing the backlash in the joints of the robot (Müller 2005).

Note also that, even if (8.145) is not valid when $\mathbf{B}$ is singular, the singularity of $\mathbf{B}$ does not impact (8.148) in which there is no inversion of that matrix.

Proof

For *PKM* with actuation redundancy, the Eqs. (8.82)–(8.84) are still valid. Let us rewrite them for reason of clarity of the demonstration. We thus still have

$$\boldsymbol{\tau} = \boldsymbol{\tau}_{ta} - \mathbf{B}^T \boldsymbol{\lambda}_1 - \mathbf{J}_{ta}^T \boldsymbol{\lambda}_2 \tag{8.149}$$

$$^0\mathbf{w}_r = \mathbf{A}_r^T \boldsymbol{\lambda}_1 - \mathbf{J}_t^T \boldsymbol{\lambda}_2 \tag{8.150}$$

$$\boldsymbol{\tau}_{td} = \mathbf{J}_{td}^T \boldsymbol{\lambda}_2. \tag{8.151}$$

In these equations, the matrices $\mathbf{B}$ and $\mathbf{J}_{td}$ are still square and invertible, as long as the robot does not encounter any singularity. Thus, they can be inverted to compute the value of the Lagrange multipliers $\boldsymbol{\lambda}_1$ and $\boldsymbol{\lambda}_2$. As a result, from (8.151), we obtain

$$\boldsymbol{\lambda}_2 = \mathbf{J}_{td}^{-T} \boldsymbol{\tau}_{td}. \tag{8.152}$$

Then, introducing (8.152) into (8.149), we find

$$\boldsymbol{\tau} = \boldsymbol{\tau}_{ta} - \mathbf{B}^T \boldsymbol{\lambda}_1 - \mathbf{J}_{ta}^T \mathbf{J}_{td}^{-T} \boldsymbol{\tau}_{td} \tag{8.153}$$

from which we can get the value of $\boldsymbol{\lambda}_1$ as

$$\boldsymbol{\lambda}_1 = -\mathbf{B}^{-T} (\boldsymbol{\tau} - \boldsymbol{\tau}_{ta} + \mathbf{J}_{ta}^T \mathbf{J}_{td}^{-T} \boldsymbol{\tau}_{td}). \tag{8.154}$$

Finally, introducing (8.152) and (8.154) into (8.150), we obtain

$$\begin{aligned} ^0\mathbf{w}_r &= -\mathbf{A}_r^T \mathbf{B}^{-T} (\boldsymbol{\tau} - \boldsymbol{\tau}_{ta} + \mathbf{J}_{ta}^T \mathbf{J}_{td}^{-T} \boldsymbol{\tau}_{td}) - \mathbf{J}_t^T \mathbf{J}_{td}^{-T} \boldsymbol{\tau}_{td} \\ &= \mathbf{J}_{inv}^T (\boldsymbol{\tau} - \boldsymbol{\tau}_{ta} + \mathbf{J}_{ta}^T \mathbf{J}_{td}^{-T} \boldsymbol{\tau}_{td}) - \mathbf{J}_t^T \mathbf{J}_{td}^{-T} \boldsymbol{\tau}_{td}. \end{aligned} \tag{8.155}$$

Then, expanding and rearranging, we get

$$\begin{aligned} ^0\mathbf{w}_r &= \mathbf{J}_{inv}^T \boldsymbol{\tau} - \mathbf{J}_{inv}^T \boldsymbol{\tau}_{ta} + (\mathbf{J}_{inv}^T \mathbf{J}_{ta}^T - \mathbf{J}_t^T \mathbf{J}_{td}^{-T}) \boldsymbol{\tau}_{td} \\ &= \mathbf{J}_{inv}^T \boldsymbol{\tau} - \mathbf{J}_{inv}^T \boldsymbol{\tau}_{ta} - \mathbf{J}_r^T \boldsymbol{\tau}_{td} \end{aligned} \tag{8.156}$$

where $\mathbf{J}_r = \mathbf{J}_{td}^{-1} (\mathbf{J}_t - \mathbf{J}_{ta} \mathbf{J}_{inv})$.

Thus, we have

$$\mathbf{J}_{inv}^T \boldsymbol{\tau} = {}^0\mathbf{w}_r + \mathbf{J}_{inv}^T \boldsymbol{\tau}_{ta} + \mathbf{J}_r^T \boldsymbol{\tau}_{td}. \tag{8.157}$$

Note that the matrix $\mathbf{J}_r$ defined in (8.147) is the matrix that allows expressing the passive joint velocities $\dot{\mathbf{q}}_d$ as a function of the platform twist ${}^0\mathbf{t}_r$ only. This can be proven as follows.

Let us recall that, from (7.91), we have

$$\mathbf{J}_t \, {}^0\mathbf{t}_r - \mathbf{J}_{ta} \dot{\mathbf{q}}_a = \mathbf{J}_{td} \dot{\mathbf{q}}_d. \tag{8.158}$$

From (7.64), we also have

$$\dot{\mathbf{q}}_a = \mathbf{J}_{inv}{}^0\mathbf{t}_r. \tag{8.159}$$

Introducing (8.159) into (8.88), we obtain

$$\mathbf{J}_{td}\dot{\mathbf{q}}_d = \mathbf{J}_t{}^0\mathbf{t}_r - \mathbf{J}_{ta}\mathbf{J}_{inv}{}^0\mathbf{t}_r = (\mathbf{J}_t - \mathbf{J}_{ta}\mathbf{J}_{inv})^0\mathbf{t}_r \tag{8.160}$$

which leads to

$$\dot{\mathbf{q}}_d = \mathbf{J}_{td}^{-1}(\mathbf{J}_t - \mathbf{J}_{ta}\mathbf{J}_{inv})^0\mathbf{t}_r = \mathbf{J}_r{}^0\mathbf{t}_r. \tag{8.161}$$

8.5.2 Direct Dynamic Model

Contrary to the *DDM* of the *PKM* without redundancy for which the input joint accelerations $\ddot{\mathbf{q}}_a$ is expressed as a function of the input efforts $\boldsymbol{\tau}$, for the *PKM* with actuation redundancy, the *DDM* expresses the platform acceleration ${}^0\dot{\mathbf{t}}_r$ as a function of the input efforts $\boldsymbol{\tau}$. It is given by

$$\boxed{{}^0\dot{\mathbf{t}}_r = \mathbf{M}_{red}^{-1}(\mathbf{x})\left(\mathbf{J}_{inv}^T\boldsymbol{\tau} - \mathbf{c}_{red}(\mathbf{x}, {}^0\mathbf{t}_r)\right)} \tag{8.162}$$

where

$$\boxed{\mathbf{M}_{red}(\mathbf{x}) = \left[\mathbf{J}_{inv}^T \; \mathbf{J}_r^T\right]\mathbf{E}_\tau^T\mathbf{M}_t\mathbf{E}_\tau\begin{bmatrix}\mathbf{J}_{inv}\\ \mathbf{J}_r\end{bmatrix} + \boldsymbol{\Psi}_t^T\,{}^0\mathbf{M}_p\,\boldsymbol{\Psi}_t} \tag{8.163}$$

is the inertia matrix of the *PKM* with actuation redundancy, and

$$\boxed{\mathbf{c}_{red}(\mathbf{x}, {}^0\mathbf{t}_r) = \left[\mathbf{J}_{inv}^T \; \mathbf{J}_r^T\right]\mathbf{E}_\tau^T\left(\mathbf{c}_t + \mathbf{M}_t\mathbf{E}_\tau\begin{bmatrix}\mathbf{a}_q\\ \mathbf{a}_r\end{bmatrix}\right) + \boldsymbol{\Psi}_t^T\left({}^0\mathbf{c}_p + {}^0\mathbf{M}_p\dot{\boldsymbol{\Psi}}_t\,{}^0\mathbf{t}_r\right)} \tag{8.164}$$

is the vector of Coriolis, centrifugal, gravity and external effects of the *PKM* with actuation redundancy, in which

$$\mathbf{a}_r = \mathbf{J}_{td}^{-1}(\mathbf{d}_c - \mathbf{J}_{ta}\mathbf{a}_q) \tag{8.165}$$

In these expressions,

- ${}^0\mathbf{M}_p$ is the inertia matrix of the free platform and ${}^0\mathbf{c}_p$ is its vector of Coriolis, centrifugal, gravity and external effects. They are defined in (8.97) and (8.98).
- $\mathbf{M}_t$ is the inertia matrix of the virtual tree structure and $\mathbf{c}_t$ is its vector of Coriolis, centrifugal, gravity and external effects. They are defined in (8.58).
- $\mathbf{E}_\tau$ is an ordering matrix defined at (8.61).
- $\mathbf{J}_r$ is defined in (8.147) and is the matrix that allows expressing the passive joint velocities $\dot{\mathbf{q}}_d$ as a function of the platform twist ${}^0\mathbf{t}_r$ only.

- $\boldsymbol{\Psi}_t$ is a $(6 \times n_{dof})$ matrix defining the independent coordinates of the platform twist (see Eq. (7.61)).
- $\mathbf{J}_{inv}$ is the *PKM* inverse kinematic Jacobian matrix defined at (7.64).
- $\mathbf{a}_q$, $\mathbf{J}_t$, $\mathbf{J}_{td}$ and $\mathbf{d}_c$ are defined in (7.164), (7.93), (7.92) and (7.182), respectively.

The present form of the *DDM* is quite unusual as it does not give the value of the acceleration of the active joint as a function of their position, velocity and input efforts, but *it gives the value of the platform acceleration as a function of the platform velocity, pose and input efforts*. All other variables can be obtained by solving the inverse geometric and kinematic problems presented in Chap. 7.

Proof

To obtain the *DDM* of the *PKM* with actuation redundancy, we first need to develop the *IDM* equations in order to obtain an expression under the form:

$$\mathbf{J}_{inv}^T \boldsymbol{\tau} = \mathbf{M}_{red}(\mathbf{x})^0\dot{\mathbf{t}}_r + \mathbf{c}_{red}(\mathbf{x}, {}^0\mathbf{t}_r). \tag{8.166}$$

So, starting from (8.71), let us decompose the expression into two sub-terms $\boldsymbol{\tau}_1$ and $\boldsymbol{\tau}_2$, such as

$$\boldsymbol{\tau} = \boldsymbol{\tau}_1 + \boldsymbol{\tau}_2 \tag{8.167}$$

with

$$\boldsymbol{\tau}_1 = \left[\mathbf{J}_{inv}^T \ \mathbf{J}_r^T\right] \mathbf{E}_\tau^T \boldsymbol{\tau}_t \tag{8.168}$$

and

$$\boldsymbol{\tau}_2 = {}^0\mathbf{w}_r. \tag{8.169}$$

Let us first concentrate on the term $\boldsymbol{\tau}_1$. Introducing (8.58) into (8.101) leads to

$$\boldsymbol{\tau}_1 = \left[\mathbf{J}_{inv}^T \ \mathbf{J}_r^T\right] \mathbf{E}_\tau^T (\mathbf{M}_t \ddot{\mathbf{q}}_t + \mathbf{c}_t). \tag{8.170}$$

Then, from (8.63), we deduce that

$$\boldsymbol{\tau}_1 = \left[\mathbf{J}_{inv}^T \ \mathbf{J}_r^T\right] \mathbf{E}_\tau^T (\mathbf{M}_t \mathbf{E}_\tau \begin{bmatrix} \ddot{\mathbf{q}}_a \\ \ddot{\mathbf{q}}_d \end{bmatrix} + \mathbf{c}_t). \tag{8.171}$$

Equation (7.183) states that

$$\ddot{\mathbf{q}}_d = \mathbf{J}_{td}^{-1}(\mathbf{J}_t \, {}^0\dot{\mathbf{t}}_r - \mathbf{J}_{ta}\ddot{\mathbf{q}}_a + \mathbf{d}_c) \tag{8.172}$$

which can be expanded by introducing (7.164) into it as

$$\begin{aligned} \ddot{\mathbf{q}}_d &= \mathbf{J}_{td}^{-1}(\mathbf{J}_t \, {}^0\dot{\mathbf{t}}_r - \mathbf{J}_{ta}(\mathbf{J}_{inv} {}^0\dot{\mathbf{t}}_r + \mathbf{a}_q) + \mathbf{d}_c) \\ &= \mathbf{J}_{td}^{-1}(\mathbf{J}_t - \mathbf{J}_{ta}\mathbf{J}_{inv})^0\dot{\mathbf{t}}_r + \mathbf{J}_{td}^{-1}(\mathbf{d}_c - \mathbf{J}_{ta}\mathbf{a}_q). \end{aligned} \tag{8.173}$$

Then, from (8.73), we deduce that

$$\ddot{\mathbf{q}}_d = \mathbf{J}_r{}^0\dot{\mathbf{t}}_r + \mathbf{a}_r \tag{8.174}$$

with $\mathbf{a}_r = \mathbf{J}_{td}^{-1}(\mathbf{d}_c - \mathbf{J}_{ta}\mathbf{a}_q)$.

Introducing (7.164) and (8.174) into (8.171), we get

$$\begin{aligned}\boldsymbol{\tau}_1 &= [\mathbf{J}_{inv}^T\ \mathbf{J}_r^T]\,\mathbf{E}_\tau^T\left(\mathbf{M}_t\mathbf{E}_\tau\left(\begin{bmatrix}\mathbf{J}_{inv}\\ \mathbf{J}_r\end{bmatrix}{}^0\dot{\mathbf{t}}_r + \begin{bmatrix}\mathbf{a}_q\\ \mathbf{a}_r\end{bmatrix}\right) + \mathbf{c}_t\right)\\ &= [\mathbf{J}_{inv}^T\ \mathbf{J}_r^T]\,\mathbf{E}_\tau^T\mathbf{M}_t\mathbf{E}_\tau\begin{bmatrix}\mathbf{J}_{inv}\\ \mathbf{J}_r\end{bmatrix}{}^0\dot{\mathbf{t}}_r + [\mathbf{J}_{inv}^T\ \mathbf{J}_r^T]\,\mathbf{E}_\tau^T\left(\mathbf{M}_t\mathbf{E}_\tau\begin{bmatrix}\mathbf{a}_q\\ \mathbf{a}_r\end{bmatrix} + \mathbf{c}_t\right).\end{aligned} \tag{8.175}$$

Let us now consider the term $\boldsymbol{\tau}_2$. Introducing (8.65)first and then (8.57) into (8.169) leads to

$$\boldsymbol{\tau}_2 = {}^0\boldsymbol{\Psi}_t^T\,{}^0\mathbf{w}_p = \boldsymbol{\Psi}_t^T\,({}^0\mathbf{M}_p{}^0\dot{\mathbf{t}}_p + {}^0\mathbf{c}_p). \tag{8.176}$$

Then, from (7.61) and (7.163), we have

$${}^0\dot{\mathbf{t}}_p = \boldsymbol{\Psi}_t\,{}^0\dot{\mathbf{t}}_r + \dot{\boldsymbol{\Psi}}_t\,{}^0\mathbf{t}_r \tag{8.177}$$

which can be introduced into (8.176) in order to obtain

$$\boldsymbol{\tau}_2 = \boldsymbol{\Psi}_t^T\,({}^0\mathbf{M}_p\boldsymbol{\Psi}_t\,{}^0\dot{\mathbf{t}}_r + {}^0\mathbf{M}_p\dot{\boldsymbol{\Psi}}_t\,{}^0\mathbf{t}_r + {}^0\mathbf{c}_p) \tag{8.178}$$

or also

$$\boldsymbol{\tau}_2 = \boldsymbol{\Psi}_t^T\,{}^0\mathbf{M}_p\boldsymbol{\Psi}_t\,{}^0\dot{\mathbf{t}}_r + \boldsymbol{\Psi}_t^T\left({}^0\mathbf{M}_p\dot{\boldsymbol{\Psi}}_t\,{}^0\mathbf{t}_r + {}^0\mathbf{c}_p\right). \tag{8.179}$$

By summing (8.175) and (8.179), we obtain

$$\mathbf{J}_{inv}^T\boldsymbol{\tau} = \mathbf{M}_{red}{}^0\dot{\mathbf{t}}_r + \mathbf{c}_{red} \tag{8.180}$$

with

$$\mathbf{M}_{red} = [\mathbf{J}_{inv}^T\ \mathbf{J}_r^T]\,\mathbf{E}_\tau^T\mathbf{M}_t\mathbf{E}_\tau\begin{bmatrix}\mathbf{J}_{inv}\\ \mathbf{J}_r\end{bmatrix} + \boldsymbol{\Psi}_t^T\,{}^0\mathbf{M}_p\boldsymbol{\Psi}_t \tag{8.181}$$

and

$$\mathbf{c}_{red} = [\mathbf{J}_{inv}^T\ \mathbf{J}_r^T]\,\mathbf{E}_\tau^T\left(\mathbf{M}_t\mathbf{E}_\tau\begin{bmatrix}\mathbf{a}_q\\ \mathbf{a}_r\end{bmatrix} + \mathbf{c}_t\right) + \left({}^0\mathbf{M}_p\dot{\boldsymbol{\Psi}}_t\,{}^0\mathbf{t}_r + {}^0\mathbf{c}_p\right). \tag{8.182}$$

Finally, the *DDM* in (8.162) can be obtained by solving Eq. (8.180).

8.5.3 Example: The DualV

8.5.3.1 Description of the DualV

The DualV (Fig. 8.15) is a prototype of a planar parallel robot with actuation redundancy developed at the LIRMM (van der Wijk et al. 2011). This robot has 3 controlled *dof* (two translations in the plane (xOy) and one rotation around the z axis) but 4 identical legs, with one actuator by leg. Thus, its degree of redundancy is equal to 1. Each leg is composed of one proximal and one distal link. The proximal link A_iB_i is attached to the base by one actuated revolute joint and to the distal link B_iC_i by one passive revolute joint. The distal link is also attached to the moving platform by one passive revolute joint.

The geometric parameters of the virtual open-loop tree structure are described in Table 8.2 using the *MDH* notation (in this table, $\gamma_{11} = 15.52°$, $\gamma_{21} = 164.48°$, $\gamma_{31} = -164.48°$ and $\gamma_{41} = -15.52°$). The platform and payload are considered as supplementary bodies, the payload being fixed on the platform. They are respectively numbered as bodies $\mathcal{B}_4$ and $\mathcal{B}_5$.

(a) (b)

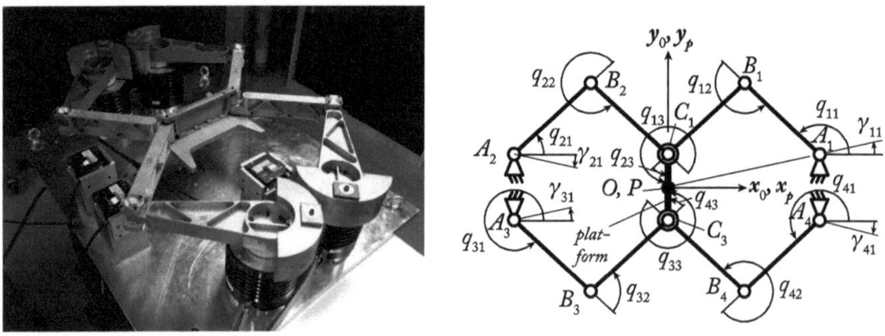

Fig. 8.15 The DualV. **a** The prototype (Courtesy of Sébastien Krut). **b** Kinematic description: in that picture, the DualV configuration is such that the base frame (x_oOy_o) coincides with the platform frame (x_pPy_p)

Table 8.2 MDH parameters for the frames corresponding to ith robot leg ($i = 1, \ldots, 4$)

ij	a_{ij}	μ_{ij}	σ_{ij}	γ_{ij}	α_{ij}	d_{ij}	θ_{ij}	r_{ij}
$i1$	0	1	0	γ_{i1}	0	$d_{i1} = l_{OA_i} = 0.41$ m	$q_{i1} - \gamma_{i1}$	0
$i2$	$i1$	0	0	0	0	$d_{i2} = l_{A_iB_i} = 0.28$ m	q_{i2}	0
$i3$	$i2$	0	0	0	0	$d_{i3} = l_{B_iC_i} = 0.28$ m	q_{i3}	0
p	$i3$	0	2	0	0	$r_p/2 = 0.11$ m	θ_{pi}	0

"p" denotes the platform frame.
$\theta_{p1} = \theta_{p2} = \pi/2, \theta_{p3} = \theta_{p4} = -\pi/2$

The DualV is actuated by four ETEL RTMB0140-100 direct drive actuators, which can deliver maximal torques of 127 Nm. The robot is able to achieve accelerations of 25 G in its workspace. The current amplifier can provide directly the measure of the input torque produced by the actuator.

8.5.3.2 Forward Kinematic Model of the Real Parallel Robot

For the DualV, the loop-closure Eqs. (7.3) can be written in the base frame $\mathscr{F}_0$ as (for $i = 1 \dots 4$):

$$\begin{aligned} 0 &= x - r_p \sin(\phi + k\pi) - x_{A_i} - d_{i2} \cos q_{i1} - d_{i3} \cos(q_{i1} + q_{i2}) \\ 0 &= y + r_p \cos(\phi + k\pi) - y_{A_i} - d_{i2} \sin q_{i1} - d_{i3} \sin(q_{i1} + q_{i2}) \\ 0 &= \phi + \bar{k}\pi - q_{i1} - q_{i2} - q_{i3} \end{aligned} \tag{8.183}$$

where

- x, y and ϕ are the platform coordinates (x and y are the position of the platform center, ϕ the platform orientation defined as the angle between the axes $\boldsymbol{x}_0$ and $\boldsymbol{x}_p$),
- r_p is the half platform length ($r_p = l_{C_1C_3}/2$), $k = 0$ ($\bar{k} = 1$) if $i = 1, 2$, $k = 1$ ($\bar{k} = 0$) if $i = 3, 4$, and
- x_{A_i} and y_{A_i} are the position coordinates along $\boldsymbol{x}_0$ and $\boldsymbol{y}_0$ axes for the point A_i.

From the two first lines of (8.183), the reduced loop-closure Eq. (7.4) that directly relate the displacements of the actuated joints to the moving platform coordinates can be obtained after deleting from these expressions the terms in $\cos q_{i2}$ or $\sin q_{i2}$ (for $i = 1 \dots 4$):

$$d_{i3}^2 = \left(x_{C_i} - x_{B_i}\right)^2 + \left(y_{C_i} - y_{B_i}\right)^2 \tag{8.184}$$

where $x_{C_i} = x - r_p \sin(\phi + k\pi)$ and $y_{C_i} = y + r_p \cos(\phi + k\pi)$ are the position coordinates along $\boldsymbol{x}_0$ and $\boldsymbol{y}_0$ axes for the point C_i, and $x_{B_i} = x_{A_i} + d_{i2} \cos q_{i1}$ and $y_{B_i} = y_{A_i} + d_{i2} \sin q_{i1}$ are the position coordinates along $\boldsymbol{x}_0$ and $\boldsymbol{y}_0$ axes for the point B_i.

Noticing that

$$\begin{aligned} x &= \frac{x_{C_1} + x_{C_3}}{2}, \quad y = \frac{y_{C_1} + y_{C_3}}{2} \\ \phi &= \tan^{-1}\left(-\frac{x_{C_1} - x_{C_3}}{y_{C_1} - y_{C_3}}\right) \end{aligned} \tag{8.185}$$

the *FGM* can be solved by finding the expressions of x_{C_j} and y_{C_j} ($j = 1, 3$) as a function of q_{i1} ($i = 1 \dots 4$). These expressions are quite simple to find as the loops formed by the legs 1 and 2, or the legs 3 and 4, are five-bar linkages. Thus, the expressions of x_{C_j} and y_{C_j} as a function of q_{i1} are (for $j = 1$ or $j = 3$):

$$x_{C_j} = f_j y_{C_j} + k_j, \quad y_{C_j} = \frac{-p_j \pm \sqrt{p_j^2 - 4g_j r_j}}{2g_j} \tag{8.186}$$

where

$$\begin{aligned} f_j &= -\frac{y_{B_{j+1}} - y_{B_j}}{x_{B_{j+1}} - x_{B_j}}, \quad g_j = f_j^2 + 1 \\ k_j &= \frac{x_{B_j}^2 + y_{B_j}^2 - x_{B_{j+1}}^2 - y_{B_{j+1}}^2}{2(x_{B_{j+1}} - x_{B_j})} \\ p_j &= 2f_j(k_j - x_{B_j}) - 2y_{B_j} \\ r_j &= x_{B_j}^2 + y_{B_j}^2 - d_{i3}^2 + k_j^2 - 2k_j x_{B_j}. \end{aligned} \tag{8.187}$$

In (8.186), the sign "±" denotes the two robot assembly modes, that are considered *a priori* fixed in the identification process as no parallel singularities are crossed.

Then, it comes easily from (8.183) that:

$$q_{i2} = \text{atan2}\left(y_{C_i} - y_{B_i}, x_{C_i} - x_{B_i}\right) - q_{i1}, \quad q_{i3} = \phi + \bar{k}\pi - q_{i1} - q_{i2}. \tag{8.188}$$

Then, differentiating (8.183) and (8.184) with respect to time, and simplifying, the matrices $\mathbf{A}_r$ and $\mathbf{B}$ of (7.62) can be found:

$$\mathbf{A}|_i = d_{i3}\left[\mathrm{c}_{12i} \;\; \mathrm{s}_{12i} \;\; -\sin(\phi + k\pi)\mathrm{s}_{12i} - \cos(\phi + k\pi)\mathrm{c}_{12i}\right] \tag{8.189}$$

where $\mathbf{A}|_i$ is the ith row of $\mathbf{A}_r$, $\mathrm{c}_{12i} = \cos(q_{i1} + q_{i2})$ and $\mathrm{s}_{12i} = \sin(q_{i1} + q_{i2})$,

$$b_{ii} = d_{i2} d_{i3} \sin q_{i2} \tag{8.190}$$

where b_{ii} is the ith term of the diagonal matrix $\mathbf{B}$.

Differentiating (8.183) w.r.t. time, we get

$$\begin{bmatrix} 1 & 0 & -r_p\cos(\phi + k\pi) \\ 0 & 1 & -r_p\sin(\phi + k\pi) \\ 0 & 0 & 1 \end{bmatrix} \begin{bmatrix} \dot{x} \\ \dot{y} \\ \dot{\phi} \end{bmatrix} - \begin{bmatrix} -d_{i2}\sin q_{i1} \\ d_{i2}\cos q_{i1} \\ 1 \end{bmatrix} \dot{q}_{i1} = \begin{bmatrix} -d_{i3}\mathrm{s}_{12i} & 0 \\ d_{i3}\mathrm{c}_{12i} & 0 \\ 1 & 1 \end{bmatrix} \begin{bmatrix} \dot{q}_{i2} \\ \dot{q}_{i3} \end{bmatrix}. \tag{8.191}$$

Now projecting (8.191) into the frame $\mathcal{F}_{i2}$ of the link $\mathcal{B}_{i2}$, we obtain

$$\begin{bmatrix} \mathrm{c}_{12i} & \mathrm{s}_{12i} & jt_{13i} \\ -\mathrm{s}_{12i} & \mathrm{c}_{12i} & jt_{23i} \\ 0 & 0 & 1 \end{bmatrix} \begin{bmatrix} \dot{x} \\ \dot{y} \\ \dot{\phi} \end{bmatrix} - \begin{bmatrix} d_{i2}\sin q_{i2} \\ d_{i2}\cos q_{i2} \\ 1 \end{bmatrix} \dot{q}_{i1} = \begin{bmatrix} 0 & 0 \\ d_{i3} & 0 \\ 1 & 1 \end{bmatrix} \begin{bmatrix} \dot{q}_{i2} \\ \dot{q}_{i3} \end{bmatrix} \tag{8.192}$$

where $jt_{13i} = -\mathrm{c}_{12i} r_p \cos(\phi + k\pi) - \mathrm{s}_{12i} r_p \sin(\phi + k\pi)$ and $jt_{23i} = \mathrm{s}_{12i} r_p \cos(\phi + k\pi) - \mathrm{c}_{12i} r_p \sin(\phi + k\pi)$.

Noticing that the first row of (8.192) does not depend on the passive joint velocities $\dot{q}_{i2}$ and $\dot{q}_{i3}$, the relation (7.87) can be identified as

$$\mathbf{J}_{t_i}^{c}\,{}^{0}\mathbf{t}_r - \mathbf{J}_{tai}\dot{\mathbf{q}}_{ai} = \mathbf{J}_{tdi}\dot{\mathbf{q}}_{di} \tag{8.193}$$

where ${}^{0}\mathbf{t}_r = \left[\dot{x}\ \dot{y}\ \dot{\phi}\right]^T$, $\dot{\mathbf{q}}_{ai} = \dot{q}_{i1}$, $\dot{\mathbf{q}}_{di}^T = \left[\dot{q}_{i2}\ \dot{q}_{i3}\right]$,

$$\mathbf{J}_{t_i}^{c} = \begin{bmatrix} -\mathrm{s}_{12i} & \mathrm{c}_{12i} & jt_{23i} \\ 0 & 0 & 1 \end{bmatrix} \tag{8.194}$$

$$\mathbf{J}_{tai} = -\begin{bmatrix} d_{i2}\cos q_{i2} \\ 1 \end{bmatrix} \tag{8.195}$$

and

$$\mathbf{J}_{tdi} = \begin{bmatrix} d_{i3} & 0 \\ 1 & 1 \end{bmatrix}. \tag{8.196}$$

Now, considering all legs, we have

$$\mathbf{J}_t\,{}^{0}\mathbf{t}_r - \mathbf{J}_{ta}\dot{\mathbf{q}}_a = \mathbf{J}_{td}\dot{\mathbf{q}}_d \tag{8.197}$$

with

$$\mathbf{J}_t = \begin{bmatrix} \mathbf{J}_{t_1}^{c} \\ \mathbf{J}_{t_2}^{c} \\ \mathbf{J}_{t_3}^{c} \\ \mathbf{J}_{t_4}^{c} \end{bmatrix} \tag{8.198}$$

an (8 × 3) matrix,

$$\mathbf{J}_{ta} = \begin{bmatrix} \mathbf{J}_{ta1} & \mathbf{0}_{2\times1} & \mathbf{0}_{2\times1} & \mathbf{0}_{2\times1} \\ \mathbf{0}_{2\times1} & \mathbf{J}_{ta2} & \mathbf{0}_{2\times1} & \mathbf{0}_{2\times1} \\ \mathbf{0}_{2\times1} & \mathbf{0}_{2\times1} & \mathbf{J}_{ta3} & \mathbf{0}_{2\times1} \\ \mathbf{0}_{2\times1} & \mathbf{0}_{2\times1} & \mathbf{0}_{2\times1} & \mathbf{J}_{ta4} \end{bmatrix} \tag{8.199}$$

an (8 × 4) matrix, and

$$\mathbf{J}_{td} = \begin{bmatrix} \mathbf{J}_{td1} & \mathbf{0}_{2\times2} & \mathbf{0}_{2\times2} & \mathbf{0}_{2\times2} \\ \mathbf{0}_{2\times2} & \mathbf{J}_{td2} & \mathbf{0}_{2\times2} & \mathbf{0}_{2\times2} \\ \mathbf{0}_{2\times2} & \mathbf{0}_{2\times2} & \mathbf{J}_{td3} & \mathbf{0}_{2\times2} \\ \mathbf{0}_{2\times2} & \mathbf{0}_{2\times2} & \mathbf{0}_{2\times2} & \mathbf{J}_{td4} \end{bmatrix} \tag{8.200}$$

an (8 × 8) matrix.

Then, all joint velocities can be computed from (7.62) and (8.88) as a function ${}^{0}\mathbf{t}_r$ or $\dot{\mathbf{q}}_a$.

Finally, the accelerations can be computed from (7.161) and (7.180) using the previous expressions and their derivatives w.r.t. time.

8.5.3.3 Computation of the *IDM*

The inverse dynamic model of the open-loop virtual structure can be obtained by noticing that each leg is indeed a planar 3*R* robot in which the last body is massless. Its inverse dynamic model is then equal to (Gautier et al. 1994):

$$\begin{aligned}
\tau_{t_{i1}} &= zz_{i1R}\ddot{q}_{i1} + zz_{i2}(\ddot{q}_{i1} + \ddot{q}_{i2}) \\
&+ d_{i2}mx_{i2}\left((2\ddot{q}_{i1} + \ddot{q}_{i2})\cos q_{i2} - \dot{q}_{i2}(2\dot{q}_{i1} + \dot{q}_{i2})\sin q_{i2}\right) \\
&+ d_{i2}my_{i2}\left((2\ddot{q}_{i1} + \ddot{q}_{i2})\sin q_{i2} + \dot{q}_{i2}(2\dot{q}_{i1} + \dot{q}_{i2})\cos q_{i2}\right) \\
&+ fs_{i1}\mathrm{sign}(\dot{q}_{i1}) + fv_{i1}\dot{q}_{i1} \qquad (8.201) \\
\tau_{t_{i2}} &= zz_{i2}(\ddot{q}_{i1} + \ddot{q}_{i2}) + d_{i2}mx_{i2}\left(\ddot{q}_{i1}\cos q_{i2} + \dot{q}_{i1}^2 \sin q_{i2}\right) \\
&+ d_{i2}my_{i2}\left(\ddot{q}_{i1}\sin q_{i2} - \dot{q}_{i1}^2\cos q_{i2}\right) \\
&+ fs_{i2}\mathrm{sign}(\dot{q}_{i2}) + fv_{i2}\dot{q}_{i2} \\
\tau_{t_{i3}} &= fs_{i3}\mathrm{sign}(\dot{q}_{i3}) + fv_{i3}\dot{q}_{i3}
\end{aligned}$$

where $zz_{i1R} = \left(zz_{i1} + Ia_{i1} + d_{i2}^2 m_{i2}\right)$ $\tau_{t_{i1}}$ is the torque of the virtual actuator located at point A_i, $\tau_{t_{i2}}$ is the torque of the virtual actuator located at point B_i, and $\tau_{t_{i3}}$ is the torque of the virtual actuator located at point C_i.

The dynamic model of the free body corresponding to the platform with the payload in the virtual system is equal to ${}^0\mathbf{w}_r = [w_1\ w_2\ w_3]^T$ with:

$$\begin{aligned}
w_1 &= (m_4 + m_5)\ddot{x} - (mx_4 + mx_5)(\ddot{\phi}\sin\phi + \dot{\phi}^2\cos\phi) \\
&+ (my_4 + my_5)(-\ddot{\phi}\cos\phi + \dot{\phi}^2\sin\phi) \qquad (8.202)
\end{aligned}$$

$$\begin{aligned}
w_2 &= (m_4 + m_5)\ddot{y} + (mx_4 + mx_5)(\ddot{\phi}\cos\phi - \dot{\phi}^2\sin\phi) \\
&- (my_4 + my_5)(\ddot{\phi}\sin\phi + \dot{\phi}^2\cos\phi) \qquad (8.203)
\end{aligned}$$

$$\begin{aligned}
w_3 &= (zz_4 + zz_5)\ddot{\phi} + (mx_4 + mx_5)(\ddot{y}\cos\phi - \ddot{x}\sin\phi) \\
&- (my_4 + my_5)(\ddot{y}\sin\phi + \ddot{x}\cos\phi). \qquad (8.204)
\end{aligned}$$

Following the procedure described in Sect. 8.5.1, the *IDM* of the DualV is given from Eq. (8.145) by:

$$\mathbf{J}_{inv}^T\,\boldsymbol{\tau} = {}^0\mathbf{w}_r + \mathbf{J}_{inv}^T\boldsymbol{\tau}_{ta} + \mathbf{J}_r^T\,\boldsymbol{\tau}_{td} \qquad (8.205)$$

with $\boldsymbol{\tau}_{ta}^T = [\tau_{t11}\ \tau_{t21}\ \tau_{t31}\ \tau_{t41}]$, $\boldsymbol{\tau}_{td}^T = [\tau_{t12}\ \tau_{t13}\ \tau_{t22}\ \tau_{t23}\ \tau_{t32}\ \tau_{t33}\ \tau_{t42}\ \tau_{t43}]$ and $\mathbf{J}_{inv} = -\mathbf{B}^{-1}\mathbf{A}_r$.

Here, we give the expression of the *IDM* only. We leave the computation of the *DDM* as training for the interested reader.

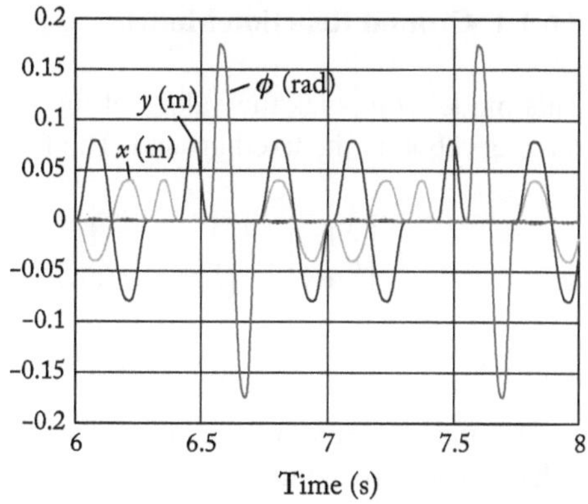

Fig. 8.16 Trajectory for validating the model of the DualV

8.5.3.4 Experimental Validation of the Model:

The dynamic parameters of the DualV have been estimated using identification procedures detailed in (Briot and Gautier 2014; Gautier et al. 2013). Their values are:

- $zz_{i1R} = 4.92 \cdot 10^{-2}$ kg m^2, $fv_{i1} = 0$ N m/s, $fs_{i1} = 0$ N m,
- $zz_{i2} = 2.23 \cdot 10^{-5}$ kg m^2, $mx_{i2} = 0.012$ kg m, $my_{i2} = 0$ kg m, $fv_{i2} = 0$ N m/s, $fs_{i2} = 0$ N m,
- $fv_{i3} = 0$ N m/s, $fs_{i3} = 0$ N m,
- $m_4 = 1.92$ kg, $zz_4 = 2.06 \cdot 10^{-2}$ kg m^2, $mx_4 = 0$ kg m, $my_4 = 0$ kg m,
- $m_5 = 5.42$ kg, $zz_5 = 1.61 \cdot 10^{-2}$ kg m^2, $mx_5 = -1.27 \cdot 10^{-1}$ kg m, $my_5 = 0$ kg m.

So now, we make the robot move on the trajectory shown in Fig. 8.16.

Then we compare the robot input torques with those predicted by the model. The comparison is shown in Fig. 8.17. It can be shown that the error between the prediction and the measure is very small: thus the model is accurate.

8.6 Other Models

8.6.1 Computation of the Ground Reactions of PKM

The *GRM* which computes the ground reactions as a function of the robot's configuration, velocity and acceleration is less known but can be used to identify its dynamic parameters (Ayusawa et al. 2008; Raucent et al. 1992) or for design purpose in shaking force and shaking moment balancing (Briot and Arakelian 2009; Briot et al. 2012a; Foucault and Gosselin 2004). An efficient way to obtain it is described thereafter.

8.6.1.1 Ground Reaction Model

This model computes the values of the force ${}^0\mathbf{f}_0$ and moment ${}^0\mathbf{m}_0$ exerted by the moving robot on the fixed ground. The force ${}^0\mathbf{f}_0$ and moment ${}^0\mathbf{m}_0$ are equal to:

$$\begin{bmatrix} {}^0\mathbf{f}_0 \\ {}^0\mathbf{m}_0 \end{bmatrix} = \begin{bmatrix} {}^0\mathbf{f}_p \\ {}^0\mathbf{m}_p + {}^0\mathbf{r}_{OP} \times {}^0\mathbf{f}_p \end{bmatrix} + \sum_{i=1}^{n} {}^{i1}\overline{\mathbf{T}}_0^T \begin{bmatrix} {}^{i1}\mathbf{f}_{i1} \\ {}^{i1}\mathbf{m}_{i1} \end{bmatrix} \tag{8.206}$$

where

- ${}^{i1}\mathbf{f}_{i1}$ and ${}^{i1}\mathbf{m}_{i1}$ are the reaction force and moments of the base on the first link of leg i. They are defined at (8.16),
- ${}^0\mathbf{m}_p$ and ${}^0\mathbf{f}_p$ are the components of the wrench ${}^0\mathbf{w}_p$ given in (8.57), and
- ${}^0\overline{\mathbf{T}}_{i1}$ is the screw transformation matrix defined at (3.13).

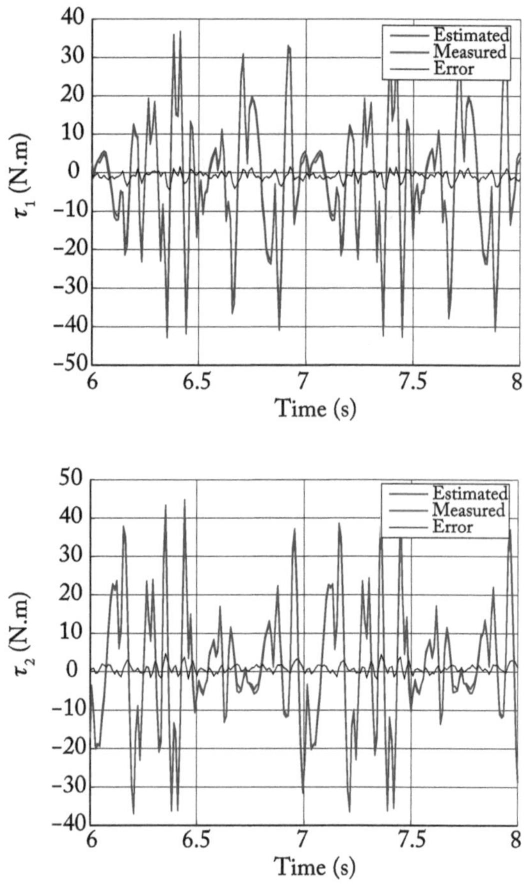

Fig. 8.17 Measured (*red lines*) and estimated (*blue lines*) input torques rebuilt using the identified parameters of the DualV

Fig. 8.17 (continued)

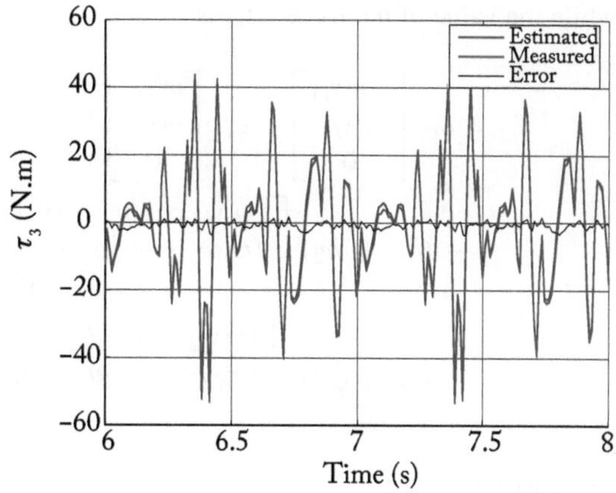

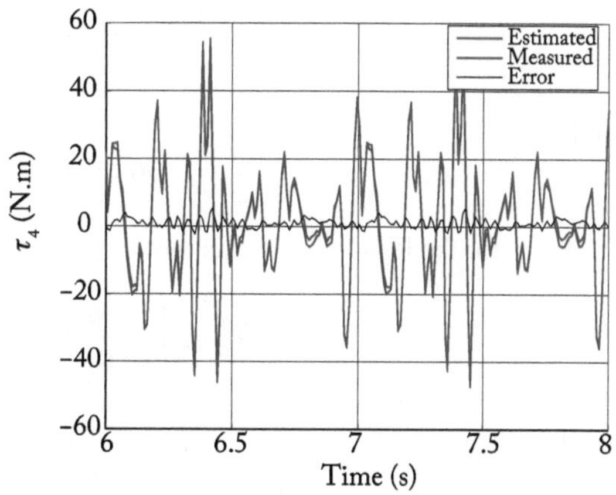

Thus, its expression can be recursively computed by using the algorithm given in Sect. 8.2.1.

8.6.1.2 Example: The Five-Bar Mechanism

For the five-bar mechanism of the Sect. 8.4.3.1, the value of the moment ${}^{i1}\mathbf{m}_{i1}$ is

$$ {}^{i1}\mathbf{m}_{i1} = \begin{bmatrix} 0 \ 0 \ m_{zi} \end{bmatrix} \tag{8.207} $$

with

$$ \begin{aligned} m_{zi} &= (zz_{i1} + d_{i2}^2 m_{i2})\ddot{q}_{i1} + zz_{i2}(\ddot{q}_{i1} + \ddot{q}_{i2}) \\ &\quad + d_{i2} mx_{i2} \left((2\ddot{q}_{i1} + \ddot{q}_{i2}) \cos q_{i2} - \dot{q}_{i2}(2\dot{q}_{i1} + \dot{q}_{i2}) \sin q_{i2}\right) \\ &\quad + d_{i2} my_{i2} \left((2\ddot{q}_{i1} + \ddot{q}_{i2}) \sin q_{i2} + \dot{q}_{i2}(2\dot{q}_{i1} + \dot{q}_{i2}) \cos q_{i2}\right) \end{aligned} \tag{8.208} $$

while the value of the force ${}^{i1}\mathbf{f}_{i1}$ is

$$
\begin{aligned}
{}^{i1}\mathbf{f}_{i1} = \ddot{q}_{i1} \begin{bmatrix} -my_{i1} \\ mx_{i1} \\ 0 \end{bmatrix} - \dot{q}_{i1}^2 \begin{bmatrix} mx_{i1} \\ my_{i1} \\ 0 \end{bmatrix} \\
+ (\ddot{q}_{i1} + \ddot{q}_{i2}) \begin{bmatrix} -my_{i2}\cos q_{i2} - \sin q_{i2}(mx_{i2} + m_{i2}d_{i2}) \\ (mx_{i2} + m_{i2}d_{i2})\cos q_{i2} - my_{i2}\sin q_{i2} \\ 0 \end{bmatrix} \\
- (\dot{q}_{i1} + \dot{q}_{i2})^2 \begin{bmatrix} (mx_{i2} + m_{i2}d_{i2})\cos q_{i2} - my_{i2}\sin q_{i2} \\ (mx_{i2} + m_{i2}d_{i2})\sin q_{i2} + my_{i2}\cos q_{i2} \\ 0 \end{bmatrix}.
\end{aligned}
\tag{8.209}
$$

Moreover, we have

$$
{}^{0}\mathbf{f}_p = m_p \begin{bmatrix} \ddot{x} \\ \ddot{y} \\ 0 \end{bmatrix} \tag{8.210}
$$

and

$$
{}^{0}\mathbf{m}_p = \mathbf{0}. \tag{8.211}
$$

Then, for the leg i, ${}^{0}\overline{\mathbf{T}}_{i1}$ is equal to

$$
{}^{i1}\overline{\mathbf{T}}_0^T = \begin{bmatrix} {}^{0}\mathbf{R}_{i1} & \mathbf{0}_3 \\ -{}^{0}\mathbf{R}_i{}^{i1}\hat{\mathbf{r}}_{i1} & {}^{0}\mathbf{R}_{i1} \end{bmatrix} \tag{8.212}
$$

with

$$
{}^{0}\mathbf{R}_{i1} = \begin{bmatrix} \cos q_{i1} & -\sin q_{i1} & 0 \\ \sin q_{i1} & \cos q_{i1} & 0 \\ 0 & 0 & 1 \end{bmatrix} \tag{8.213}
$$

and

$$
\mathbf{r}_{i1} = k \begin{bmatrix} d_{i1} \\ 0 \\ 0 \end{bmatrix} \tag{8.214}
$$

where $k = -1$ if $i = 1$, $k = +1$ if $i = 2$.

Then, the ground reaction model of the five-bar mechanism can be obtained from

$$
\begin{bmatrix} {}^{0}\mathbf{f}_0 \\ {}^{0}\mathbf{m}_0 \end{bmatrix} = \begin{bmatrix} {}^{0}\mathbf{f}_p \\ {}^{0}\mathbf{m}_p + {}^{0}\mathbf{r}_{OP} \times {}^{0}\mathbf{f}_p \end{bmatrix} + \sum_{i=1}^{n} {}^{i1}\overline{\mathbf{T}}_0^T \begin{bmatrix} {}^{i1}\mathbf{f}_{i1} \\ {}^{i1}\mathbf{m}_{i1} \end{bmatrix} \tag{8.215}
$$

8.6.2 Energy Models of PKM

Energy models can be used to identify robot's dynamic parameters (Gautier 1997; Gautier and Briot 2013) or for design purpose or trajectory planning to reduce robot's energy consumption (Ur-Rehman et al. 2009). An efficient way to obtain them is described thereafter.

8.6.2.1 Calculation of the Energy

From (6.8), the kinetic energy of the body $\mathcal{B}_{ij}$ is given by

$$
\begin{aligned}
E_{ij} &= \frac{1}{2}\begin{bmatrix} {}^{ij}\mathbf{v}_{ij}^T & {}^{ij}\boldsymbol{\omega}_{ij}^T \end{bmatrix} \begin{bmatrix} m_{ij}\mathbf{1}_3 & {}^{ij}\widehat{\mathbf{ms}}_{ij}^T \\ {}^{ij}\widehat{\mathbf{ms}}_{ij} & {}^{ij}\mathbf{I}_{O_{ij}} \end{bmatrix} \begin{bmatrix} {}^{ij}\mathbf{v}_{ij} \\ {}^{ij}\boldsymbol{\omega}_{ij} \end{bmatrix} \\
&= \frac{1}{2} {}^{ij}\mathbf{t}_{ij}^T \, {}^{ij}\mathbf{M}_{ij} \, {}^{ij}\mathbf{t}_{ij} \qquad (8.216) \\
&= \frac{1}{2}(m_{ij} \, {}^{ij}\mathbf{v}_{ij}^T \, {}^{ij}\mathbf{v}_{ij} + {}^{ij}\boldsymbol{\omega}_{ij}^T \, {}^{ij}\mathbf{I}_{O_{ij}} \, {}^{ij}\boldsymbol{\omega}_{ij} + 2^{ij}\mathbf{ms}_{ij}^T({}^{ij}\mathbf{v}_{ij} \times {}^{ij}\boldsymbol{\omega}_{ij})).
\end{aligned}
$$

Moreover, the kinetic energy due to the inertia of the actuator k is

$$
E_k = \frac{1}{2} I a_k \dot{q}_k^2 \qquad (8.217)
$$

where $\dot{q}_k$ is the velocity of the actuator k.

As a result, the total kinetic energy of the parallel robot including the end-effector is

$$
\begin{aligned}
E &= E_p + \sum_{i=1}^{n}\sum_{j=1}^{m_i} E_{ij} + \sum_{k=1}^{n_a} E_k \\
&= \frac{1}{2}(m_p \, {}^p\mathbf{v}_p^T \, {}^p\mathbf{v}_p + {}^p\boldsymbol{\omega}_p^T \, {}^p\mathbf{I}_p \, {}^p\boldsymbol{\omega}_p + 2^p\mathbf{ms}_p^T({}^p\mathbf{v}_p \times {}^p\boldsymbol{\omega}_p) + \sum_{k=1}^{n_a} I a_k \dot{q}_k^2 \qquad (8.218) \\
&\quad + \sum_{i=1}^{n}\sum_{j=1}^{m_i}(m_{ij} \, {}^{ij}\mathbf{v}_{ij}^T \, {}^{ij}\mathbf{v}_{ij} + {}^{ij}\boldsymbol{\omega}_{ij}^T \, {}^{ij}\mathbf{I}_{O_{ij}} \, {}^{ij}\boldsymbol{\omega}_{ij} + 2^{ij}\mathbf{ms}_{ij}^T({}^{ij}\mathbf{v}_{ij} \times {}^{ij}\boldsymbol{\omega}_{ij}))).
\end{aligned}
$$

From (6.20), the potential energy of the body $\mathcal{B}_{ij}$ is given by

$$
U_{ij} = -\begin{bmatrix} {}^0\mathbf{g}^T & 0 \end{bmatrix} {}^0\mathbf{T}_{ij}(\mathbf{q}) \begin{bmatrix} {}^{ij}\mathbf{ms}_{ij} \\ m_{ij} \end{bmatrix}. \qquad (8.219)
$$

As a result, the total potential energy of the parallel robot including the end-effector is

$$U = U_p + \sum_{i=1}^{n}\sum_{j=1}^{m_i} U_{ij}$$
$$= -\left[{}^0\mathbf{g}^T\ 0\right]{}^0\mathbf{T}_p(\mathbf{q})\begin{bmatrix}{}^p\mathbf{ms}_p\\ m_p\end{bmatrix} - \sum_{i=1}^{n}\sum_{j=1}^{m_i}\left(\left[{}^0\mathbf{g}^T\ 0\right]{}^0\mathbf{T}_{ij}(\mathbf{q})\begin{bmatrix}{}^{ij}\mathbf{ms}_{ij}\\ m_{ij}\end{bmatrix}\right). \tag{8.220}$$

Expressions (8.218) and (8.220) are valid for *PKM* with or without redundancy and can be recursively computed by using the algorithm for the velocity calculation given in Sect. 8.2.1.

8.6.2.2 Example: The Five-Bar Mechanism

For the Five-bar mechanism of the Sect. 8.4.3.1, the link velocities can be obtained as

$${}^{i1}\mathbf{v}_{i1} = \mathbf{0},\ {}^{i1}\boldsymbol{\omega}_{i1} = \begin{bmatrix}0\\0\\ \dot{q}_{i1}\end{bmatrix} \tag{8.221}$$

$${}^{i1}\mathbf{v}_{i2} = d_{i2}\dot{q}_{i1}\begin{bmatrix}-\sin q_{i2}\\ \cos q_{i2}\\ 0\end{bmatrix},\ {}^{i2}\boldsymbol{\omega}_{i2} = \begin{bmatrix}0\\0\\ \dot{q}_{i1}+\dot{q}_{i2}\end{bmatrix} \tag{8.222}$$

while the end-effector velocity is

$${}^0\mathbf{v}_p = \begin{bmatrix}\dot{x}\\ \dot{y}\\ 0\end{bmatrix},\ {}^0\boldsymbol{\omega}_p = \begin{bmatrix}0\\0\\ \dot{q}_{21}+\dot{q}_{22}\end{bmatrix}. \tag{8.223}$$

As a result, the kinetic energy of the links $\mathscr{B}_{i1}$ and $\mathscr{B}_{i2}$ is

$$E_{i1} = \frac{1}{2}zz_{i1}\dot{q}_{i1}^2 \tag{8.224}$$

$$E_{i2} = \frac{1}{2}(m_{i2}d_{i2}^2\dot{q}_{i1}^2 + zz_{i2}(\dot{q}_{i1}+\dot{q}_{i2})^2 + 2d_{i2}\dot{q}_{i1}(\dot{q}_{i1}+\dot{q}_{i2})(mx_{i2}\cos q_{i2} + my_{i2}\sin q_{i2})) \tag{8.225}$$

and the kinetic energy of the end-effector is

$$E_p = \frac{1}{2}m_p(\dot{x}^2+\dot{y}^2). \tag{8.226}$$

The kinetic energy of the actuators is

$$E_{act} = \frac{1}{2}(Ia_{11}\dot{q}_{11}^2 + Ia_{21}\dot{q}_{21}^2). \tag{8.227}$$

Thus the total kinetic energy is $E = E_p + E_{act} + \sum_{i=1}^{2}(E_{i1} + E_{i2})$.

If the gravity field ${}^0\mathbf{g}$ is directed along z_0, so the potential energy is constant (which is the case for the example proposed in Sect. 8.4.3.1).

If the gravity field ${}^0\mathbf{g}$ is not directed along z_0 (for instance, ${}^0\mathbf{g}^T = [0\ g\ 0]$), so the potential energy of the links $\mathcal{B}_{i1}$ and $\mathcal{B}_{i2}$ is

$$U_{i1} = -(mx_{i1} \sin q_{i1} + my_{i1} \cos q_{i1})g \tag{8.228}$$

$$U_{i2} = -(mx_{i2} \sin(q_{i1} + q_{i2}) + my_{i2} \cos(q_{i1} + q_{i2}) + d_{i2} m_{i2} \sin q_{i1})g \tag{8.229}$$

and the potential energy of the end-effector is

$$U_p = -m_p y g. \tag{8.230}$$

Thus the total potential energy is $U = U_p + \sum_{i=1}^{2}(U_{i1} + U_{i2})$.

8.7 Computation of the Base Dynamic Parameters

In this section, we study the concept of base dynamic parameters or identifiable parameters. We develop a straightforward numerical method to determine them. These parameters constitute the minimum set of inertial parameters that are needed to compute the dynamic model of a robot (Mayeda et al. 1990). The use of the base dynamic parameters in a customized Newton-Euler algorithm reduces its computational cost (Khalil and Kleinfinger 1987; Khalil et al. 1986). The determination of the base parameters is also essential for identification of the dynamic parameters (Briot and Gautier 2014; Gautier 1991; Gautier and Khalil 1990; Khalil and Dombre 2002), since they constitute the only identifiable parameters. The base dynamic parameters can be deduced from standard parameters by eliminating those that have no effect on the dynamic model and by grouping some others.

The determination of the base parameters of tree structure robots and some particular closed-loop architectures can be obtained using symbolic algorithms (Khalil and Bennis 1995). For *PKM*, the numerical method developed in (Briot and Gautier 2014; Gautier 1991; Khalil and Guégan 2004) can be used for this purpose. This method is presented below.

8.7.1 Expressing the Dynamic Model Linearly as a Function of the Standard Dynamic Parameters

In this section, we will demonstrate that the energy of the robot, and also the *IDM*, is linear with respect to the dynamic parameters.

8.7.2 Linearity of the Energy w.r.t. the Inertial Parameters

Let us consider the energy of the body $\mathscr{B}_{ij}$. From (8.216), we can easily prove that:

$$\begin{aligned} E_{ij} &= \frac{1}{2}(m_{ij}\ {}^{ij}\mathbf{v}_{ij}^T\ {}^{ij}\mathbf{v}_{ij} + {}^{ij}\boldsymbol{\omega}_{ij}^T\ {}^{ij}\mathbf{I}_{O_{ij}}\ {}^{ij}\boldsymbol{\omega}_{ij} + 2^{ij}\mathbf{ms}_{ij}^T({}^{ij}\mathbf{v}_{ij} \times {}^{ij}\boldsymbol{\omega}_{ij})) \\ &= \mathbf{e}_{ij}\boldsymbol{\chi}_{ij} \end{aligned} \tag{8.231}$$

where $\boldsymbol{\chi}_{ij}$ is the vector of the link inertial parameters that have been defined in Sect. 6.1.2,

$$\boldsymbol{\chi}_{ij} = \begin{bmatrix} xx_{ij} & xy_{ij} & xz_{ij} & yy_{ij} & yz_{ij} & zz_{ij} & mx_{ij} & my_{ij} & mz_{ij} & m_{ij} \end{bmatrix}^T \tag{8.232}$$

and $\mathbf{e}_{ij}$ is a vector of function also called the Jacobian matrix of the kinetic energy w.r.t. the inertial parameters of the link $\mathscr{B}_{ij}$,

$$\begin{aligned} \mathbf{e}_{ij} &= \frac{\partial E_{ij}}{\partial \boldsymbol{\chi}_{ij}} \\ &= \begin{bmatrix} e_{xx_{ij}} & e_{xy_{ij}} & e_{xz_{ij}} & e_{yy_{ij}} & e_{yz_{ij}} & e_{zz_{ij}} & e_{mx_{ij}} & e_{my_{ij}} & e_{mz_{ij}} & e_{m_{ij}} \end{bmatrix} \end{aligned} \tag{8.233}$$

with

$$e_{xx_{ij}} = \frac{1}{2}\omega_{1ij}\omega_{1ij} \tag{8.234}$$

$$e_{xy_{ij}} = \omega_{1ij}\omega_{2ij} \tag{8.235}$$

$$e_{xz_{ij}} = \omega_{1ij}\omega_{3ij} \tag{8.236}$$

$$e_{yy_{ij}} = \frac{1}{2}\omega_{2ij}\omega_{2ij} \tag{8.237}$$

$$e_{yz_{ij}} = \omega_{2ij}\omega_{3ij} \tag{8.238}$$

$$e_{zz_{ij}} = \frac{1}{2}\omega_{3ij}\omega_{3ij} \tag{8.239}$$

$$e_{mx_{ij}} = \omega_{3ij}v_{2ij} - \omega_{2ij}v_{3ij} \tag{8.240}$$

$$e_{my_{ij}} = \omega_{1ij}v_{3ij} - \omega_{3ij}v_{1ij} \tag{8.241}$$

$$e_{mz_{ij}} = \omega_{2ij}v_{1ij} - \omega_{1ij}v_{2ij} \tag{8.242}$$

$$e_{m_{ij}} = \frac{1}{2}{}^{ij}\mathbf{v}_{ij}^T\,{}^{ij}\mathbf{v}_{ij} \tag{8.243}$$

in which ${}^{ij}\boldsymbol{\omega}_{ij}^T = [\omega_{1ij}\ \omega_{2ij}\ \omega_{3ij}]$ and ${}^{ij}\mathbf{v}_{ij}^T = [v_{1ij}\ v_{2ij}\ v_{3ij}]$.

As a result, the platform kinetic energy E_p can also be written as a function of its inertial parameters $\boldsymbol{\chi}_p$ such that

$$E_p = \mathbf{e}_p \boldsymbol{\chi}_p. \tag{8.244}$$

Finally, from (8.218), the total kinetic energy can be written as

$$\begin{aligned} E &= E_p + \sum_{i=1}^{n}\sum_{j=1}^{m_i} E_{ij} \\ &= \mathbf{e}_p \boldsymbol{\chi}_p + \sum_{i=1}^{n}\sum_{j=1}^{m_i} \mathbf{e}_{ij}\boldsymbol{\chi}_{ij} \\ &= \left[\mathbf{e}_p\ \mathbf{e}_{11}\ \mathbf{e}_{12}\ \ldots\ \mathbf{e}_{n,m_n}\right] \begin{bmatrix} \boldsymbol{\chi}_p \\ \boldsymbol{\chi}_t \end{bmatrix} \\ &= \left[\mathbf{e}_p\ \mathbf{e}_t\right] \begin{bmatrix} \boldsymbol{\chi}_p \\ \boldsymbol{\chi}_t \end{bmatrix} \end{aligned} \tag{8.245}$$

where

- $\mathbf{e}_t$ is the kinetic energy function of the tree-structure,
- $\boldsymbol{\chi}_t^T = [\boldsymbol{\chi}_{11}^T\ \boldsymbol{\chi}_{12}^T \cdots \boldsymbol{\chi}_{n,m_n}^T]$ is the vector of standard dynamic parameters of the tree structure.

Similarly, the potential energy of the body $\mathscr{B}_{ij}$ can be linearly written as a function of the vector of the link inertial parameters $\boldsymbol{\chi}_{ij}$. From (8.219), we can prove that:

$$\begin{aligned} U_{ij} &= -\left[{}^0\mathbf{g}^T\ 0\right] {}^0\mathbf{T}_{ij}(\mathbf{q}) \begin{bmatrix} {}^{ij}\mathbf{ms}_{ij} \\ m_{ij} \end{bmatrix} \\ &= \mathbf{u}_{ij}\boldsymbol{\chi}_{ij} \end{aligned} \tag{8.246}$$

where

$$\begin{aligned} \mathbf{u}_{ij} &= \frac{\partial U_{ij}}{\partial \boldsymbol{\chi}_{ij}} \\ &= \left[u_{xx_{ij}}\ u_{xy_{ij}}\ u_{xz_{ij}}\ u_{yy_{ij}}\ u_{yz_{ij}}\ u_{zz_{ij}}\ u_{mx_{ij}}\ u_{my_{ij}}\ u_{mz_{ij}}\ u_{m_{ij}}\right] \end{aligned} \tag{8.247}$$

with

$$u_{xx_{ij}} = u_{xy_{ij}} = u_{xz_{ij}} = u_{yy_{ij}} = u_{yz_{ij}} = u_{zz_{ij}} = 0 \tag{8.248}$$

$$u_{mx_{ij}} = -\left[{}^0\mathbf{g}^T\ 0\right] {}^0\tilde{\mathbf{s}}_{ij} \tag{8.249}$$

$$u_{my_{ij}} = -\left[{}^0\mathbf{g}^T\ 0\right] {}^0\tilde{\mathbf{n}}_{ij} \tag{8.250}$$

$$u_{mz_{ij}} = -\left[{}^0\mathbf{g}^T\ 0\right]{}^0\tilde{\mathbf{a}}_{ij} \tag{8.251}$$

$$u_{m_{ij}} = -\left[{}^0\mathbf{g}^T\ 0\right]{}^0\tilde{\mathbf{r}}_{ij} \tag{8.252}$$

in which

- ${}^0\tilde{\mathbf{s}}_{ij}$ is the first column of matrix ${}^0\mathbf{T}_{ij}$,
- ${}^0\tilde{\mathbf{n}}_{ij}$ is the second column of matrix ${}^0\mathbf{T}_{ij}$,
- ${}^0\tilde{\mathbf{a}}_{ij}$ is the third column of matrix ${}^0\mathbf{T}_{ij}$,
- ${}^0\tilde{\mathbf{r}}_{ij}$ is the fourth column of matrix ${}^0\mathbf{T}_{ij}$.

As a result, the platform potential energy U_p can also be written as a function of its inertial parameters $\boldsymbol{\chi}_p$ such that

$$U_p = \mathbf{u}_p \boldsymbol{\chi}_p. \tag{8.253}$$

Finally, from (8.220), the total kinetic energy can be written as

$$\begin{aligned} U &= U_p + \sum_{i=1}^{n}\sum_{j=1}^{m_i} U_{ij} \\ &= \mathbf{u}_p\boldsymbol{\chi}_p + \sum_{i=1}^{n}\sum_{j=1}^{m_i} \mathbf{u}_{ij}\boldsymbol{\chi}_{ij} \\ &= \left[\mathbf{u}_p\ \mathbf{u}_{11}\ \mathbf{u}_{12}\ \dots\ \mathbf{u}_{n,m_n}\right]\begin{bmatrix}\boldsymbol{\chi}_p \\ \boldsymbol{\chi}_t\end{bmatrix} \\ &= \left[\mathbf{u}_p\ \mathbf{u}_t\right]\begin{bmatrix}\boldsymbol{\chi}_p \\ \boldsymbol{\chi}_t\end{bmatrix} \end{aligned} \tag{8.254}$$

where $\mathbf{u}_t = \left[\mathbf{u}_{11}\ \mathbf{u}_{12}\ \dots\ \mathbf{u}_{n,m_n}\right]$ is the potential energy function of the tree-structure.

As a result, the Lagrangian of the system can be linearly written as a function of the robot inertial parameters $\boldsymbol{\chi}_{st}$ such that

$$\begin{aligned} L &= E - U \\ &= \left[\mathbf{e}_p\ \mathbf{e}_t\right]\begin{bmatrix}\boldsymbol{\chi}_p \\ \boldsymbol{\chi}_t\end{bmatrix} - \left[\mathbf{u}_p\ \mathbf{u}_t\right]\begin{bmatrix}\boldsymbol{\chi}_p \\ \boldsymbol{\chi}_t\end{bmatrix} \\ &= \left[\mathbf{e}_p - \mathbf{u}_p\ \mathbf{e}_t - \mathbf{u}_t\right]\begin{bmatrix}\boldsymbol{\chi}_p \\ \boldsymbol{\chi}_t\end{bmatrix} \\ &= \left[\mathbf{l}_p\ \mathbf{l}_t\right]\begin{bmatrix}\boldsymbol{\chi}_p \\ \boldsymbol{\chi}_t\end{bmatrix} \end{aligned} \tag{8.255}$$

where $\mathbf{l}_p = \mathbf{e}_p - \mathbf{u}_p$ is the Lagrangian function of the platform and $\mathbf{l}_t = \mathbf{e}_t - \mathbf{u}_t$ is the Lagrangian function of the tree-structure.

8.7.3 Linearity of the IDM *w.r.t. the Dynamic Parameters*

From (8.59), (8.60) and (8.255), we can see that the input efforts of the virtual tree-structure can be linearly written as a function of the robot inertial parameters:

$$\begin{aligned}\boldsymbol{\tau}_{ta} &= \frac{\mathrm{d}}{\mathrm{d}t}\left(\frac{\partial L_t}{\partial \dot{\mathbf{q}}_a}\right)^T - \left(\frac{\partial L_t}{\partial \mathbf{q}_a}\right)^T \\ &= \frac{\mathrm{d}}{\mathrm{d}t}\left(\frac{\partial \mathbf{l}_t \boldsymbol{\chi}_t}{\partial \dot{\mathbf{q}}_a}\right)^T - \left(\frac{\partial \mathbf{l}_t \boldsymbol{\chi}_t}{\partial \mathbf{q}_a}\right)^T \\ &= \frac{\mathrm{d}}{\mathrm{d}t}\left(\frac{\partial \mathbf{l}_t}{\partial \dot{\mathbf{q}}_a}\right)^T \boldsymbol{\chi}_t - \left(\frac{\partial \mathbf{l}_t}{\partial \mathbf{q}_a}\right)^T \boldsymbol{\chi}_t \\ &= \left(\frac{\mathrm{d}}{\mathrm{d}t}\left(\frac{\partial \mathbf{l}_t}{\partial \dot{\mathbf{q}}_a}\right)^T - \left(\frac{\partial \mathbf{l}_t}{\partial \mathbf{q}_a}\right)^T\right)\boldsymbol{\chi}_t \\ &= \mathbf{D}_a \boldsymbol{\chi}_t \end{aligned} \tag{8.256}$$

where

$$\mathbf{D}_a = \left[\mathbf{D}_a^{11}\ \mathbf{D}_a^{12} \ldots \mathbf{D}_a^{n,m_n}\right]. \tag{8.257}$$

Thus,

$$\begin{aligned}\boldsymbol{\tau}_{ta} &= \left[\mathbf{D}_a^{11}\ \mathbf{D}_a^{12} \ldots \mathbf{D}_a^{n,m_n}\right]\boldsymbol{\chi}_t \\ &= \sum_{i=1}^{n}\sum_{j=1}^{m_i} \mathbf{D}_a^{ij}\boldsymbol{\chi}_{ij}. \end{aligned} \tag{8.258}$$

Similarly,

$$\begin{aligned}\boldsymbol{\tau}_{td} &= \mathbf{D}_d \boldsymbol{\chi}_{st} \\ &= \left[\mathbf{D}_d^{11}\ \mathbf{D}_d^{12} \ldots \mathbf{D}_d^{n,m_n}\right]\boldsymbol{\chi}_t \\ &= \sum_{i=1}^{n}\sum_{j=1}^{m_i} \mathbf{D}_d^{ij}\boldsymbol{\chi}_{ij} \end{aligned} \tag{8.259}$$

where

$$\mathbf{D}_d = \left[\mathbf{D}_d^{11}\ \mathbf{D}_d^{12} \ldots \mathbf{D}_d^{n,m_n}\right]. \tag{8.260}$$

In these expressions,

- $\mathbf{D}_a^{ij}$ is the Jacobian matrix of the effort $\boldsymbol{\tau}_{ta}$ w.r.t. the inertial parameters $\boldsymbol{\chi}_{ij}$ of the body $\mathscr{B}_{ij}$,
- $\mathbf{D}_d^{ij}$ is the Jacobian matrix of the effort $\boldsymbol{\tau}_{td}$ w.r.t. the inertial parameters $\boldsymbol{\chi}_{ij}$ of the body $\mathscr{B}_{ij}$.

Previous expressions do not take into account the inertia of the robot actuators and the friction effects into the joints (see Sects. 8.2.2 and 8.2.3). Let us recall that, for the joint ij, when the inertia of the robot actuators and the friction effects are taken into account, the input effort becomes:

$$\tau_{t_{ij}}^{n} = \tau_{t_{ij}} + Ia_{ij}\ddot{q}_{ij} + fs_{ij}\mathrm{sign}(\dot{q}_{ij}) + fv_{ij}\dot{q}_{ij} \tag{8.261}$$

where $\tau_{t_{ij}}$ is the component of $\boldsymbol{\tau}_{ta}$ or $\boldsymbol{\tau}_{td}$ defined at (8.258) and (8.259) corresponding to the joint ij.

As a result, $\boldsymbol{\tau}_{ta}$ and $\boldsymbol{\tau}_{td}$ can be rewritten as

$$\begin{aligned}\boldsymbol{\tau}_{ta} &= \sum_{i=1}^{n}\sum_{j=1}^{m_i} \mathbf{D}_a^{ij}\boldsymbol{\chi}_{ij} + \mathrm{diag}(\ddot{\mathbf{q}}_a)\mathbf{ia}^a + \mathrm{diag}(\dot{\mathbf{q}}_a)\mathbf{fv}^a + \mathrm{diag}(\mathrm{sign}(\dot{\mathbf{q}}_a))\mathbf{fs}^a \\ &= \sum_{i=1}^{n}\sum_{j=1}^{m_i} \mathbf{D}_a^{ij\,st}\boldsymbol{\chi}_{ij}^{st} \\ &= \mathbf{D}_a^{tot}\boldsymbol{\chi}_t^{tot}\end{aligned} \tag{8.262}$$

with

- $\mathbf{D}_a^{ij\,st} = \begin{bmatrix}\mathbf{D}_a^{ij} & \ddot{q}_{ij} & \dot{q}_{ij} & \mathrm{sign}(\dot{q}_{ij})\end{bmatrix}$
- $\mathbf{D}_a^{tot} = \begin{bmatrix}\mathbf{D}_a^{11\,st} & \mathbf{D}_a^{12\,st} & \ldots & \mathbf{D}_a^{n,m_n\,st}\end{bmatrix}$.

Moreover,

$$\begin{aligned}\boldsymbol{\tau}_{td} &= \sum_{i=1}^{n}\sum_{j=1}^{m_i} \mathbf{D}_d^{ij}\boldsymbol{\chi}_{ij} + \mathrm{diag}(\ddot{\mathbf{q}}_d)\mathbf{ia}^d + \mathrm{diag}(\dot{\mathbf{q}}_d)\mathbf{fv}^a + \mathrm{diag}(\mathrm{sign}(\dot{\mathbf{q}}_d))\mathbf{fs}^a \\ &= \sum_{i=1}^{n}\sum_{j=1}^{m_i} \mathbf{D}_d^{ij\,st}\boldsymbol{\chi}_{ij}^{st} \\ &= \mathbf{D}_d^{tot}\boldsymbol{\chi}_t^{tot}\end{aligned} \tag{8.263}$$

with

- $\mathbf{D}_d^{ij\,st} = \begin{bmatrix}\mathbf{D}_d^{ij} & \ddot{q}_{ij} & \dot{q}_{ij} & \mathrm{sign}(\dot{q}_{ij})\end{bmatrix}$
- $\mathbf{D}_d^{tot} = \begin{bmatrix}\mathbf{D}_d^{11\,st} & \mathbf{D}_d^{12\,st} & \ldots & \mathbf{D}_d^{n,m_n\,st}\end{bmatrix}$.

In all the previous expressions

- for a vector $\mathbf{s} = [s_1\ s_2 \ldots s_n]^T$, $\mathrm{diag}(\mathbf{s}) = \begin{bmatrix} s_1 & 0 & \ldots & 0 \\ 0 & s_2 & \ldots & 0 \\ \vdots & \vdots & \ddots & \vdots \\ 0 & 0 & \ldots & s_n \end{bmatrix}$,

- $\mathbf{ia}^a$ ($\mathbf{ia}^d$) is a $(n_a \times 1)$ vector ($(n_d \times 1)$, resp.) grouping the inertia of the actuators of the virtual structure in the joints corresponding to the active (passive, resp.) joints of the real robot,
- $\mathbf{fv}^a$ ($\mathbf{fv}^d$) is a $(n_a \times 1)$ vector ($(n_d \times 1)$, resp.) grouping the viscous friction terms of the virtual structure in the joints corresponding to the active (passive, resp.) joints of the real robot,
- $\mathbf{fs}^a$ ($\mathbf{fs}^d$) is a $(n_a \times 1)$ vector ($(n_d \times 1)$, resp.) grouping the Coulomb friction terms of the virtual structure in the joints corresponding to the active (passive, resp.) joints of the real robot,
- $\boldsymbol{\chi}_t^{tot\,T} = [\boldsymbol{\chi}_{11}^{st\,T}\ \boldsymbol{\chi}_{12}^{st\,T} \ldots \boldsymbol{\chi}_{n,m_n}^{st\,T}]$,
- $\boldsymbol{\chi}_{ij}^{st\,T} = [xx_{ij}\ xy_{ij}\ xz_{ij}\ yy_{ij}\ yz_{ij}\ zz_{ij}\ mx_{ij}\ my_{ij}\ mz_{ij}\ m_{ij}\ Ia_{ij}\ fv_{ij}\ fs_{ij}]$ is called the vector of standard dynamic parameters of the link $\mathscr{B}_{ij}$,
- $\mathbf{D}_a^{ij\,st}$ the Jacobian matrix of the effort $\boldsymbol{\tau}_{ta}$ w.r.t. the standard dynamic parameters $\boldsymbol{\chi}_{ij}^{st}$ of the body $\mathscr{B}_{ij}$,
- $\mathbf{D}_d^{ij\,st}$ the Jacobian matrix of the effort $\boldsymbol{\tau}_{td}$ w.r.t. the standard dynamic parameters $\boldsymbol{\chi}_{ij}^{st}$ of the body $\mathscr{B}_{ij}$.

In a fully similar way to what is written above, the vector ${}^0\mathbf{w}_r$ of (8.71) can be written linearly as a function of the end-effector inertial parameters $\boldsymbol{\chi}_p$ such that

$$ {}^0\mathbf{w}_r = \mathbf{D}_p \boldsymbol{\chi}_p. \tag{8.264} $$

Finally, from (8.71), (8.262) and (8.263), the dynamic model of the real robot can be linearly written w.r.t. the standard dynamic parameters $\boldsymbol{\chi}_{tot}$ such as:

$$ \begin{aligned} \boldsymbol{\tau} &= \boldsymbol{\tau}_{ta} + \mathbf{J}^{T\,0}\mathbf{w}_r + \mathbf{J}_d^T \boldsymbol{\tau}_{td} \\ &= \mathbf{D}_a^{tot}\boldsymbol{\chi}_t^{tot} + \mathbf{J}^T\mathbf{D}_p\boldsymbol{\chi}_p + \mathbf{J}_d^T\mathbf{D}_d^{tot}\boldsymbol{\chi}_t^{tot} \\ &= \mathbf{J}^T\mathbf{D}_p\boldsymbol{\chi}_p + \left(\mathbf{D}_a^{tot} + \mathbf{J}_d^T\mathbf{D}_d^{tot}\right)\boldsymbol{\chi}_t^{tot} \\ &= \left[\mathbf{J}^T\mathbf{D}_p\ \left(\mathbf{D}_a^{tot} + \mathbf{J}_d^T\mathbf{D}_d^{tot}\right)\right]\begin{bmatrix}\boldsymbol{\chi}_p \\ \boldsymbol{\chi}_t^{tot}\end{bmatrix} \\ &= \mathbf{D}_{tot}(\mathbf{q}_a, \dot{\mathbf{q}}_a, \ddot{\mathbf{q}}_a)\boldsymbol{\chi}_{tot} \end{aligned} \tag{8.265} $$

where $\mathbf{D}_{tot}(\mathbf{q}_a, \dot{\mathbf{q}}_a, \ddot{\mathbf{q}}_a) = \left[\mathbf{J}^T\mathbf{D}_p\ \left(\mathbf{D}_a^{tot} + \mathbf{J}_d^T\mathbf{D}_d^{tot}\right)\right]$ and $\boldsymbol{\chi}_{tot}^T = \left[\boldsymbol{\chi}_p^T\ \boldsymbol{\chi}_t^{tot\,T}\right]$.

This demonstration has been done for the *IDM* of the *PKM* without actuation redundancy, but similar relations can be obtained for *PKM* with actuation redundancy.

8.7.4 Numerical Method Based on a QR *Decomposition*

The symbolic approach of computing the base parameters is based on determining the independent elements of the *IDM* represented by the row matrix $\mathbf{D}_{tot}$ (Eq. (8.265)).

Numerically this problem is equivalent to study of the space spanned by the columns of a matrix $\mathbf{W}_{tot}$ formed from $\mathbf{D}_{tot}$ using r random values of $\mathbf{q}_a, \dot{\mathbf{q}}_a$ and $\ddot{\mathbf{q}}_a$. This study can be carried out using the singular value decomposition (*SVD*) or the *QR* decomposition of $\mathbf{W}_{tot}$ (Gautier 1991). In this section, we develop the numerical method that is based on the *QR* decomposition of a matrix $\mathbf{W}_{tot}$, which is derived from the *IDM*.

To determine the base parameters, we construct a matrix $\mathbf{W}_{tot}$ by calculating the rows of $\mathbf{D}_{tot}$ for r random values of joint positions, velocities and acceleration $\mathbf{q}_a^{(i)}$, $\dot{\mathbf{q}}_a^{(i)}$ and $\ddot{\mathbf{q}}_a^{(i)}$ (the upperscript (i) denotes the ith random value of the concerned vector $\mathbf{q}_a$, $\dot{\mathbf{q}}_a$ or $\ddot{\mathbf{q}}_a$) for $r >> \dim \boldsymbol{\chi}_{tot} = n_{st}$ and $i = 1, \ldots, r$. The matrix $\mathbf{W}_{tot}$ with dimension $((r \times n_a) \times n_{st})$:

$$\mathbf{W}_{tot} = \begin{bmatrix} \mathbf{W}_{tot}^{(1)} \\ \vdots \\ \mathbf{W}_{tot}^{(n_a)} \end{bmatrix} \tag{8.266}$$

where

$$\mathbf{W}_{tot}^{(j)} = \begin{bmatrix} \mathbf{d}_{tot}^{(j)}(\mathbf{q}_a^{(1)}, \dot{\mathbf{q}}_a^{(1)}, \ddot{\mathbf{q}}_a^{(1)}) \\ \mathbf{d}_{tot}^{(j)}(\mathbf{q}_a^{(2)}, \dot{\mathbf{q}}_a^{(2)}, \ddot{\mathbf{q}}_a^{(2)}) \\ \vdots \\ \mathbf{d}_{tot}^{(j)}(\mathbf{q}_a^{(r)}, \dot{\mathbf{q}}_a^{(r)}, \ddot{\mathbf{q}}_a^{(r)}) \end{bmatrix} \tag{8.267}$$

in which $\mathbf{d}_{tot}^{(j)}$ is the jth line of the matrix $\mathbf{D}_{tot}$.

An inertial parameter has no effect on the dynamic model if the elements of its corresponding column in $\mathbf{W}_{tot}$ are all equal to zero. By eliminating such parameters and the corresponding columns, the matrix $\mathbf{W}_{tot}$ is reduced to a matrix $\mathbf{W}_{tot}^r$ with c columns and r rows. However, some columns in $\mathbf{W}_{tot}^r$ are linearly dependent and their corresponding parameters can be grouped in order to obtain a set of n_b base parameters ($n_b \leq n_{st}$), i.e. the base parameters are those corresponding to n_b independent columns of $\mathbf{W}_{tot}^r$ (and thus of $\mathbf{W}_{tot}$).

Application of the foregoing statements can be achieved by the use of the *QR* decomposition of $\mathbf{W}_{tot}^r$, which is given by:

$$\mathbf{Q}^T \mathbf{W}_{tot}^r = \begin{bmatrix} \mathbf{R} \\ \mathbf{0}_{(r-c)\times c} \end{bmatrix} \tag{8.268}$$

where $\mathbf{Q}$ is an $(r \times r)$ orthogonal matrix, $\mathbf{R}$ is a $(c \times c)$ upper-triangular matrix, and $\mathbf{0}_{(r-c)\times c}$ is the $((r-c) \times c)$ matrix of zeros.

Theoretically, the non-identifiable parameters are those whose corresponding elements on the diagonal of the matrix $\mathbf{R}$ are zero (Forsythe et al. 1977; Golub and van Loan 1983). Let α be the numerical zero which is different from 0 because of round-off errors (α can be chosen such that $\alpha = \varepsilon \max \left|R_{jj}\right|$, where ε is a small coefficient depending on the level of perturbations in $\mathbf{W}^r_{tot}$ (due to noise measurement—in the case of a model identification procedure—and/or error modeling) and max $\left|R_{jj}\right|$ is the largest diagonal absolute value of $\mathbf{R}$ (Gautier 1991).

If the absolute value $|R_{kk}|$ of the kth component located on the diagonal of $\mathbf{R}$ is inferior to α, the kth column $\mathbf{W}^r_{tot_k}$ of $\mathbf{W}^r_{tot}$ can be deleted because it depends on the other.

On the contrary, if $|R_{kk}| > \alpha$, then the corresponding column in $\mathbf{W}^r_{tot_k}$ is independent and constitutes a base of the space span by $\mathbf{W}^r_{tot_k}$ (and thus $\mathbf{W}_{tot_k}$). Let the n_b independent columns be collected in the matrix $\mathbf{W}_1$, and the corresponding parameters be collected in the vector $\boldsymbol{\chi}_1$. The other columns and parameters are represented by $\mathbf{W}_2$ and $\boldsymbol{\chi}_2$ respectively, such that:

$$\mathbf{W}^r_{tot}\boldsymbol{\chi} = \begin{bmatrix}\mathbf{W}_1 \ \mathbf{W}_2\end{bmatrix}\begin{bmatrix}\boldsymbol{\chi}_1\\ \boldsymbol{\chi}_2\end{bmatrix}. \tag{8.269}$$

The matrix $\mathbf{W}_2$ which groups the non independent columns of $\mathbf{W}^r_{tot}$ can be written in terms of $\mathbf{W}_1$ as:

$$\mathbf{W}_2 = \mathbf{W}_1\boldsymbol{\beta}. \tag{8.270}$$

Consequently:

$$\mathbf{W}^r_{tot}\boldsymbol{\chi} = \begin{bmatrix}\mathbf{W}_1 \ \mathbf{W}_2\end{bmatrix}\begin{bmatrix}\boldsymbol{\chi}_b\\ \mathbf{0}_{(n_{st}-n_b)\times 1}\end{bmatrix} \tag{8.271}$$

where $\boldsymbol{\chi}_b$ is the vector of the base parameter given by

$$\boldsymbol{\chi}_b = \boldsymbol{\chi}_1 + \boldsymbol{\beta}\boldsymbol{\chi}_2. \tag{8.272}$$

Thus, the matrix $\boldsymbol{\beta}$ allows to obtain the grouping equations of the parameters $\boldsymbol{\chi}_2$ with $\boldsymbol{\chi}_1$. The matrix $\boldsymbol{\beta}$, can be computed using (8.270) by:

$$\boldsymbol{\beta} = \mathbf{W}_1^{+}\mathbf{W}_2 \tag{8.273}$$

Because the QR algorithm starts from the last columns to the first of $\mathbf{W}_{tot}$, the $(n_{st} - n_b)$ standard parameters to delete are dependent on the ordering of the columns of that matrix. For serial robots, the matrix $\mathbf{W}_{tot}$ is build such that the columns with smaller indices are those corresponding to the links closest from the base. Thus, using the previous algorithm, the parameters with the smallest influence (those of the wrist) are eliminated from the base parameters.

For parallel robots, to take into account the symmetry in the leg dynamic parameters, it is preferable to order the columns of $\mathbf{W}_{tot}$ such that

$$\mathbf{W}_{tot} = \begin{bmatrix}\mathbf{W}^p_{tot} \ \mathbf{W}^{\chi_1,1:n}_{tot} \ \mathbf{W}^{\chi_2,1:n}_{tot} \ \ldots \ \mathbf{W}^{\chi_{n_{st_{leg}}},1:n}_{tot}\end{bmatrix} \tag{8.274}$$

where $n_{st_{leg}}$ is the number of standard parameters for one leg, matrix $\mathbf{W}_{tot}^{p}$ is the observation matrix corresponding to the platform inertial parameters and matrices $\mathbf{W}_{tot}^{\chi_k,1:n}$ concatenates the columns of matrix $\mathbf{W}_{tot}$ corresponding to the parameters χ_k that are *a priori* identical for the n robot legs. Then, $(n_{st} - n_b)$ columns of $\mathbf{W}_{tot}$ can be deleted using the previous approach based on the *QR* factorization to obtain a new observation matrix $\mathbf{W}_1$ associated with a set of symmetrical base parameters denoted as $\boldsymbol{\chi}_b$.

8.7.5 Examples

8.7.5.1 The Five-Bar Mechanism

For the five-bar mechanism presented in Sect. 8.4.3.1, and using the proposed procedure for computation of the base parameters, we get the following grouping relations for the parameters in $\boldsymbol{\chi}_b$:

$$\begin{aligned} zz_{i1R} &= zz_{i1} + Ia_{i1} - 0.24158\, mx_{i2} + 0.045369\, m_{i2} \\ zz_{i2R} &= zz_{i2} - 0.1878\, mx_{i2} \end{aligned} \tag{8.275}$$

for the legs $i = 1, 2$, and also

$$m_{3R} = m_3 + 5.3248\, mx_{12} + 5.3248\, mx_{22} \tag{8.276}$$

for the end-effector parameters.

8.7.5.2 The Orthoglide

For the Orthoglide presented in Sect. 8.4.3.2, and using the proposed procedure for computation of the base parameters, we get the following grouping relations for the parameters in $\boldsymbol{\chi}_b$:

$$m_{6R} = m_6 + 3.2258\, mx_{13} + m_{14} + 3.2258\, mx_{23} + m_{24} + 3.2258\, mx_{33} + m_{34} \tag{8.277}$$

for the end-effector,

$$\begin{aligned} m_{11R} &= m_{11} + Ia_{11} + m_{12} - 3.2258\, mx_{13} + m_{14} \\ m_{21R} &= m_{21} + Ia_{21} + m_{22} - 3.2258\, mx_{23} + m_{24} \\ m_{31R} &= m_{31} + m_{32} - 3.2258 mx_{33} + m_{34} \end{aligned} \tag{8.278}$$

for the actuated joints, and

$$
\begin{aligned}
zz_{i2R} &= zz_{i2} + yy_{i3} - 0.31\, mx_{i3} + yy_{i4} \\
my_{i2R} &= my_{i2} + mz_{i3} + mz_{i4} \\
fv_{i2R} &= fv_{i2} + fv_{i5} \\
fs_{i2R} &= fs_{i2} + fs_{i5} \\
xx_{i3R} &= xx_{i3} - yy_{i3} + 0.31 mx_{i3} \\
xz_{i3R} &= xz_{i3} - 0.31\, mz_{i4} \\
zz_{i3R} &= zz_{i3} - 0.31\, mx_{i3} \\
fv_{i3R} &= fv_{i3} + fv_{i4} \\
fs_{i3R} &= fs_{i3} + fs_{i4}
\end{aligned} \tag{8.279}
$$

for the other bodies of the legs.

As can be observed, the same grouping relations appear for each robot leg. There is a small difference concerning the grouped parameter m_{i1R}: as actuated joint 31 is vertical, contrary to joints 11 and 21 that are horizontal, this joint must support the gravity effects applied on the foot of leg 3 (link 31). This is not the case for joints 11 and 21. Thus, this is the reason why parameters m_{31R} and Ia_{31} are not grouped together.

Chapter 9
Analysis of the Degeneracy Conditions for the Dynamic Model of Parallel Robots

Abstract The conditions of degeneracy of the dynamic model of parallel robots have been little investigated. In this chapter, we present an exhaustive study of the conditions of degeneracy of the dynamics models of *PKM*. We will show that two types of singularity impact their dynamics: Type 2 singularities and *LPJTS* singularities. We define criteria, based on analyses of dynamic models, that allows a definition of optimal trajectories that can avoid the degeneracy of the dynamic model and make it possible to cross the singularities. Simulations and experiments show the efficiency of such types of trajectories.

9.1 Introduction

Parallel robots have been used increasingly for a few decades. This is due to their main advantages over serial counterparts that are: (i) higher intrinsic rigidity, (ii) larger payload-to-weight ratio, and (iii) higher velocity and acceleration capacities (Merlet 2006b). However, their main drawback is probably the presence of singularities in the workspace which divide their workspace into different aspects (each aspect corresponding to one or more assembly modes (Merlet 2006b)) and near which their performance are drastically reduced.

As already mentioned, various types of singularity exist. In general, singularities lead to two different types of phenomena (that can be combined at the same robot configuration):

1. The loss of the ability for the robot to move along one given direction (instantaneously or not): this is the case of the so-called *Type 1 singularities* (Gosselin and Angeles 1990) which correspond to workspace boundaries,
2. The gain of some uncontrollable robot motions (instantaneously or not): the so-called *Type 2 singularities* (Gosselin and Angeles 1990) and *constraint singularities* (Zlatanov et al. 2002) belong to this category. Another type of singularity, which is less known than the two previously mentioned ones, also belongs to this category: the *LPJTS* singularities presented in Sect. 7.5.2 which are due to the degeneracy of the leg passive joint twist system. Near these configurations, the robot stiffness is considerably decreased and the robot capabilities in terms of effort transmission deteriorates.

S. Briot and W. Khalil, *Dynamics of Parallel Robots*,
Mechanisms and Machine Science 35, DOI 10.1007/978-3-319-19788-3_9

It should be mentioned that, historically, the first designed parallel robots were made of quite simple legs (in terms of joint arrangement) and encountered only Type 1 and Type 2 singularities. However, due to the problem of the non homogeneity of the performance inside the robot workspace, designers have tried to propose mechanisms with more complex leg architectures but with better performance distribution all along the workspace, such as the decoupled robots (Carricato and Parenti Castelli 2002; Gogu 2004; Kong and Gosselin 2002) which are fully isotropic with regard to their input/output kinematic performance. However, the increased complexity of the leg arrangement has led to the appearance of other kinds of singularities, such as the *LPJTS* singularities. The best known examples of decoupled robots whose legs encounter *LPJTS* singularities are the Tripteron-like or Isoglide-like robots from three to six degrees of freedom (Carricato and Parenti-Castelli 2002; Gogu 2004; Gosselin 2009; Kong and Gosselin 2002, 2011b; Rizk et al. 2007; Seward and Bonev 2014).

In order to increase the workspace size, several approaches have been envisaged in the literature, such as:

- The design of parallel robots without singularities. This can be done by using the optimal design approach (Briot et al. 2010; Liu et al. 2006) or by creating fully-isotropic mechanisms (Carricato and Parenti-Castelli 2002; Gogu 2004; Kong and Gosselin 2002) (which have no Type 2 singularities but usually have *LPJTS* singularities). This solution is the most usual one, but it often leads to the design of robots with a small ratio workspace size/robot footprint or to the design of robot architectures with very low practicability.
- The use of redundancy (Kotlarski et al. 2010; Kurtz and Hayward 1992; Müller 2005; Yi et al. 1994) or the use of mechanisms with variable actuation modes (Arakelian et al. 2008; Rakotomanga et al. 2008). These mechanisms can change the actuation mode without adding additional actuators, but this change can only be carried out when the mechanism is stopped, thus increasing the time necessary to perform the task.
- Planning working mode changing trajectories. The main way to proceed is to cross a Type 1 singularity by reaching the workspace boundary and changing the leg configuration (Bourbonais et al. 2014). By changing the leg configuration, the singularity loci appearing in the workspace for the initial configuration disappear and are replaced by other singularity loci linked to the new leg configuration. Thus, the robot is able to access new workspace zones (Campos et al. 2010). It should be mentioned that:

  - Type 1 singularities are a special type of serial singularities (Conconi and Carricato 2009) due to the degeneracy of the leg twist systems including active joint twists.
  - For the moment, changing the leg configuration by crossing a Leg Passive Joint Twist System (*LPJTS*) singularity has not been investigated, even if this process could allow accessing new workspace zones.

- Planning assembly mode changing trajectories. A first way to do this is to bypass a cusp point (Zein et al. 2008). However, this solution is hardly practical for two main reasons: (*i*) it forces the mechanism to follow a particular trajectory, which can be very different from the desired one; (*ii*) only a few mechanisms have cusp points. A second solution is to go directly through a Type 2 singularity (Briot and Arakelian 2008; Hesselbach et al. 2004; Ider 2005).

The last two solutions (i.e. changing the leg configuration or changing the assembly mode) are promising, since they can considerably increase the workspace size of any parallel mechanism by using only trajectory planning approaches. However, as shown in (Briot and Arakelian 2008) for changing assembly modes by passing through the Type 2 singularities, a physical criterion based on the analysis of the degeneracy conditions of the *PKM IDM* must be respected. We will show below that another criterion must be satisfied if we need to cross the *LPJTS* singularities.

In the next section, the degeneracy conditions of the *IDM* of *PKM* are disclosed and optimal trajectory planning approaches through singularities are proposed.

9.2 Analysis of the Degeneracy Conditions of the *IDM* of *PKM*

In the following section, we will only focus on the analysis of the degeneracy conditions of the *IDM* of *PKM* without redundancy as *PKM* with redundancy are especially designed for avoiding the presence of singularities inside their workspace.

To assure clarity, let us rewrite here the equations of the dynamic model including the Lagrange multipliers developed in Sect. 8.4 and defined in Eqs. (8.82)–(8.84):

$$\boldsymbol{\tau} = \boldsymbol{\tau}_{ta} - \mathbf{B}^T \boldsymbol{\lambda}_1 - \mathbf{J}_{ta}^T \boldsymbol{\lambda}_2 \tag{9.1}$$

$${}^0\mathbf{w}_r = \mathbf{A}_r^T \boldsymbol{\lambda}_1 - \mathbf{J}_t^T \boldsymbol{\lambda}_2 \tag{9.2}$$

$$\boldsymbol{\tau}_{td} = \mathbf{J}_{td}^T \boldsymbol{\lambda}_2. \tag{9.3}$$

Analysing the Eqs. (9.1)–(9.3), we get:

- if the matrix $\mathbf{A}_r$ is rank deficient, which appears in Type 2 singularities (see Sect. 7.5.1), or
- if the matrix $\mathbf{J}_{td}$ is rank deficient, which appears in *LPJTS* singularities (see Sect. 7.5.2),

the Lagrange multipliers cannot be computed and the dynamic model degenerates.

The conditions of rank-deficiency of matrices $\mathbf{A}_r$ and $\mathbf{J}_{td}$ have been presented in Sect. 7.5. The impact of their degeneracy on the robot input efforts is disclosed thereafter.

9.2.1 Degeneracy Conditions of the IDM Due to the Matrix $\mathbf{A}_r$

As already mentioned in Sect. 7.5, the matrix $\mathbf{A}_r$ becomes rank-deficient if and only if the robot is in a *Type 2* singularity. In Type 2 singularities, at least one platform motion becomes uncontrollable. Moreover, Type 2 singularities separates the workspace aspects (Merlet 2006b) and prevent the robot from reaching all possible workspace configurations.

From Eq. (9.2), it can be deduced that, when matrix $\mathbf{A}_r$ becomes rank-deficient in Type 2 singularities, a non-null vector $\boldsymbol{\lambda}_1$ corresponding to a null value of ${}^0\mathbf{w}_r + \mathbf{J}_t^T \boldsymbol{\lambda}_2$ can exist. This also means that there is an infinity of solutions for $\boldsymbol{\lambda}_1$ and that the robot platform is not in equilibrium. Another consequence is that in the neighborhood of the Type 2 singularities, the active joint effort $\boldsymbol{\tau}$ may increase considerably as their expression is proportional to the inverse of the determinant of $\mathbf{A}_r$, which is close to zero in that area. Such singularity may thus lead to a breakdown of the mechanism (if the joints cannot support the load) or to the impossibility of tracking the desired trajectory due to the technological limitations in terms of maximal input efforts for the actuators.

9.2.2 Degeneracy Conditions of the IDM Due to the Matrix $\mathbf{J}_{td}$

For reason of simplicity, but without loss of generality, let us consider that the robot legs are made of serial chains. In that case, as explained previously (see Sect. 7.3.3), the matrix $\mathbf{J}_{td}$ is block-diagonal. As a result, $\mathbf{J}_{td}$ is rank-deficient if and only if at least one block $\mathbf{J}_{tdi}$ on the diagonal is rank deficient.

If the ith block $\mathbf{J}_{tdi}$ is rank-deficient (let us recall that $\mathbf{J}_{tdi}$ is the kinematic Jacobian matrix that relates the twist of the last link of the leg i to the passive joint velocities $\dot{\mathbf{q}}_i$ of the same leg), then the sub-chain composed of the passive joints of the leg i is in a singular configuration. Such kind of singularity has been described in Sect. 7.5.2 and is called in this book a *LPJTS* singularity (Leg Passive Joint Twist System singularity). In *LPJTS* singularities, at least one leg gets an internal and uncontrollable motion while the platform remains rigid. Moreover, *LPJTS* singularities separate the passive joint space aspects and thus prevent the leg from reaching all the possible joint configurations (Conconi and Carricato 2009).

As mentioned in the introduction, *LPJTS* singularities are encountered in numerous robot architectures among which the best known examples are probably the Tripteron-like or Isoglide-like robots (Carricato and Parenti-Castelli 2002; Gogu 2004; Gosselin 2009; Kong and Gosselin 2002, 2011b; Rizk et al. 2007; Seward and Bonev 2014).

From Eq. (8.85), it can be deduced that, when matrix $\mathbf{J}_{tdi}$ (and consequently the matrix $\mathbf{J}_{td}$) becomes rank-deficient in *LPJTS* singularities, there can be a non-null vector $\boldsymbol{\lambda}_2$ corresponding to a null value of $\boldsymbol{\tau}_{td}$. This also means that there is an infinity of solutions for $\boldsymbol{\lambda}_2$ and that the leg i is not in equilibrium. Another consequence is that in the neighborhood of the *LPJTS* singularities, the value of $\boldsymbol{\lambda}_2$, and as a result the active joint efforts $\boldsymbol{\tau}$, may increase considerably as its expression is proportional to the inverse of the determinant of $\mathbf{J}_{td}$, which is close to zero in that area. As for Type 2 singularities, a *LPJTS* singularity may thus lead to a breakdown of the mechanism (if the joints cannot support the load) or to the impossibility of tracking the desired trajectory due to the technological limitations in terms of maximal input efforts for the actuators.

9.3 Avoiding Infinite Input Efforts While Crossing Type 2 or *LPJTS* Singularities Thanks to an Optimal Trajectory Planning

In this section, conditions for avoiding infinite input efforts while approaching and crossing the Type 2 or *LPJTS* singularities are disclosed.

9.3.1 Optimal Trajectory Planning Through Type 2 Singularities

In this section, we consider that only the matrix $\mathbf{A}_r$ is not invertible. So, let us rewrite (9.2) such as

$$\mathbf{A}_r^T \boldsymbol{\lambda}_1 = \mathbf{w}_d \tag{9.4}$$

where $\mathbf{w}_d$ is equal to

$$\mathbf{w}_d = {}^0\mathbf{w}_r + \mathbf{J}_t^T \boldsymbol{\lambda}_2 = {}^0\mathbf{w}_r + \mathbf{J}_t^T \mathbf{J}_{td}^{-T} \boldsymbol{\tau}_{td}. \tag{9.5}$$

As previously explained, Eq. (9.4) represents the platform equilibrium so that the loops of the parallel robot can be closed. As a result, the term $\mathbf{w}_d$ represents the sum of:

- The inertial/gravitational effects and external efforts applied on the platform,
- The reactions applied by the legs on the robot platform.

Let us also express (9.1) again such as

$$\boldsymbol{\tau} = \mathbf{w}_b - \mathbf{B}^T \boldsymbol{\lambda}_1, \tag{9.6}$$

where $\mathbf{w}_b$ is equal to

$$\mathbf{w}_b = \boldsymbol{\tau}_{ta} - \mathbf{J}_{ta}^T \boldsymbol{\lambda}_2 = \boldsymbol{\tau}_{ta} - \mathbf{J}_{ta}^T \mathbf{J}_{td}^{-T} \boldsymbol{\tau}_{td}. \tag{9.7}$$

As a result, the dynamic model is parameterized by the system of equations:

$$\begin{cases} \boldsymbol{\tau} = \mathbf{w}_b - \mathbf{B}^T \boldsymbol{\lambda}_1 \\ \mathbf{A}_r^T \boldsymbol{\lambda}_1 = \mathbf{w}_d. \end{cases} \tag{9.8}$$

If $\mathbf{A}_r$ is rank deficient, so a non-null vector $\mathbf{t}_s$ exists, defined such as

$$\mathbf{A}_r \mathbf{t}_s = \mathbf{0} \Leftrightarrow \mathbf{t}_s^T \mathbf{A}_r^T = \mathbf{0}. \tag{9.9}$$

From (7.58), $\mathbf{t}_s$[1]

- is a twist reciprocal to all the wrenches defining the rows of the matrix $\mathbf{A}_r$ (see Eqs. (7.58) and (7.62)), and
- describes the uncontrollable motion of the platform inside the Type 2 singularity (Briot and Arakelian 2008; Merlet 2006b).

Multiplying the left side of (9.4) by $\mathbf{t}_s^T$, one obtains

$$\mathbf{t}_s^T \mathbf{A}_r^T \boldsymbol{\lambda}_1 = 0. \tag{9.10}$$

As a result, for the *IDM* to be consistent, the right-hand part of (9.4) must strictly follow the condition

$$\mathbf{t}_s^T \mathbf{w}_d = 0 \tag{9.11}$$

which involves that:
For avoiding infinite input efforts while crossing a Type 2 singularity, the sum of the wrenches applied on the platform by the legs, inertia/gravitational effects and external environment $\mathbf{w}_d$ must be reciprocal to the uncontrollable motion of the platform inside the singularity $\mathbf{t}_s$ (in other words, the power of these wrenches along the platform uncontrollable motion must be null).

This physical criterion was first provided in (Briot and Arakelian 2008) and can be satisfied through a proper robot trajectory planning.

However, to better understand the phenomenon, let us consider the five-bar mechanism depicted in Fig. 9.1. As already mentioned, the five-bar mechanism is a planar parallel mechanism composed of two actuators located at the revolute joints located at points A_{11} and A_{21} and three passive revolute joints at points A_{12}, A_{22} and $A_{13} \equiv A_{23} \equiv P$.

Considering that the mechanism is not moving and that the gravity effects are canceled. A force $\mathbf{f}$ is applied on the end-effector. An analysis of the effort transmission shows that the reactions in the passive joints located at points A_{12} and P (A_{22} and P, resp.) are collinear to the vector $\overrightarrow{A_{12}P}$ ($\overrightarrow{A_{22}P}$, resp.) for any mechanism configurations and that $\mathbf{f} = \mathbf{r}_1 + \mathbf{r}_2$ (with $\mathbf{r}_i$ the force in the joint of the leg i).

[1]In this chapter, we consider rank-deficiency of matrices $\mathbf{A}_r$ and $\mathbf{J}_{td}$ of order 1, as this is the most usual case. However, the methodology could be extended to matrices with higher order of rank-deficiency.

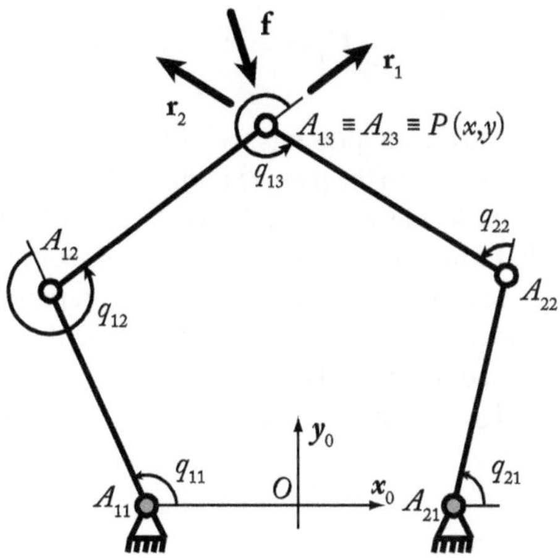

Fig. 9.1 Kinematic chain of the five-bar mechanism

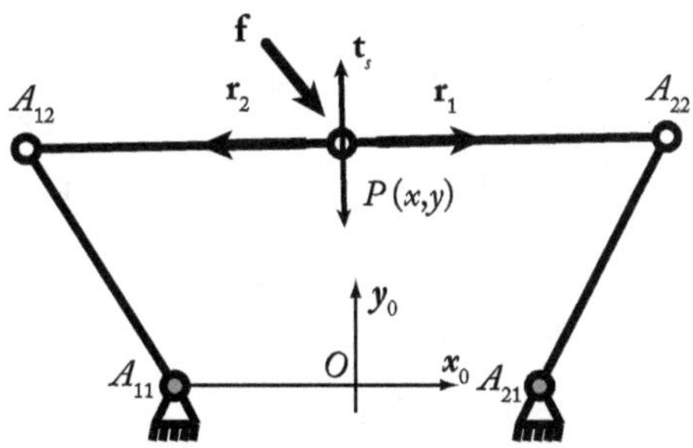

Fig. 9.2 The five-bar mechanism in a Type 2 singularity: the uncontrollable motion is described by the vector $\mathbf{t}_s$

In Type 2 singularity, $\overrightarrow{A_{12}P}$ is collinear to $\overrightarrow{A_{22}P}$ and, as a result, $\mathbf{r}_1$ is collinear to $\mathbf{r}_2$. It can be proven that, in such a case, the robot gets an uncontrollable motion along the vector $\mathbf{t}_s$ which is perpendicular to $\overrightarrow{A_{12}P}$ and $\overrightarrow{A_{22}P}$ (Fig. 9.2). To compensate a force $\mathbf{f}$ *which is not collinear* to $\mathbf{r}_1$ and $\mathbf{r}_2$, (i.e. for which the criterion (9.11) is not satisfied as $\mathbf{t}_s^T\mathbf{f}$ will be different from zero in this case), the reactions $\mathbf{r}_1$ and $\mathbf{r}_2$ must have infinite norms. If the force $\mathbf{f}$ is collinear to $\mathbf{r}_1$ and $\mathbf{r}_2$ (i.e. the criterion (9.11) is respected as $\mathbf{t}_s^T\mathbf{f} = 0$ in this case), the reactions $\mathbf{r}_1$ and $\mathbf{r}_2$ will have finite norms.

This simplified problem gives an insight into the general theory presented in this section.

9.3.2 *Optimal Trajectory Planning Through* LPJTS *Singularities*

In this section, we consider that only the matrix $\mathbf{J}_{td}$ is not invertible. To be clear, let us rewrite here Eq. (9.3):

$$\mathbf{J}_{td}^T \boldsymbol{\lambda}_2 = \boldsymbol{\tau}_{td} \text{ where } \boldsymbol{\tau}_{td} = \frac{\mathrm{d}}{\mathrm{d}t}\left(\frac{\partial L}{\partial \dot{\mathbf{q}}_d}\right)^T - \left(\frac{\partial L}{\partial \mathbf{q}_d}\right)^T. \tag{9.12}$$

Thus, $\boldsymbol{\tau}_{td}$ represents the virtual input efforts in the joints of the virtual system that correspond to the passive joints of the real robot. Moreover, as previously mentioned, $\boldsymbol{\lambda}_2$ stacks the wrenches $\boldsymbol{\lambda}_2^1$ to $\boldsymbol{\lambda}_2^n$ (Fig. 8.5) applied by the virtual tree structure on the platform at points $A_{k\,m_k}$, so that this virtual structure can have the same motion as the real parallel robot. Then, (9.12) represents the equations of the dynamics of the passive legs in contact with the external environment (here the platform on which is applied the wrenches $\boldsymbol{\lambda}_2$).

If $\mathbf{J}_{td}$ is rank deficient, then a non-null vector $\dot{\mathbf{q}}_d^s$ exists, defined as

$$\mathbf{J}_{td}\dot{\mathbf{q}}_d^s = \mathbf{0} \Leftrightarrow \dot{\mathbf{q}}_d^{s\,T} \mathbf{J}_{td}^T = \mathbf{0}. \tag{9.13}$$

Thus, $\dot{\mathbf{q}}_d^s$ represents the passive joint velocities describing the uncontrolled motion of the legs inside the *LPJTS* singularity.

Multiplying the left-hand side of (9.12) by $\dot{\mathbf{q}}_d^{s\,T}$, one obtains

$$\dot{\mathbf{q}}_d^{s\,T} \mathbf{J}_{td}^T \boldsymbol{\lambda}_2 = 0. \tag{9.14}$$

As a result, for the *IDM* to be consistent, the right-hand part of (9.12) must strictly follow the condition

$$\dot{\mathbf{q}}_d^{s\,T} \boldsymbol{\tau}_{td} = 0 \tag{9.15}$$

which involves that:

For avoiding infinite input efforts while crossing a LPJTS singularity, the input efforts of the virtual system in the joints that correspond to the passive joints of the real robot must be reciprocal to the uncontrollable motion of the passive legs inside the singularity (in other words, the power of these efforts along the leg uncontrollable motion must be null).

As for crossing the Type 2 singularities, we will show thereafter that the criterion (9.15) can be respected through a proper robot trajectory planning.

To better understand the phenomenon, let us consider the Tripteron proposed by Gosselin et al. (2002) depicted in Fig. 9.3. The robot is composed of three identical legs made of an active prismatic ($\underline{P}$) joint mounted onto the base and followed by a serial 3R passive chain. In each leg, all $\underline{P}$ and R joint axes are collinear (i.e. the 3R chain is planar and its displacement is orthogonal to the one of the $\underline{P}$ joint). The legs are mounted so that each leg is orthogonal to the two others.

This special arrangement of the leg leads to a fully-isotropic robot with 3 translational degrees of freedom, i.e.

$$\dot{\mathbf{q}}_a = {}^0\mathbf{v}_p \tag{9.16}$$

where $\dot{\mathbf{q}}_a$ are the input velocities and ${}^0\mathbf{v}_p$ is the platform translational velocity. As a result, the Jacobian matrix is the identity matrix of dimension 3. From Sect. 6.4, if a

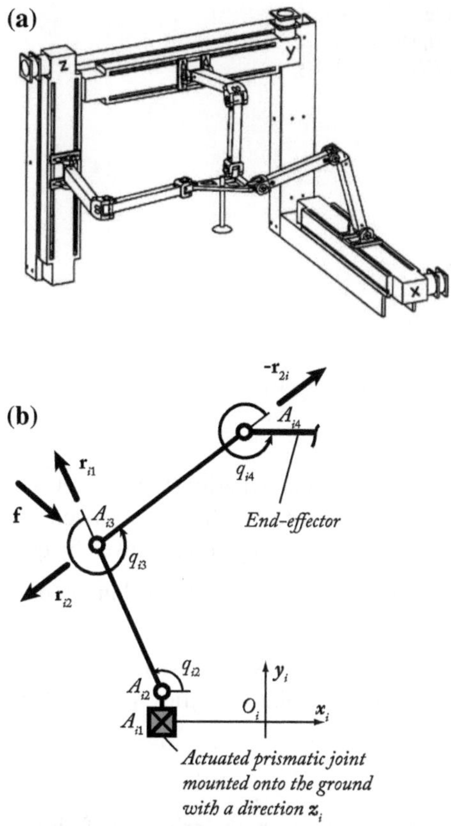

Fig. 9.3 The Tripteron (Kong and Gosselin 2002). **a** CAD view (courtesy of C.M. Gosselin). **b** Kinematic architecture of the leg *i*

force $\mathbf{f}_p$ is applied on any point of the platform (and in absence of any other effects), the robot input efforts $\boldsymbol{\tau}$ are equal to

$$\boldsymbol{\tau} = \mathbf{f}_p. \tag{9.17}$$

It is considered in this example that the mechanism is not moving and that the gravity effects are canceled. A force $\mathbf{f}$ (which could represent any kind of effort applied on the leg, e.g. gravity, inertia, interaction with the environment, etc.) is applied on leg 1 at point A_{13} (Fig. 9.3b). A simple analysis of the effort transmission shows that the reactions in the passive joints located at points A_{12} and A_{13} (A_{13} and A_{14}, resp.) is collinear to the vector $\overrightarrow{A_{12}A_{13}}$ ($\overrightarrow{A_{13}A_{14}}$, resp.) for any robot configurations and that $\mathbf{f} = \mathbf{r}_{11} + \mathbf{r}_{12}$ (with $\mathbf{r}_{1j}$ the force in the joint of the element j of the leg 1). Moreover, as the force $-\mathbf{r}_{12}$ is applied on the platform through the passive joint located at A_{14}, from (9.17), we have

$$\boldsymbol{\tau} = -\mathbf{r}_{12}. \tag{9.18}$$

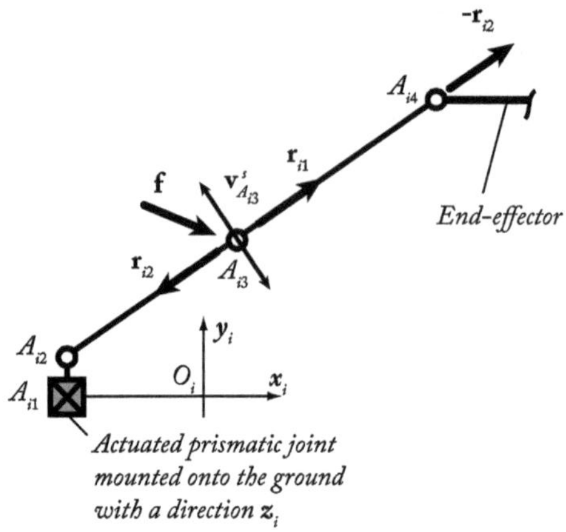

Fig. 9.4 Leg i of the Tripteron in a *LPJTS* singularity

In a *LPJTS* singularity (Fig. 9.4), $\overrightarrow{A_{12}A_{13}}$ is collinear to $\overrightarrow{A_{13}A_{14}}$ and, as a result, $\mathbf{r}_{11}$ is collinear to $\mathbf{r}_{12}$. It can be proven (see Sect. 9.5) that, in such a case, the robot gets an uncontrollable motion given by $\dot{\mathbf{q}}_d^s$ that produces a displacement $\mathbf{v}^s_{A_{13}}$ of point A_{13} (Fig. 9.4). $\mathbf{v}^s_{A_{13}}$ is contained in the plane $x_i O_i y_i$ and is perpendicular to $\overrightarrow{A_{12}A_{13}}$ and $\overrightarrow{A_{13}A_{14}}$.

Let us denote as $\mathbf{J}_{A_{13}}$ the Jacobian matrix linking the velocity $\mathbf{v}_{A_{13}}$ of point A_{13} to the passive joint velocities $\dot{\mathbf{q}}_d$ such that:

$$\mathbf{v}_{A_{13}} = \mathbf{J}_{A_{13}}\dot{\mathbf{q}}_d. \tag{9.19}$$

As a result, from the principle of virtual powers, $\boldsymbol{\tau}_{td}$ is the vector of the efforts in the virtual structure defined such that

$$\boldsymbol{\tau}_{td} = \mathbf{J}^T_{A_{13}}\mathbf{f} \Rightarrow \mathbf{f} = \mathbf{J}^{-T}_{A_{13}}\boldsymbol{\tau}_{td}. \tag{9.20}$$

The virtual power due to $\mathbf{f}$ and the displacement of the point A_{13} is thus equal to

$$\mathbf{v}^T_{A_{13}}\mathbf{f} = \dot{\mathbf{q}}^T_d\mathbf{J}^T_{A_{13}}\mathbf{J}^{-T}_{A_{13}}\boldsymbol{\tau}_{td} = \dot{\mathbf{q}}^T_d\boldsymbol{\tau}_{td} \tag{9.21}$$

To compensate a force $\mathbf{f}$ which is not collinear to $\mathbf{r}_{11}$ and $\mathbf{r}_{12}$, and thus not reciprocal to $\mathbf{v}_{A_{13}}$, (i.e. for which the criterion (9.15) is not respected as $\dot{\mathbf{q}}_d^{s\,T}\boldsymbol{\tau}_{td} = \mathbf{v}^{s\,T}_{A_{13}}\mathbf{f}$ will be different from zero in this case), the reactions $\mathbf{r}_{11}$ and $\mathbf{r}_{12}$ must have infinite norms, thus leading to infinite input efforts from (9.18). If the force $\mathbf{f}$ is collinear to $\mathbf{r}_{11}$ and $\mathbf{r}_{12}$, and thus reciprocal to $\mathbf{v}_{A_{13}}$, (i.e. the criterion (9.15) is respected as $\dot{\mathbf{q}}_d^{s\,T}\boldsymbol{\tau}_{td} = \mathbf{v}^{s\,T}_{A_{13}}\mathbf{f} = 0$ in this case), the reactions $\mathbf{r}_{11}$ and $\mathbf{r}_{12}$ will have finite norms, and the input efforts $\boldsymbol{\tau}$ will also remain finite.

This simplified problem gives an insight into the general theory presented in this section.

Examples of trajectories for crossing Type 2 or *LPJTS* singularities are shown in the next section.

9.4 Example 1: The Five-Bar Mechanism Crossing a Type 2 Singularity

9.4.1 Trajectory Planning Through the Type 2 Singularities

From the analysis of matrices $\mathbf{A}_r$ and $\mathbf{J}_{td}$, the five-bar mechanism encounters only Type 2 singularities, but no *LPJTS* singularity. So, let us analyze the degeneracy conditions of the expression (9.4), and first, let us compute the term $\mathbf{w}_d$ of (9.5).

For that, let us rewrite the vector $\boldsymbol{\tau}_{td}$ in the form:

$$\boldsymbol{\tau}_{td} = \mathbf{M}_{td}(\mathbf{q_t}) \begin{bmatrix} \ddot{\mathbf{q}}_a \\ \ddot{\mathbf{q}}_d \end{bmatrix} + \mathbf{c}_{td}(\mathbf{q}_t, \dot{\mathbf{q}}_t) \tag{9.22}$$

where $\mathbf{M}_{td}$ can be defined by using the Eq. (8.127) as

$$\mathbf{M}_{td} = \begin{bmatrix} M_{11} & 0 & zz_{21} & 0 & 0 \\ 0 & 0 & 0 & 0 & 0 \\ 0 & M_{54} & 0 & 0 & zz_{22} \end{bmatrix} \tag{9.23}$$

and $\mathbf{c}_{td} = \begin{bmatrix} c_2 \; c_4 \; c_5 \end{bmatrix}^T$ where the terms c_2, c_4 and c_5 are defined in Eqs. (8.133), (8.135) and (8.136) and can be rewritten as:

$$\begin{aligned}\mathbf{c}_{td} &= \begin{bmatrix} d_{12}mx_{12}\sin q_{12} - d_{12}my_{12}\cos q_{12} & 0 \\ 0 & 0 \\ 0 & d_{22}mx_{22}\sin q_{22} - d_{22}my_{22}\cos q_{22} \end{bmatrix} \begin{bmatrix} \dot{q}_{11}^2 \\ \dot{q}_{21}^2 \end{bmatrix} \\ &+ \begin{bmatrix} fv_{12} & 0 & 0 \\ 0 & fv_{13} & 0 \\ 0 & 0 & fv_{22} \end{bmatrix} \begin{bmatrix} \dot{q}_{12} \\ \dot{q}_{13} \\ \dot{q}_{22} \end{bmatrix} + \begin{bmatrix} fs_{12}\mathrm{sign}(\dot{q}_{12}) \\ fs_{13}\mathrm{sign}(\dot{q}_{13}) \\ fs_{22}\mathrm{sign}(\dot{q}_{22}) \end{bmatrix} \\ &= \mathbf{C}_{td}^r \begin{bmatrix} \dot{q}_{11}^2 \\ \dot{q}_{21}^2 \end{bmatrix} + \mathbf{F}_{vd}\dot{\mathbf{q}}_d + \mathbf{f}_{sd}. \end{aligned} \tag{9.24}$$

Now, let us express the values of $\ddot{\mathbf{q}}_a$ and $\ddot{\mathbf{q}}_d$ as a function of ${}^0\dot{\mathbf{t}}_r$. From (7.187), we obtain

$$\ddot{\mathbf{q}}_a = \mathbf{J}_{inv}{}^0\dot{\mathbf{t}}_r + \mathbf{a}_q. \tag{9.25}$$

Then, introducing (9.25) into (7.189), we get

$$\ddot{\mathbf{q}}_d = \mathbf{J}_r{}^0\dot{\mathbf{t}}_r + \mathbf{a}_r \tag{9.26}$$

with

$$\mathbf{J}_r = \mathbf{J}_{td}^{-1}(\mathbf{J}_t - \mathbf{J}_{ta}\mathbf{J}_{inv}) \tag{9.27}$$

and

$$\mathbf{a}_r = \mathbf{J}_{td}^{-1}(\mathbf{d}_c - \mathbf{J}_{ta}\mathbf{a}_q). \tag{9.28}$$

Introducing (9.25)–(9.28) into (9.23), simplifying and skipping all mathematical derivations, we get

$$\tau_{td} = \mathbf{M}_d^{x\,0}\dot{\mathbf{t}}_r + \mathbf{c}_d^x \tag{9.29}$$

where

$$\mathbf{M}_d^x = \mathbf{M}_d \begin{bmatrix} \mathbf{J}_{inv} \\ \mathbf{J}_r \end{bmatrix} \tag{9.30}$$

and

$$\mathbf{c}_d^x = \mathbf{M}_d \begin{bmatrix} \mathbf{a}_q \\ \mathbf{a}_r \end{bmatrix} + \mathbf{c}_{td} \tag{9.31}$$

with

$$\mathbf{c}_{td} = \mathbf{C}_{td}^r \begin{bmatrix} (\mathbf{j}_{inv}^{(1)}\,{}^0\mathbf{t}_r)^2 \\ (\mathbf{j}_{inv}^{(2)}\,{}^0\mathbf{t}_r)^2 \end{bmatrix} + \mathbf{F}_{vd}\mathbf{J}_{inv}\,{}^0\mathbf{t}_r + \mathbf{f}_{sd} \tag{9.32}$$

in which $\mathbf{j}_{inv}^{(j)}$ corresponds to the jth row of the matrix $\mathbf{J}_{inv}$.

Then, introducing (9.29) into (9.5), we obtain

$$\begin{aligned} \mathbf{w}_d &= {}^0\mathbf{w}_r + \mathbf{J}_t^T\mathbf{J}_{td}^{-T}\tau_{td} \\ &= m_4\,{}^0\dot{\mathbf{t}}_r + \mathbf{J}_t^T\mathbf{J}_{td}^{-T}(\mathbf{M}_d^{x\,0}\dot{\mathbf{t}}_r + \mathbf{c}_d^x) \end{aligned} \tag{9.33}$$

which, for one given robot configuration, is a function of ${}^0\dot{\mathbf{t}}_r$ and ${}^0\mathbf{t}_r$ only.

From the degeneracy analysis of matrix $\mathbf{A}_r$ of (7.113) (see Sect. 7.5.6.1), the gained motion inside the Type 2 singularity can be expressed as:

$$\mathbf{t}_s = \begin{bmatrix} -\sin(q_{i1}+q_{i2}) \\ \cos(q_{i1}+q_{i2}) \end{bmatrix}. \tag{9.34}$$

Thus, the criterion (9.11) to respect in order to cross the Type 2 singularity takes the general form

$$\mathbf{t}_s^T\mathbf{w}_d = \left[-\sin(q_{i1}+q_{i2})\ \cos(q_{i1}+q_{i2})\right](m_4\,{}^0\dot{\mathbf{t}}_r + \mathbf{J}_t^T\mathbf{J}_{td}^{-T}(\mathbf{M}_d^{x\,0}\dot{\mathbf{t}}_r + \mathbf{c}_d^x)) = 0 \tag{9.35}$$

which, for one given singularity configuration, is a function of ${}^0\dot{\mathbf{t}}_r$ and ${}^0\mathbf{t}_r$ only. Therefore, it is possible to plan, for one given singularity configuration, a Cartesian trajectory which respects (9.35).

9.4.2 Simulations and Experimental Results

In order to validate the theoretical results presented above, we will test the proposed criterion for crossing the Type 2 singularities on the five-bar mechanism prototype designed at the Institut Pascal from Clermont-Ferrand (France) and presented in Sect. 8.4.3.1.

Taking into account the real identified parameters of the robot given in Sect. 8.4.3.1, the following model fully describes the robot dynamics of the studied mechanism:

$$\boldsymbol{\tau} = \mathbf{w}_b - \mathbf{B}^T \boldsymbol{\lambda}_1, \tag{9.36}$$

$$\mathbf{A}_r^T \boldsymbol{\lambda}_1 = \mathbf{w}_d \tag{9.37}$$

with

$$\begin{aligned} \mathbf{w}_d &= m_3 \begin{bmatrix} \ddot{x} \\ \ddot{y} \end{bmatrix}, \\ \mathbf{w}_b &= \begin{bmatrix} zz_{11R}\, \ddot{q}_{11} \\ zz_{21R}\, \ddot{q}_{21} \end{bmatrix} + \begin{bmatrix} f_{v11}\dot{q}_{11} \\ f_{v21}\dot{q}_{21} \end{bmatrix} + \begin{bmatrix} f_{s11}\mathrm{sign}(\dot{q}_{11}) \\ f_{s21}\mathrm{sign}(\dot{q}_{21}) \end{bmatrix}. \end{aligned} \tag{9.38}$$

From (9.35) and (9.37), the criterion for crossing the Type 2 singularities becomes

$$\begin{aligned} \mathbf{t}_s^T \mathbf{w}_d &= \left[-\sin(q_{i1}+q_{i2}) \; \cos(q_{i1}+q_{i2}) \right] m_4 {}^0\dot{\mathbf{t}}_r \\ &= m_4(-\sin(q_{i1}+q_{i2})\ddot{x} + \sin(q_{i1}+q_{i2})\ddot{y}) = 0 \end{aligned} \tag{9.39}$$

or also

$$\ddot{y} = \ddot{x}\tan(q_{i1}+q_{i2}). \tag{9.40}$$

Then, let us define two different types of trajectory with a duration $t_f = 1.5$ s between the points P_0 ($\mathbf{x}_{P_0} = \left[x_{p0}\ y_{p0}\right]^T = [0; 0.338]^T$ m) and P_f ($\mathbf{x}_{P_f} = \left[x_{pf}\ y_{pf}\right]^T = [0.1; 0.1]^T$ m) which are separated by a Type 2 singularity (Fig. 9.5):

- A trajectory defined using a fifth-degree polynomial which can fix the position, velocity and acceleration of the robot at the trajectory extremities only; for this polynomial, those conditions are:

$$x(t=0) = x_{p0},\ \dot{x}(t=0) = 0,\ \ddot{x}(t=0) = 0 \tag{9.41}$$
$$y(t=0) = y_{p0},\ \dot{y}(t=0) = 0,\ \ddot{y}(t=0) = 0 \tag{9.42}$$
$$x(t=t_f) = x_{pf},\ \dot{x}(t=t_f) = 0,\ \ddot{x}(t=t_f) = 0 \tag{9.43}$$
$$y(t=t_f) = y_{pf},\ \dot{y}(t=t_f) = 0,\ \ddot{y}(t=t_f) = 0 \tag{9.44}$$

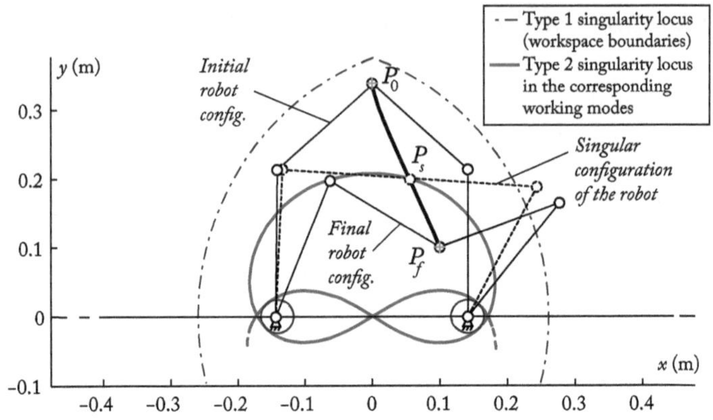

Fig. 9.5 Starting point P_0 and ending point P_f of the Type 2 singularity crossing trajectory

which lead to the following polynomials for x and y:

$$x(t) = 0.296296t^3 - 0.296296t^4 + 0.079012t^5 \tag{9.45}$$

$$y(t) = 0.338175 - 0.705704t^3 + 0.705704t^4 - 0.188188t^5. \tag{9.46}$$

Note that for the reader who doesn't know how to plan a trajectory between two points, the Appendix D explains the procedure.

- A trajectory using an eighth-degree polynomial law which can fix the position, velocity and acceleration of the robot at the trajectory extremity plus the position and acceleration of the robot in the singular configuration; for this polynomial, those conditions are fixed at

$$x(t=0) = 0, \quad \dot{x}(t=0) = 0, \quad \ddot{x}(t=0) = 0 \tag{9.47}$$

$$y(t=0) = 0.3381, \quad \dot{y}(t=0) = 0, \quad \ddot{y}(t=0) = 0 \tag{9.48}$$

$$x(t=t_f) = 0.1, \quad \dot{x}(t=t_f) = 0, \quad \ddot{x}(t=t_f) = 0 \tag{9.49}$$

$$y(t=t_f) = 0.1, \quad \dot{y}(t=t_f) = 0, \quad \ddot{y}(t=t_f) = 0 \tag{9.50}$$

$$x(t=t_s) = 0.0543, \quad \ddot{x}(t=t_s) = 6.8e^{-4} \tag{9.51}$$

$$y(t=t_s) = 0.2, \quad \ddot{y}(t=t_s) = -0.01 \tag{9.52}$$

with $t_s = 0.75$ s the time at which the robot must cross the singularity and $\ddot{x}(t=t_s)$ and $\ddot{y}(t=t_s)$ values that respects the criterion (9.40). These conditions lead to the following polynomials for x and y:

$$\begin{aligned} x(t) =& 0.03062t^3 + 0.36498t^4 - 0.08964t^5 \\ &- 0.63889t^6 + 0.55539t^7 - 0.13198t^8 \end{aligned} \tag{9.53}$$

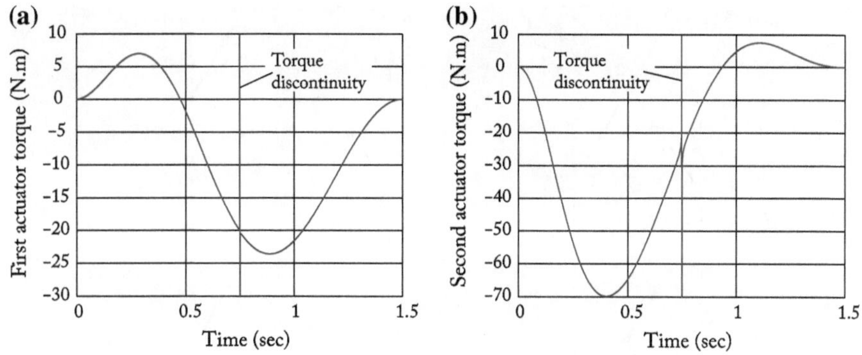

Fig. 9.6 Input torques simulated for the five-bar mechanism crossing the Type 2 singularity locus at $t_s = 0.75$ s without respecting the dynamic criterion. **a** Actuator 1. **b** Actuator 2 (9.40)

$$\begin{aligned} y(t) =& 0.33818 + 0.40308t^3 - 1.95392t^4 + 0.1490t^5 \\ &+ 3.13259t^6 - 2.56936t^7 + 0.60008t^8. \end{aligned} \tag{9.54}$$

First, let us simulate the behavior of the robot when perfectly tracking the two different trajectories. The input torques for both trajectories are shown in Figs. 9.6 and 9.7. It can be observed that, for the trajectory that respects the criterion (9.40) (Fig. 9.7), the input torques remain finite while in the other case (Fig. 9.6), they tend to infinity when crossing the singularity at $t_s = 0.75$ s.

Now, let us launch each trajectory on the five-bar mechanism prototype. The results in terms of

- robot displacements are shown in Figs. 9.8 and 9.9
- input torques are shown in Figs. 9.10 and 9.11.

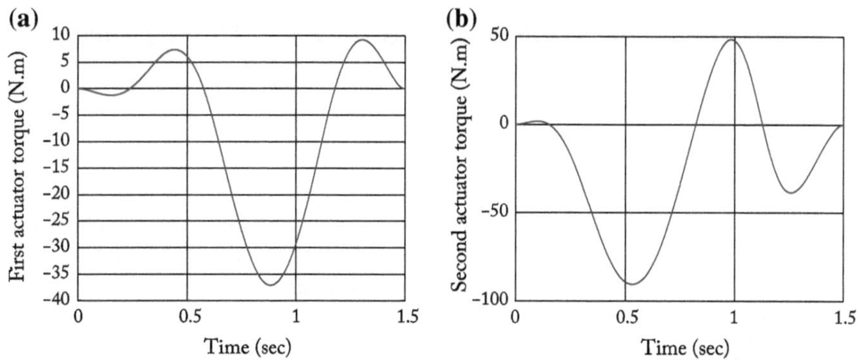

Fig. 9.7 Input torques simulated for the five-bar mechanism crossing the Type 2 singularity locus at $t_s = 0.75$ s with respecting the dynamic criterion. **a** Actuator 1. **b** Actuator 2 (9.40)

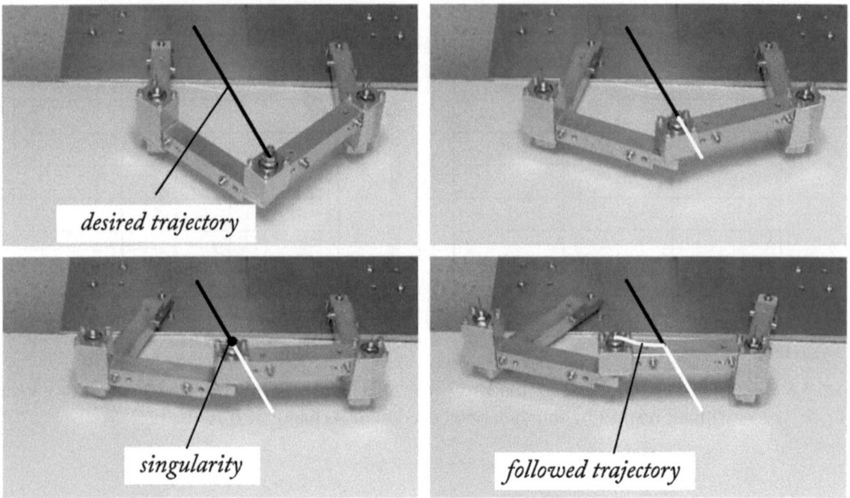

Fig. 9.8 The five-bar mechanism tracking the trajectory which does not respect the dynamic criterion (9.40)

It can be observed that for the trajectory that satisfies the criterion (9.40), the robot can cross the singularity with finite torques while in the other case, it stays blocked in it. Note that:

- when the robot fails to cross the singularity, the data are not recorded after 0.7 s because we activated the emergency stop,
- experimental results in terms of input torques are different from the simulated ones because the robot is not able to perfectly track the desired trajectory.

9.5 Example 2: The Tripterion Crossing a *LPJTS* Singularity

9.5.1 Geometric Description of the Tripteron

As already mentioned, the Tripteron is a spatial parallel mechanism with three degrees of freedom composed of three actuators located at the prismatic joints attached to the ground and three passive revolute joints per leg at points A_{i2}, A_{13} and A_{14}. The *MDH* parameters of the virtual open-loop tree structure are described in Tables 9.1 and 9.2 and Figs. 9.3b and 9.12. The end-effector is considered as a supplementary body numbered as body 5.

The gravity field is directed along z_0.

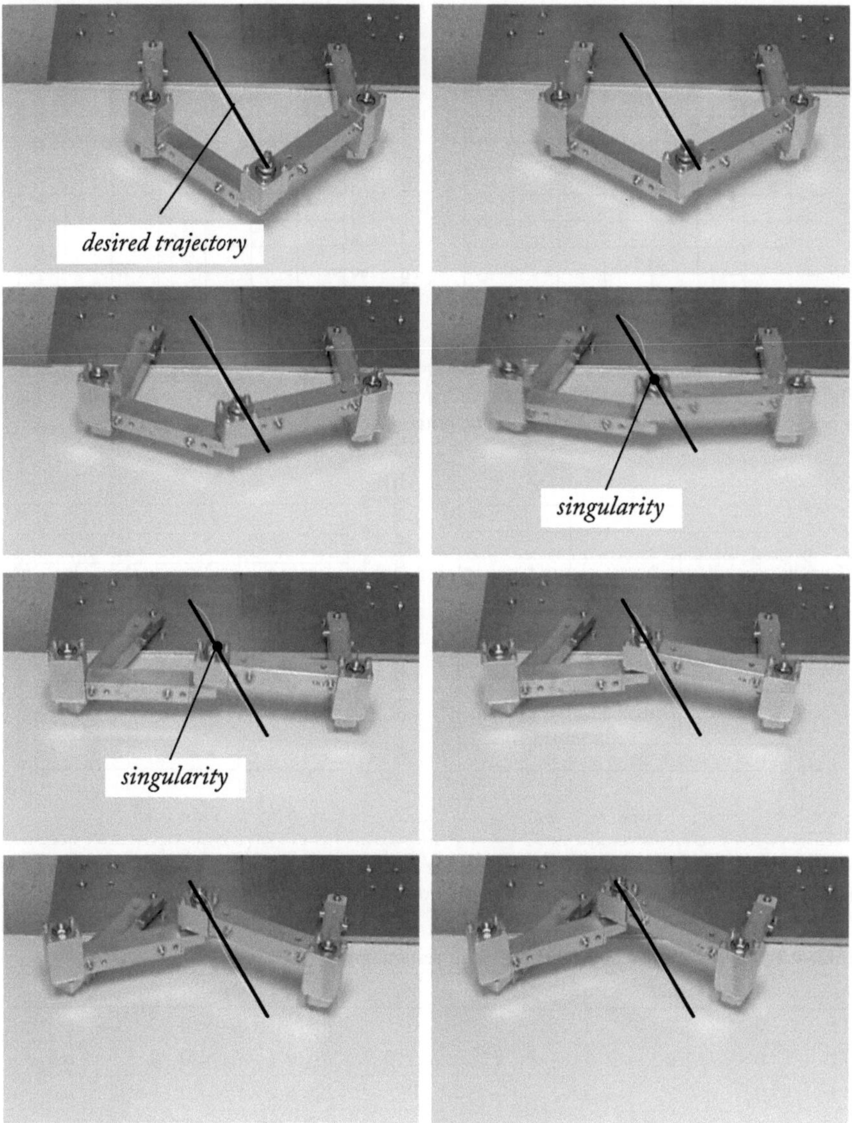

Fig. 9.9 The five-bar mechanism tracking the trajectory respecting the dynamic criterion (9.40)

For this mechanism:

- the end-effector coordinates are $\mathbf{x}^T = [x\ y\ z]$,
- the active joint coordinates are $\mathbf{q}_a^T = [q_{11}\ q_{21}\ q_{31}]$,
- the passive joint coordinates are $\mathbf{q}_d^T = [\mathbf{q}_{d1}^T\ \mathbf{q}_{d2}^T\ \mathbf{q}_{d3}^T]$ with $\mathbf{q}_{di}^T = [q_{i2}\ q_{i3}\ q_{i4}]$ $(i = 1, 2, 3)$.

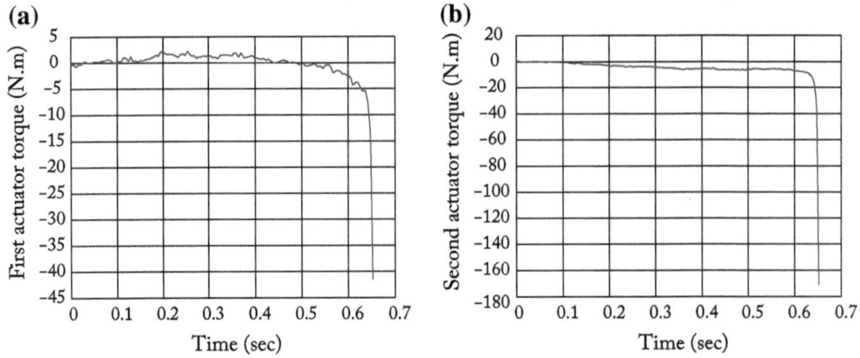

Fig. 9.10 Input torques measured for the five-bar mechanism crossing the Type 2 singularity locus at $t_s = 0.75$ s without respecting the dynamic criterion (9.40). **a** Actuator 1. **b** Actuator 2

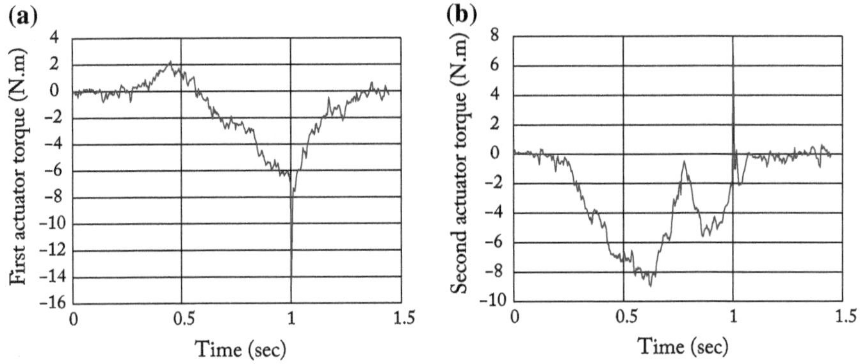

Fig. 9.11 Input torques measured for the five-bar mechanism crossing the Type 2 singularity locus at $t_s = 0.75$ s with respecting the dynamic criterion (9.40). **a** Actuator 1. **b** Actuator 2

Table 9.1 *MDH* parameters for the frames corresponding to robot active joints

ij	a_{ij}	μ_{ij}	σ_{ij}	α_{ij}	γ_{ij}	b_{ij}	d_{ij}	θ_{ij}	r_{ij}
11	0	1	1	0	0	b_{11}	$d_{i1} = 0$	0	q_{11}
12	0	1	1	$\pi/2$	$\pi/2$	$b_{12} = a$	$d_{i2} = 0$	0	$q_{21} - a$
13	0	1	1	$-\pi/2$	0	$b_{13} = a$	$d_{i3} = 0$	$-\pi/2$	$q_{31} + a$

9.5.2 Kinematics of the Tripteron

For the Tripteron, the loop-closure Eq. (7.4) can be expanded in the leg i frame (Fig. 9.3b) as

$$\begin{aligned} 0 &= {}^i x_{A_{i4}} - {}^i x_{A_{i1}} - d_{i2}\cos q_{i2} - d_{i3}\cos(q_{i2} + q_{i3}) \\ 0 &= {}^i y_{A_{i4}} - {}^i y_{A_{i1}} - d_{i2}\sin q_{i2} - d_{i3}\sin(q_{i2} + q_{i3}) \\ 0 &= {}^i z_{A_{i4}} - r_{i1} \end{aligned} \tag{9.55}$$

Table 9.2 *MDH* parameters for the frames corresponding to the passive joints of the ith robot leg ($i = 1, \ldots, 3$)

ij	a_{ij}	μ_{ij}	σ_{ij}	γ_{ij}	d_{ij}	θ_{ij}	r_{ij}
$i2$	$i1$	0	0	0	$d_{i2} = 0$	q_{i2}	0
$i3$	$i2$	0	0	0	$d_{i3} = l_{A_{i2}A_{i3}}$	q_{i3}	0
$i4$	$i3$	0	0	0	$d_{i4} = l_{A_{i3}A_{i4}}$	q_{i4}	0

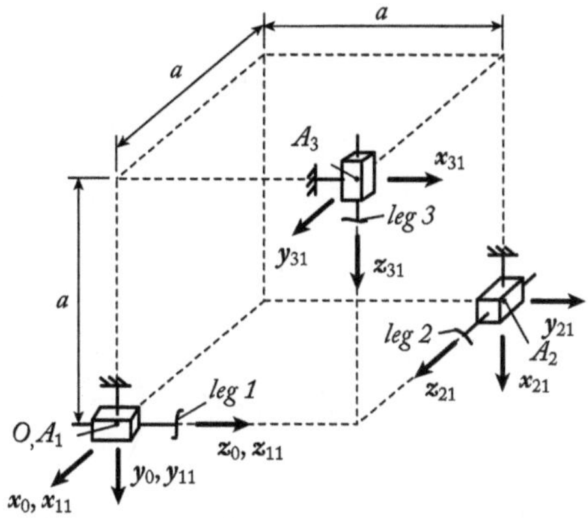

Fig. 9.12 Kinematic description of the actuated prismatic joint arrangement for the Tripteron.eps

and

$$0 = q_{i2} + q_{i3} + q_{i4} \tag{9.56}$$

where ${}^i x_{A_{i4}}$, ${}^i y_{A_{i4}}$ and ${}^i z_{A_{i4}}$ are the point A_{i4} coordinates expressed in the frame of the leg i,

$$^1x_{A14} = x,\ ^1y_{A14} = y,\ ^1z_{A14} = z \tag{9.57}$$

$$^2x_{A24} = y,\ ^2y_{A24} = z,\ ^2z_{A24} = x \tag{9.58}$$

$$^3x_{A34} = z,\ ^3y_{A34} = x,\ ^3z_{A34} = y. \tag{9.59}$$

${}^i x_{A_{i1}}$, ${}^i y_{A_{i1}}$ and ${}^i z_{A_{i1}}$ are the point A_{i1} coordinates (also regrouped in the vector $\mathbf{x}_{A_{i1}}$) expressed in the frame of the leg i, $\mathbf{x}_{A_{i4}P} = \overrightarrow{A_{i4}P}$ (P is the platform center) and r_{i1} is defined in the Table 9.1.

From the last line of (9.55), we directly get:

$$\begin{aligned} x &= q_{12} - a \\ y &= q_{13} + a \\ z &= q_{11}. \end{aligned} \tag{9.60}$$

From (9.55), by deleting the terms in $\cos(q_{i2}+q_{i3})$ or $\sin(q_{i2}+q_{i3})$, it is possible to obtain (for $i = 1\dots 3$):

$$d_{i4}^2 = \left(x_{A_{i2}A_{i4}} - d_{i3}\cos q_{i2}\right)^2 + \left(y_{A_{i2}A_{i4}} - d_{i3}\sin q_{i2}\right)^2 \tag{9.61}$$

where $x_{A_{i1}A_{i4}} = {}^i x_{A_{i4}} - {}^i x_{A_{i1}}$ and $y_{A_{i1}A_{i4}} = {}^i y_{A_{i4}} - {}^i y_{A_{i1}}$.

Then, expanding (9.61),

$$0 = A_{i1}\cos q_{i2} + A_{i2}\sin q_{i2} + A_{i3} \tag{9.62}$$

where

$$\begin{aligned} A_{i1} &= -2d_{i3}x_{A_{i1}A_{i4}} \\ A_{i2} &= -2d_{i3}y_{A_{i1}A_{i4}} \\ A_{i3} &= x_{A_{i1}A_{i4}}^2 + y_{A_{i1}A_{i4}}^2 + d_{i3}^2 - d_{i4}^2 \end{aligned} \tag{9.63}$$

Finally, by using the tangent half-angle formula, we can obtain

$$q_{i2} = 2\tan^{-1}\left(\frac{-A_{i2} \pm \sqrt{A_{i2}^2 - A_{i3}^2 + A_{i1}^2}}{A_{i3} - A_{i1}}\right). \tag{9.64}$$

In (9.64), the sign "±" denotes the two robot leg working modes.

Then, we get easily from (9.55) and (9.56) that:

$$q_{i3} = \tan^{-1}\left(\frac{{}^i y_{A_{i4}} - {}^i y_{A_{i3}}}{{}^i x_{A_{i4}} - {}^i x_{A_{i3}}}\right), \tag{9.65}$$

with ${}^i x_{A_{i3}} = {}^i x_{A_{i1}} + d_{i2}\cos q_{i2}$, ${}^i y_{A_{i3}} = {}^i y_{A_{i1}} + d_{i2}\sin q_{i2}$, and

$$q_{i4} = -q_{i2} - q_{i3}. \tag{9.66}$$

Now, differentiating (9.60) with respect to time, and simplifying, the matrices $\mathbf{A}_r$ and $\mathbf{B}$ of (7.62) can be found:

$$\mathbf{A}_r = \mathbf{1}_3,\ \mathbf{B} = \begin{bmatrix} 0 & 1 & 0 \\ 0 & 0 & 1 \\ 1 & 0 & 0 \end{bmatrix} \tag{9.67}$$

where $\mathbf{1}_3$ is the identity matrix of dimension 3 leading thus to

$$\mathbf{A}_r \begin{bmatrix} \dot{x} \\ \dot{y} \\ \dot{z} \end{bmatrix} + \mathbf{B} \begin{bmatrix} \dot{q}_{11} \\ \dot{q}_{12} \\ \dot{q}_{13} \end{bmatrix} = \begin{bmatrix} \dot{x} \\ \dot{y} \\ \dot{z} \end{bmatrix} - \begin{bmatrix} \dot{q}_{11} \\ \dot{q}_{12} \\ \dot{q}_{13} \end{bmatrix} = \mathbf{0} \tag{9.68}$$

from which we get:

$$^0\mathbf{t}_r = \mathbf{J}\dot{\mathbf{q}}_a. \tag{9.69}$$

with $^0\mathbf{t}_r = [\dot{x}\ \dot{y}\ \dot{z}]^T$, $\dot{\mathbf{q}}_a = [\dot{q}_{11}\ \dot{q}_{21}\ \dot{q}_{31}]^T$ and $\mathbf{J} = \mathbf{1}_3$.

Differentiating (9.69) w.r.t. time, we also get:

$$^0\dot{\mathbf{t}}_r = \mathbf{J}\ddot{\mathbf{q}}_a \tag{9.70}$$

Now, differentiating (9.55) and (9.56) with respect to time, it can be found that:

$$\begin{aligned} 0 &= {}^i\dot{x}_{A_{i4}} + d_{i2}\sin q_{i2}\dot{q}_{i2} + d_{i3}\sin(q_{i2}+q_{i3})(\dot{q}_{i2}+\dot{q}_{i3}) \\ 0 &= {}^i\dot{y}_{A_{i4}} - d_{i2}\cos q_{i2}\dot{q}_{i2} - d_{i3}\cos(q_{i2}+q_{i3})(\dot{q}_{i2}+\dot{q}_{i3}) \\ 0 &= {}^i\dot{z}_{A_{i4}} - \dot{q}_{i1} \end{aligned} \tag{9.71}$$

$$0 = \dot{q}_{i2} + \dot{q}_{i3} + \dot{q}_{i4} \tag{9.72}$$

for $i = 1, 2, 3$ and where ${}^i\dot{x}_{A_{i4}}$, ${}^i\dot{y}_{A_{i4}}$ and ${}^i\dot{z}_{A_{i4}}$ are the point A_{i4} velocities along the axes of the frame of the leg i,

$${}^1\dot{x}_{A_{14}} = \dot{x},\ {}^1\dot{y}_{A_{14}} = \dot{y},\ {}^1\dot{z}_{A_{14}} = \dot{z} \tag{9.73}$$

$${}^2\dot{x}_{A_{24}} = \dot{y},\ {}^2\dot{y}_{A_{24}} = \dot{z},\ {}^2\dot{z}_{A_{24}} = \dot{x} \tag{9.74}$$

$${}^3\dot{x}_{A_{34}} = \dot{z},\ {}^3\dot{y}_{A_{34}} = \dot{x},\ {}^3\dot{z}_{A_{34}} = \dot{y}. \tag{9.75}$$

Combining (9.71), (9.72) and (9.75) and noticing that the last line of (9.71) can be disregarded as the velocities of the passive joints are not included in this equation, we get

$$\mathbf{J}_{ti}\begin{bmatrix}\dot{x}\\ \dot{y}\\ \dot{z}\end{bmatrix} = \begin{bmatrix}0\\0\\0\end{bmatrix}\dot{q}_{i1} + \begin{bmatrix} d_{i2}\sin q_{i2} + d_{i3}\sin(q_{i2}+q_{i3}) & d_{i3}\sin(q_{i2}+q_{i3}) & 0 \\ -d_{i2}\cos q_{i2} - d_{i3}\cos(q_{i2}+q_{i3}) & -d_{i3}\cos(q_{i2}+q_{i3}) & 0 \\ 1 & 1 & 1 \end{bmatrix}\begin{bmatrix}\dot{q}_{i2}\\ \dot{q}_{i3}\\ \dot{q}_{i4}\end{bmatrix} \tag{9.76}$$

which can be rewritten as

$$\mathbf{J}_{ti}\,{}^0\mathbf{t}_r - \mathbf{J}_{tai}\dot{q}_{i1} - \mathbf{J}_{tdi}\dot{\mathbf{q}}_{di} = \mathbf{0} \tag{9.77}$$

with

$$\mathbf{J}_{t1} = \begin{bmatrix}1&0&0\\0&1&0\\0&0&0\end{bmatrix},\ \mathbf{J}_{t2} = \begin{bmatrix}0&1&0\\0&0&1\\0&0&0\end{bmatrix},\ \mathbf{J}_{t3} = \begin{bmatrix}0&0&1\\1&0&0\\0&0&0\end{bmatrix} \tag{9.78}$$

$$\mathbf{J}_{tai} = [0\ 0\ 0]^T \tag{9.79}$$

$$\mathbf{J}_{tdi} = \begin{bmatrix} d_{i2}\sin q_{i2} + d_{i3}\sin(q_{i2}+q_{i3}) & d_{i3}\sin(q_{i2}+q_{i3}) & 0 \\ -d_{i2}\cos q_{i2} - d_{i3}\cos(q_{i2}+q_{i3}) & -d_{i3}\cos(q_{i2}+q_{i3}) & 0 \\ 1 & 1 & 1 \end{bmatrix} \quad (9.80)$$

and ${}^{0}\mathbf{t}_r^T = [\dot{x}\ \dot{y}\ \dot{z}]$, and $\dot{\mathbf{q}}_{di}^T = [\dot{q}_{i2}\ \dot{q}_{i3}\ \dot{q}_{i4}]$.

Now, considering the legs 1 to 3, we obtain

$$\mathbf{J}_t\, {}^{0}\mathbf{t}_r - \mathbf{J}_{ta}\dot{\mathbf{q}}_a - \mathbf{J}_{td}\dot{\mathbf{q}}_d = \mathbf{0} \quad (9.81)$$

with

$$\mathbf{J}_t = \begin{bmatrix} \mathbf{J}_{t1} \\ \mathbf{J}_{t2} \\ \mathbf{J}_{t3} \end{bmatrix} \quad (9.82)$$

$$\mathbf{J}_{ta} = \mathbf{0}_{9\times 3} \quad (9.83)$$

with $\mathbf{0}_{9\times 3}$ a (9×3) zero matrix and

$$\mathbf{J}_{td} = \begin{bmatrix} \mathbf{J}_{td1} & \mathbf{0}_{3\times 3} & \mathbf{0}_{3\times 3} \\ \mathbf{0}_{3\times 3} & \mathbf{J}_{td2} & \mathbf{0}_{3\times 3} \\ \mathbf{0}_{3\times 3} & \mathbf{0}_{3\times 3} & \mathbf{J}_{td3} \end{bmatrix} \quad (9.84)$$

with $\mathbf{0}_{3\times 3}$ a (3×3) zero matrix and $\dot{\mathbf{q}}_d^T = [\dot{\mathbf{q}}_{d1}^T\ \dot{\mathbf{q}}_{d2}^T\ \dot{\mathbf{q}}_{d3}^T]$.

Now, introducing (9.69) into (9.81), we obtain:

$$\dot{\mathbf{q}}_d = \mathbf{J}_r\, {}^{0}\mathbf{t}_r \quad (9.85)$$

with

$$\mathbf{J}_r = \mathbf{J}_{td}^{-1}(\mathbf{J}_t - \mathbf{J}_{ta}) = \mathbf{J}_{td}^{-1}\mathbf{J}_t. \quad (9.86)$$

Finally, after differentiating (9.81) w.r.t. time and by introducing (9.70) in it, we obtain

$$\ddot{\mathbf{q}}_d = \mathbf{J}_r\, {}^{0}\dot{\mathbf{t}}_r + \mathbf{J}_r^d\, {}^{0}\mathbf{t}_r \quad (9.87)$$

with

$$\mathbf{J}_r^d = -\mathbf{J}_{td}^{-1}\dot{\mathbf{J}}_{td}\mathbf{J}_r. \quad (9.88)$$

From (9.67) and (9.84), it is possible to observe that

- matrix $\mathbf{J}_{td}$ is singular if one block matrix $\mathbf{J}_{tdi}$ is singular; $\mathbf{J}_{tdi}$ is singular if and only if $q_{i3} = 0$ or π (i.e. $\boldsymbol{x}_{i2}$ is collinear to $\boldsymbol{x}_{i3}$—Fig. 9.4),
- matrix $\mathbf{A}_r$ is constant and never singular; as a result, the robot does not encounter Type 2 singularities.

9.5.3 Full IDM of the Tripteron

As mentioned above, the Tripteron encounters only *LPJTS* singularities. Thus, let us now compute the criterion (9.15).

The inverse dynamic model of the open loop virtual structure of the Tripteron can be obtained by noticing that each leg is composed

- of a first active prismatic joint,
- followed by a planar $3R$ robot in which the last body is massless.

The inverse dynamic model of the leg i is:

$$\tau_{t_{i1}} = (m_{i1} + m_{i2} + m_{i3} + Ia_{i1})\ddot{q}_{i1} + fs_{i1}\text{sign}(\dot{q}_{i1}) + fv_{i1}\dot{q}_{i1} + \tau_{g_{i1}} \tag{9.89}$$

$$\begin{aligned}\tau_{t_{i2}} = &\left(zz_{i2} + d_{i3}^2 m_{i3}\right)\ddot{q}_{i2} + zz_{i3}(\ddot{q}_{i2} + \ddot{q}_{i3})\\ &+ d_{i3}mx_{i3}\left((2\ddot{q}_{i2} + \ddot{q}_{i3})\cos q_{i3} - \dot{q}_{i3}(2\dot{q}_{i2} + \dot{q}_{i3})\sin q_{i3}\right)\\ &+ d_{i3}my_{i3}\left((2\ddot{q}_{i2} + \ddot{q}_{i3})\sin q_{i3} + \dot{q}_{i3}(2\dot{q}_{i2} + \dot{q}_{i3})\cos q_{i3}\right)\\ &+ fs_{i2}\text{sign}(\dot{q}_{i2}) + fv_{i2}\dot{q}_{i2} + \tau_{g_{i2}}\end{aligned} \tag{9.90}$$

$$\begin{aligned}\tau_{t_{i3}} = &zz_{i3}(\ddot{q}_{i2} + \ddot{q}_{i3}) + d_{i3}mx_{i3}\left(\ddot{q}_{i2}\cos q_{i3} + \dot{q}_{i2}^2\sin q_{i3}\right)\\ &+ d_{i3}my_{i3}\left(\ddot{q}_{i2}\sin q_{i3} - \dot{q}_{i2}^2\cos q_{i3}\right)\\ &+ fs_{i3}\text{sign}(\dot{q}_{i3}) + fv_{i3}\dot{q}_{i3} + \tau_{g_{i3}}\end{aligned} \tag{9.91}$$

$$\tau_{t_{i4}} = fs_{i4}\text{sign}(\dot{q}_{i4}) + fv_{i4}\dot{q}_{i4} \tag{9.92}$$

where

$$\tau_{g_{11}} = g\,(m_{11} + m_{21} + m_{31}),\ \tau_{g_{12}} = \tau_{g_{13}} = 0 \tag{9.93}$$

$$\tau_{g_{21}} = 0,\ \tau_{g_{i2}} = g\,(mx_{i2} + m_{i3}d_{i3})\cos q_{i2} - g\,my_{i2}\sin q_{i2} + \tau_{g_{i3}} \text{ for } i = 2, 3 \tag{9.94}$$

$$\tau_{g_{31}} = 0,\ \tau_{g_{i3}} = g\,mx_{i3}\cos(q_{i2} + q_{i3}) - g\,my_{i3}\sin(q_{i2} + q_{i3}) \text{ for } i = 2, 3 \tag{9.95}$$

and

- parameters q_{ij} and length d_{i3} are defined in Tables 9.1, 9.2 and Figs. 9.3b and 9.12 ($j = 1 \ldots 4$),
- $\tau_{t_{i1}}$ is the torque of the virtual actuator located in the prismatic pair, $\tau_{t_{i2}}$ is the torque of the virtual actuator located at point A_{i2}, $\tau_{t_{i3}}$ is the torque of the virtual actuator located at point A_{i3} and $\tau_{t_{i4}}$ is the torque of the virtual actuator located at point A_{i4}.

The vector $\boldsymbol{\tau}_{ta}$ of (8.59) stacks all vectors $\boldsymbol{\tau}_{ta} = \begin{bmatrix}\tau_{t_{11}} & \tau_{t_{12}} & \tau_{t_{13}}\end{bmatrix}^T$ while the vector $\boldsymbol{\tau}_{td}$ of (8.60) stacks all vectors $\boldsymbol{\tau}_{td} = \begin{bmatrix}\tau_{t_{d1}} & \tau_{t_{d2}} & \tau_{t_{d3}}\end{bmatrix}^T$ with $\tau_{t_{di}} = \begin{bmatrix}\tau_{t_{i2}} & \tau_{t_{i3}} & \tau_{t_{i4}}\end{bmatrix}^T$.

The inverse dynamic model of the free body corresponding to the end-effector (body 5) in the virtual system is

$$\begin{aligned} w_1 &= m_5\ddot{x} \\ w_2 &= m_5\ddot{y} \\ w_3 &= m_5(\ddot{z}+g) \end{aligned} \tag{9.96}$$

with w_j being the jth components of the vector ${}^0\mathbf{w}_r$ of (8.65); m_5 is the end-effector mass.

Combining these expressions into the equations of Sect. 8.4, the inverse dynamic model of the Tripteron can be straightforwardly computed.

9.5.4 *Trajectory Planning Through the* LPJTS *Singularities*

Let us rewrite the vector $\boldsymbol{\tau}_{td}$ in the form:

$$\boldsymbol{\tau}_{td} = \mathbf{M}_{td}(\mathbf{q_t})\ddot{\mathbf{q}}_t + \mathbf{c}_{td}(\mathbf{q}_t, \dot{\mathbf{q}}_t) \tag{9.97}$$

where

$$\mathbf{M}_d = \begin{bmatrix} \mathbf{0}_{3\times3} & \mathbf{M}_{d1} & \mathbf{0}_{3\times3} & \mathbf{0}_{3\times3} \\ \mathbf{0}_{3\times3} & \mathbf{0}_{3\times3} & \mathbf{M}_{d2} & \mathbf{0}_{3\times3} \\ \mathbf{0}_{3\times3} & \mathbf{0}_{3\times3} & \mathbf{0}_{3\times3} & \mathbf{M}_{d3} \end{bmatrix} \tag{9.98}$$

and

$$\mathbf{c}_d = \begin{bmatrix} \mathbf{c}_{d1} \\ \mathbf{c}_{d2} \\ \mathbf{c}_{d3} \end{bmatrix} \tag{9.99}$$

in which

$$\mathbf{M}_{di} = \begin{bmatrix} m_{di}^{11} & m_{di}^{12} & 0 \\ m_{di}^{12} & zz_{i3} & 0 \\ 0 & 0 & 0 \end{bmatrix} \tag{9.100}$$

with

- $m_{di}^{11} = zz_{i2} + d_{i3}^2 m_{i3} + zz_{i3} + 2d_{i3}(mx_{i3}\cos q_{i3} + my_{i3}\sin q_{i3})$,
- $m_{di}^{12} = zz_{i3} + d_{i3}(mx_{i3}\cos q_{i3} + my_{i3}\sin q_{i3})$,

and

$$\begin{aligned}
\mathbf{c}_{di} &= \begin{bmatrix} 0 & c_{di}^{12} & 2c_{di}^{12} \\ d_{i3}mx_{i3} - d_{i3}my_{i3}\cos q_{i3}\sin q_{i3} & 0 & 0 \\ 0 & 0 & 0 \end{bmatrix} \begin{bmatrix} \dot{q}_{i2}^2 \\ \dot{q}_{i3}^2 \\ \dot{q}_{i3}\dot{q}_{i2} \end{bmatrix} \\
&+ \begin{bmatrix} fv_{i2} & 0 & 0 \\ 0 & fv_{i3} & 0 \\ 0 & 0 & fv_{i4} \end{bmatrix} \begin{bmatrix} \dot{q}_{i2} \\ \dot{q}_{i3} \\ \dot{q}_{i4} \end{bmatrix} + \begin{bmatrix} fs_{i2}\mathrm{sign}(\dot{q}_{i2}) \\ fs_{i3}\mathrm{sign}(\dot{q}_{i3}) \\ fs_{i4}\mathrm{sign}(\dot{q}_{i4}) \end{bmatrix} \\
&= \mathbf{C}_{di}^{r} \begin{bmatrix} \dot{q}_{i2}^2 \\ \dot{q}_{i3}^2 \\ \dot{q}_{i3}\dot{q}_{i2} \end{bmatrix} + \mathbf{F}_{v_{di}}\dot{\mathbf{q}}_{di} + \mathbf{f}_{s_{di}}
\end{aligned} \tag{9.101}$$

with $c_{di}^{12} = -d_{i3}(mx_{i3}\sin q_{i3} + my_{i3}\cos q_{i3})$.

Introducing (7.64), (7.91), (7.164) and (7.183) into (9.97), simplifying and skipping all mathematical derivations, we get

$$\boldsymbol{\tau}_{td} = \mathbf{M}_d^x(\mathbf{x}, \mathbf{q}_t)\,{}^0\dot{\mathbf{t}}_r + \mathbf{c}_d^x(\mathbf{x}, \mathbf{q}_t, {}^0\mathbf{t}_r) \tag{9.102}$$

where

$$\mathbf{M}_d^x = \mathbf{M}_d \begin{bmatrix} \mathbf{J}^{-1} \\ \mathbf{J}_r \end{bmatrix} \tag{9.103}$$

and

$$\mathbf{c}_d^x = \mathbf{M}_d \begin{bmatrix} \mathbf{0}_{3\times 3} \\ \mathbf{J}_r^d \end{bmatrix} {}^0\mathbf{t}_r + \begin{bmatrix} \mathbf{c}_{d1}^x \\ \mathbf{c}_{d2}^x \\ \mathbf{c}_{d3}^x \end{bmatrix} \tag{9.104}$$

with

$$\mathbf{c}_{di}^x = \mathbf{C}_{di}^r \begin{bmatrix} (\mathbf{j}_{ri}^1\,{}^0\mathbf{t}_r)^2 \\ (\mathbf{j}_{ri}^2\,{}^0\mathbf{t}_r)^2 \\ (\mathbf{j}_{ri}^1\,{}^0\mathbf{t}_r)(\mathbf{j}_{di}^2\,{}^0\mathbf{t}_r) \end{bmatrix} + \mathbf{F}_{v_{di}}\mathbf{J}_{q_{di}}\,{}^0\mathbf{t}_r + \mathbf{f}_{s_{di}} \tag{9.105}$$

in which

- $\mathbf{J}_r$ and $\mathbf{J}_r^d$ are three matrices defined at (9.86) and (9.88),
- $\mathbf{j}_{ri}^j$ the line of the matrix $\mathbf{J}_r$ corresponding to the variable $\dot{q}_{d_{ij}}$, i.e. the jth component of the vector $\dot{\mathbf{q}}_{di}$.

Thus, for one given robot configuration, $\boldsymbol{\tau}_{td}$ is a function of ${}^0\dot{\mathbf{t}}_r$ and ${}^0\mathbf{t}_r$ only.

From the degeneracy analysis of matrix $\mathbf{J}_{tdi}$ of (9.80), the gained motion inside the *LPJTS* singularity of the leg i can be expressed as:

$$\dot{\mathbf{q}}_{di}^s = \begin{bmatrix} d_{i3} \\ -(d_{i2} + d_{i3}) \\ d_{i2} \end{bmatrix}. \tag{9.106}$$

Thus,

- If the leg 1 encounters a *LPJTS* singularity, $\dot{\mathbf{q}}_d^{s\,T} = [\dot{\mathbf{q}}_{d1}^{s\,T}\ \mathbf{0}_3\ \mathbf{0}_3]$,
- If the leg 2 encounters a *LPJTS* singularity, $\dot{\mathbf{q}}_d^{s\,T} = [\mathbf{0}_3\ \dot{\mathbf{q}}_{d2}^{s\,T}\ \mathbf{0}_3]$,
- If the leg 3 encounters a *LPJTS* singularity, $\dot{\mathbf{q}}_d^{s\,T} = [\mathbf{0}_3\ \mathbf{0}_3\ \dot{\mathbf{q}}_{d3}^{s\,T}]$,

where $\mathbf{0}_3$ is a zero vector of dimension 3.

Thus, the criterion (9.15) to satisfy in order to cross the *LPJTS* singularity of the leg i takes the general form

$$\dot{\mathbf{q}}_d^{s\,T} \boldsymbol{\tau}_{t_d} = 0 = \dot{\mathbf{q}}_d^{s\,T} (\mathbf{M}_d^{x\,0}\dot{\mathbf{t}}_r + \mathbf{c}_d^x) \tag{9.107}$$

which, for one given singularity configuration, is a function of ${}^0\dot{\mathbf{t}}_r$ and ${}^0\mathbf{t}_r$ only. Therefore, it is possible to define, for one given singularity configuration, a Cartesian trajectory which respects (9.107).

9.5.5 Simulations and Experimental Results

For the simulations, we have decided to simulate the behavior of a Tripteron during the crossing of a *LPJTS* singularity for the leg 1 with the following hypothesis which does not affect the genericity of the example: we consider that only the elements of the leg 1 have mass and inertia properties (all other terms are canceled).

This hypothesis which may seem strong does not affect the problem because, when crossing the leg 1 *LPJTS* singularity, from the equations of the sections above, it can be seen that only the mass and inertia parameters of the legs can make the dynamic model degenerate. Moreover, this hypothesis brings the following main advantage: we do not have any Tripteron prototype in our laboratory, but we will be able to experimentally simulate the Tripteron behavior during the *LPJTS* singularity crossing by using the five-bar mechanism prototype presented in Sect. 8.4.3.1. Indeed, this experimental simulation can be done by taking into account that:

- the passive planar $3R$ serial chain $A_{12}A_{13}A_{14}$ of the leg 1 of the Tripteron is equivalent to the passive chain $A_{12}PA_{22}$ of the five-bar mechanism (see Figs. 9.1 and 9.3b);
- if we brake the active joint of the five-bar mechanism prototype located at A_{11}, the joint A_{12} of the five-bar prototype mechanism is equivalent to the passive joint A_{12} of the Tripteron (Fig. 9.13)
- then, the crossing of the singularity of the chain $A_{12}A_{13}A_{14}$ of the leg 1 of the Tripteron which is equivalent to the passive chain $A_{12}PA_{22}$ of the five-bar mechanism can be driven by the active link $A_{21}A_{22}$ of the five-bar mechanism prototype that will simulate the end-effector displacement of the Tripteron when motors 2 and 3 are moving (see Figs. 9.2 and 9.4).

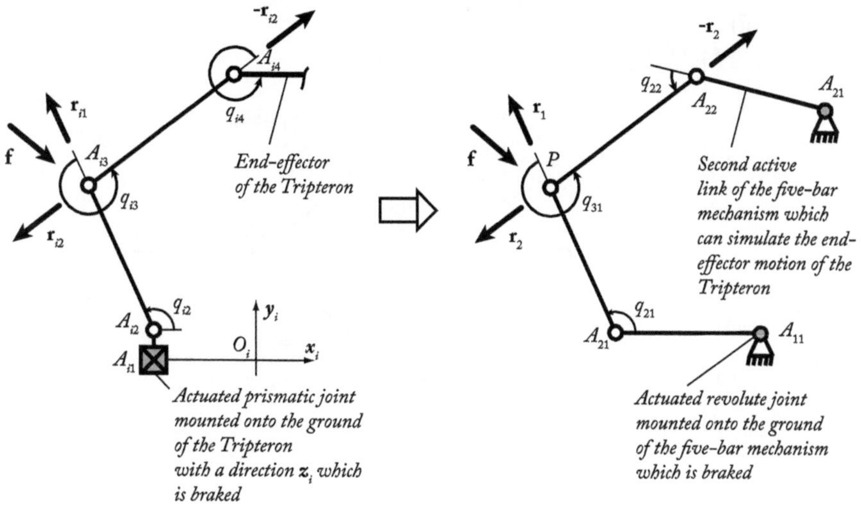

Fig. 9.13 Equivalence between the leg i of the Tripteron and the five-bar mechanism architecture

Due to this analogy, the mass and inertia parameters of the leg 1 of the Tripteron must be equal to:

- $m_{13} = 0.40 \pm 0.02$ kg, $m_{11} = m_{12} = 0$ kg,
- $Ia_{i1} = zz_{i2} = zz_{i3} = 0\,\mathrm{kg\,m^2}$,
- $mx_{i2} = mx_{i3} = my_{i2} = my_{13} = 0\,\mathrm{kg\,m}$,
- $fs_{i2} = fs_{13} = fs_{i4} = 0\,\mathrm{N\,m}$,
- $fv_{i2} = fv_{13} = fv_{i4} = 0\,\mathrm{N\,m/rad}$,

while the length parameters are $d_{13} = 0.1888$ m and $d_{14} = 0.1878$ m in order to fit to the five-bar mechanism prototype parameters.

From (9.107) and (9.102) and by using the parameters given above, the criterion for crossing the *LPJTS* singularities of the leg 1 becomes

$$\dot{\mathbf{q}}_d^{s\,T} \tau_{t_d} = m_{31} d_{i3}^3 \ddot{q}_{21}^2 = 0 \Rightarrow \ddot{q}_{12}^2 = 0 = \mathbf{j}_{d1}^1\, {}^0\dot{\mathbf{t}}_r + \mathbf{J}_d^{d1}\, {}^0\mathbf{t}_r \tag{9.108}$$

where $\mathbf{J}_d^{d1}$ is the first line of the matrix $\mathbf{J}_d^d$.

Let us now define for the point A_{13} of the leg 1 two different types of trajectory with a duration $t_f = 1$ s between the points A_{130} ($\mathbf{x}_{A130} = [x_{A130}\; y_{A130}]^T = [0\,0.338]^T$m) and A_{13f} ($\mathbf{x}_{A13f} = [x_{A13f}\; y_{A13f}]^T = [0\,0.0878]^T$m) which are separated by a *LPJTS* singularity (Fig. 9.14):

- A trajectory defined using fifth-degree polynomial which can fix the position, velocity and acceleration of the robot at the trajectory extremity only; for this polynomial, those conditions are:

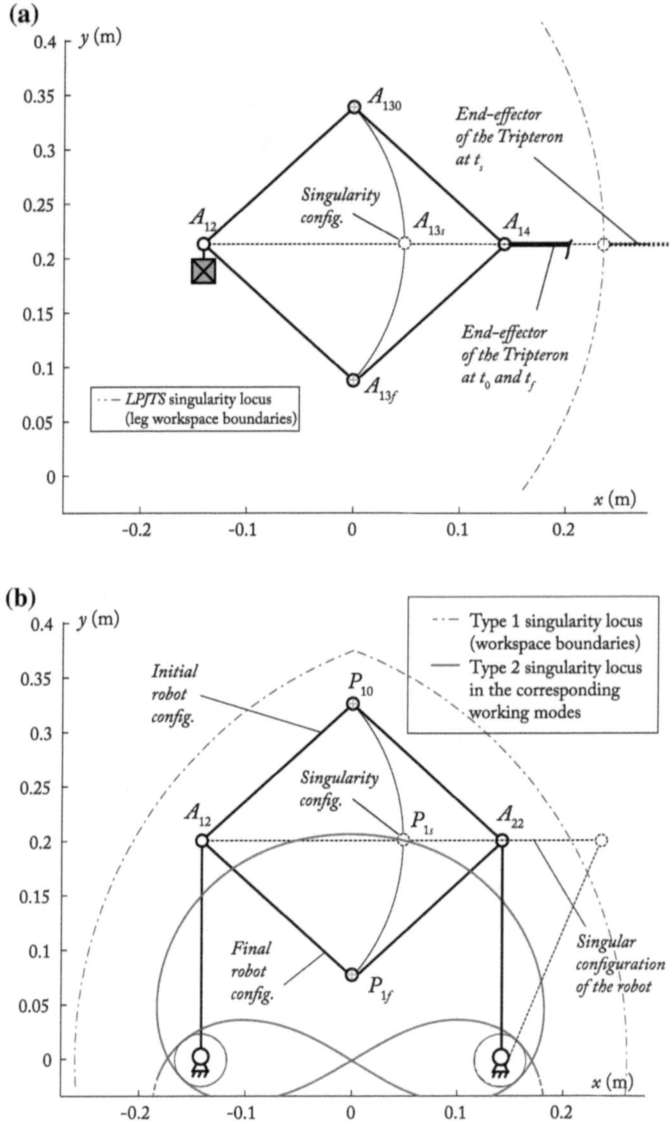

Fig. 9.14 Starting point A_{120} and ending point A_{12f} of the *LPJTS* singularity crossing trajectory for the Tripteron. **a** Trajectory for the Tripteron leg. **b** Equivalent trajectory of the five-bar mechanism

$$y_{A_{13}}(t=0) = y_{A_{130}}, \quad \dot{y}_{A_{13}}(t=0) = 0, \quad \ddot{y}_{A_{13}}(t=0) = 0 \tag{9.109}$$

$$y_{A_{13}}(t=t_f) = y_{A_{13f}}, \quad \dot{y}_{A_{13}}(t=t_f) = 0, \quad \ddot{y}_{A_{13}}(t=t_f) = 0 \tag{9.110}$$

$$x_{A_{13}}(t) = x_{A_{12}} + \sqrt{d_{13}^2 + (y_{A_{13}}(t) - y_{A_{12}})^2} \tag{9.111}$$

which lead to the following polynomials for y (see Appendix D):

$$y_{A_{13}}(t) = 0.33818 - 2.50350t^3 + 3.75525t^4 - 1.50210t^5. \tag{9.112}$$

- A trajectory using an eighth-degree polynomial law which can fix the position, velocity and acceleration of the robot at the trajectory extremity plus the position and acceleration of the robot in the singular configuration; for this polynomial, those conditions are:

$$y_{A_{13}}(t=0) = 0.3381, \quad \dot{y}_{A_{13}}(t=0) = 0, \quad \ddot{y}_{A_{13}}(t=0) = 0 \tag{9.113}$$

$$y_{A_{13}}(t=t_f) = 0.0878, \quad \dot{y}_{A_{13}}(t=t_f) = 0, \quad \ddot{y}_{A_{13}}(t=t_f) = 0 \tag{9.114}$$

$$y_{A_{13}}(t=t_s) = 0.2021, \quad \dot{y}_{A_{13}}(t=t_s) = 0.147, \quad \ddot{y}_{A_{13}}(t=t_s) = -0.693 \tag{9.115}$$

$$x_{A_{13}}(t) = x_{A_{12}} + \sqrt{d_{13}^2 + (y_{A_{13}}(t) - y_{A_{12}})^2} \tag{9.116}$$

which lead to the following polynomials for y (see Appendix D):

$$\begin{aligned} y_{A_{13}}(t) = 0.33818 + 3.05172t^3 - 23.59052t^4 + 43.55897t^5 \\ - 26.66084t^6 - 0.25459t^7 + 3.64490t^8. \end{aligned} \tag{9.117}$$

First, let us simulate the behavior of the robot when following the two different trajectories. The input torques for both trajectories are shown in Figs. 9.15 and 9.16 (τ_1 is not shown as it is null at any time). It can be observed that, for the trajectory that respects the criterion (9.108), the input torques remain finite while in the other case, they tend to infinity.

Now, let us launch each trajectory on the prototype. The results in terms of five-bar mechanism displacement are shown in Figs. 9.17 and 9.18. The torque in the

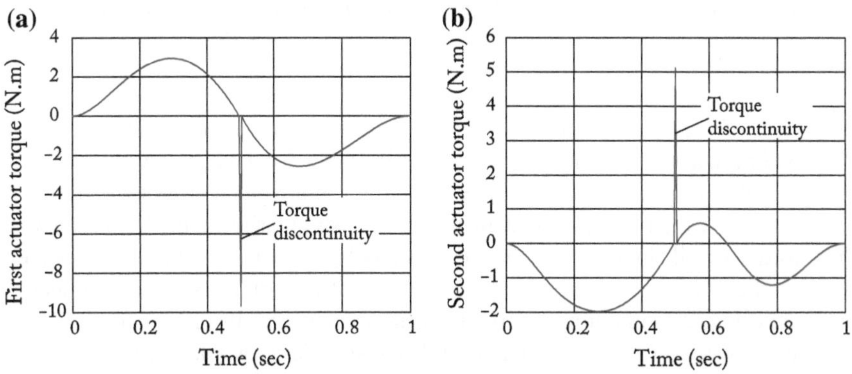

Fig. 9.15 Input torques simulated for the Tripteron crossing the *LPJTS* singularity locus at $t_s = 0.5$ s without respecting the dynamic criterion (9.108). **a** Actuator 2. **b** Actuator 3

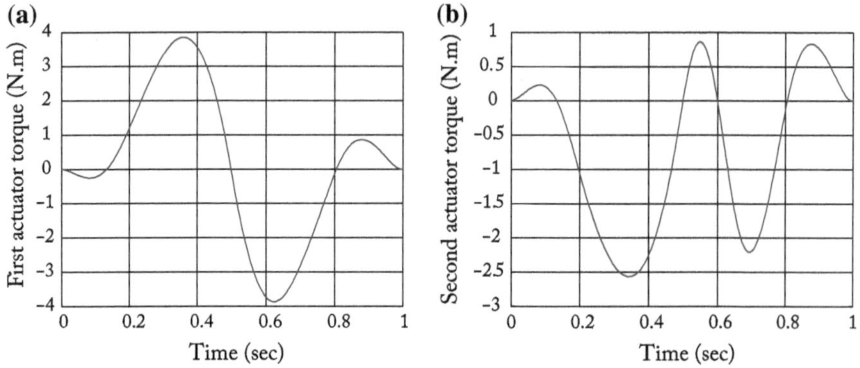

Fig. 9.16 Input torques simulated for the Tripteron crossing the *LPJTS* singularity locus at $t_s = 0.5$ s with respecting the dynamic criterion (9.108). **a** Actuator 2. **b** Actuator 3

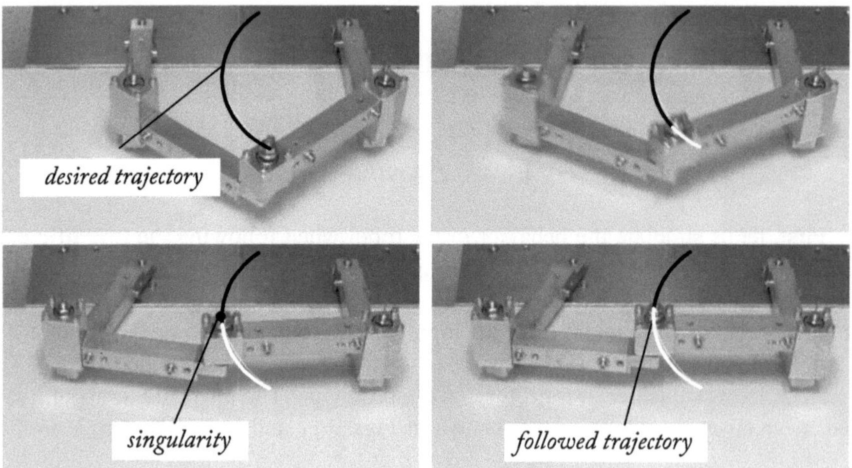

Fig. 9.17 The five-bar mechanism tracking the trajectory which does not respect the dynamic criterion (9.108)

actuator 2 of the five-bar mechanism prototype are given as information, to show their evolution and check their degeneracy (Fig. 9.19). It can be observed that for the trajectory that respects the criterion (9.108), the robot leg can cross the singularity configuration which is equivalent to the *LPJTS* singularity of the Tripteron with finite torques while in the other case, it stays blocked in it at 0.7 s (in order to prevent harming the mechanism, a security stopped the mechanism). Thus, with the trajectory defined at (9.112) without respecting the criterion (9.108), the Tripteron would not be able to cross the *LPJTS* singularity while the singularity would have been crossed by using the trajectory defined at (9.117) without respecting the criterion (9.108).

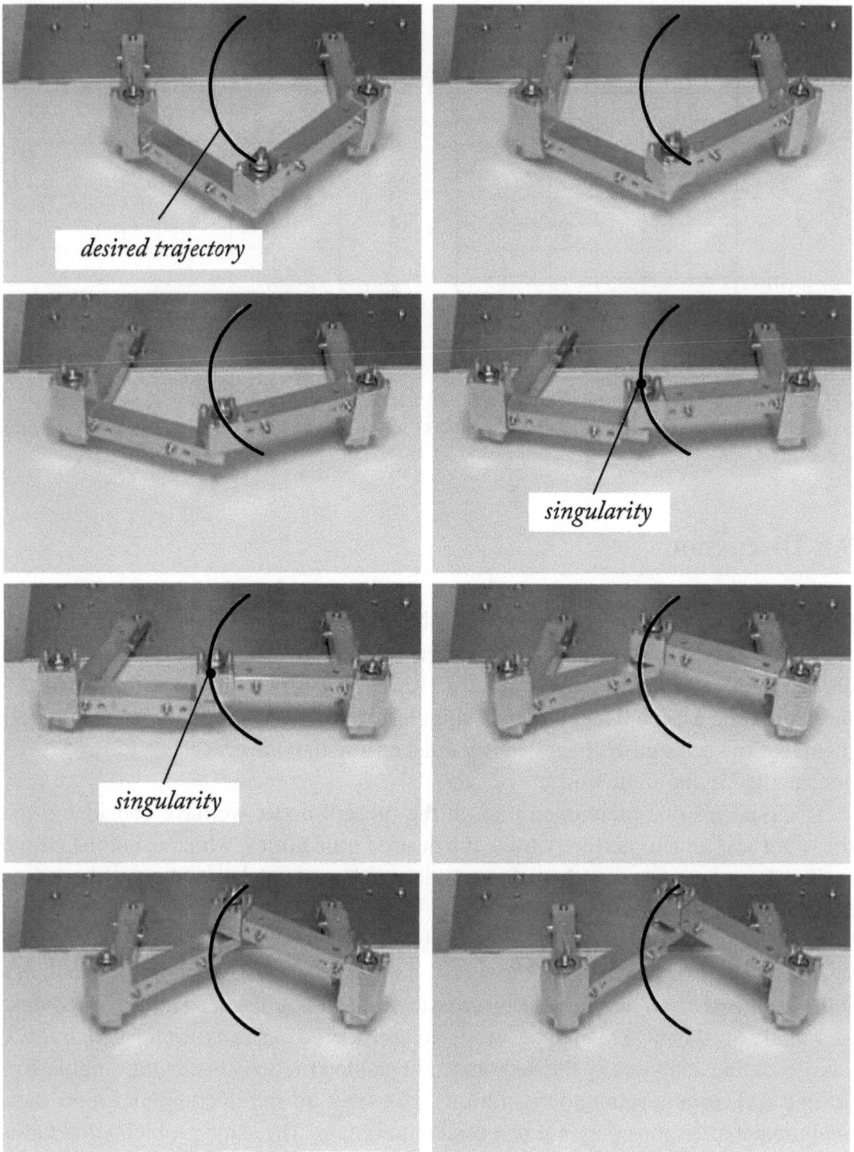

Fig. 9.18 The five-bar mechanism tracking the trajectory respecting the dynamic criterion (9.108)

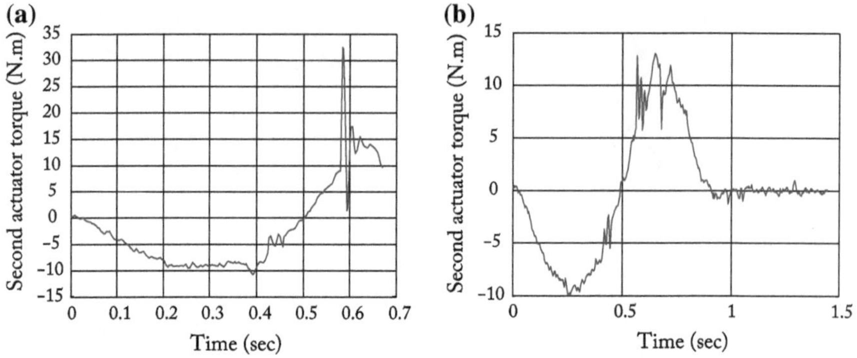

Fig. 9.19 Input torques of the actuator 2 of the five-bar mechanism when tracking two types of trajectories. **a** Without respecting the dynamic criterion (9.108). **b** With respecting the dynamic criterion (9.108)

9.6 Discussion

In this book, we have shown that it is possible to cross the Type 2 and *LPJTS* singularities of the robots without degeneracy of the robot input efforts. We have deliberately chosen to treat each problem separately. However, it is of course possible to cross at the same time a Type 2 and a *LPJTS* singularity, whereas we cannot show it experimentally. In such a case, the trajectory must ensure that the criteria (9.11) and (9.15) are satisfied in the same time.

It should also be mentioned that, in the present book, we have considered that the robot was able to perfectly track the desired trajectories, which is not the case in reality. In order to deal with such a problem, adequate and robust controllers must be developed, such as the one proposed in (Pagis et al. 2014, 2015) for crossing Type 2 singularities. However, developing adequate controllers for crossing Type 2 or *LPJTS* singularities is out of the scope of the present book, even if this problem is interesting and can make singularity crossing more attractive in an industrial context.

Finally, in the present chapter, we have considered case studies for which it was easy to define analytically the vectors of the gained motions inside the singularities ($\mathbf{t}_s$ and $\dot{\mathbf{q}}_d^s$) for any robot configurations. However, for more complex robots these analytical expressions may not be possible to obtain. The same problem could arise for the definition of the singularity loci that the robot will cross. In such a case, a numerical approach must be used which could, for instance for crossing a Type 2 singularity, follow the steps:

1. numerically find a robot singular configuration through which it must pass.
2. then, numerically compute the value of the vector $\mathbf{t}_s$. For that, we can use the "eig" function in *Matlab*, as $\mathbf{t}_s$ is an eigenvector of the matrix $\mathbf{A}_r$ corresponding to a null eigenvalue α. Of course, due to numerical problems, the corresponding

eigenvalue α may not be null, but its absolute value will be lower than a small number ϵ, i.e. $|\alpha| < \epsilon$.

3. compute the value of $\mathbf{w}_d$ as a function of $\mathbf{x}_s$, $\dot{\mathbf{x}}_s$ and $\ddot{\mathbf{x}}_s$, where $\mathbf{x}_s$ represents ($\dot{\mathbf{x}}_s$, $\ddot{\mathbf{x}}_s$, resp.) the platform pose (velocity, acceleration, resp.) at the singular configuration. $\mathbf{x}_s$ can be numerically defined at the first step, while $\dot{\mathbf{x}}_s$ and $\ddot{\mathbf{x}}_s$ must be for the moment taken as variables. At the end of this step, the expressions of each term of $\mathbf{w}_d$ can be obtained under the form of polynomials in $\dot{\mathbf{x}}_s$ and $\ddot{\mathbf{x}}_s$ whose coefficients can be numerically computed.
4. finally, define the criterion (9.11) which must have the form of a polynomial in $\dot{\mathbf{x}}_s$ and $\ddot{\mathbf{x}}_s$ whose coefficients can be numerically computed.

Obviously, to avoid numerical issues due to round-off problems, the maximal computational accuracy in Matlab must be used.

The reader interested in these results can find videos of some robots crossing singularities in the following webpages:

- http://www.irccyn.ec-nantes.fr/~briot/Books.html
- http://www.irccyn.ec-nantes.fr/~briot/SingControl.html
- http://www.irccyn.ec-nantes.fr/~briot/SingExit.html

Moreover, note that the case of flexible *PKM* crossing Type 2 singularities has been investigated in (Briot and Arakelian 2010, 2011).

Part III
Dynamics of Flexible Parallel Robots

Chapter 10
Elastodynamic Modeling of Parallel Robots

Abstract The present chapter deals with computation of the dynamic model of flexible parallel robots. In order to obtain the elastodynamic model, as in the rigid case, all closed loops must be virtually opened to make the platform virtually disassembled from the rest of the structure which becomes a tree structure with all joints actuated. The elastodynamic model of the tree structure and of the free platform is then computed using a systematic procedure based on the generalized Newton-Euler principle, that makes it possible to reduce the computational complexity of the model. Then, the loops are closed using the principle of virtual powers. As a matter of fact, after an introductory section, this chapter will introduce an effective way to compute the elastodynamic models of a single clamped-free flexible link and of a tree structure robot. Then, the computation of the elastodynamic model of parallel robots is investigated. The simulation results obtained with the dynamic model of a flexible parallel robot are compared with experiments.

10.1 Introduction

Many studies have been devoted to computation of the full dynamic model of rigid parallel robots, however there are still many open questions concerning the computation of their elastodynamic model. One of them concerns reduction of the computational time that is generally huge and prevents the use of such models in many applications, such as real-time control, design optimization process, etc. To decrease the computational cost, it is either (i) possible to decrease the number of variables (using model reduction methods (Briot et al. 2011; Craig 1981; Craig and Bampton 1968) and truncated series of shape functions (Blevins 2001)) or (ii) to efficiently compute the symbolic model in order to minimize the number of operators (similarly to what has been done for rigid robots (Khalil and Dombre 2002), robots with lumped springs (Khalil and Gautier 2000) or for serial robots with distributed flexibilities (Boyer and Khalil 1998)). Both methods can obviously be combined.

For the computation of the elastodynamic models of robots, two main approaches are generally proposed (see (Dwivedy and Eberhard 2006b) for a large literature review): (i) lumped modeling (Khalil and Gautier 2000; Kruszewski et al. 1975;

S. Briot and W. Khalil, *Dynamics of Parallel Robots*,
Mechanisms and Machine Science 35, DOI 10.1007/978-3-319-19788-3_10

Wittbrodt et al. 2006) and (ii) modeling using distributed flexibilities (Bauchau 2011; Boyer and Coiffet 1996; Boyer and Khalil 1998; de Jalon and Bayo 1994; Rognant et al. 2010; Shabana 2005; Stachera and Schumacher 2008). Lumped modeling is generally simpler to use by non-experts in finite element methods but, to obtain a model with sufficient accuracy, a higher number of elements is required, thus increasing the computational time. The most relevant works in lumped modeling methods are probably (Khalil and Gautier 2000) (for serial robots) and (Wittbrodt et al. 2006) (for any type of robots). In Khalil and Gautier (2000), the flexibilities are modeled by one *DOF* springs and a systematic procedure for the symbolic computation of the model is proposed. This procedure allows minimization of the number of operators in the model. In Wittbrodt et al. (2006), springs of higher dimension are used, but it is shown that, to obtain good accuracy, the number of elements must be high, thus leading to longer computational time.

Contrary to lumped modeling, using distributed flexibilities allows improvement of the model accuracy. However, such methods require highly-skilled users. In Bauchau (2011), de Jalon and Bayo (1994), Rognant et al. (2010), Shabana (2005), some general methodologies based on the Lagrange principle that can be applied to any system are proposed. In the case of closed-loop mechanisms, some Jacobian matrices are computed that allow taking into account the kinematic dependencies. The work (Stachera and Schumacher 2008) combines the Lagrange principle and the principle of virtual work for computing the elastodynamic model of parallel robots. However, the main drawback of such general methodologies is that they are not specifically designed for parallel robots and that they do not guarantee minimization of the number of operators for the symbolic computation of the model. A first approach for systematic computation of the required Jacobian matrices has been proposed in Bouzgarrou et al. (2005). However, this approach was not designed so that a minimal number of operators for the model symbolic computation can be obtained. The objective of this chapter is to present a systematic procedure to compute the elastodynamic model (using distributed flexibilities) of parallel robot with a minimal numbers of operators. This model is useful for several different reasons:

- In design optimization processes, optimization algorithms that test thousands of robot parameters are used. If the computational time required for the calculation of one iteration of the elastodynamic model of the robot is not minimized, several days, and even months, can pass before the results are obtained.
- Symbolic expressions, with a minimized number of variables and operators, are requested for computing the identification model, in order to decrease the risk of error propagation due to noisy measured data.

In order to minimize the number of operations, a generalized *NE* model (which is known to reduce the number of operators (Boyer and Khalil 1998; Boyer et al. 2007; Khalil and Dombre 2002; Shabana 1990)) is used and combined with the *PVP*. The Jacobian matrices defined in the *PVP* are computed using recursive algorithms that decrease the number of operators.

In this part, we will consider a parallel robot composed of one rigid fixed base (denoted as the element 0), one rigid moving platform and n legs, each leg being a serial kinematic chain composed of $m_i - 1$ bodies linked by m_i joints (revolute, prismatic or even fixed joints—$i = 1, \ldots, n$) (Fig. 8.1a). The actuated variables are denoted by $\mathbf{q}_a$ (of size n_a) and the leg passive variables by $\mathbf{q}_d$ (of size n_d). The platform coordinates are denoted as $\mathbf{x}_p$.

To obtain the desired equations for the *IDM* and *DDM*, we will use a method similar to the one used in Chap. 8 for the rigid case. The problem will be once again divided into two steps (Briot and Khalil 2014a):

1. first, all closed loops are virtually opened to make the platform virtually disassembled from the rest of the structure (Fig. 8.1b) so that the robot becomes a tree structure and a free body: the platform; The leg joints are virtually considered actuated (even for unactuated actual joints). The dynamic model of the tree structure (being composed of flexible bodies) and of the free platform is then computed using a systematic procedure based on the Newton-Euler principle, which makes it possible to obtain

$$\begin{bmatrix} \boldsymbol{\tau}_t \\ \mathbf{0}_{n_e} \end{bmatrix} = \mathbf{dm}_t(\ddot{\mathbf{q}}_t, \dot{\mathbf{q}}_t, \mathbf{q}_t, \ddot{\mathbf{q}}_e, \dot{\mathbf{q}}_e, \mathbf{q}_e, \mathbf{w}_t) \tag{10.1}$$

$$\mathbf{w}_p = \mathbf{idm}_p(\dot{\mathbf{t}}_p, \mathbf{t}_p, \mathbf{x}_p, \mathbf{w}_e) \tag{10.2}$$

   where $\mathbf{dm}_t$ represents the dynamic model of the flexible tree structure, $\mathbf{idm}_p$ the *IDM* of the platform, $\mathbf{q}_t$ are the joint coordinates of the tree structure, $\mathbf{q}_e$ is the vector of the elastic coordinates of dimension n_e, and $\mathbf{t}_p, \mathbf{x}_p$ are the platform twist and pose, $\mathbf{w}_t$ is the system of wrenches applied by the tree-structure robot on the environment and $\mathbf{w}_e$ is the system of wrenches applied by the platform on the environment.
2. Then, the loops are closed using the *PVP*.

As a result, the present chapter is organized as follows:

- Section 10.2 presents the generalized *NE* equations for the considered body. The generalized *NE* equations are optimized so that the number of operators "+", "−", "×" and "/" used for computing the models is minimized.
- Sections 10.3–10.5 present the computation of the dynamic models of

1. the mechanism composed of the virtual tree-structure and the free body corresponding to the platform,
2. the flexible *PKM*.

- Section 10.6 treats the practical implementation of the algorithm.
- Section 10.7 shows a case study.

10.2 Generalized Newton-Euler Equations of a Flexible Link

This section aims at presenting the generalized *NE* model of a flexible clamped-free body (Boyer and Coiffet 1996; Boyer et al. 2007; Shabana 1990; Sharf and Damaren 1992).

Remark

The goal of this chapter and of the following is not to provide a complete lecture on mechanics of deformable bodies, but to show for people having some basic knowledge in this field how to obtain the dynamics of a flexible *PKM* starting from basic considerations in mechanics of deformable bodies (such as the fact that the deformation of a body can be parameterized as a truncated series of Rayleigh-Ritz shape functions, or such as the expression of the potential energy due to elastic deformations, etc.).

If the reader requires additional information on mechanics of deformable bodies, he/she is invited to read the reference books of Shabana (2005) and of Bauchau (2011).

10.2.1 Geometry and First-Order Kinematics of a Clamped-Free Flexible Body

In this part, we consider the model of the flexible body depicted in Fig. 10.1. This body, denoted as body $\mathscr{B}_j$, is supposed to have small elastic deformations around its reference (rigid) configuration $O_{0j}A_{0j}$. The body is clamped at O_j such that $O_{0j} \equiv O_j$. The position of any point M_j belonging to $\mathscr{B}_j$ can be computed as the superposition of

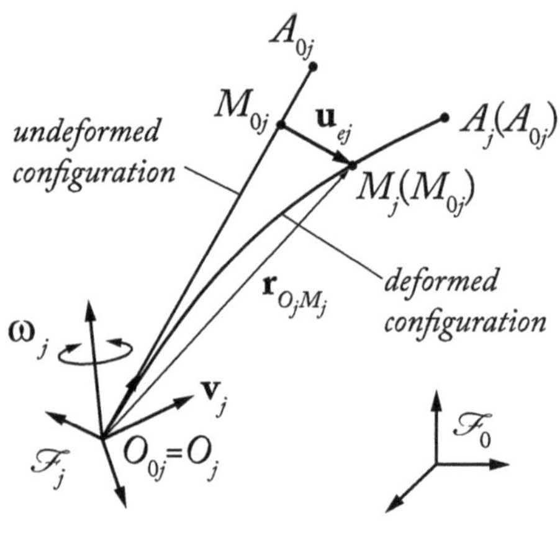

Fig. 10.1 Schematics of one flexible link $\mathscr{B}_j$

- a rigid state $\mathbf{r}_{O_j M_{0j}}$ characterizing the position of the point M_j with respect to the local frame $\mathscr{F}_j$ in absence of deformation (in that case, $M_j \equiv M_{0j}$) and,
- an elastic deformation $\mathbf{u}_{e_j}(M_{0j})$ which can be parameterized as truncated series of Rayleigh-Ritz shape functions (Blevins 2001) such that:

$$\mathbf{u}_{e_j}(M_{0j}) = \mathbf{\Phi}_{d_j}(M_{0j})\mathbf{q}_{e_j} \tag{10.3}$$

with

- $\mathbf{\Phi}_{d_j} = \left[\mathbf{\Phi}_{d1_j}, \ldots, \mathbf{\Phi}_{dN_{j_j}}\right]$, $\mathbf{\Phi}_{dk_j}(M_{0j})$ being the kth shape functions for the displacement of the flexible body expressed at point M_{0j},
- $\mathbf{q}_{e_j}^T = \left[q_{e1_j}, \ldots, q_{eN_{j_j}}\right]$, q_{ek_j} being the kth generalized elastic coordinate of the body $\mathscr{B}_j$ and N_j the number of considered shape functions.

As a result, the vector $\mathbf{r}_{O_j M_j}$ characterizing the position of any point M_j belonging to the body $\mathscr{B}_j$ is given by:

$$\mathbf{r}_{O_j M_j} = \mathbf{r}_{O_j M_{0j}} + \mathbf{u}_{e_j}(M_{0j}) = \mathbf{r}_{O_j M_{0j}} + \mathbf{\Phi}_{d_j}(M_{0j})\mathbf{q}_{e_j}. \tag{10.4}$$

Moreover, the orientation of the body $\mathscr{B}_j$ at any point M_j belonging to this body can also be computed as the superposition of

- a rigid state defined by the orientation matrix ${}^0\mathbf{R}_j^r = {}^0\mathbf{R}_j$ characterizing the rigid orientation of the point M_j with respect to the global frame $\mathscr{F}_0$ in absence of deformation (in that case, $M_j \equiv M_{0j}$) and,
- an elastic deformation state parameterized by the three angles $\phi_j^e(M_{0j})$, $\theta_j^e(M_{0j})$ and $\psi_j^e(M_{0j})$ characterizing the respective rotations around $\boldsymbol{x}_j, \boldsymbol{y}_j$ and $\boldsymbol{z}_j$[1] which can be defined as truncated series of Rayleigh-Ritz shape functions (Blevins 2001) such that:

$$\begin{bmatrix} \phi_j^e(M_{0j}) \\ \theta_j^e(M_{0j}) \\ \psi_j^e(M_{0j}) \end{bmatrix} = \boldsymbol{\vartheta}_{e_j}(M_{0j}) = \mathbf{\Phi}_{r_j}(M_{0j})\mathbf{q}_{e_j} \tag{10.5}$$

with $\mathbf{\Phi}_{r_j} = \left[\mathbf{\Phi}_{r1_j}, \ldots, \mathbf{\Phi}_{rN_{j_j}}\right]$, $\mathbf{\Phi}_{rk_j}(M_{0j})$ being the kth shape functions for the rotation of the flexible body expressed at point M_{0j}.

As a result, the rotation matrix ${}^0\mathbf{R}_j(M_j)$ characterizing the orientation of the body $\mathscr{B}_j$ at any point M_j is given by:

$${}^0\mathbf{R}_j(M_j) = {}^0\mathbf{R}_j^r\ \mathbf{rot}(x, \phi_j^e(M_{0j}))\ \mathbf{rot}(y, \theta_j^e(M_{0j}))\ \mathbf{rot}(z, \psi_j^e(M_{0j})). \tag{10.6}$$

Based on these considerations, the twist of any free flexible body $\mathscr{B}_j$ at any point M_j can always be expressed as (Boyer et al. 2007):

[1] The order of the rotations is not important as we consider small deformations.

$$\begin{bmatrix} \mathbf{v}_{M_j} \\ \boldsymbol{\omega}_{M_j} \end{bmatrix} = \mathbf{t}_j + \begin{bmatrix} \boldsymbol{\omega}_j \times \mathbf{r}_{O_j M_j} \\ \mathbf{0} \end{bmatrix} + \begin{bmatrix} \mathbf{v}_{e_j}(M_j) \\ \boldsymbol{\omega}_{e_j}(M_j) \end{bmatrix} \tag{10.7}$$

where

- $\mathbf{t}_j = \left[\mathbf{v}_j^T, \boldsymbol{\omega}_j^T\right]^T$ is the twist of the local frame $\mathscr{F}_j$ fixed on the body $\mathscr{B}_j$ expressed at point O_j, $\mathbf{v}_j$ and $\boldsymbol{\omega}_j$ being the translational and rotational velocities, respectively,
- $\mathbf{v}_{e_j}(M_j)$ and $\boldsymbol{\omega}_{e_j}(M_j)$ are the translational and rotational velocities due to the body elasticity.

$\mathbf{v}_{e_j}(M_j)$ and $\boldsymbol{\omega}_{e_j}(M_j)$ can be obtained such that:

$$\begin{bmatrix} \mathbf{v}_{e_j}(M_j) \\ \boldsymbol{\omega}_{e_j}(M_j) \end{bmatrix} = \begin{bmatrix} \boldsymbol{\Phi}_{d_j}(M_{0j}) \\ \boldsymbol{\Phi}_{r_j}(M_{0j}) \end{bmatrix} \dot{\mathbf{q}}_{e_j} = \boldsymbol{\Phi}_j(M_{0j})\dot{\mathbf{q}}_{e_j} \tag{10.8}$$

with $\dot{\mathbf{q}}_{e_j}^T = \left[\dot{q}_{e1_j}, \ldots, \dot{q}_{eN_{j_j}}\right]$, $\dot{q}_{ek_j}$ being the kth elastic generalized velocity of the body and N_j the number of considered shape functions.

Thus, (10.7) can be rewritten as:

$$\begin{bmatrix} \mathbf{v}_{M_j} \\ \boldsymbol{\omega}_{M_j} \end{bmatrix} = \mathbf{t}_j + \begin{bmatrix} \boldsymbol{\omega}_j \times (\mathbf{r}_{O_j M_{0j}} + \boldsymbol{\Phi}_{d_j}(M_{0j})\mathbf{q}_{e_j}) \\ \mathbf{0} \end{bmatrix} + \begin{bmatrix} \boldsymbol{\Phi}_{d_j}(M_{0j}) \\ \boldsymbol{\Phi}_{r_j}(M_{0j}) \end{bmatrix} \dot{\mathbf{q}}_{e_j}. \tag{10.9}$$

Equations (10.4)–(10.7) define the kinematic model of the flexible body $\mathscr{B}_j$. This model is thus parameterized by the following set of variables:

- $\mathbf{t}_j$ that are the Euler variables characterizing the rigid velocity of the body $\mathscr{B}_j$ at the origin of the local frame,
- $\mathbf{q}_{e_j}$ that are the Lagrange variables characterizing the elastic displacement of the body $\mathscr{B}_j$.

Thus, the kinematics model of a flexible free body is parameterized by N_j supplementary variables $\mathbf{q}_{e_j}$ and N_j shape functions compared to the free rigid body case.

It should be noted that this description can be applied to both robot segments and joints, as long as all the shape functions can be defined. Many expressions of shape functions can be found in Blevins (2001). Moreover, the shape functions for 3D Bernoulli beams are given in Appendix F.

10.2.2 Computation of the Elastodynamic Model of the Flexible Free Body Using the PVP

In Boyer et al. (2007), the *PVP* is used for computing the elastodynamic model of a free flexible body. This principle, which has been given in Sect. 6.3 in the case of a rigid link, can be rewritten for a flexible body as:

$$P^*_{acc} = P^*_{int} + P^*_{ext} \tag{10.10}$$

where P^*_{acc} is the virtual power due to the acceleration quantities, P^*_{int} is the virtual power due to the internal elastic efforts and P^*_{ext} is the virtual power of the external efforts. Let us recall that the superscript '$*$' stands for a virtual quantity.

10.2.2.1 Computation of the Virtual Power Due to the Acceleration Quantities

By definition, the virtual power due to the acceleration quantities is equal to (Boyer et al. 2007),

$$P^*_{acc} = \int_{\mathscr{B}_j} \dot{\mathbf{v}}^T_{M_j} \mathbf{v}^*_{M_j} \mathrm{d}m \tag{10.11}$$

where $\mathscr{B}_j$ is the body under consideration, dm is a small quantity of mass and

$$\begin{aligned} \dot{\mathbf{v}}_{M_j} &= \dot{\mathbf{v}}_j + \mathbf{\Phi}_{d_j}(M_{0j})\ddot{\mathbf{q}}_{e_j} + 2\boldsymbol{\omega}_j \times \mathbf{\Phi}_{d_j}(M_{0j})\dot{\mathbf{q}}_{e_j} \\ &\quad + \boldsymbol{\omega}_j \times \left(\boldsymbol{\omega}_j \times \mathbf{r}_{O_j M_j}\right) + \dot{\boldsymbol{\omega}}_j \times \mathbf{r}_{O_j M_j} \end{aligned} \tag{10.12}$$

is the translational acceleration of the point M_j with respect to the base frame, with $\dot{\boldsymbol{\omega}}_j$ the rotational acceleration of the considered body at point O_j with respect to the base frame. The expression of $\dot{\mathbf{v}}_{M_j}$ can be obtained through the differentiation w.r.t. time of the expression of $\mathbf{v}_{M_j}$ given in (10.7).

The expression of $\dot{\mathbf{v}}_{M_j}$ can be straightforwardly obtained by differentiating w.r.t. time the upper part of the Eq. (10.9) which states that:

$$\mathbf{v}_{M_j} = \mathbf{v}_j + \boldsymbol{\omega}_j \times (\mathbf{r}_{O_j M_{0j}} + \mathbf{\Phi}_{d_j}(M_{0j})\mathbf{q}_{e_j}) + \mathbf{\Phi}_{d_j}(M_{0j})\dot{\mathbf{q}}_{e_j} \tag{10.13}$$

which implies that

$$\begin{aligned} \dot{\mathbf{v}}_{M_j} &= \dot{\mathbf{v}}_j + \mathbf{\Phi}_{d_j}(M_{0j})\ddot{\mathbf{q}}_{e_j} + \boldsymbol{\omega}_j \times \mathbf{\Phi}_{d_j}(M_{0j})\dot{\mathbf{q}}_{e_j} \\ &\quad + \dot{\boldsymbol{\omega}}_j \times (\mathbf{r}_{O_j M_{0j}} + \mathbf{\Phi}_{d_j}(M_{0j})\mathbf{q}_{e_j}) \\ &\quad + \boldsymbol{\omega}_j \times (\boldsymbol{\omega}_j \times \mathbf{r}_{O_j M_{0j}} + \boldsymbol{\omega}_j \times \mathbf{\Phi}_{d_j}(M_{0j})\mathbf{q}_{e_j} + \mathbf{\Phi}_{d_j}(M_{0j})\dot{\mathbf{q}}_{e_j}). \end{aligned} \tag{10.14}$$

Equation (10.12) can be thus obtained by factorizing the expression (10.14).

Introducing (10.12) into (10.11), it turns out that:

$$\begin{aligned} P^*_{acc} = &\left(\int_{\mathscr{B}_j} \dot{\mathbf{v}}_{M_j} \mathrm{d}m\right)^T \mathbf{v}^*_j + \\ &\left(\int_{\mathscr{B}_j} \mathbf{r}_{O_j M_j} \times \dot{\mathbf{v}}_{M_j} \mathrm{d}m\right)^T \boldsymbol{\omega}^*_j + \end{aligned}$$

$$\left(\int_{\mathscr{B}_j} \mathbf{\Phi}_{d_j}^T(M_{0j})\dot{\mathbf{v}}_{M_j}\mathrm{d}m\right)^T \dot{\mathbf{q}}_{e_j}^* \tag{10.15}$$

By introducing the expression (10.12) in (10.15) and after simplifications, detailed in Appendix E, we get

$$P_{acc}^* = \mathbf{f}_{acc_1}^T \mathbf{v}_j^* + \mathbf{f}_{acc_2}^T \boldsymbol{\omega}_j^* + \mathbf{f}_{acc_3}^T \dot{\mathbf{q}}_{e_j}^* \tag{10.16}$$

in which

$$\begin{aligned}
\mathbf{f}_{acc_1} = & \int_{\mathscr{B}_j} \mathrm{d}m\ \dot{\mathbf{v}}_j + \int_{\mathscr{B}_j} \mathbf{\Phi}_{d_j}(M_{0j})\mathrm{d}m\ \ddot{\mathbf{q}}_{e_j} \\
& + \dot{\boldsymbol{\omega}}_j \times \left(\int_{\mathscr{B}_j} \mathbf{r}_{O_jM_j}\mathrm{d}m\right) \\
& + 2\boldsymbol{\omega}_j \times \left(\int_{\mathscr{B}_j} \mathbf{\Phi}_{d_j}(M_{0j})\mathrm{d}m\ \dot{\mathbf{q}}_{e_j}\right) \\
& + \boldsymbol{\omega}_j \times \left(\boldsymbol{\omega}_j \times \left(\int_{\mathscr{B}_j} \mathbf{r}_{O_jM_j}\mathrm{d}m\right)\right)
\end{aligned} \tag{10.17}$$

$$\begin{aligned}
\mathbf{f}_{acc_2} = & \left(\int_{\mathscr{B}_j} \mathbf{r}_{O_jM_j}\mathrm{d}m\right) \times\ \dot{\mathbf{v}}_j + \int_{\mathscr{B}_j} \hat{\mathbf{r}}_{O_jM_j}\mathbf{\Phi}_{d_j}(M_{0j})\mathrm{d}m\ \ddot{\mathbf{q}}_{e_j} \\
& + \int_{\mathscr{B}_j} \hat{\mathbf{r}}_{O_jM_j}^T\hat{\mathbf{r}}_{O_jM_j}\mathrm{d}m\ \dot{\boldsymbol{\omega}}_j \\
& + \boldsymbol{\omega}_j \times \left(\left(\left(\int_{\mathscr{B}_j} \hat{\mathbf{r}}_{O_jM_j}^T\hat{\mathbf{r}}_{O_jM_j}\mathrm{d}m\right)\boldsymbol{\omega}_j\right)\right) \\
& + 2\sum_{k=1}^{N_j}\left(\int_{\mathscr{B}_j} \hat{\mathbf{r}}_j^T(M_{0j})\hat{\mathbf{\Phi}}_{dk_j}(M_{0j})\mathrm{d}m\right)\boldsymbol{\omega}_j\dot{q}_{ek_j} \\
& + 2\sum_{i,k=1}^{N_j}\left(\int_{\mathscr{B}_j} \hat{\mathbf{\Phi}}_{di_j}^T(M_{0j})\hat{\mathbf{\Phi}}_{dk_j}(M_{0j})\mathrm{d}m\right)\boldsymbol{\omega}_j q_{ei_j}\dot{q}_{ek_j}
\end{aligned} \tag{10.18}$$

in which $\hat{\mathbf{\Phi}}_{dk_j}$ is the cross-product matrix associated with the vector $\mathbf{\Phi}_{dk_j}$, and

$$\mathbf{f}_{acc_3}|_i = \left(\int_{\mathscr{B}_j} \mathbf{\Phi}_{d_j}(M_{0j})\mathrm{d}m\right)^T |_i\ \dot{\mathbf{v}}_j$$

$$
\begin{aligned}
&+\left(\int_{\mathscr{B}_j} \hat{\mathbf{r}}_{O_jM_j}\boldsymbol{\Phi}_{d_j}(M_{0j})\mathrm{d}m\right)^T|_i\,\dot{\boldsymbol{\omega}}_j \\
&+\left(\int_{\mathscr{B}_j} \boldsymbol{\Phi}_{d_j}^T(M_{0j})\boldsymbol{\Phi}_{d_j}(M_{0j})\mathrm{d}m\right)|_i\,\ddot{\mathbf{q}}_{e_j} \\
&+2\sum_{k=1}^{N_j}\left(\int_{\mathscr{B}_j} \boldsymbol{\Phi}_{dk_j}(M_{0j})\times\boldsymbol{\Phi}_{di_j}(M_{0j})\mathrm{d}m\right)^T\boldsymbol{\omega}_j\dot{q}_{ek_j} \\
&-\boldsymbol{\omega}_j^T\left(\int_{\mathscr{B}_j} \hat{\mathbf{r}}_j^T(M_{0j})\hat{\boldsymbol{\Phi}}_{di_j}(M_{0j})\mathrm{d}m\right)^T\boldsymbol{\omega}_j \\
&-\sum_{k=1}^{N_j}\boldsymbol{\omega}_j^T\left(\int_{\mathscr{B}_j} \hat{\boldsymbol{\Phi}}_{di_j}^T(M_{0j})\hat{\boldsymbol{\Phi}}_{dk_j}(M_{0j})\mathrm{d}m\right)\boldsymbol{\omega}_j q_{ek_j}
\end{aligned}
\tag{10.19}
$$

where the symbol '$|_i$' denotes the line i of the considered vector or matrix.

10.2.2.2 Computation of the Virtual Power of External Efforts

The virtual power of external efforts can be divided into two parts

$$P_{ext}^* = P_{grav}^* + P_{reac}^* \tag{10.20}$$

where

$$P_{grav}^* = \int_{\mathscr{B}_j}\left(\mathbf{v}_{M_j}^*\right)^T\mathbf{g}\,\mathrm{d}m \tag{10.21}$$

is the virtual power of the gravity field $\mathbf{g}$, and

$$P_{reac}^* = \mathbf{f}_{O_j}^T\mathbf{v}_j^* + \mathbf{m}_{O_j}^T\boldsymbol{\omega}_j^* - \mathbf{f}_{A_j}^T\mathbf{v}_{A_j}^* - \mathbf{m}_{A_j}^T\boldsymbol{\omega}_{A_j}^* \tag{10.22}$$

is the virtual power due to the reactions at point O_j and A_j (Fig. 10.1), where $\mathbf{f}_{O_j}$ and $\mathbf{m}_{O_j}$ are the force and moment applied at point O_j while $\mathbf{f}_{A_j}$ and $\mathbf{m}_{A_j}$ are the force and moment applied at point A_j.

Thus, introducing (10.7) in the previous expressions, we get

$$
\begin{aligned}
P_{grav}^* = &\left(\int_{\mathscr{B}_j}\mathbf{g}\,\mathrm{d}m\right)^T\mathbf{v}_j^* + \\
&\left(\int_{\mathscr{B}_j}\mathbf{r}_{O_jM_j}\times\mathbf{g}\,\mathrm{d}m\right)^T\boldsymbol{\omega}_j^* +
\end{aligned}
$$

$$\left(\int_{\mathscr{B}_j} \mathbf{g}^T \mathbf{\Phi}_{d_j}(M_{0j})\mathrm{d}m\right)\dot{\mathbf{q}}_{e_j}^* \tag{10.23}$$

which can be expanded as

$$\begin{aligned} P_{grav}^* = &\left(\int_{\mathscr{B}_j} \mathrm{d}m\right)\mathbf{g}^T\ \mathbf{v}_j^* + \\ &\left(\left(\int_{\mathscr{B}_j} \mathbf{r}_{O_jM_j}\ \mathrm{d}m\right)\times \mathbf{g}\right)^T\ \boldsymbol{\omega}_j^* + \\ &\left(\mathbf{g}^T \int_{\mathscr{B}_j} \mathbf{\Phi}_{d_j}(M_{0j})\mathrm{d}m\right)\dot{\mathbf{q}}_{e_j}^* \end{aligned} \tag{10.24}$$

and

$$\begin{aligned} P_{reac}^* = &\left(\mathbf{f}_{O_j} - \mathbf{f}_{A_j}\right)^T\ \mathbf{v}_j^* + \\ &\left(\mathbf{m}_{O_j} - \mathbf{m}_{A_j} - \mathbf{r}_{O_jA_j}\times \mathbf{f}_{A_j}\right)^T\ \boldsymbol{\omega}_j^* - \\ &\left(\mathbf{f}_{A_j}^T \mathbf{\Phi}_{d_j}(A_{0j}) + \mathbf{m}_{A_j}^T \mathbf{\Phi}_{r_j}(A_{0j})\right)\dot{\mathbf{q}}_{e_j}^*. \end{aligned} \tag{10.25}$$

10.2.2.3 Computation of the Virtual Power Due to Internal Elastic Efforts

For computing this quantity, it is first necessary to compute the elastic potential energy of the body $\mathscr{B}_j$.

The elastic potential energy of any body is given by (Shabana 2005):

$$U_{e_j} = \frac{1}{2}\int_{V_j} \boldsymbol{\sigma}_j^T \mathbf{I}_t \boldsymbol{\varepsilon}_j \mathrm{d}V \tag{10.26}$$

where

- V_j is the volume of the body $\mathscr{B}_j$,
- $\boldsymbol{\sigma}_j$ and $\boldsymbol{\varepsilon}_j$ are the six-dimensional stress and strain vectors due to the small elastic displacement $\mathbf{u}_{e_j}(M_{0j}) = \mathbf{\Phi}_{d_j}(M_{0j})\mathbf{q}_{e_j}$ in body $\mathscr{B}_j$,
- $\mathbf{I}_t$ is a (6×6) diagonal matrix. The first three diagonal terms are equal to one, whereas the last three diagonal terms are equal to two, because of the two multipliers associated with the shear strains (Shabana 2005), i.e.

$$\mathbf{I}_t = \begin{bmatrix} \mathbf{1}_3 & \mathbf{0}_{3\times 3} \\ \mathbf{0}_{3\times 3} & 2\,\mathbf{1}_3 \end{bmatrix}. \tag{10.27}$$

The strain vector is defined as $\boldsymbol{\varepsilon}_j = \begin{bmatrix} \varepsilon_{j11} & \varepsilon_{j22} & \varepsilon_{j33} & \varepsilon_{j12} & \varepsilon_{j13} & \varepsilon_{j23} \end{bmatrix}^T$, where:

$$\begin{bmatrix} \varepsilon_{j11} & \varepsilon_{j12} & \varepsilon_{j13} \\ \varepsilon_{j12} & \varepsilon_{j22} & \varepsilon_{j23} \\ \varepsilon_{j13} & \varepsilon_{j23} & \varepsilon_{j33} \end{bmatrix} = \frac{1}{2}\left(\nabla\,\mathbf{u}_{e_j}(M_{0j}) + \left(\nabla\,\mathbf{u}_{e_j}(M_{0j})\right)^T\right) \tag{10.28}$$

with

$$\begin{aligned} \nabla\,\mathbf{u}_{e_j}(M_{0j}) &= \begin{bmatrix} \dfrac{\partial \mathbf{u}_{e_j}}{\partial x}(M_{0j}) & \dfrac{\partial \mathbf{u}_{e_j}}{\partial y}(M_{0j}) & \dfrac{\partial \mathbf{u}_{e_j}}{\partial z}(M_{0j}) \end{bmatrix} \\ &= \begin{bmatrix} \dfrac{\partial \boldsymbol{\Phi}_{d_j}^{(1)}}{\partial x}(M_{0j})\mathbf{q}_{e_j} & \dfrac{\partial \boldsymbol{\Phi}_{d_j}^{(1)}}{\partial y}(M_{0j})\mathbf{q}_{e_j} & \dfrac{\partial \boldsymbol{\Phi}_{d_j}^{(1)}}{\partial z}(M_{0j})\mathbf{q}_{e_j} \\ \dfrac{\partial \boldsymbol{\Phi}_{d_j}^{(2)}}{\partial x}(M_{0j})\mathbf{q}_{e_j} & \dfrac{\partial \boldsymbol{\Phi}_{d_j}^{(2)}}{\partial y}(M_{0j})\mathbf{q}_{e_j} & \dfrac{\partial \boldsymbol{\Phi}_{d_j}^{(2)}}{\partial z}(M_{0j})\mathbf{q}_{e_j} \\ \dfrac{\partial \boldsymbol{\Phi}_{d_j}^{(3)}}{\partial x}(M_{0j})\mathbf{q}_{e_j} & \dfrac{\partial \boldsymbol{\Phi}_{d_j}^{(3)}}{\partial y}(M_{0j})\mathbf{q}_{e_j} & \dfrac{\partial \boldsymbol{\Phi}_{d_j}^{(3)}}{\partial z}(M_{0j})\mathbf{q}_{e_j} \end{bmatrix} \end{aligned} \tag{10.29}$$

where $\boldsymbol{\Phi}_{d_j}^{(k)}$ corresponds to the kth row of matrix $\boldsymbol{\Phi}_{d_j}$, $k = 1, 2, 3$. As a result,

$$\boldsymbol{\varepsilon}_j = \begin{bmatrix} \dfrac{\partial \boldsymbol{\Phi}_{d_j}^{(1)}}{\partial x}(M_{0j}) \\ \dfrac{\partial \boldsymbol{\Phi}_{d_j}^{(2)}}{\partial y}(M_{0j}) \\ \dfrac{\partial \boldsymbol{\Phi}_{d_j}^{(3)}}{\partial z}(M_{0j}) \\ \dfrac{1}{2}\left(\dfrac{\partial \boldsymbol{\Phi}_{d_j}^{(1)}}{\partial y}(M_{0j}) + \dfrac{\partial \boldsymbol{\Phi}_{d_j}^{(2)}}{\partial x}(M_{0j})\right) \\ \dfrac{1}{2}\left(\dfrac{\partial \boldsymbol{\Phi}_{d_j}^{(1)}}{\partial z}(M_{0j}) + \dfrac{\partial \boldsymbol{\Phi}_{d_j}^{(3)}}{\partial x}(M_{0j})\right) \\ \dfrac{1}{2}\left(\dfrac{\partial \boldsymbol{\Phi}_{d_j}^{(2)}}{\partial z}(M_{0j}) + \dfrac{\partial \boldsymbol{\Phi}_{d_j}^{(3)}}{\partial y}(M_{0j})\right) \end{bmatrix} \mathbf{q}_{e_j} = \boldsymbol{\Phi}_{\varepsilon_j}\mathbf{q}_{e_j}. \tag{10.30}$$

The stress vector is expressed as:

$$\boldsymbol{\sigma}_j = \begin{bmatrix} \sigma_{j11} & \sigma_{j22} & \sigma_{j33} & \sigma_{j12} & \sigma_{j13} & \sigma_{j23} \end{bmatrix}^T = \mathbf{H}_j\boldsymbol{\varepsilon}_j \tag{10.31}$$

where matrix $\mathbf{H}_j$ is given by Hooke's law (Shabana 2005), and is equal to, for an isotropic elastic material with elastic modulus E_j and Poisson coefficient ν_j,

$$\mathbf{H}_j = \frac{E_j}{(1+\nu_j)(1-2\nu_j)} \begin{bmatrix} \mathbf{H}_j^{(11)} & \mathbf{0}_{3\times 3} \\ \mathbf{0}_{3\times 3} & \frac{1-2\nu_j}{2}\mathbf{1}_3 \end{bmatrix} \tag{10.32}$$

in which

$$\mathbf{H}_j^{(11)} = \begin{bmatrix} 1-\nu_j & \nu_j & \nu_j \\ \nu_j & 1-\nu_j & \nu_j \\ \nu_j & \nu_j & 1-\nu_j \end{bmatrix}. \tag{10.33}$$

Thus, introducing (10.30) and (10.31) into (10.26) leads to:

$$U_{e_j} = \frac{1}{2}\mathbf{q}_{e_j}^T \mathbf{K}_{ee_j} \mathbf{q}_{e_j} \tag{10.34}$$

where $\mathbf{K}_{ee_j}$ is the stiffness matrix of body $\mathscr{B}_j$ and takes the form:

$$\mathbf{K}_{ee_j} = \int_{V_j} \mathbf{\Phi}_{\varepsilon_j}^T \mathbf{H}_j^T \mathbf{I}_t \mathbf{\Phi}_{\varepsilon_j} \mathrm{d}V. \tag{10.35}$$

The resulting expression of the virtual power due to internal elastic efforts is thus given by (Boyer et al. 2007):

$$P_{int}^* = -\left(\frac{\partial U_{e_j}}{\partial \mathbf{q}_{e_j}}\right)^T \dot{\mathbf{q}}_{e_j}^* = -\mathbf{q}_{e_j}^T \mathbf{K}_{ee_j} \dot{\mathbf{q}}_{e_j}^*. \tag{10.36}$$

The stiffness matrices for 3D Bernoulli beams are given in Appendix F.

10.2.2.4 Computation of Equilibrium Equations

Thus, introducing (10.16), (10.24), (10.25) and (10.36) into (10.10), developing and simplifying the expressions, and taking into account the fact that the virtual velocities $\mathbf{v}_j^*$, $\boldsymbol{\omega}_j^*$ and $\dot{q}_{ei_j}^*$ are independent, three sets of equilibrium equations can be obtained:

$$\begin{aligned} \Sigma\mathbf{f}_j =& m_j\left(\dot{\mathbf{v}}_j - \mathbf{g}\right) + \mathbf{MS}_{de_j}\ddot{\mathbf{q}}_{e_j} + 2\boldsymbol{\omega}_j \times \mathbf{MS}_{de_j}\dot{\mathbf{q}}_{e_j} \\ & - \mathbf{ms}_j \times \dot{\boldsymbol{\omega}}_j + \boldsymbol{\omega}_j \times \left(\boldsymbol{\omega}_j \times \mathbf{ms}_j\right) \end{aligned} \tag{10.37}$$

$$\Sigma\mathbf{m}_j = \mathbf{ms}_j \times \left(\dot{\mathbf{v}}_j - \mathbf{g}\right) + \mathbf{MS}_{re_j}\ddot{\mathbf{q}}_{e_j}$$

$$+2\sum_{i,k=1}^{N_j} \mathbf{I}_{ee_{ikj}} \boldsymbol{\omega}_j q_{ei_j} \dot{q}_{ek_j} + \mathbf{I}_{O_j} \dot{\boldsymbol{\omega}}_j$$
$$+\boldsymbol{\omega}_j \times \left(\mathbf{I}_{O_j} \boldsymbol{\omega}_j\right) + 2\sum_{k=1}^{N_j} \mathbf{I}_{re_{kj}} \boldsymbol{\omega}_j \dot{q}_{ek_j} \tag{10.38}$$

and

$$\begin{aligned}\Sigma \mathbf{s}_j|_i =& \mathbf{MS}_{de_j}^T|_i \ (\dot{\mathbf{v}}_j - \mathbf{g}) + \mathbf{MS}_{re_j}^T|_i \ \dot{\boldsymbol{\omega}}_j + \mathbf{M}_{ee_j}|_i \ \ddot{\mathbf{q}}_{e_j} \\ &+2\sum_{k=1}^{N_j} \boldsymbol{\lambda}_{ki_j}^T \boldsymbol{\omega}_j \dot{q}_{ek_j} - \boldsymbol{\omega}_j^T \mathbf{I}_{re_{ij}}^T \boldsymbol{\omega}_j \\ &-\sum_{k=1}^{N_j} \boldsymbol{\omega}_j^T \mathbf{I}_{ee_{ikj}} \boldsymbol{\omega}_j q_{ek_j} + \mathbf{K}_{ee_j}|_i \ \mathbf{q}_{e_j}\end{aligned} \tag{10.39}$$

where the symbol '$|_i$' denotes the row i of the considered vector or matrix. Equations (10.37)–(10.39) represent respectively the linear rigid equilibrium, the angular rigid equilibrium and the elastic equilibrium.

In the expressions (10.37)–(10.39), m_j is the mass of the body $\mathscr{B}_j$ and the terms $\mathbf{ms}_j$, $\mathbf{MS}_{de_j}$, $\mathbf{MS}_{re_j}$, $\mathbf{I}_{O_j}$, $\mathbf{I}_{re_{ij}}$, $\mathbf{I}_{ee_{ikj}}$, $\mathbf{M}_{ee_j}$, $\Sigma \mathbf{f}_j$, $\Sigma \mathbf{m}_j$ and $\Sigma \mathbf{s}_j$ are defined in the following expressions:

$$\begin{aligned}\mathbf{ms}_j &= \int_{\mathscr{B}_j} \mathbf{r}_{O_j M_j} \mathrm{d}m \\ &= \int_{\mathscr{B}_j} \mathbf{r}_{O_j M_{0j}} \mathrm{d}m + \int_{\mathscr{B}_j} \boldsymbol{\Phi}_{d_j}(M_{0j}) \mathrm{d}m \ \mathbf{q}_{e_j} \\ &= \mathbf{ms}_{r_j} + \mathbf{MS}_{de_j} \mathbf{q}_{e_j}\end{aligned} \tag{10.40}$$

is the global vector of the first moments of inertia in which

$$\mathbf{ms}_{r_j} = \int_{\mathscr{B}_j} \mathbf{r}_{O_j M_{0j}} \mathrm{d}m \tag{10.41}$$

is the (3×1) constant vector of the first moments of inertia of the rigid link,

$$\mathbf{MS}_{de_j} = \begin{bmatrix} msdex_j \\ msdey_j \\ msdez_j \end{bmatrix} = \int_{\mathscr{B}_j} \boldsymbol{\Phi}_{d_j}(M_{0j}) \mathrm{d}m \tag{10.42}$$

is a $(3 \times N_j)$ constant matrix when expressed in the frame $\mathscr{F}_{ij}$,

$$\begin{aligned}\mathbf{MS}_{re_j} &= \int_{\mathscr{B}_j} \hat{\mathbf{r}}_{O_jM_j} \mathbf{\Phi}_{d_j}(M_{0j})\mathrm{d}m \\ &= \int_{\mathscr{B}_j} \hat{\mathbf{r}}_{O_jM_{0j}} \mathbf{\Phi}_{d_j}\mathrm{d}m + \int_{\mathscr{B}_j} (\widehat{\mathbf{\Phi}_{d_j}(M_{0j})}\mathbf{q}_{e_j})\mathbf{\Phi}_{d_j}(M_{0j})\mathrm{d}m \\ &= \left[\boldsymbol{\beta}_{1_j} + \textstyle\sum_{k=1}^{N_j} \boldsymbol{\lambda}_{k1_j} q_{ek_j} \ \dots \ \boldsymbol{\beta}_{N_j,j} + \textstyle\sum_{k=1}^{N_j} \boldsymbol{\lambda}_{kN_{j},j} q_{ek_j}\right]\end{aligned} \tag{10.43}$$

where

$$\boldsymbol{\beta}_{k_j} = \int_{\mathscr{B}_j} \mathbf{r}_{O_jM_{0j}} \times \mathbf{\Phi}_{dk_j}(M_{0j})\mathrm{d}m \tag{10.44}$$

and

$$\boldsymbol{\lambda}_{ki_j} = \int_{\mathscr{B}_j} \mathbf{\Phi}_{dk_j}(M_{0j}) \times \mathbf{\Phi}_{di_j}(M_{0j})\mathrm{d}m \tag{10.45}$$

are two (3×1) constant vectors when expressed in the frame $\mathscr{F}_{ij}$ in which, let us recall,

- $\mathbf{\Phi}_{di_j}$ corresponds to the ith column of the matrix $\mathbf{\Phi}_{d_j}$,
- $\mathbf{\Phi}_{dk_j}$ corresponds to the kth column of the matrix $\mathbf{\Phi}_{d_j}$,

$$\begin{aligned}\mathbf{I}_{O_j} &= \int_{\mathscr{B}_j} \hat{\mathbf{r}}^T_{O_jM_j} \hat{\mathbf{r}}_{O_jM_j}\mathrm{d}m \\ &= \int_{\mathscr{B}_j} \hat{\mathbf{r}}^T_{O_jM_{0j}} \hat{\mathbf{r}}_{O_jM_{0j}}\mathrm{d}m + \int_{\mathscr{B}_j} \sum_{k=1}^{N_j} \hat{\mathbf{r}}^T_{O_jM_{0j}} \hat{\mathbf{\Phi}}_{dk_j}(M_{0j}) q_{ek_j}\mathrm{d}m \\ &+ \int_{\mathscr{B}_j} \sum_{k=1}^{N_j} \hat{\mathbf{\Phi}}^T_{dk_j}(M_{0j}) \hat{\mathbf{r}}_{O_jM_{0j}} q_{ek_j}\mathrm{d}m \\ &+ \int_{\mathscr{B}_j} \sum_{i,k=1}^{N_j} \hat{\mathbf{\Phi}}^T_{di_j}(M_{0j}) \hat{\mathbf{\Phi}}_{dk_j}(M_{0j}) q_{ei_j} q_{ek_j}\mathrm{d}m \\ &= \mathbf{I}_{rr_j} + \sum_{k=1}^{N_j} \left(\mathbf{I}_{rek_j} + \mathbf{I}^T_{rekj}\right) q_{ek_j} + \sum_{i,k=1}^{N_j} \mathbf{I}_{eeikj} q_{ei_j} q_{ek_j}\end{aligned} \tag{10.46}$$

is the global matrix of the second moments of inertia, in which:

$$\mathbf{I}_{rr_j} = \int_{\mathscr{B}_j} \hat{\mathbf{r}}^T_{O_jM_{0j}} \hat{\mathbf{r}}_{O_jM_{0j}}\mathrm{d}m \tag{10.47}$$

is the (3×3) constant inertia matrix of the rigid body when expressed in the frame $\mathscr{F}_{ij}$,

$$\mathbf{I}_{re_{kj}} = \int_{\mathscr{B}_j} \hat{\mathbf{r}}^T_{O_j M_{0j}} \hat{\mathbf{\Phi}}_{dk_j}(M_{0j})\mathrm{d}m \tag{10.48}$$

and

$$\mathbf{I}_{ee_{ikj}} = \int_{\mathscr{B}_j} \hat{\mathbf{\Phi}}^T_{di_j}(M_{0j})\hat{\mathbf{\Phi}}_{dk_j}(M_{0j})\mathrm{d}m \tag{10.49}$$

are (3×3) constant matrices when expressed in the frame $\mathscr{F}_{ij}$,

$$\mathbf{M}_{ee_j} = \int_{\mathscr{B}_j} \mathbf{\Phi}^T_{d_j}(M_{0j})\mathbf{\Phi}_{d_j}(M_{0j})\mathrm{d}m \tag{10.50}$$

is a $(N_j \times N_j)$ constant matrix when expressed in the frame $\mathscr{F}_{ij}$, and

$$\Sigma \mathbf{f}_j = \mathbf{f}_{O_j} - \mathbf{f}_{A_j} \tag{10.51}$$

and

$$\Sigma \mathbf{m}_j = \mathbf{m}_{O_j} - \mathbf{m}_{A_j} - \mathbf{r}_{O_j A_j} \times \mathbf{f}_{A_j} \tag{10.52}$$

represent the total external forces and moments at point O_j, and

$$\Sigma \mathbf{s}_j = \mathbf{f}^T_{A_j}\mathbf{\Phi}_{d_j}(A_{0j}) + \mathbf{m}^T_{A_j}\mathbf{\Phi}_{r_j}(A_{0j}), \tag{10.53}$$

represents the generalized elastic forces and moments at A_{0j}.

10.2.2.5 Remark on Computation of the Terms $\boldsymbol{\beta}_{k_j}$, $\boldsymbol{\lambda}_{ki_j}$, $\mathbf{I}_{re_{kj}}$, $\mathbf{I}_{ee_{ikj}}$ and $\mathbf{M}_{ee_j}$

The elements of the matrices and vectors $\boldsymbol{\beta}_{k_j}$, $\boldsymbol{\lambda}_{ki_j}$, $\mathbf{I}_{re_{kj}}$, $\mathbf{I}_{ee_{ikj}}$ and $\mathbf{M}_{ee_j}$ are not independent and can be computed via the use of a limited set of parameters.

Let us consider the (3×3) matrix defined by

$$\mathbf{J}_{ee_{ikj}} = \begin{bmatrix} jexx_{ikj} & jexy_{ikj} & jexz_{ikj} \\ jeyx_{ikj} & jeyy_{ikj} & jeyz_{ikj} \\ jezx_{ikj} & jezy_{ikj} & jezz_{ikj} \end{bmatrix} = \int_{\mathscr{B}_j} \mathbf{\Phi}_{di_j}(M_{0j})\mathbf{\Phi}^T_{dk_j}(M_{0j})\mathrm{d}m. \tag{10.54}$$

By definition, we have $\mathbf{J}_{ee_{ikj}} = \mathbf{J}^T_{ee_{kij}}$.

As a result, it turns out that the element on the ith row and kth column of the matrix $\mathbf{M}_{ee_j}$ (denoted as the element $\mathbf{M}_{ee_j}(i,k)$) is equal to:

$$\mathbf{M}_{ee_j}(i,k) = jexx_{ikj} + jeyy_{ikj} + jezz_{ikj}. \tag{10.55}$$

Moreover, we have also

$$\boldsymbol{\lambda}_{ki_j} = \begin{bmatrix} jezy_{ikj} - jeyz_{ikj} \\ jexz_{ikj} - jezx_{ikj} \\ jeyx_{ikj} - jexy_{ikj} \end{bmatrix} \tag{10.56}$$

and

$$\mathbf{I}_{ee_{ikj}} = \begin{bmatrix} jeyy_{ikj} + jezz_{ikj} & -jeyx_{ikj} & -jezx_{ikj} \\ -jexy_{ikj} & jexx_{ikj} + jezz_{ikj} & -jezy_{ikj} \\ -jexz_{ikj} & -jeyz_{ikj} & jexx_{ikj} + jeyy_{ikj} \end{bmatrix}. \tag{10.57}$$

Now, let us consider the (3 × 3) matrix defined by

$$\mathbf{J}_{re_{kj}} = \begin{bmatrix} jrxx_{kj} & jrxy_{kj} & jrxz_{kj} \\ jryx_{kj} & jryy_{kj} & jryz_{kj} \\ jrzx_{kj} & jrzy_{kj} & jrzz_{kj} \end{bmatrix} = \int_{\mathscr{B}_j} \mathbf{r}^T_{O_j M_{0j}} \boldsymbol{\Phi}^T_{dk_j}(M_{0j}) \mathrm{d}m. \tag{10.58}$$

As a result, it turns out that the matrix $\mathbf{I}_{re_{kj}}$ is equal to:

$$\mathbf{I}_{re_{kj}} = \begin{bmatrix} jryy_{kj} + jrzz_{kj} & -jryx_{kj} & -jrzx_{kj} \\ -jrxy_{kj} & jrxx_{kj} + jrzz_{kj} & -jrzy_{kj} \\ -jrxz_{kj} & -jryz_{kj} & jrxx_{kj} + jryy_{kj} \end{bmatrix}. \tag{10.59}$$

Moreover, we have

$$\boldsymbol{\beta}_{k_j} = \begin{bmatrix} jryz_{kj} - jrzy_{kj} \\ jrzx_{kj} - jrxz_{kj} \\ jrxy_{kj} - jryx_{kj} \end{bmatrix}. \tag{10.60}$$

10.2.3 Matrix Form of the Generalized Newton-Euler Model for a Flexible Clamped-Free Body

The generalized *NE* model of a flexible free body presented in the previous section can be put in the following matrix form:

$$\begin{bmatrix} \Sigma\mathbf{f}_j \\ \Sigma\mathbf{m}_j \\ \Sigma\mathbf{s}_j \end{bmatrix} = \begin{bmatrix} m_j\mathbf{1}_3 & \widehat{\mathbf{ms}}_j^T & \mathbf{MS}_{de_j} \\ \widehat{\mathbf{ms}}_j & \mathbf{I}_{O_j} & \mathbf{MS}_{re_j} \\ \mathbf{MS}^T_{de_j} & \mathbf{MS}^T_{re_j} & \mathbf{M}_{ee_j} \end{bmatrix} \begin{bmatrix} \dot{\mathbf{v}}_j \\ \dot{\boldsymbol{\omega}}_j \\ \ddot{\mathbf{q}}_{e_j} \end{bmatrix} + \begin{bmatrix} \mathbf{f}_{in_j} \\ \mathbf{c}_{in_j} \\ \mathbf{s}_{in_j} \end{bmatrix} + \begin{bmatrix} \mathbf{0} \\ \mathbf{0} \\ \mathbf{K}_{ee_j}\mathbf{q}_{e_j} \end{bmatrix} + \begin{bmatrix} \mathbf{f}_{g_j} \\ \mathbf{c}_{g_j} \\ \mathbf{s}_{g_j} \end{bmatrix}$$
$$= \mathbf{M}_j \begin{bmatrix} \dot{\mathbf{t}}_j \\ \ddot{\mathbf{q}}_{e_j} \end{bmatrix} + \mathbf{c}_j \tag{10.61}$$

where

- $\widehat{\mathbf{ms}}_j$ is the (3 × 3) cross-product matrix corresponding to the vector $\mathbf{ms}_j$ defined at (10.40),
- $\dot{\mathbf{t}}_j = \left[\dot{\mathbf{v}}_j^T, \dot{\boldsymbol{\omega}}_j^T\right]^T$ is the acceleration of frame $\mathscr{F}_j$ expressed at point O_j with $\dot{\mathbf{v}}_j$ and $\dot{\boldsymbol{\omega}}_j$ the translational and rotational accelerations of the local frame fixed on the body $\mathscr{B}_j$ at point O_j, respectively,
- $\mathbf{f}_{in_j}$ and $\mathbf{c}_{in_j}$ are vectors of the inertial force and torques, respectively,

$$\mathbf{f}_{in_j} = \boldsymbol{\omega}_j \times \left(\widehat{\mathbf{ms}}_j^T \boldsymbol{\omega}_j + 2\mathbf{MS}_{de_j}\dot{\mathbf{q}}_{e_j}\right) \tag{10.62}$$

$$\begin{aligned}\mathbf{c}_{in_j} =& \boldsymbol{\omega}_j \times \left(\mathbf{I}_{O_j}\boldsymbol{\omega}_j\right) + 2\sum_{k=1}^{N_j} \mathbf{I}_{rek_j}\boldsymbol{\omega}_j \dot{q}_{ek_j} \\ &+ 2\sum_{i,k=1}^{N_j} \mathbf{I}_{eeik_j}\boldsymbol{\omega}_j q_{ei_j}\dot{q}_{ek_j}\end{aligned} \tag{10.63}$$

- $\mathbf{s}_{in_j}$ is the vector of the generalized elastic forces,

$$\begin{aligned}\mathbf{s}_{in_j}|_i =& 2\sum_{k=1}^{N_j} \boldsymbol{\lambda}_{ki_j}\boldsymbol{\omega}_j \dot{q}_{ek_j} - \boldsymbol{\omega}_j^T \mathbf{I}_{rei_j}^T \boldsymbol{\omega}_j \\ &- \sum_{k=1}^{N_j} \boldsymbol{\omega}_j^T \mathbf{I}_{eeik_j}\boldsymbol{\omega}_j q_{ek_j}\end{aligned} \tag{10.64}$$

- $\mathbf{f}_{g_j}$ and $\mathbf{c}_{g_j}$ are vectors of the gravity force and torques plus the other external forces, respectively,

$$\mathbf{f}_{g_j} = -m_j \mathbf{g} \tag{10.65}$$

$$\mathbf{c}_{g_j} = -\widehat{\mathbf{ms}}_j \mathbf{g} \tag{10.66}$$

- $\mathbf{s}_{g_j}$ is the vector of the generalized elastic forces due to gravity,

$$\mathbf{s}_{g_j} = -\mathbf{MS}_{de_j}^T \mathbf{g} \tag{10.67}$$

- $\mathbf{M}_j = \begin{bmatrix} m_j \mathbf{1}_3 & \widehat{\mathbf{ms}}_j^T & \mathbf{MS}_{de_j} \\ \widehat{\mathbf{ms}}_j & \mathbf{I}_{O_j} & \mathbf{MS}_{re_j} \\ \mathbf{MS}_{de_j}^T & \mathbf{MS}_{re_j}^T & \mathbf{M}_{ee_j} \end{bmatrix}$ is the generalized inertia matrix of the body $\mathscr{B}_j$,

- $\mathbf{c}_j = \begin{bmatrix} \mathbf{f}_{in_j} \\ \mathbf{c}_{in_j} \\ \mathbf{s}_{in_j} \end{bmatrix} + \begin{bmatrix} \mathbf{0} \\ \mathbf{0} \\ \mathbf{K}_{ee_j}\mathbf{q}_{e_j} \end{bmatrix} + \begin{bmatrix} \mathbf{f}_{g_j} \\ \mathbf{c}_{g_j} \\ \mathbf{s}_{g_j} \end{bmatrix}$ is the global vector of the centrifugal, Coriolis, gravity and elastic forces of the body $\mathscr{B}_j$.

The Eq. (10.61) generalizes for a flexible body the *NE* model of the rigid case. In fact, by eliminating the elastic terms in (10.61), we get the *NE* model for the rigid body defined in (6.35).

Note that we can see that the gravity effects can be automatically taken into account in (10.61) by adding $-\mathbf{g}$ to the acceleration $\dot{\mathbf{v}}_j$. As a result, the Eq. (10.61) becomes:

$$\begin{bmatrix} \Sigma\mathbf{f}_j \\ \Sigma\mathbf{m}_j \\ \Sigma\mathbf{s}_j \end{bmatrix} = \begin{bmatrix} m_j\mathbf{1}_3 & \widehat{\mathbf{ms}}_j^T & \mathbf{MS}_{de_j} \\ \widehat{\mathbf{ms}}_j & \mathbf{I}_{O_j} & \mathbf{MS}_{re_j} \\ \mathbf{MS}_{de_j}^T & \mathbf{MS}_{re_j}^T & \mathbf{M}_{ee_j} \end{bmatrix} \begin{bmatrix} \dot{\mathbf{v}}_j - \mathbf{g} \\ \dot{\boldsymbol{\omega}}_j \\ \ddot{\mathbf{q}}_{e_j} \end{bmatrix} + \begin{bmatrix} \mathbf{f}_{in_j} \\ \mathbf{c}_{in_j} \\ \mathbf{s}_{in_j} \end{bmatrix} + \begin{bmatrix} \mathbf{0} \\ \mathbf{0} \\ \mathbf{K}_{ee_j}\mathbf{q}_{e_j} \end{bmatrix}. \quad (10.68)$$

This formulation is better in terms of computational cost.

This generalized *NE* model is known to reduce the number of operators necessary for computation of the elastodynamics behavior of a flexible link (Boyer and Khalil 1998).

10.3 Dynamic Model of Virtual Flexible Systems

In this section, we compute the elastodynamic model of the virtual tree-structure and of the free moving platform by applying the *PVP*.

Let us consider a parallel robot composed of one rigid fixed base (denoted as the element 0), one rigid moving platform and n legs, each leg being a serial kinematic chain composed of $m_i - 1$ bodies linked by m_i joints (revolute, prismatic or even fixed joints—$i = 1, \ldots, n$) (Fig. 8.1a). The actuated variables are denoted by $\mathbf{q}_a$ and the leg passive variables by $\mathbf{q}_d$. The platform coordinates are denoted as $\mathbf{x}_p$. The size n_a of $\mathbf{q}_a$ must be equal or superior to the number of *DOF* of the parallel robot. The number of shape functions by element is denoted as N_{ij} ($j = 1, \ldots, m_i - 1$). As a result, there are $n_e = \sum_{i=1}^{n}\sum_{j=1}^{m_i-1} N_{ij}$ elastic variables grouped in the vector $\mathbf{q}_e$. All the active, passive and elastic variables are grouped into the vector $\mathbf{q}_t^T = \left[\mathbf{q}_a^T, \mathbf{q}_d^T, \mathbf{q}_e^T\right]$.

10.3.1 *Application of the* PVP

Considering the jth body of leg i (denoted in the following as the body $\mathscr{B}_{ij}$), the *PVP* states that:

$$\begin{bmatrix} \mathbf{v}_{ij}^{*T} & \boldsymbol{\omega}_{ij}^{*T} & \dot{\mathbf{q}}_{e_{ij}}^{*T} \end{bmatrix} \begin{bmatrix} \Sigma\mathbf{f}_{ij} \\ \Sigma\mathbf{m}_{ij} \\ \Sigma\mathbf{s}_{ij} \end{bmatrix} = \dot{\mathbf{q}}_t^{*T} \begin{bmatrix} \boldsymbol{\tau}_t^{(ij)} \\ \mathbf{0}_{n_e} \end{bmatrix} \quad (10.69)$$

where $\boldsymbol{\tau}_t^{(ij)}$ is the vector of the virtual input torques of the tree structure (Fig. 8.1b) due to the movement of the link $\mathscr{B}_{ij}$ and $\mathbf{0}_{n_e}$ a null vector of dimension n_e.

The twist $\mathbf{t}_{ij}^{*T} = \begin{bmatrix} \mathbf{v}_{ij}^{*T} & \boldsymbol{\omega}_{ij}^{*T} \end{bmatrix}$ and generalized elastic velocities $\dot{\mathbf{q}}_{e_{ij}}^*$ are linked to the generalized velocities $\dot{\mathbf{q}}_t^*$ by the relation:

$$\begin{bmatrix} \mathbf{t}_{ij} \\ \dot{\mathbf{q}}_{e_{ij}} \end{bmatrix} = \mathbf{J}_{ij} \dot{\mathbf{q}}_t \tag{10.70}$$

where $\mathbf{J}_{ij}$ is the Jacobian matrix of the body $\mathscr{B}_i j$ whose expression will be given in the following section.

Equation (10.69) can thus be rewritten as:

$$\dot{\mathbf{q}}_t^{*T} \mathbf{J}_{ij}^T \begin{bmatrix} \Sigma \mathbf{f}_{ij} \\ \Sigma \mathbf{m}_{ij} \\ \Sigma \mathbf{s}_{ij} \end{bmatrix} = \dot{\mathbf{q}}_t^{*T} \begin{bmatrix} \boldsymbol{\tau}_t^{(ij)} \\ \mathbf{0}_{n_e} \end{bmatrix} \tag{10.71}$$

which leads to, for any virtual velocity $\dot{\mathbf{q}}_t^*$:

$$\begin{bmatrix} \boldsymbol{\tau}_t^{(ij)} \\ \mathbf{0}_{n_e} \end{bmatrix} = \mathbf{J}_{ij}^T \begin{bmatrix} \Sigma \mathbf{f}_{ij} \\ \Sigma \mathbf{m}_{ij} \\ \Sigma \mathbf{s}_{ij} \end{bmatrix}. \tag{10.72}$$

Thus, now considering all the links of the robot, it turns out that

$$\begin{aligned} \begin{bmatrix} \boldsymbol{\tau}_t = \sum_{i,j} \boldsymbol{\tau}_t^{(ij)} \\ \mathbf{0}_{n_e} \end{bmatrix} &= \sum_{i,j} \mathbf{J}_{ij}^T \begin{bmatrix} \Sigma \mathbf{f}_{ij} \\ \Sigma \mathbf{m}_{ij} \\ \Sigma \mathbf{s}_{ij} \end{bmatrix} \\ &= \sum_{i,j} \mathbf{J}_{ij}^T \left(\mathbf{M}_{ij} \begin{bmatrix} \dot{\mathbf{t}}_{ij} \\ \ddot{\mathbf{q}}_{e_{ij}} \end{bmatrix} + \mathbf{c}_{ij} \right) \end{aligned} \tag{10.73}$$

where $\boldsymbol{\tau}_t$ is the vector of the tree-structure input efforts.

In the next section, recursive algorithms for the computation of the vectors $\mathbf{t}_{ij}$, $\dot{\mathbf{t}}_{ij}$ and of the Jacobian matrices $\mathbf{J}_{ij}$ are developed.

10.3.2 Recursive Computation of Velocities and Jacobian Matrices

Let us consider Fig. 10.2 describing the displacement of the body $\mathscr{B}_{il}$. From (10.7), we get that

$$\begin{bmatrix} \mathbf{v}_{A_{il}} \\ \boldsymbol{\omega}_{A_{il}} \end{bmatrix} = \mathbf{t}_{il} + \begin{bmatrix} \boldsymbol{\omega}_{il} \times \mathbf{r}_{O_{il}A_{il}} \\ \mathbf{0} \end{bmatrix} + \begin{bmatrix} \mathbf{v}_{e_{il}}(A_{il}) \\ \boldsymbol{\omega}_{e_{il}}(A_{il}) \end{bmatrix}. \tag{10.74}$$

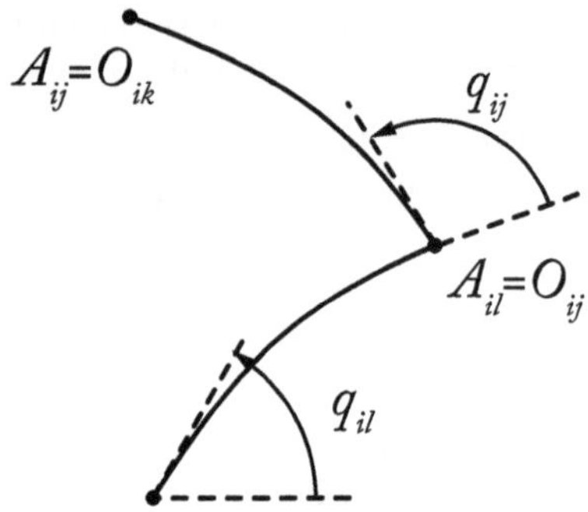

Fig. 10.2 Assembly of two flexible links $\mathscr{B}_{il}$ and $\mathscr{B}_{ij}$

If a body $\mathscr{B}_{ij}$ is linked at A_{il} by an actuated joint moving at a velocity $\dot{q}_{ij}$ (Fig. 10.2—if the joint is a fixed joint, $\dot{q}_{ij} = 0$), we get that:

$$\mathbf{t}_{ij} = \begin{bmatrix} \mathbf{v}_{A_{il}} \\ \boldsymbol{\omega}_{A_{il}} \end{bmatrix} + \dot{q}_{ij} \begin{bmatrix} \sigma_{ij}\mathbf{a}_{ij} \\ \bar{\sigma}_{ij}\mathbf{a}_{ij} \end{bmatrix} \tag{10.75}$$

where $\mathbf{a}_{ij}$ is the unit vector direction along the joint axis (see Sect. 5.2.4).

As a result, for the element $\mathscr{B}_{ij}$ of the global robot, it can be demonstrated that (in the following expressions, the preceding superscript indicates the frame in which the vector expression is given and it is considered that the body $\mathscr{B}_{il}$ is antecedent to the body $\mathscr{B}_{ij}$, i.e. $il = a(ij)$) (Boyer and Khalil 1998):

$$^{ij}\mathbf{t}_{ij} = {}^{ij}\overline{\mathbf{T}}_{il}{}^{il}\mathbf{t}_{il} + {}^{ij}\overline{\mathbf{R}}_{il}\boldsymbol{\Phi}_{il}(O_{ij})\dot{\mathbf{q}}_{e_{il}} + \dot{q}_{ij}{}^{ij}\bar{\mathbf{a}}_{ij} \tag{10.76}$$

where

$$^{ij}\bar{\mathbf{a}}_{ij} = \begin{bmatrix} \sigma_{ij}{}^{ij}\mathbf{a}_{ij} \\ \bar{\sigma}_{ij}{}^{ij}\mathbf{a}_{ij} \end{bmatrix}.$$

Equation (10.76) can also be written as:

$$^{ij}\mathbf{t}_{ij} = \mathbf{J}_{t_{ij}}\dot{\mathbf{q}}_t \tag{10.77}$$

with

$$\mathbf{J}_{t_{ij}} = {}^{ij}\overline{\mathbf{T}}_{il}\mathbf{J}_{t_{il}} + \boldsymbol{\Phi}_{\mathbf{q}_{eij}} + \mathbf{A}_{ij} \tag{10.78}$$

where

$$\begin{aligned} \boldsymbol{\Phi}_{\mathbf{q}_{eij}} &= \begin{bmatrix} \mathbf{0} \cdots {}^{ij}\overline{\mathbf{R}}_{il}\boldsymbol{\Phi}_{il}(O_{ij}) \cdots \mathbf{0} \end{bmatrix} \\ \mathbf{A}_{ij} &= \begin{bmatrix} \mathbf{0} \cdots {}^{ij}\bar{\mathbf{a}}_{ij} \cdots \mathbf{0} \end{bmatrix}. \end{aligned} \tag{10.79}$$

In the matrix $\mathbf{\Phi}_{\mathbf{q}_{eij}}$, the term ${}^{ij}\overline{\mathbf{R}}_{il}\mathbf{\Phi}_{il}(O_{ij})$ is located at the columns corresponding to the variables $\dot{\mathbf{q}}_{e_{il}}$ and, for the matrix $\mathbf{A}_{ij}$, the term ${}^{ij}\bar{\mathbf{a}}_{ij}$ is located at the column corresponding to the variable $\dot{q}_{ij}$.

In the previous expressions, matrix $\mathbf{\Phi}_{il}$ is the matrix containing all shape functions for the element $\mathscr{B}_{il}$ and ${}^{ij}\overline{\mathbf{T}}_{il}$ and ${}^{ij}\overline{\mathbf{R}}_{il}$ are defined at Eqs. (3.13) and (3.25) in Chap. 3. However, the reader must not forget to take into account the rigid and elastic displacements of each body in the computation of these matrices.

Finally, the global Jacobian matrix $\mathbf{J}_{ij}$ of (10.70) can be computed as:

$$\mathbf{J}_{ij} = \begin{bmatrix} \mathbf{J}_{t_{ij}} \\ \mathbf{O}_{\mathbf{q}_{eij}} \end{bmatrix} \tag{10.80}$$

where $\mathbf{O}_{\mathbf{q}_{eij}}$ is defined such that

$$\dot{\mathbf{q}}_{e_{ij}} = \mathbf{O}_{\mathbf{q}_{eij}}\dot{\mathbf{q}}_{t}.$$

10.3.3 Recursive Computation of the Accelerations

Differentiating (10.74), it can be shown that (Boyer and Khalil 1998):

$${}^{ij}\dot{\mathbf{t}}_{ij} = {}^{ij}\overline{\mathbf{T}}_{il}{}^{il}\dot{\mathbf{t}}_{il} + \ddot{q}_{ij}{}^{ij}\bar{\mathbf{a}}_{ij} + {}^{ij}\mathbf{h}_{ij} + {}^{ij}\overline{\mathbf{R}}_{il}\mathbf{\Phi}_{il}(O_{ij})\ddot{\mathbf{q}}_{e_{il}} \tag{10.81}$$

where:

$$\begin{aligned}
{}^{ij}\mathbf{h}_{ij} &= \begin{bmatrix} {}^{ij}\mathbf{h}_{lin_{ij}} \\ {}^{ij}\mathbf{h}_{ang_{ij}} \end{bmatrix} \\
{}^{ij}\mathbf{h}_{lin_{ij}} &= {}^{ij}\mathbf{R}_{il}{}^{il}\boldsymbol{\omega}_{il} \times {}^{ij}\mathbf{R}_{il}\left(2\mathbf{\Phi}_{d_{il}}(O_{ij})\dot{\mathbf{q}}_{e_{il}} + {}^{il}\boldsymbol{\omega}_{il} \times {}^{il}\mathbf{r}_{O_{il}O_{ij}}\right) \\
&\quad + \dot{q}_{ij}\sigma_{ij}{}^{ij}\mathbf{R}_{il}{}^{il}\left((2\boldsymbol{\omega}_{il} + \mathbf{\Phi}_{r_{il}}(O_{ij})\dot{\mathbf{q}}_{e_{il}}) \times {}^{ij}\mathbf{a}_{ij}\right) \\
{}^{ij}\mathbf{h}_{ang_{ij}} &= {}^{ij}\mathbf{R}_{il}\left(\dot{q}_{ij}({}^{il}\boldsymbol{\omega}_{il} + \mathbf{\Phi}_{r_{il}}(O_{ij})\dot{\mathbf{q}}_{e_{il}}) \times \bar{\sigma}_{ij}{}^{ij}\mathbf{a}_{ij}\right) \\
&\quad + {}^{ij}\mathbf{R}_{il}\left({}^{il}\boldsymbol{\omega}_{il} \times {}^{ij}\mathbf{R}_{il}\mathbf{\Phi}_{r_{il}}(O_{ij})\dot{\mathbf{q}}_{e_{il}}\right).
\end{aligned} \tag{10.82}$$

Equation (10.81) can be then put into the form

$${}^{ij}\dot{\mathbf{t}}_{ij} = \mathbf{J}_{t_{ij}}\ddot{\mathbf{q}}_{t} + {}^{ij}\mathbf{g}_{ij} \tag{10.83}$$

with

$${}^{ij}\mathbf{g}_{ij} = {}^{ij}\mathbf{h}_{ij} + {}^{ij}\mathbf{T}_{il}{}^{il}\mathbf{g}_{il} \tag{10.84}$$

initialized with ${}^{i0}\mathbf{g}_{i0} = \mathbf{0}$ if the base is fixed.[2] Thus,

$$\begin{bmatrix} {}^{ij}\dot{\mathbf{t}}_{ij} \\ \ddot{\mathbf{q}}_{e_{ij}} \end{bmatrix} = \begin{bmatrix} \mathbf{J}_{t_{ij}} \\ \mathbf{O}_{\mathbf{q}_{eij}} \end{bmatrix} \ddot{\mathbf{q}}_t + \begin{bmatrix} {}^{ij}\mathbf{g}_{ij} \\ \mathbf{0} \end{bmatrix} = \mathbf{J}_{ij}\ddot{\mathbf{q}}_t + {}^{ij}\mathbf{g}^s_{ij}. \tag{10.85}$$

Proof

From (10.75), we have the velocity expressed in a fixed frame:

$$\mathbf{v}_{ij} = \mathbf{v}_{il} + \boldsymbol{\omega}_{il} \times \mathbf{r}_{O_{il}O_{ij}} + \dot{q}_{ij}\sigma_{ij}\mathbf{a}_{ij} + \mathbf{v}_{e_{il}}(O_{ij}) \tag{10.86}$$

$$\boldsymbol{\omega}_{ij} = \boldsymbol{\omega}_{il} + \dot{q}_{ij}\bar{\sigma}_{ij}\mathbf{a}_{ij} + \boldsymbol{\omega}_{e_{il}}(O_{ij}). \tag{10.87}$$

Differentiating (10.86) and (10.87) w.r.t. time, it turns out that

$$\begin{aligned} \dot{\mathbf{v}}_{ij} =& \dot{\mathbf{v}}_{il} + \dot{\boldsymbol{\omega}}_{il} \times \mathbf{r}_{O_{il}O_{ij}} + \boldsymbol{\omega}_{il} \times \frac{\mathrm{d}}{\mathrm{d}t}\left(\mathbf{r}_{O_{il}O_{ij}}\right) \\ & + \ddot{q}_{ij}\sigma_{ij}\mathbf{a}_{ij} + \dot{q}_{ij}\sigma_{ij}\frac{\mathrm{d}\mathbf{a}_{ij}}{\mathrm{d}t} + \frac{\mathrm{d}\,\mathbf{v}_{e_{il}}(O_{ij})}{\mathrm{d}t} \end{aligned} \tag{10.88}$$

$$\dot{\boldsymbol{\omega}}_{ij} = \dot{\boldsymbol{\omega}}_{il} + \ddot{q}_{ij}\bar{\sigma}_{ij}\mathbf{a}_{ij} + \dot{q}_{ij}\bar{\sigma}_{ij}\frac{\mathrm{d}\mathbf{a}_{ij}}{\mathrm{d}t} + \frac{\mathrm{d}\,\boldsymbol{\omega}_{e_{il}}(O_{ij})}{\mathrm{d}t}. \tag{10.89}$$

Expanding (10.88) and from the fact that:

$$\begin{bmatrix} \mathbf{v}_{e_{il}}(O_{ij}) \\ \boldsymbol{\omega}_{e_{il}}(O_{ij}) \end{bmatrix} = \begin{bmatrix} \boldsymbol{\Phi}_{d_{il}}(O_{ij}) \\ \boldsymbol{\Phi}_{r_{il}}(O_{ij}) \end{bmatrix} \dot{\mathbf{q}}_{e_{il}} \tag{10.90}$$

we get:

$$\begin{aligned} \dot{\mathbf{v}}_{ij} =& \dot{\mathbf{v}}_{il} + \dot{\boldsymbol{\omega}}_{il} \times \mathbf{r}_{O_{il}O_{ij}} \\ & + \boldsymbol{\omega}_{il} \times \left(\boldsymbol{\omega}_{il} \times \mathbf{r}_{O_{il}O_{ij}} + \dot{q}_{ij}\sigma_{ij}\mathbf{a}_{ij}\right) \\ & + \boldsymbol{\omega}_{il} \times \boldsymbol{\Phi}_{d_{il}}(O_{ij})\dot{\mathbf{q}}_{e_{il}} \\ & + \ddot{q}_{ij}\sigma_{ij}\mathbf{a}_{ij} + \dot{q}_{ij}\sigma_{ij}\left(\left(\boldsymbol{\omega}_{il} + \boldsymbol{\omega}_{e_{il}}\right) \times \mathbf{a}_{ij}\right) \\ & + \boldsymbol{\omega}_{il} \times \left(\boldsymbol{\Phi}_{d_{il}}(O_{ij})\dot{\mathbf{q}}_{e_{il}}\right) \\ & + \boldsymbol{\Phi}_{d_{il}}(O_{ij})\ddot{\mathbf{q}}_{e_{il}} \end{aligned} \tag{10.91}$$

or after simplifications and expressing it in the frame $\mathscr{F}_{ij}$,

[2] Note that, as mentioned in Sect. 10.2.3, in order to decrease the computational cost, it is better to initialize the algorithm with ${}^{i0}\mathbf{g}_{i0}^T = [-\mathbf{g}^T\ \mathbf{0}]$. In such a case, the model (10.68) must be used instead of (10.61).

$$\begin{aligned}
{}^{ij}\dot{\mathbf{v}}_{ij} = {}^{ij}\mathbf{R}_{il}({}^{il}\dot{\mathbf{v}}_{il} &+ {}^{il}\dot{\boldsymbol{\omega}}_{il} \times {}^{il}\mathbf{r}_{O_{il}O_{ij}} \\
&+ {}^{il}\boldsymbol{\omega}_{il} \times \left({}^{il}\boldsymbol{\omega}_{il} \times {}^{il}\mathbf{r}_{O_{il}O_{ij}} + \dot{q}_{ij}\sigma_{ij}{}^{il}\mathbf{a}_{ij}\right) \\
&+ {}^{il}\boldsymbol{\omega}_{il} \times \boldsymbol{\Phi}_{d_{il}}(O_{ij})\dot{\mathbf{q}}_{e_{il}} \\
&+ \ddot{q}_{ij}\sigma_{ij}{}^{il}\mathbf{a}_{ij} + \dot{q}_{ij}\sigma_{ij}\left(({}^{il}\boldsymbol{\omega}_{il} + {}^{il}\boldsymbol{\omega}_{e_{il}}) \times {}^{il}\mathbf{a}_{ij}\right) \\
&+ {}^{il}\boldsymbol{\omega}_{il} \times \left(\boldsymbol{\Phi}_{d_{il}}(O_{ij})\dot{\mathbf{q}}_{e_{il}}\right) \\
&+ \boldsymbol{\Phi}_{d_{il}}(O_{ij})\ddot{\mathbf{q}}_{e_{il}})
\end{aligned} \tag{10.92}$$

or also

$$\begin{aligned}
{}^{ij}\dot{\mathbf{v}}_{ij} = {}^{ij}\dot{\mathbf{v}}_{il} &- {}^{ij}\hat{\mathbf{r}}_{O_{il}O_{ij}}{}^{ij}\dot{\boldsymbol{\omega}}_{il} \\
&+ \ddot{q}_{ij}\sigma_{ij}{}^{ij}\mathbf{a}_{ij} + {}^{ij}\mathbf{R}_{il}\boldsymbol{\Phi}_{d_{il}}(O_{ij})\ddot{\mathbf{q}}_{e_{il}} \\
&+ {}^{ij}\boldsymbol{\omega}_{il} \times {}^{ij}\boldsymbol{\omega}_{il} \times {}^{ij}\mathbf{r}_{O_{il}O_{ij}} \\
&+ \dot{q}_{ij}\sigma_{ij}\left((2{}^{ij}\boldsymbol{\omega}_{il} + {}^{ij}\boldsymbol{\omega}_{e_{il}}) \times {}^{ij}\mathbf{a}_{ij}\right) \\
&+ 2{}^{ij}\boldsymbol{\omega}_{il} \times \left({}^{ij}\mathbf{R}_{il}\boldsymbol{\Phi}_{d_{il}}(O_{ij})\dot{\mathbf{q}}_{e_{il}}\right).
\end{aligned} \tag{10.93}$$

Now, expanding (10.89) and using (10.90), we get

$$\begin{aligned}
{}^{ij}\dot{\boldsymbol{\omega}}_{ij} = {}^{ij}\dot{\boldsymbol{\omega}}_{il} &+ \ddot{q}_{ij}\bar{\sigma}_{ij}{}^{ij}\mathbf{a}_{ij} \\
&+ \dot{q}_{ij}\bar{\sigma}_{ij}\left(({}^{ij}\boldsymbol{\omega}_{il} + {}^{ij}\boldsymbol{\omega}_{e_{il}}) \times {}^{ij}\mathbf{a}_{ij}\right) \\
&+ {}^{ij}\boldsymbol{\omega}_{il} \times \left({}^{ij}\mathbf{R}_{il}\boldsymbol{\Phi}_{r_{il}}(O_{ij})\dot{\mathbf{q}}_{e_{il}}\right) \\
&+ {}^{ij}\mathbf{R}_{il}\boldsymbol{\Phi}_{r_{il}}(O_{ij})\ddot{\mathbf{q}}_{e_{il}}.
\end{aligned} \tag{10.94}$$

Finally, introducing the facts that

$$ {}^{ij}\boldsymbol{\omega}_{il} = {}^{ij}\mathbf{R}_{il}{}^{il}\boldsymbol{\omega}_{il} \tag{10.95}$$

$$ {}^{ij}\boldsymbol{\omega}_{e_{il}} = {}^{ij}\mathbf{R}_{il}\,\boldsymbol{\Phi}_{r_{il}}(O_{ij})\dot{\mathbf{q}}_{e_{il}} \tag{10.96}$$

$$ {}^{ij}\overline{\mathbf{R}}_{il} = \begin{bmatrix} {}^{ij}\mathbf{R}_{il} & \mathbf{0}_3 \\ \mathbf{0}_3 & {}^{ij}\mathbf{R}_{il} \end{bmatrix} \tag{10.97}$$

and

$$ {}^{ij}\overline{\mathbf{T}}_{il} = \begin{bmatrix} {}^{ij}\mathbf{R}_{il} & -{}^{ij}\mathbf{R}_{il}{}^{il}\hat{\mathbf{r}}_{ij} \\ \mathbf{0}_3 & {}^{ij}\mathbf{R}_{il} \end{bmatrix} \tag{10.98}$$

in which $\mathbf{r}_{ij} = \mathbf{r}_{O_{il}O_{ij}}$, Eq. (10.81) can be obtained.

10.3.4 Elastodynamic Model of the Virtual System

Introducing (10.85) into (10.73) leads to:

$$\begin{bmatrix} \boldsymbol{\tau}_t \\ \mathbf{0}_{n_e} \end{bmatrix} = \sum_{i,j} \mathbf{J}_{ij}^T \mathbf{M}_{ij} \mathbf{J}_{ij} \ddot{\mathbf{q}}_t + \mathbf{c}_{ij}^s \tag{10.99}$$

where

$$\mathbf{c}_{ij}^s = \mathbf{J}_{ij}^T \left(\mathbf{c}_{ij} + \mathbf{M}_{ij} {}^{ij}\mathbf{g}_{ij}^s \right) \tag{10.100}$$

in which $\mathbf{M}_{ij}$ and $\mathbf{c}_{ij}$ are projected into $\mathscr{F}_{ij}$.

The *NE* equations for a rigid moving platform are given by (from Eq. (8.57)):

$${}^0\mathbf{w}_p = {}^0\mathbf{M}_p {}^0\dot{\mathbf{t}}_p + {}^0\mathbf{c}_p \tag{10.101}$$

where ${}^0\mathbf{w}_p$ is the platform reaction wrench, ${}^0\mathbf{M}_p$ is the platform mass matrix expressed in the base frame and is defined at (8.97), ${}^0\dot{\mathbf{t}}_p$ is the platform acceleration and ${}^0\mathbf{c}_p$ the centrifugal, Coriolis, gravity effects and external efforts applied on the platform defined at (8.98).

Finally, the global elastodynamic model of the virtual structure can be put into the form:

$$\begin{aligned} \begin{bmatrix} \begin{bmatrix} \boldsymbol{\tau}_t \\ \mathbf{0}_{n_e} \end{bmatrix} \\ {}^0\mathbf{w}_p \end{bmatrix} &= \begin{bmatrix} \sum_{i,j} \mathbf{J}_{ij}^T \mathbf{M}_{ij} \mathbf{J}_{ij} & \mathbf{0} \\ \mathbf{0} & {}^0\mathbf{M}_p \end{bmatrix} \begin{bmatrix} \ddot{\mathbf{q}}_t \\ {}^0\dot{\mathbf{t}}_p \end{bmatrix} + \begin{bmatrix} \mathbf{c}_{ij}^s \\ {}^0\mathbf{c}_p \end{bmatrix} \\ &= \mathbf{M}_t \begin{bmatrix} \ddot{\mathbf{q}}_t \\ {}^0\dot{\mathbf{t}}_p \end{bmatrix} + \mathbf{c}_t \end{aligned} \tag{10.102}$$

where

- $\mathbf{M}_t = \begin{bmatrix} \sum_{i,j} \mathbf{J}_{ij}^T \mathbf{M}_{ij} \mathbf{J}_{ij} & \mathbf{0} \\ \mathbf{0} & {}^0\mathbf{M}_p \end{bmatrix}$,
- $\mathbf{c}_t = \begin{bmatrix} \mathbf{c}_{ij}^s \\ {}^0\mathbf{c}_p \end{bmatrix}$.

Adding the contributions of the motor inertia and friction effects (from Sect. 8.2):

$$\begin{aligned} \begin{bmatrix} \begin{bmatrix} \boldsymbol{\tau}_t \\ \mathbf{0}_{n_e} \end{bmatrix} \\ {}^0\mathbf{w}_p \end{bmatrix} &= \mathbf{M}_t \begin{bmatrix} \ddot{\mathbf{q}}_t \\ {}^0\dot{\mathbf{t}}_p \end{bmatrix} + \mathbf{c}_t + \begin{bmatrix} \mathbf{I}_t & \mathbf{0} \\ \mathbf{0} & \mathbf{0} \end{bmatrix} \begin{bmatrix} \ddot{\mathbf{q}}_t \\ {}^0\dot{\mathbf{t}}_p \end{bmatrix} + \begin{bmatrix} \mathbf{F}_v \dot{\mathbf{q}}_t \\ \mathbf{0} \end{bmatrix} + \begin{bmatrix} \mathbf{f}_s sign\left(\dot{\mathbf{q}}_t\right) \\ \mathbf{0} \end{bmatrix} \\ &= \mathbf{M}_{tot} \ddot{\mathbf{q}}_{tot} + \mathbf{c}_{tot} \end{aligned} \tag{10.103}$$

where

- $\mathbf{I}_t$ is a diagonal matrix whose jth element corresponds to the value of the inertia of the rotor of joint j (the jth element of $\mathbf{I}_t$ is equal to zero if the joint is passive or if it corresponds to an elastic coordinate),
- $\mathbf{F}_v$, ($\mathbf{f}_s$, resp.) a diagonal matrix (a vector, resp.) of viscous (Coulomb, resp.) friction parameters,
- $\mathbf{M}_{tot} = \mathbf{M}_t + \begin{bmatrix} \mathbf{I}_t & \mathbf{0} \\ \mathbf{0} & \mathbf{0} \end{bmatrix}$ is the global inertia matrix of the virtual system,
- $\mathbf{c}_{tot} = \mathbf{c}_t + \begin{bmatrix} \mathbf{F}_v \dot{\mathbf{q}}_t \\ \mathbf{0} \end{bmatrix} + \begin{bmatrix} \mathbf{f}_s sign\left(\dot{\mathbf{q}}_t\right) \\ \mathbf{0} \end{bmatrix}$ is the vector of Coriolis, centrifugal/ gravity/ friction/elastic effects of the virtual system.

10.4 Dynamic Model of a Flexible Parallel Robot

The model of the virtual tree structure and of the free moving platform does not consider the closed-loop kinematic chains. As a matter of fact, the n_{tot} components of the generalized velocity vector $\dot{\mathbf{q}}_{tot}^T = [\dot{\mathbf{q}}_t^T \;\; {}^0\mathbf{t}_p^T]$ are dependent. The independent components are gathered in vector $\dot{\mathbf{q}}$ ($\dim \dot{\mathbf{q}} = n_q < n_{tot}$) and their determination is described thereafter.

10.4.1 Determination of the Joint and Platform Velocities as a Function of the Generalized Velocities $\dot{\mathbf{q}}$ of the Parallel Robot

For determining one possible subset of generalized coordinates for the parallel robot, let us express the relations between the vector of generalized velocities of the tree structure $\dot{\mathbf{q}}_t$ and the twist of the last element m_i for each leg i. Using (10.70) to compute the twist ${}^{i,m_i}\mathbf{t}_{i,m_i}$ of the extremity of each leg, it turns out that:

$$ {}^{i,m_i}\mathbf{t}_{i,m_i} = \mathbf{J}^i_{t_{i,m_i}} \dot{\mathbf{q}}_{t_i} \tag{10.104} $$

where $\mathbf{J}^i_{t_{i,m_i}}$ is expressed in the frame $\mathscr{F}_{i,m_i}$ (the preceding superscript is omitted for reason of clarity) and can be obtained from $\mathbf{J}_{t_{i,m_i}}$ by extracting the columns corresponding to the vector $\dot{\mathbf{q}}_{t_i}^T = \left[\dot{\mathbf{q}}_{a_i}^T, \dot{\mathbf{q}}_{d_i}^T, \dot{\mathbf{q}}_{e_i}^T\right]$, i.e. the vector concatenating all active, passive and elastic variables of the leg i.

As the leg extremity is also linked to the platform, which is supposed to be rigid (for flexible platform, see Long et al. 2014), its twist can be related to the platform twist ${}^0\mathbf{t}_p$ via the rigid displacement relation:

$$ {}^{i,m_i}\mathbf{t}_{i,m_i} = \mathbf{J}_p^i \, {}^0\mathbf{t}_p \text{ , where } \mathbf{J}_p^i = {}^{i,m_i}\overline{\mathbf{R}}_0 \begin{bmatrix} \mathbf{1}_3 & -{}^0\hat{\mathbf{r}}_{A_{i,m_i}P} \\ \mathbf{0} & \mathbf{1}_3 \end{bmatrix} \tag{10.105} $$

in which $\mathbf{J}_p^i$ is a (6×6) matrix expressed in the frame $\mathscr{F}_{i,m_i}$ (the preceding superscript is omitted for reason of clarity), ${}^0\hat{\mathbf{r}}_{A_{i,m_i}P}$ is the cross product matrix of vector ${}^0\mathbf{r}_{A_{i,m_i}P}$ that characterizes the position of the attachment point A_{i,m_i} with respect to the platform center position P (Fig. 8.1a) and ${}^{i,m_i}\overline{\mathbf{R}}_0$ is the (6×6) rotation matrix between the global frame and the local frame attached to body $\mathscr{B}_{i,m_i}$.

Thus, expressing the twist ${}^{i,m_i}\mathbf{t}_{i,m_i}$ for each leg in relation with the platform twist ${}^0\mathbf{t}_p$ and generalized coordinates $\dot{\mathbf{q}}_{t_i}$, the following set of equations is obtained:

$$ \begin{bmatrix} \mathbf{J}^1_{t1,m_1} & \cdots & \mathbf{0} \\ \vdots & \ddots & \vdots \\ \mathbf{0} & \cdots & \mathbf{J}^n_{tn,mn} \end{bmatrix} \begin{bmatrix} \dot{\mathbf{q}}_{t1} \\ \vdots \\ \dot{\mathbf{q}}_{tn} \end{bmatrix} - \begin{bmatrix} \mathbf{J}^1_p \\ \vdots \\ \mathbf{J}^n_p \end{bmatrix} {}^0\mathbf{t}_p = \mathbf{0} \tag{10.106} $$

or also

$$ \mathbf{J}_t\dot{\mathbf{q}}_t - \mathbf{J}_p\,{}^0\mathbf{t}_p = \begin{bmatrix} \mathbf{J}_t & -\mathbf{J}_p \end{bmatrix} \begin{bmatrix} \dot{\mathbf{q}}_t \\ {}^0\mathbf{t}_p \end{bmatrix} = \mathbf{J}_{tot}\dot{\mathbf{q}}_{tot} = \mathbf{0} \tag{10.107} $$

where $\mathbf{J}_{tot}$ is a $(c \times n_{tot})$ matrix $(c = r\,n)$, $n_{tot} > c$ ($r = 6$ for a spatial robot, $r = 3$ for a planar robot). This means that $\dot{\mathbf{q}}_{tot}$ contains a subset of c dependent variables $\dot{\mathbf{q}}_{dep}$. This subset is not unique. An idea could be to put all passive joints and platform variables in this subset, i.e., $\dot{\mathbf{q}}^T_{dep} = \begin{bmatrix} \dot{\mathbf{q}}^T_d & {}^0\mathbf{t}^T_p \end{bmatrix}$. However, for over-constrained parallel robots, $\dim\left(\dot{\mathbf{q}}_{dep}\right) < c$. As a result, this vector should be completed using some other elastic variables that could be chosen arbitrarily. Meanwhile, it must be mentioned that most parallel robots have identical legs and that such a methodology will lead to an asymmetrical description of the leg variables, which is not ideal. In order to avoid this problem, we better put in $\dot{\mathbf{q}}_{dep}$ the last r components $\dot{\mathbf{q}}^f_{t_i}$ of each vector $\dot{\mathbf{q}}_{t_i}$ that is now divided into two parts: $\dot{\mathbf{q}}^T_{t_i} = \begin{bmatrix} \dot{\mathbf{q}}^{0T}_{t_i} & \dot{\mathbf{q}}^{fT}_{t_i} \end{bmatrix}$. Thus, variables $\dot{\mathbf{q}}^f_{t_i}$ are related to the others using (10.107):

$$ -\begin{bmatrix} \mathbf{J}^{f1}_{t1,m_1} & \cdots & \mathbf{0} \\ \vdots & \ddots & \vdots \\ \mathbf{0} & \cdots & \mathbf{J}^{fn}_{tn,mn} \end{bmatrix} \begin{bmatrix} \dot{\mathbf{q}}^f_{t1} \\ \vdots \\ \dot{\mathbf{q}}^f_{tn} \end{bmatrix} = \begin{bmatrix} \mathbf{J}^{01}_{t1,m_1} & \cdots & \mathbf{0} & -\mathbf{J}^1_p \\ \vdots & \ddots & \vdots & \vdots \\ \mathbf{0} & \cdots & \mathbf{J}^{0n}_{tn,mn} & -\mathbf{J}^n_p \end{bmatrix} \begin{bmatrix} \dot{\mathbf{q}}^0_{t1} \\ \vdots \\ \dot{\mathbf{q}}^0_{tn} \\ {}^0\mathbf{t}_p \end{bmatrix} \tag{10.108} $$

which can be written in the form

$$ -\mathbf{J}^f_t \begin{bmatrix} \dot{\mathbf{q}}^f_{t1} \\ \vdots \\ \dot{\mathbf{q}}^f_{tn} \end{bmatrix} = \begin{bmatrix} \mathbf{J}^0_t & -\mathbf{J}_p \end{bmatrix} \begin{bmatrix} \dot{\mathbf{q}}^0_{t1} \\ \vdots \\ \dot{\mathbf{q}}^0_{tn} \\ {}^0\mathbf{t}_p \end{bmatrix} \tag{10.109} $$

or also

$$\begin{bmatrix} \dot{\mathbf{q}}^f_{t_1} \\ \vdots \\ \dot{\mathbf{q}}^f_{t_n} \end{bmatrix} = \dot{\mathbf{q}}_{dep} = -\left(\mathbf{J}^f_t\right)^{-1}\left[\mathbf{J}^0_t - \mathbf{J}_p\right]\begin{bmatrix} \dot{\mathbf{q}}^0_{t_1} \\ \vdots \\ \dot{\mathbf{q}}^0_{t_n} \\ {}^0\mathbf{t}_p \end{bmatrix} = \begin{bmatrix} \mathbf{J}_{d_{1,1}} & \cdots & \mathbf{J}_{d_{1,n}} & \mathbf{J}_{d_{1,n+1}} \\ \vdots & \ddots & \vdots & \vdots \\ \mathbf{J}_{d_{n,1}} & \cdots & \mathbf{J}_{d_{n,n}} & \mathbf{J}_{d_{n,n+1}} \end{bmatrix} \dot{\mathbf{q}} = \mathbf{J}_d \dot{\mathbf{q}} \tag{10.110}$$

where

- $\mathbf{J}^{0i}_{t_{i,m_i}}$ ($\mathbf{J}^{fi}_{t_{i,m_i}}$, resp.) are the columns of matrix $\mathbf{J}^i_{t_{i,m_i}}$ corresponding to variables $\dot{\mathbf{q}}^0_{t_i}$ ($\dot{\mathbf{q}}^f_{t_i}$, resp.);
- $\mathbf{J}_{d_{ij}}$ is the matrix that relates the variable $\dot{\mathbf{q}}^f_{t_i}$ to $\dot{\mathbf{q}}^0_{t_j}$, $j = 1, \ldots, n$;
- $\mathbf{J}_{d_{i,n+1}}$ is the matrix that relates the variable $\dot{\mathbf{q}}^f_{t_i}$ to ${}^0\mathbf{t}_p$.

It is noteworthy that the inversion of matrix $\mathbf{J}^f_t$ involves only inversion of the n $(r \times r)$ matrices $\mathbf{J}^{fj}_{t_{1,m_1}}$, which is more efficient in terms of computational time. Moreover, when 3D beam elements are used for leg i, if the coordinates $\mathbf{q}^f_{t_i}$ are the elastic coordinates of uth element of this leg (denoted as body $\mathscr{B}_{iu}$), it can be proven from Sect. 5.2 and Appendix F that the kth column of matrix $\mathbf{J}^{fi}_{t_{iu}}$ corresponds to a unit twist that describes the displacement of the leg extremity due to the kth coordinate of vector $\mathbf{q}^f_{t_i}$. Taking into account the fact that, for a 3D beam, the first three components of $\mathbf{q}^f_{t_i}$ correspond to translations along $\boldsymbol{x}_{iu}$, $\boldsymbol{y}_{iu}$ and $\boldsymbol{z}_{iu}$ and that the last three components of $\mathbf{q}^f_{t_i}$ correspond to rotations around $\boldsymbol{x}_{iu}$, $\boldsymbol{y}_{iu}$ and $\boldsymbol{z}_{iu}$, the matrix $\mathbf{J}^{fi}_{t_{iu}}$ is equal to:

$$\mathbf{J}^{fi}_{t_{iu}} = {}^{i,m_i}\overline{\mathbf{R}}_{iu} \begin{bmatrix} \mathbf{1}_3 & -{}^{iu}\hat{\mathbf{r}}_{O_{iu}O_{i,m_i}} \\ \mathbf{0} & \mathbf{1}_3 \end{bmatrix} \tag{10.111}$$

or also

$$\mathbf{J}^{fi}_{t_{iu}} = \begin{bmatrix} {}^{i,m_i}\mathbf{R}_{iu} & -{}^{i,m_i}\mathbf{R}_{iu}\,{}^{iu}\hat{\mathbf{r}}_{O_{i,m_i}O_{iu}} \\ \mathbf{0} & {}^{i,m_i}\mathbf{R}_{iu} \end{bmatrix} \tag{10.112}$$

where ${}^{i,m_i}\mathbf{R}_{iu}$ (${}^{i,m_i}\overline{\mathbf{R}}_{iu}$, resp.) is the (3×3) ((6×6), resp.) rotation matrix between the local frame $\mathscr{F}_{i,m_i}$ and the local frame $\mathscr{F}_{iu}$ and ${}^{iu}\hat{\mathbf{r}}_{O_{iu}O_{i,m_i}}$ is the cross-product matrix of the vector ${}^{iu}\mathbf{r}_{O_{iu}O_{i,m_i}}$ that characterizes the position of the leg extremity with respect to the frame $\mathscr{F}_{iu}$. Thus its inverse is equal to

$$\left(\mathbf{J}^{fi}_{t_{iu}}\right)^{-1} = \begin{bmatrix} {}^{i,m_i}\mathbf{R}^T_{iu} & {}^{iu}\hat{\mathbf{r}}_{O_{iu}O_{i,m_i}}\,{}^{i,m_i}\mathbf{R}^T_{iu} \\ \mathbf{0} & {}^{i,m_i}\mathbf{R}^T_{iu} \end{bmatrix} \tag{10.113}$$

which does not require much calculation. If 2D beam elements are used, some similar relations can be obtained.

Finally, the Jacobian matrix relating all variables $\dot{\mathbf{q}}_{tot}$ to the configuration variables $\dot{\mathbf{q}}^T = \left[\dot{\mathbf{q}}_{t_1}^{0T} \cdots \dot{\mathbf{q}}_{t_n}^{0T}\ {}^0\mathbf{t}_p\right]$ can be obtained as:

$$\dot{\mathbf{q}}_{tot} = \begin{bmatrix} \dot{\mathbf{q}}_{t_1}^0 \\ \dot{\mathbf{q}}_{t_1}^f \\ \vdots \\ \dot{\mathbf{q}}_{t_n}^0 \\ \dot{\mathbf{q}}_{t_n}^f \\ {}^0\mathbf{t}_p \end{bmatrix} = \begin{bmatrix} \mathbf{1}_{c_1} & \cdots & \mathbf{0} & \mathbf{0} \\ \mathbf{J}_{d_{1,1}} & \cdots & \mathbf{J}_{d_{1,n}} & \mathbf{J}_{d_{1,n+1}} \\ \vdots & \cdots & \vdots & \vdots \\ \mathbf{0} & \cdots & \mathbf{1}_{c_n} & \mathbf{0} \\ \mathbf{J}_{d_{n,1}} & \cdots & \mathbf{J}_{d_{n,n}} & \mathbf{J}_{d_{n,n+1}} \\ \mathbf{0} & \cdots & \mathbf{0} & \mathbf{1}_6 \end{bmatrix} \begin{bmatrix} \dot{\mathbf{q}}_{t_1}^0 \\ \vdots \\ \dot{\mathbf{q}}_{t_n}^0 \\ {}^0\mathbf{t}_p \end{bmatrix} = \mathbf{J}\dot{\mathbf{q}} \tag{10.114}$$

where $\mathbf{1}_{c_i}$ is the $(c_i \times c_i)$ identity matrix, c_i being the size of vector $\dot{\mathbf{q}}_{t_i}^0$.

10.4.2 Determination of Joint and Platform Accelerations as a Function of the Generalized Accelerations $\ddot{\mathbf{q}}$ of the Parallel Robot

Expressing the acceleration ${}^{i,m_i}\dot{\mathbf{t}}_{i,m_i}$ of the extremity of each leg using (10.85) and combining this expression (7.173), it turns out that:

$${}^{i,m_i}\dot{\mathbf{t}}_{i,m_i} = \mathbf{J}_{t_{i,m_i}}^i \ddot{\mathbf{q}}_{ti} + {}^{i,m_i}\mathbf{g}_{i,m_i} = \mathbf{J}_{p_i}\,{}^0\dot{\mathbf{t}}_p + \mathbf{d}_i. \tag{10.115}$$

Thus, considering all the robot legs,

$$\begin{bmatrix} \mathbf{J}_{t_{1,m_1}}^1 & \cdots & \mathbf{0} \\ \vdots & \ddots & \vdots \\ \mathbf{0} & \cdots & \mathbf{J}_{t_{n,m_n}}^n \end{bmatrix} \begin{bmatrix} \ddot{\mathbf{q}}_{t_1} \\ \vdots \\ \ddot{\mathbf{q}}_{t_n} \end{bmatrix} - \begin{bmatrix} \mathbf{J}_p^1 \\ \vdots \\ \mathbf{J}_p^n \end{bmatrix} {}^0\dot{\mathbf{t}}_p + \begin{bmatrix} {}^{1,m_1}\mathbf{g}_{1,m_1} - \mathbf{d}_1 \\ \vdots \\ {}^{n,m_n}\mathbf{g}_{n,m_n} - \mathbf{d}_n \end{bmatrix} = \mathbf{0} \tag{10.116}$$

or also

$$\mathbf{J}_t\ddot{\mathbf{q}}_t - \mathbf{J}_p\,{}^0\dot{\mathbf{t}}_p + \mathbf{b}' = \left[\mathbf{J}_t\ -\mathbf{J}_p\right] \begin{bmatrix} \ddot{\mathbf{q}}_t \\ {}^0\dot{\mathbf{t}}_p \end{bmatrix} + \mathbf{b}' = \mathbf{0} \tag{10.117}$$

where $\mathbf{b}' = \begin{bmatrix} {}^{1,m_1}\mathbf{g}_{1,m_1} - \mathbf{d}_1 \\ \vdots \\ {}^{n,m_n}\mathbf{g}_{n,m_n} - \mathbf{d}_n \end{bmatrix}$.

As a result, by analogy with (10.110), it can be demonstrated that

$$\ddot{\mathbf{q}}_d = \mathbf{J}_d\ddot{\mathbf{q}} + \left(\mathbf{J}_{t_{iu}}^f\right)^{-1}\mathbf{b}' = \mathbf{J}_d\ddot{\mathbf{q}} + \mathbf{b}_d \tag{10.118}$$

where $\mathbf{b}_d = \left(\mathbf{J}_{t_{iu}}^f\right)^{-1}\mathbf{b}'$.

Finally, similarly to expression (10.114),

$$\ddot{\mathbf{q}}_{tot} = \begin{bmatrix} \ddot{\mathbf{q}}_{t_1}^0 \\ \ddot{\mathbf{q}}_{t_1}^f \\ \vdots \\ \ddot{\mathbf{q}}_{t_n}^0 \\ \ddot{\mathbf{q}}_{t_n}^f \\ {}^0\dot{\mathbf{t}}_p \end{bmatrix} = \mathbf{J}\ddot{\mathbf{q}} + \begin{bmatrix} \mathbf{0} \\ \mathbf{b}_{d1} \\ \vdots \\ \mathbf{0} \\ \mathbf{b}_{dn} \\ \mathbf{0} \end{bmatrix} = \mathbf{J}\ddot{\mathbf{q}} + \mathbf{b} \tag{10.119}$$

where $\mathbf{b}_{di}$ is the part of the vector $\mathbf{b}_d$ ($\mathbf{b}_d^T = \begin{bmatrix}\mathbf{0}\ \mathbf{b}_{d1}^T \ldots \mathbf{0}\ \mathbf{b}_{dn}^T\ \mathbf{0}\end{bmatrix}$) corresponding to the accelerations $\ddot{\mathbf{q}}_{t_i}^f$.

10.4.3 Elastodynamic Model of the Actual Parallel Robot

Considering the actual robot, the *PVP* states that:

$$\dot{\mathbf{q}}_{tot}^{T*}\left(\mathbf{M}_{tot}\ddot{\mathbf{q}}_{tot} + \mathbf{c}_{tot}\right) = \dot{\mathbf{q}}_r^{T*}\begin{bmatrix}\boldsymbol{\tau} \\ \mathbf{0}_{n_c}\end{bmatrix} \tag{10.120}$$

where

- $\dot{\mathbf{q}}_r = \mathbf{E}_q\dot{\mathbf{q}}$, in which $\mathbf{E}_q$ is a matrix that makes it possible to sort vector $\dot{\mathbf{q}}$ in such a manner that $\dot{\mathbf{q}}_r^T = [\dot{\mathbf{q}}_a^T\ \dot{\mathbf{q}}_c^T]$ in which the first n_a rows of $\dot{\mathbf{q}}_r$ correspond to the vector $\dot{\mathbf{q}}_a$ of the actual active variables, and the last $n_c = n_q - n_a$ rows of $\dot{\mathbf{q}}_r$ correspond to the vector $\dot{\mathbf{q}}_c$ of the actual non constrained variables,
- $\boldsymbol{\tau}$ is the vector of the actual actuator input efforts and $\mathbf{0}_{n_c}$ is a zero vector of dimension n_c.

Introducing (10.114) into (10.120), we get,

$$\dot{\mathbf{q}}_r^{T*}\mathbf{E}_q\mathbf{J}^T\left(\mathbf{M}_{tot}\ddot{\mathbf{q}}_{tot} + \mathbf{c}_{tot}\right) = \dot{\mathbf{q}}_r^{T*}\begin{bmatrix}\boldsymbol{\tau} \\ \mathbf{0}_{n_c}\end{bmatrix}. \tag{10.121}$$

Equation (10.121) must be true for any value of $\dot{\mathbf{q}}_r^{T*}$, thus

$$\begin{bmatrix}\boldsymbol{\tau} \\ \mathbf{0}_{n_c}\end{bmatrix} = \mathbf{E}_q\mathbf{J}^T\left(\mathbf{M}_{tot}\ddot{\mathbf{q}}_{tot} + \mathbf{c}_{tot}\right). \tag{10.122}$$

Finally, introducing (10.119) into (10.122) leads to:

$$\begin{aligned}\begin{bmatrix}\boldsymbol{\tau} \\ \mathbf{0}_{n_c}\end{bmatrix} &= \mathbf{E}_q\mathbf{J}^T\mathbf{M}_{tot}\mathbf{J}\mathbf{E}_q^T\ddot{\mathbf{q}}_r + \mathbf{E}_q\mathbf{J}^T\left(\mathbf{M}_{tot}\mathbf{b} + \mathbf{c}_{tot}\right) \\ &= \mathbf{M}\ddot{\mathbf{q}}_r + \mathbf{c}\end{aligned} \tag{10.123}$$

where $\mathbf{M} = \mathbf{E}_q\mathbf{J}^T\mathbf{M}_{tot}\mathbf{J}\mathbf{E}_q^T$ and $\mathbf{c} = \mathbf{E}_q\mathbf{J}^T\,(\mathbf{M}_{tot}\mathbf{b} + \mathbf{c}_{tot})$.

Equation (10.123) is the full elastodynamic model of the parallel robot, and it can also be rewritten as

$$\begin{bmatrix} \boldsymbol{\tau} \\ \mathbf{0}_{n_c} \end{bmatrix} = \begin{bmatrix} \mathbf{M}_{aa} & \mathbf{M}_{ac} \\ \mathbf{M}_{ac}^T & \mathbf{M}_{cc} \end{bmatrix} \begin{bmatrix} \ddot{\mathbf{q}}_a \\ \ddot{\mathbf{q}}_c \end{bmatrix} + \begin{bmatrix} \mathbf{c}_a \\ \mathbf{c}_c \end{bmatrix} \tag{10.124}$$

with

- $\mathbf{M} = \begin{bmatrix} \mathbf{M}_{aa} & \mathbf{M}_{ac} \\ \mathbf{M}_{ac}^T & \mathbf{M}_{cc} \end{bmatrix}$ its generalized inertia matrix, in which $\mathbf{M}_{aa}$ is a $(n_a \times n_a)$ matrix, $\mathbf{M}_{ac}$ is a $(n_a \times n_c)$ matrix, and $\mathbf{M}_{cc}$ is a $(n_c \times n_c)$ matrix.
- $\mathbf{c} = \begin{bmatrix} \mathbf{c}_a \\ \mathbf{c}_c \end{bmatrix}$ the global vector of the centrifugal, Coriolis, gravity and elastic effects, in which $\mathbf{c}_a$ is a $(n_a \times 1)$ vector and $\mathbf{c}_c$ is a $(n_c \times 1)$ vector.

Inverse Problem

Giving $\mathbf{q}_a$, $\dot{\mathbf{q}}_a$ and $\ddot{\mathbf{q}}_a$, but also $\mathbf{q}_c$, $\dot{\mathbf{q}}_c$ at the previous step of the iteration, first calculate $\ddot{\mathbf{q}}_c$ from the second row of (10.124) by inverting $\mathbf{M}_{cc}$. Then, by injecting it in the first row of (10.124), calculate $\boldsymbol{\tau}$.

Direct Problem

Giving τ, $\dot{\mathbf{q}}_a$ and $\ddot{\mathbf{q}}_a$, but also $\mathbf{q}_c$, $\dot{\mathbf{q}}_c$, calculate $\ddot{\mathbf{q}}_r$ from (10.123) by inverting $\mathbf{M}$.

10.5 Including the Actuator Elasticity

The presence of actuated joint flexibility is a common feature of many current industrial robots. The joint elasticity may arise from several sources, such as elasticity in the gears, transmission belts, harmonic drives, etc. It follows that a time-varying displacement is introduced between the position of the driving actuator and the joint position. The joint elasticity is modeled as a linear torsional spring for revolute joints and a linear spring for prismatic joints (Khalil and Dombre 2002; Spong 1987) (Fig. 10.3). Consequently, the vector of joint deformations is given by $(\mathbf{q}_a - \mathbf{q}_M)$, where $\mathbf{q}_M$ denotes the vector of the actuator positions, while $\mathbf{q}_a$ represents the vector of the actuated link positions.

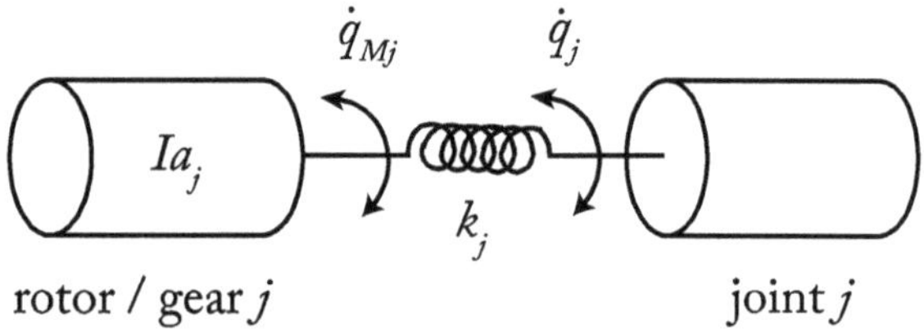

Fig. 10.3 Modeling of joint flexibility

The dynamic equations (10.124) becomes:

$$\begin{bmatrix} \tau \\ \mathbf{0}_{n_a} \\ \mathbf{0}_{n_c} \end{bmatrix} = \begin{bmatrix} \mathbf{I}_a & \mathbf{0}_{(n_a \times n_a)} & \mathbf{0}_{(n_a \times n_c)} \\ \mathbf{0}_{(n_a \times n_a)} & \mathbf{M}_{aa} & \mathbf{M}_{ac} \\ \mathbf{0}_{(n_c \times n_a)} & \mathbf{M}_{ac}^T & \mathbf{M}_{cc} \end{bmatrix} \begin{bmatrix} \ddot{\mathbf{q}}_M \\ \ddot{\mathbf{q}}_a \\ \ddot{\mathbf{q}}_c \end{bmatrix} + \begin{bmatrix} \mathbf{F}_{vM}\, \dot{\mathbf{q}}_M + \mathbf{f}_{sM} - \mathbf{K}_a(\mathbf{q}_a - \mathbf{q}_M) \\ \mathbf{c}_a + \mathbf{K}_a(\mathbf{q}_a - \mathbf{q}_M) \\ \mathbf{c}_c \end{bmatrix} \tag{10.125}$$

where $\mathbf{I}_a = diag([Ia_1 \dots Ia_{n_a}])$ is the $(n_a \times n_a)$ diagonal matrix containing the inertia of the rotors, $\mathbf{F}_{vM}$ is the $(n_a \times n)$ matrix containing the viscous friction parameters of the actuators and transmissions referred to the joint side and $\mathbf{f}_{sM}$ is a $(n_a \times 1)$ vector containing Coulomb friction parameters.

Note that in the case where all robot elements are considered rigid, while the elasticity is located only in the actuated joints, the model becomes:

$$\begin{bmatrix} \tau \\ \mathbf{0}_{n_a} \end{bmatrix} = \begin{bmatrix} \mathbf{I}_a & \mathbf{0}_{(n_a \times n_a)} \\ \mathbf{0}_{(n_a \times n_a)} & \mathbf{M}_{aa} \end{bmatrix} \begin{bmatrix} \ddot{\mathbf{q}}_M \\ \ddot{\mathbf{q}}_a \end{bmatrix} + \begin{bmatrix} \mathbf{F}_{vM}\, \dot{\mathbf{q}}_M + \mathbf{f}_{sM} - \mathbf{K}_a(\mathbf{q}_a - \mathbf{q}_M) \\ \mathbf{c}_a + \mathbf{K}_a(\mathbf{q}_a - \mathbf{q}_M) \end{bmatrix} \tag{10.126}$$

where:

- the matrix $\mathbf{M}_{aa}$ becomes the inertia matrix of the rigid *PKM* and is defined at (8.93) in the case of a *PKM* without redundancy,
- the vector $\mathbf{c}_a$ becomes the vector of Coriolis, centrifugal/gravity/friction effects of the rigid *PKM* and is defined at (8.94) in the case of a *PKM* without redundancy.

10.6 Practical Implementation of the Algorithm

Finally, in order to obtain symbolic equations for the model with a minimum number of operations, the following method is used. First, the rigid kinematics of each element are modeled using the modified Denavit-Hartenberg notations (Sect. 4.2). If body $\mathcal{B}_{ij}$ is flexible, N_{ij} supplementary elastic variables $\mathbf{q}_{e_{ij}}$ are introduced in combination with N_{ij} shape functions. Then, the previously developed equations are used in the following sequence:

- **Step 0**: *Initialization of the algorithm*
  Variables considered known: $\mathbf{q}_{tot}$, $\dot{\mathbf{q}}_{tot}$. They constitute the state variables of the robot
  Computation of:
  - $\mathbf{\Phi}_{d_{il}}(O_{ij})$, $\mathbf{\Phi}_{r_{il}}(O_{ij})$, $\mathbf{\Phi}_{il}(O_{ij})$ from (10.8); ${}^{il}\mathbf{r}_{O_{il}O_{ij}}$ from (10.4);
  - $\mathbf{M}_{ij}$, $\mathbf{K}_{ee_{ij}}$, $\mathbf{f}_{g_{ij}}$, $\mathbf{c}_{g_{ij}}$, $\mathbf{s}_{g_{ij}}$ from (10.61);
  - ${}^0\mathbf{M}_p$, ${}^0\mathbf{c}_p$ from (10.101);
  - ${}^{ij}\overline{\mathbf{R}}_{il}$, ${}^{ij}\overline{\mathbf{T}}_{il}$ from (10.97) and (10.98); ${}^{ij}\bar{\mathbf{a}}_{ij}$ from (10.75);
  - $\mathbf{A}_{ij}$, $\mathbf{\Phi}_{\mathbf{q}_{eij}}$ from (10.79);

- **Step 1**: *Forward recurrence (computation of the twist and acceleration of each body, and computation of the Jacobian matrices required for applying the PVP on the virtual tree structure)*
  Computation of:
  - ${}^{ij}\mathbf{t}_{ij}$ from (10.76);
  - $\mathbf{f}_{in_{ij}}$, $\mathbf{c}_{in_{ij}}$, $\mathbf{s}_{in_{ij}}$, $\mathbf{c}_{ij}$ from (10.61);
  - $\mathbf{J}_{t_{ij}}$ from (10.78); $\mathbf{J}_{ij}$ from (10.80);
  - ${}^{ij}\mathbf{h}_{ij}$ from (10.82); ${}^{ij}\mathbf{g}_{ij}$ from (10.84); $\mathbf{g}^{s}_{ij}$ from (10.85);
- **Step 2**: *Computation of the global inertia matrix and global vector of the centrifugal, Coriolis, gravity and elastic forces of the virtual tree structure*
  Computation of:
  - $\mathbf{M}_t$, $\mathbf{c}_t$ from (10.102);
  - $\mathbf{M}_{tot}$, $\mathbf{c}_{tot}$ from (10.103);
- **Step 3**: *Computation of the global inertia matrix and global vector of the centrifugal, Coriolis, gravity and elastic forces of the parallel robot*
  Computation of:
  - $\mathbf{J}_t$, $\mathbf{J}_p$ from (10.107);
  - $\mathbf{J}^f_t$ from (10.108); $\left(\mathbf{J}^f_t\right)^{-1}$ from (10.113);
  - $\mathbf{J}_d$ from (10.110); $\mathbf{J}$ from (10.114);
  - $\dot{\mathbf{J}}^i_p$ from (10.115);
  - $\mathbf{b}'$ from (10.117); $\mathbf{b}_d$ from (10.118); $\mathbf{b}$ from (10.119);
  - $\mathbf{M}$, $\mathbf{c}$ from (10.123);
- **Step 4**: *Solving the model*
  - Inverse dynamic model: computation of $\boldsymbol{\tau}$ and $\ddot{\mathbf{q}}_e$ as a function of $\mathbf{q}$, $\dot{\mathbf{q}}$ and $\ddot{\mathbf{q}}_a$.
  - Forward dynamic model: computation of $\ddot{\mathbf{q}}$ as a function of $\mathbf{q}$, $\dot{\mathbf{q}}$ and $\boldsymbol{\tau}$.

To reduce the number of operations of this model, customized symbolic techniques based on the use of intermediate variables can be used (Khalil and Creusot 1997). For each computation, the elements of a vector or a matrix containing at least one mathematical operation are replaced by an intermediate variable. This variable is written in an output file which contains the model. The elements that do not contain any operations are not modified. The obtained vectors and matrices are propagated in the subsequent equations. Consequently, at the end, the dynamic model is obtained as a set of intermediate variables. Those that have no effect on the desired output ($\boldsymbol{\tau}$ and the n_c last values of $\ddot{\mathbf{q}}_r$ corresponding to the elastic variables in the case of the inverse model, $\ddot{\mathbf{q}}_r$ for the direct model) can be eliminated by scanning the intermediate variables from the end to the beginning. With this procedure, it is also possible to know the exact number of operators necessary for the computation of the model. This algorithm has been successfully implemented with Mathematica and is used in the next section for computing the elastodynamic model of a flexible planar parallel robot.

10.7 Case Study: The DualEMPS

The previous equations are used to compute the elastodynamic model of the DualEMPS, a prototype of flexible planar $\underline{P}RRR\underline{P}$ robot designed and manufactured at IRCCyN (Fig. 10.4). This robot is actuated by two rotary actuators controlled by a dSPACE card in which simple PD control laws are introduced (the cut-off pulsation is set to 100 rad/s). The actuator movements are transmitted to the prismatic pairs via the use of ball screws (the stroke of the prismatic pairs is of 25 cm). The two prismatic pairs are parallel and are linked to the two legs via passive revolute joints.

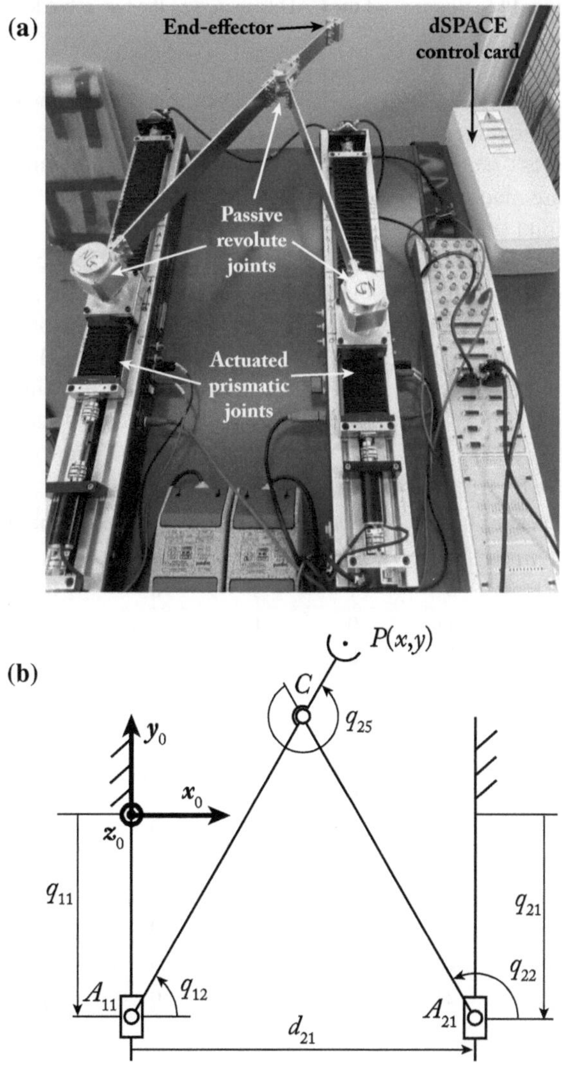

Fig. 10.4 The DualEMPS flexible parallel robot. **a** Picture of the prototype. **b** Schematics of the kinematic chain

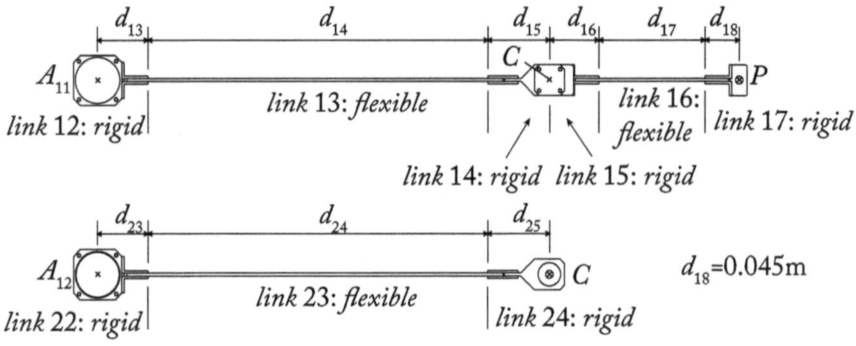

Fig. 10.5 *Top view* of the CAD drawings of the two robot legs

Each leg can be divided into a succession of rigid and elastic links (Fig. 10.5) attached all together. The elastic links are very thin when compared to the rigid links. Their cross-sections are rectangular with a width of 4 mm and height of 50 mm and they are made of Duralumin (AU4G). The two legs are connected via a passive revolute joint (Fig. 10.4).

The modified Denavit-Hartenberg parameters corresponding to the architectural description of the Fig. 10.5 are given in Table 10.1, where the parameters d_{ij} and q_{ij} are detailed in Fig. 10.4. The gravity is directed along z_0. The *a priori* rigid dynamic parameters of the links have been extracted from *CAD* and are described in Table 10.2. In this table, m_{ij} is the mass of the link $\mathscr{B}_{ij}$, mx_{ij} its first moment of inertia around the axes of the plane $(\mathbf{y}_{ij} O_{ij} \mathbf{z}_{ij})$, zz_{ij} the second moment of inertia around $\mathbf{z}_{ij}$, fv_{ij}

Table 10.1 Modified Denavit-Hartenberg parameters of the DualEMPS robot

ij	$a(ij)$	μ_{ij}	σ_{ij}	α_{ij}	d_{ij}	θ_{ij}	r_{ij}
11	0	1	1	0	0	0	q_{11}
12	11	0	0	$\pi/2$	0	q_{12}	0
13	12	0	2	0	$d_{13} = 0.0675$ m	0	0
14	13	0	2	0	$d_{14} = 0.4505$ m	0	0
15	14	0	2	0	$d_{15} = 0.082$ m	0	0
16	15	0	2	0	$d_{16} = 0.065$ m	0	0
17	16	0	2	0	$d_{17} = 0.14$ m	0	0
18	17	0	2	0	$d_{18} = 0.035$ m	0	0
21	0	1	1	0	$d_{21} = 0.4$ m	0	q_{21}
22	21	0	0	$\pi/2$	0	q_{22}	0
23	22	0	0	0	$d_{23} = 0.0675$ m	0	0
24	23	0	0	0	$d_{24} = 0.4505$ m	0	0
25	24	0	0	0	$d_{25} = 0.082$ m	q_{25}	0

Table 10.2 Rigid dynamic parameters for the links of the DualEMPS robot (SI units)

ij	m_{ij}	mx_{ij}	zz_{ij}	fs_{ij}	fv_{ij}
11	95.196	0.000	0.000	25	250
12	0.843	0.004	$7.06e-4$	0.02	0.08
13	0.252	0.057	0.017	0.00	0.00
14	0.396	0.030	0.003	0.00	0.00
15	0.000	0.000	0.000	0.00	0.00
16	0.078	0.005	$5.12e-4$	0.00	0.00
17	0.177	0.006	$2.56e-4$	0.00	0.00
18	0.00	0.00	0.00	0.00	0.00
21	95.196	0.000	0.000	25	250
22	0.843	0.004	$7.06e-4$	0.02	0.08
23	0.252	0.057	0.017	0.00	0.00
24	0.190	0.011	$9.44e-4$	0.00	0.00
25	0.000	0.000	0.000	0.01	0.005

the viscous friction parameter and fs_{ij} the Coulomb inertia parameter. Note that the values of the friction terms have been identified so that the obtained results best fit with experimental data. The elastic links are modeled as planar beam finite elements (one element by elastic link, i.e. there are 3 elastic coordinates per flexible link giving 9 elastic coordinates in total). Their corresponding elastic dynamic parameters can be computed using the formulas given in Sect. 10.2 applied to the beam shape functions given in Appendix F and are given in Tables 10.3, 10.4, 10.5 and 10.6.

For this mechanism,

- the active joint coordinates are $\mathbf{q}_a^T = [q_{11}\ q_{21}]$ (Fig. 10.4b).

Table 10.3 Terms of the vectors $\mathbf{MS}_{de_{ij}}$ for the flexible links of the DualEMPS robot (SI units)

ij	$msdex_{ij}$	$msdex_{ij}$	$msdex_{ij}$
13	0.1261	0.1261	−0.0095
16	0.392	0.392	$-9.1467e-4$
23	0.1261	0.1261	−0.0095

Table 10.4 Non zero terms of the matrices $\mathbf{J}_{re_{kij}}$ for the flexible links of the DualEMPS robot (SI units)

ij	$jrxx_{1ij}$	$jrxy_{1ij}$	$jryx_{1ij}$	$jrxy_{3ij}$
13	0.0379	0.0398	$-7.4667e-7$	−0.0026
16	0.0037	0.0038	$-7.4667e-7$	$-7.6832e-5$
23	0.0379	0.0398	$-7.4667e-7$	−0.0026

Table 10.5 Non zero terms of the matrices $\mathbf{J}_{ee_{lkij}}$ for the flexible links of the DualEMPS robot (SI units)

ij	$jexx_{11ij}$	$jexy_{12ij}$	$jexy_{13ij}$	$jeyx_{21ij}$	$jexx_{22ij}$	$jeyy_{22ij}$	$jexx_{23ij}$
13	0.0841	0.0883	−0.0057	0.0883	$1.9889e-6$	0.0937	$-7.4667e-8$
16	0.0261	0.0274	$-5.4880e-4$	0.0274	$6.4000e-6$	0.0291	$-7.4667e-8$
23	0.0841	0.0883	−0.0057	0.0883	$1.9889e-6$	0.0937	$-7.4667e-8$
ij	$jeyy_{23ij}$	$jeyx_{31ij}$	$jexx_{32ij}$	$jeyy_{32ij}$	$jexx_{33ij}$	$jeyy_{33ij}$	
13	−0.0060	−0.0057	$-7.4667e-8$	−0.0060	$4.4850-e8$	$4.4850e-4$	
16	$-5.7493e-4$	$5.4880e-4$	$-7.4667e-8$	$-5.7493e-4$	$1.3938e-8$	$1.4635e-5$	
23	−0.0060	−0.0057	$-7.4667e-8$	−0.0060	$4.4850-e8$	$4.4850e-4$	

Table 10.6 Non zero terms of the matrices $\mathbf{K}_{ee_{ij}}$ for the flexible links of the DualEMPS robot (SI units)

ij	k_{11ij}	k_{22ij}	k_{23ij}	k_{32ij}	k_{33ij}
13	3.2852e7	2.5900e3	−583.3942	−583.3942	175.2127
16	1.0571e8	8.6297e4	−6.0403e3	−6.0403e3	563.8095
23	3.2852e7	2.5900e3	−583.3942	−583.3942	175.2127

k_{uvij} represents the element on the uth row and vth column of the matrix $\mathbf{K}_{ee_{ij}}$
$k_{23ij} = k_{32ij}$ because the stiffness matrices are symmetric

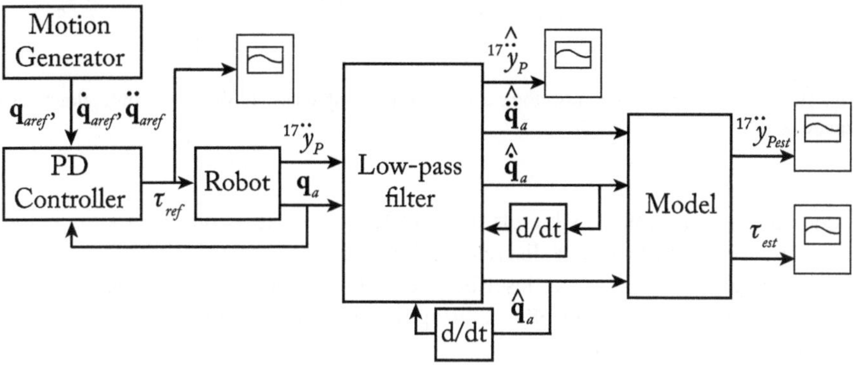

Fig. 10.6 Measured and estimated data for the experimental tests

- The passive joint coordinates are $\mathbf{q}_d^T = [q_{12}\ q_{22}\ q_{25}]$ (Fig. 10.4b).
- The platform coordinates are $\mathbf{x}^T = [x\ y]$ (Fig. 10.4b).
- The elastic coordinates are $\mathbf{q}_e^T = [\mathbf{q}_{e13}^T\ \mathbf{q}_{e16}^T\ \mathbf{q}_{e23}^T]$, in which $\mathbf{q}_{e_{ij}}^T = [q_{eij1}\ q_{eij2}\ q_{eij2}]$ is a vector of dimension 3 parameterizing the deformation of the body $\mathscr{B}_{ij}$ (Fig. 10.5); the two first components of $\mathbf{q}_{e_{ij}}$ represent the translations along the $\boldsymbol{x}_{ij}$ and $\boldsymbol{y}_{ij}$ axes of the frame $\mathscr{F}_{ij}$ while the last component represents the rotation around z_{ij}.
- The vector $\mathbf{q}_r$ which represents the generalized coordinates of the parallel robot and which is defined in (10.121) is equal to $\mathbf{q}_r^T = [\mathbf{q}_a^T\ \mathbf{q}_c^T]$, with $\mathbf{q}_c^T = [q_{12}\ q_{22}\ q_{25}\ x\ y\ q_{e131}\ q_{e161}\ q_{e162}\ q_{e163}]$.
- The vector $\mathbf{q}_{dep}$ which represents the dependent coordinates of the parallel robot and which is defined in (10.110) is equal to $\mathbf{q}_{dep}^T = [q_{e132}\ q_{e133}\ q_{e231}\ q_{e232}\ q_{e233}]$.

The model is thus calculated with Mathematica applying the proposed methods and then included in a S-function solved using Matlab/Simulink. The model includes 1041 intermediate variables and 1287 '+' or '−' and 1555 '∗' or '/' operators. The model is given on the webpage:
http://www.irccyn.ec-nantes.fr/~briot/Books.html

For reasons of comparison, an Adams model interfaced with Simulink via the use of the module Adams/Controls is also created. In this model, the elastic links are

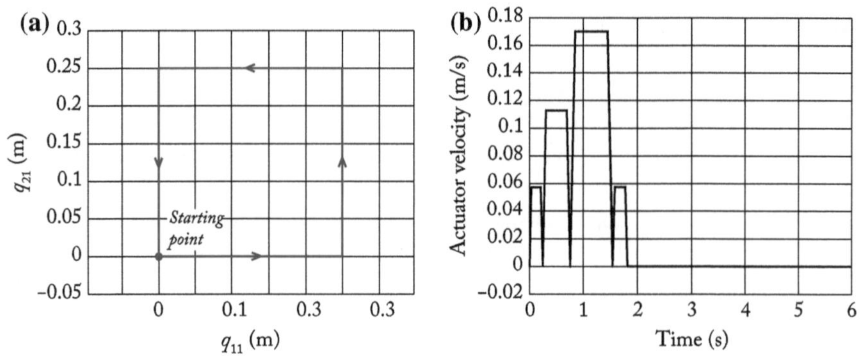

Fig. 10.7 The first reference trajectory. **a** Path in the joint space. **b** Velocity profile along each square edge

modeled using discrete flexible elements (after several tests on the Adams model accuracy, we decided to use 8 discrete flexible element by elastic link).

The experimental tests are described hereafter. Some reference trajectories are introduced in the robot controller. During the robot displacements, three types of data are recorded (Fig. 10.6):

- the reference input torques $\boldsymbol{\tau}_{ref}$ given by the controller to the actuators;
- the actuator displacements $\mathbf{q}_a$ given by the robot encoders;
- the acceleration [17] $\ddot{y}_P$ of the point P (in the local $\mathbf{y}$ direction) measured via the use of a uniaxial accelerometer with a sensitivity of 995 mV/g.

Then, to make sure that the models give a correct estimation of the robot input torques and elastic displacements, the real measured actuator displacements are given as the model inputs, as well as the computed actuator velocities and accelerations, estimated

Fig. 10.8 The reference trajectories in the workspace (scaled)

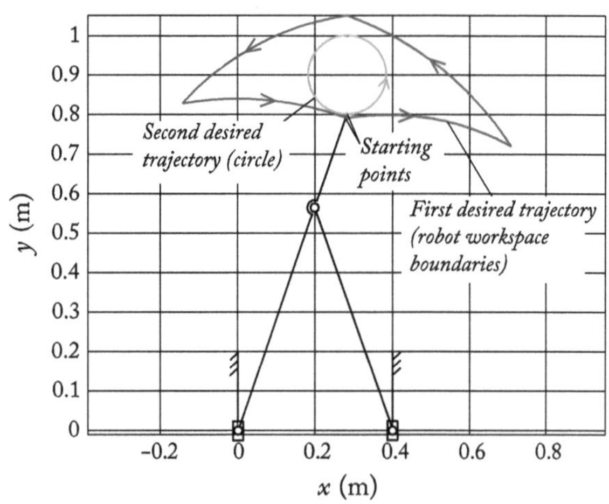

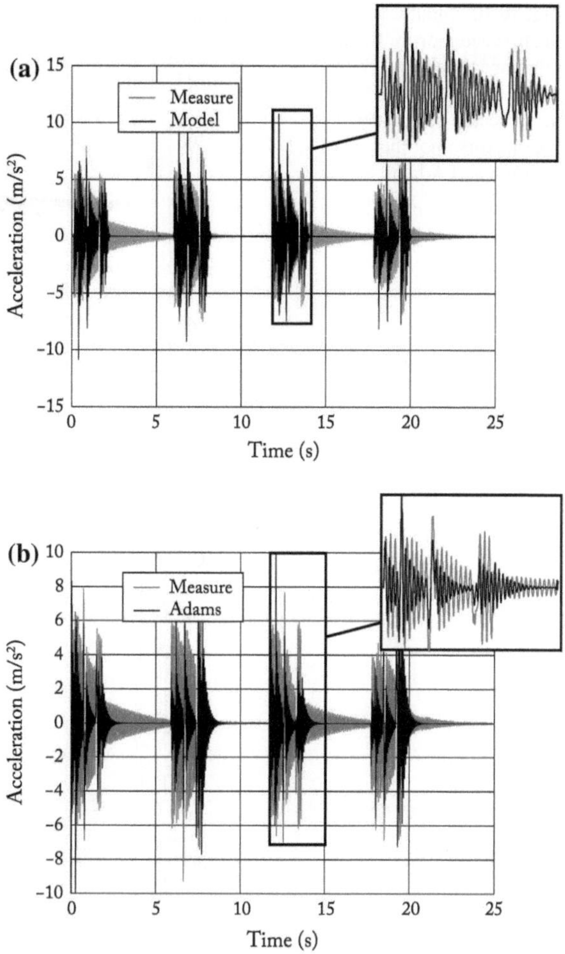

Fig. 10.9 Point P acceleration for the first reference trajectory. **a** Comparison of the measure and our model. **b** Comparison of the measure and Adams

from differentiation and low-pass filtering of the encoders data (Fig. 10.6). Results in terms of end-effector accelerations ${}^{17}\ddot{y}_{Pest}$ and computed input torques $\boldsymbol{\tau}_{est}$ are finally compared with the real robot.

The first reference trajectory is a square in the joint space (Fig. 10.7a). The resulting displacement of the end-effector is a path following the workspace boundaries (Fig. 10.8). On each square edge, the velocity profile represented in Fig. 10.7b is applied: the actuator moves during 1.8 s and then stops so that the free vibration of the robot can be observed. The measured acceleration ${}^{17}\ddot{y}_P$ of point P, low-pass filtered at 100 Hz in order to suppress high-frequency noise, is plotted at Fig. 10.9 and compared with the acceleration computed with our model and with Adams. There are very good correlations between the measurements and the simulations (with better results for our model), even if the damping is higher in the simulations which can be due both to solver problems and to too large model approximation for the dissipative

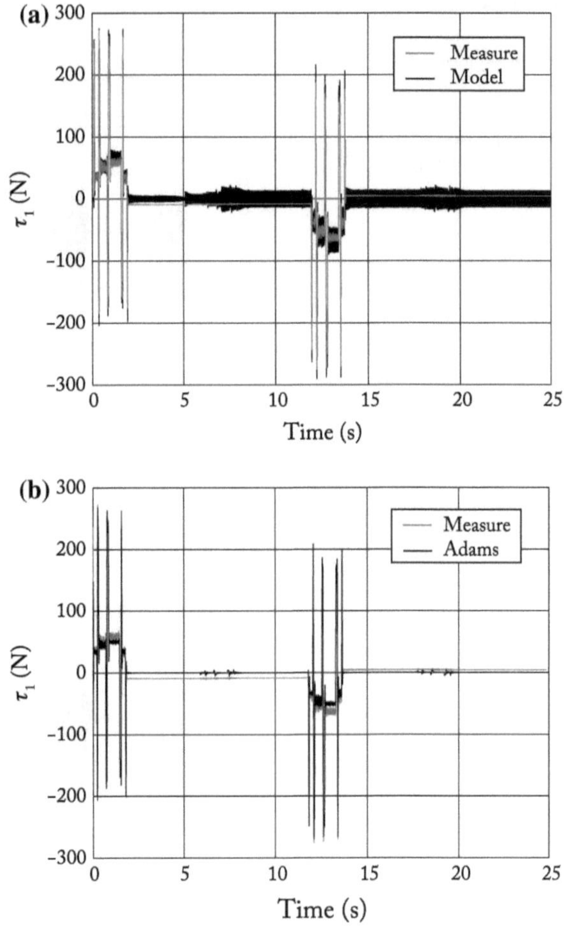

Fig. 10.10 Input torque of the first actuator for the first reference trajectory. **a** Comparison of the measure and our model. **b** Comparison of the measure and Adams

terms. For the simulations, Adams gives the result after several hours of computation while our model sends the results in 6′05 min (for a Pentium 4 2.70 GHz, 8 Go of RAM).

The motor torques prediction (in prismatic joint side) is also compared with the measured actuator torques in Figs. 10.10 and 10.11. The torque prediction is better for our model even if there are some problems of noise when the actuator velocity is very low (mainly due to problem of Coulomb friction modeling with the '*sign*' function).

A second reference trajectory is introduced in the robot controller. This is a circle of radius 0.1 m centered in [0.28, 0.90] m along which is applied a constant velocity profile of 0.20 m/s (Fig. 10.8). The predictions of the acceleration ${}^{17}\ddot{y}_P$ of point P and of the input torques using our model are shown in Figs. 10.12, 10.13 and 10.14. There is still a very good correlation between the measurements and the simulations. Note

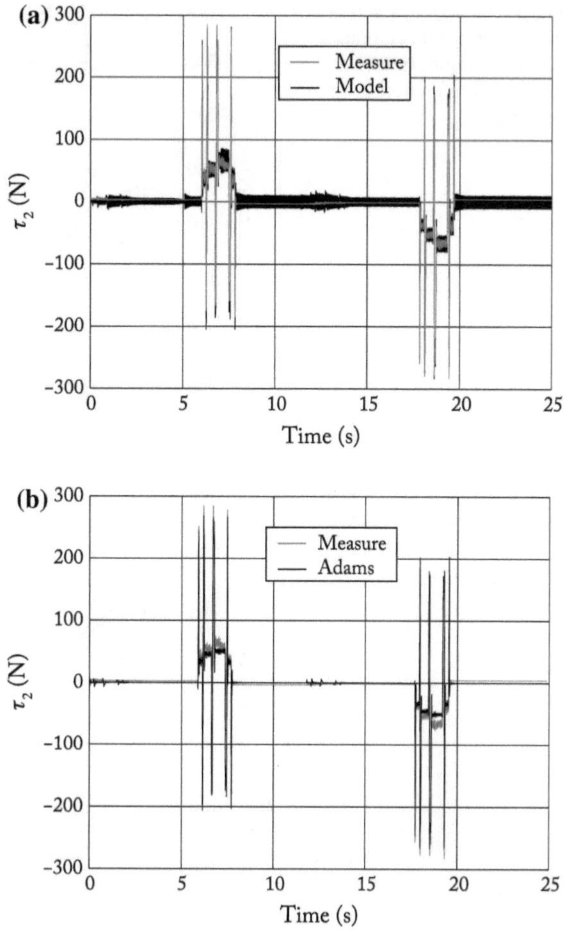

Fig. 10.11 Input torque of the second actuator for the first reference trajectory. **a** Comparison of the measure and our model. **b** Comparison of the measure and Adams

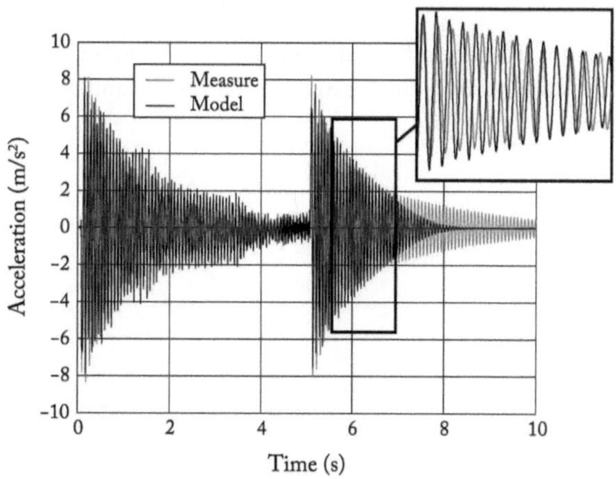

Fig. 10.12 Comparison of the measured and computed acceleration of point P for the second reference trajectory

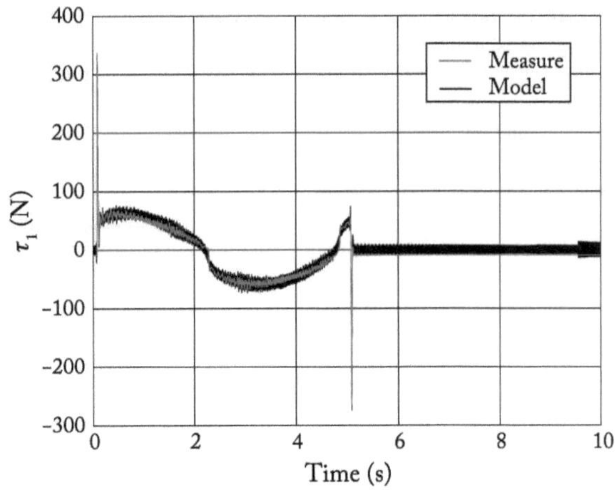

Fig. 10.13 Comparison of the measured and computed actuator 1 input torque for the second reference trajectory

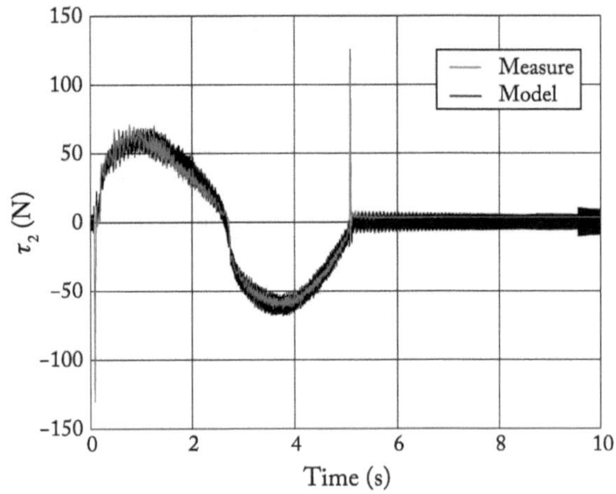

Fig. 10.14 Comparison of the measured and computed actuator 2 input torque for the second reference trajectory

that for this second simulation, data provided by Adams are not presented as we have just shown that Adams gives similar results and requires much longer computational time.

Chapter 11
Computation of Natural Frequencies

Abstract The frequencies at which a *PKM* tends to vibrate when hit, struck or somehow disturbed are known as the natural frequencies. The determination of the natural frequencies of *PKM* is needed for both design and control purposes. In this chapter, we present the computation of natural frequencies of parallel robots. We adapt the algorithm presented in Chap. 10 to make it efficient for calculation of the stiffness and inertia matrices of a parallel robot, matrices which are necessary for computing natural frequencies. The simulation results obtained through the use of natural frequency models of two flexible parallel robots are compared with experiments.

11.1 Introduction

In many applications of robot design (Bouzgarrou et al. 2005; Briot et al. 2009; Voglewede and Ebert-Uphoff 2005) and control (Pelaez et al. 2005; Singer and Seering 1988; Singh and Singhose 2002), the computation of the full elastodynamic model of a robot is not necessary, while the knowledge of its natural frequencies is required.

The natural frequencies are evaluated for a given robot configuration around its undeformed state (Wittbrodt et al. 2006) and they can be obtained through the use of the expression:

$$f_i = \frac{1}{2\pi}\sqrt{\text{eig}\left(\mathbf{M}^{-1}\mathbf{K}\right)} \tag{11.1}$$

where

- f_i is the ith natural frequency of the robot,
- eig($\mathbf{Q}$) is the *Matlab* function which returns the eigenvalues of the matrix $\mathbf{Q}$,
- $\mathbf{M}$ is the robot inertia matrix evaluated at the undeformed configuration, i.e. $\mathbf{q}_e = \mathbf{0}$,
- $\mathbf{K}$ the robot stiffness matrix evaluated at the undeformed configuration, i.e. $\mathbf{q}_e = \mathbf{0}$.

The aim of the present chapter is to provide an efficient method for calculation of the matrices $\mathbf{K}$ and $\mathbf{M}$ evaluated for $\mathbf{q}_e = \mathbf{0}$ (Briot and Khalil 2014b).

S. Briot and W. Khalil, *Dynamics of Parallel Robots*,
Mechanisms and Machine Science 35, DOI 10.1007/978-3-319-19788-3_11

11.2 Stiffness and Inertia Matrices of the Virtual System

Once again, let us consider a parallel robot composed of a rigid fixed base (denoted as the element 0), a rigid moving platform and n legs, each leg being a serial kinematic chain composed of $m_i - 1$ bodies linked by m_i joints (revolute, prismatic or even fixed joints – $i = 1, ..., n$) (Fig. 8.1a). The actuated variables are denoted by $\mathbf{q}_a$ (of size n_a) and the leg passive variables by $\mathbf{q}_d$ (of size n_d). The platform coordinates are denoted as $\mathbf{x}_p$.

The size n_a of the vector $\mathbf{q}_a$ must be greater than or equal to the number of degrees of freedom of the parallel robot.

The number of shape functions for the body $\mathscr{B}_{ij}$ is denoted by N_{ij} ($j = 1, ..., m_i - 1$). As a result, the dimension n_e of the vector of elastic variables $\mathbf{q}_e$ is equal to $\sum_{i=1}^{n} \sum_{j=1}^{m_i-1} N_{ij}$.

As mentioned in the introduction of the present chapter, the natural frequencies are evaluated for a given robot configuration parameterized by $\mathbf{q}_a$, $\mathbf{q}_d$ and $\mathbf{q}_e = \mathbf{0}$ around its undeformed state (Wittbrodt et al. 2006). As a result, under the assumption of an elastic deformation, the variations in the joint and platform variables (which represent the coordinates of the free displacement of robot bodies around the nominal configuration $\mathbf{q}_a$, $\mathbf{q}_d$ and $\mathbf{q}_e = \mathbf{0}$.) are denoted by $\delta\mathbf{q}_a$, $\delta\mathbf{q}_d$ and $\delta\mathbf{x}_p$, respectively.

The vector of generalized coordinates of the tree-structure is thus defined $\mathbf{q}_t^T = [\delta\mathbf{q}_a^T, \delta\mathbf{q}_d^T, \mathbf{q}_e^T]$, where $\delta\mathbf{q}_a^T = [\delta q_{a_1} \dots \delta q_{a_{n_a}}]$, $\delta\mathbf{q}_d^T = [\delta q_{d_1} \dots \delta q_{d_{n_d}}]$. $\delta\mathbf{q}_a$, $\delta\mathbf{q}_d$ and $\mathbf{q}_e$ are the vectors of the actuated, passive and elastic generalized coordinates for the tree structure and represent the coordinates of the free displacement of the robot bodies around the nominal configuration parameterized by $\mathbf{q}_a$, $\mathbf{q}_d$ and $\mathbf{q}_e = \mathbf{0}$.

Finally, it should be mentioned that the vector of generalized velocities of the tree-structure is now given by $\dot{\mathbf{q}}_t^T = [\dot{\mathbf{q}}_a^T, \dot{\mathbf{q}}_d^T, \dot{\mathbf{q}}_e^T]$, where $\dot{\mathbf{q}}_a = \mathrm{d}(\delta\mathbf{q}_a)/\mathrm{d}t$, $\dot{\mathbf{q}}_d = \mathrm{d}(\delta\mathbf{q}_d)/\mathrm{d}t$. Moreover, we denote as ${}^0\mathbf{t}_p = \mathrm{d}(\delta\mathbf{x}_p)/\mathrm{d}t$.

To compute the inertia and stiffness matrices of the virtual tree structure, we need to compute its kinetic and potential elastic energies. The elastic potential energy of the tree structure system can be expressed as:

$$U_{et} = \sum_{i,j} U_{e_{ij}} \tag{11.2}$$

with $U_{e_{ij}}$ the elastic potential energy of the body $\mathscr{B}_{ij}$ defined at (10.34), and its kinetic energy is given by

$$E_t = \sum_{i,j} E_{ij} \tag{11.3}$$

where E_{ij} is the kinetic energy of the body $\mathscr{B}_{ij}$.

11.2.1 Kinetic Energy and Elastic Potential Energy of the Body $\mathscr{B}_{ij}$

The kinetic energy of the body $\mathscr{B}_{ij}$ is defined from (6.6) by

$$E_{ij} = \frac{1}{2}\int_{\mathscr{B}_{ij}} \mathbf{v}_{M_{ij}}^T \mathbf{v}_{M_{ij}} \mathrm{d}m \tag{11.4}$$

where $\mathbf{v}_{M_{ij}}$ is the velocity of a point M_j belonging to $\mathscr{B}_{ij}$ given from (10.7) by:

$$\mathbf{v}_{M_{ij}} = \mathbf{v}_{ij} + \boldsymbol{\omega}_{ij} \times \mathbf{r}_{O_{ij}M_{ij}} + \boldsymbol{\Phi}_{d_{ij}}(M_{0ij})\dot{\mathbf{q}}_{e_{ij}} \tag{11.5}$$

Equation (11.5) can be put into a matrix form such that:

$$\mathbf{v}_{M_{ij}} = \begin{bmatrix} \mathbf{1}_3 & \hat{\mathbf{r}}_{O_{ij}M_{ij}}^T & \boldsymbol{\Phi}_{d_{ij}}(M_{0ij}) \end{bmatrix} \begin{bmatrix} \mathbf{v}_{ij} \\ \boldsymbol{\omega}_{ij} \\ \dot{\mathbf{q}}_{e_{ij}} \end{bmatrix}. \tag{11.6}$$

Introducing (11.6) in (11.4), we obtain

$$E_{ij} = \frac{1}{2}\int_{\mathscr{B}_{ij}} \begin{bmatrix} \mathbf{v}_{ij} \\ \boldsymbol{\omega}_{ij} \\ \dot{\mathbf{q}}_{e_{ij}} \end{bmatrix}^T \begin{bmatrix} \mathbf{1}_3 \\ \hat{\mathbf{r}}_{O_{ij}M_{ij}} \\ \boldsymbol{\Phi}_{d_{ij}}^T(M_{0ij}) \end{bmatrix} \begin{bmatrix} \mathbf{1}_3 & \hat{\mathbf{r}}_{O_{ij}M_{ij}}^T & \boldsymbol{\Phi}_{d_{ij}}(M_{0ij}) \end{bmatrix} \begin{bmatrix} \mathbf{v}_{ij} \\ \boldsymbol{\omega}_{ij} \\ \dot{\mathbf{q}}_{e_{ij}} \end{bmatrix} \mathrm{d}m \tag{11.7}$$

or also,

$$E_{ij} = \frac{1}{2}\begin{bmatrix} \mathbf{t}_{ij}^T & \dot{\mathbf{q}}_{e_{ij}}^T \end{bmatrix} \mathbf{M}_{ij} \begin{bmatrix} \mathbf{t}_{ij} \\ \dot{\mathbf{q}}_{e_{ij}} \end{bmatrix} \tag{11.8}$$

where

$$\mathbf{M}_{ij} = \int_{\mathscr{B}_{ij}} \begin{bmatrix} \mathbf{1}_3 & \hat{\mathbf{r}}_{O_{ij}M_{ij}}^T & \boldsymbol{\Phi}_{d_{ij}}(M_{0ij}) \\ \hat{\mathbf{r}}_{O_{ij}M_{ij}} & \hat{\mathbf{r}}_{O_{ij}M_{ij}}^T \hat{\mathbf{r}}_{O_{ij}M_{ij}} & \hat{\mathbf{r}}_{O_{ij}M_{ij}} \boldsymbol{\Phi}_{d_{ij}}(M_{0ij}) \\ \boldsymbol{\Phi}_{d_{ij}}^T(M_{0ij}) & \boldsymbol{\Phi}_{d_{ij}}^T(M_{0ij})\hat{\mathbf{r}}_{O_{ij}M_{ij}}^T & \boldsymbol{\Phi}_{d_{ij}}^T(M_{0ij})\boldsymbol{\Phi}_{d_{ij}}(M_{0ij}) \end{bmatrix} \mathrm{d}m. \tag{11.9}$$

By identification with the expressions defined in Sect. 10.2.2.4, we have

$$\mathbf{M}_{ij} = \begin{bmatrix} m_{ij}\mathbf{1}_3 & \widehat{\mathbf{ms}}_{ij}^T & \mathbf{MS}_{de_{ij}} \\ \widehat{\mathbf{ms}}_{ij} & \mathbf{I}_{ij} & \mathbf{MS}_{re_{ij}} \\ \mathbf{MS}_{de_{ij}}^T & \mathbf{MS}_{re_{ij}}^T & \mathbf{M}_{ee_{ij}} \end{bmatrix} \tag{11.10}$$

where

- $\mathbf{ms}_{ij}$ is defined at (10.40),
- $\mathbf{MS}_{de_{ij}}$ is defined at (10.42),
- $\mathbf{MS}_{re_{ij}}$ is defined at (10.43),
- $\mathbf{I}_{ij}$ is defined at (10.46),
- $\mathbf{M}_{ee_{ij}}$ is defined at (10.50).

As mentioned in Sect. 11.1, the natural frequencies are evaluated for a robot around an undeformed state for which $\mathbf{q}_{e_{ij}} = \mathbf{0}$. In such a case, the matrix $\mathbf{M}_{ij}$ becomes

$$\mathbf{M}_{ij}(\mathbf{q}_{e_{ij}} = \mathbf{0}) = \mathbf{M}_{ij0} = \begin{bmatrix} m_{ij}\mathbf{1}_3 & \widehat{\mathbf{ms}}_{r_{ij}}^T & \mathbf{MS}_{de_{ij}} \\ \widehat{\mathbf{ms}}_{r_{ij}} & \mathbf{I}_{rr_{ij}} & \mathbf{MS}_{re_{ij}}^r \\ \mathbf{MS}_{de_{ij}}^T & \mathbf{MS}_{re_{ij}}^{r\,T} & \mathbf{M}_{ee_{ij}} \end{bmatrix} \tag{11.11}$$

where

$$\mathbf{MS}_{re_{ij}}^r = \left[\boldsymbol{\beta}_{1_{ij}} \cdots \boldsymbol{\beta}_{N_{ij,ij}}\right] \tag{11.12}$$

and

- $\mathbf{ms}_{r_{ij}}$ is defined at (10.41),
- $\mathbf{I}_{rr_{ij}}$ is defined at (10.47),
- $\boldsymbol{\beta}_{k_{ij}}$ is defined at (10.44).

Matrix $\mathbf{M}_{ij0}$ is the inertia matrix of the body $\mathcal{B}_{ij}$ in the undeformed state.

The elastic potential energy of the body $\mathcal{B}_{ij}$ given in Sect. 10.2.2.3 is equal to

$$U_{e_{ij}} = \frac{1}{2}\mathbf{q}_{e_{ij}}^T \mathbf{K}_{ee_{ij}} \mathbf{q}_{e_{ij}} \tag{11.13}$$

where $\mathbf{K}_{ee_{ij}}$ is the stiffness matrix of body $\mathcal{B}_{ij}$ defined in (10.35).

The expressions of the matrices $\mathbf{M}_{ij0}$ and $\mathbf{K}_{ee_{ij}}$ in the case of a flexible 3D Bernoulli beam are given in Appendix F.

11.2.2 Kinetic Energy and Elastic Potential Energy of the Virtual Tree Structure

Introducing (11.8) and (11.13) into (11.2) and (11.3), we obtain

$$U_{et} = \frac{1}{2}\sum_{i,j} \mathbf{q}_{e_{ij}}^T \mathbf{K}_{ee_{ij}} \mathbf{q}_{e_{ij}} \tag{11.14}$$

and

$$E_t = \frac{1}{2}\sum_{i,j}\begin{bmatrix}{}^{ij}\mathbf{t}_{ij}^T & \dot{\mathbf{q}}_{e_{ij}}^T\end{bmatrix}\mathbf{M}_{ij0}\begin{bmatrix}{}^{ij}\mathbf{t}_{ij}\\ \dot{\mathbf{q}}_{e_{ij}}\end{bmatrix} \tag{11.15}$$

in which $\mathbf{M}_{ij0}$ is projected into $\mathscr{F}_{ij}$.

We need to rewrite these expressions as functions of the tree structure generalized coordinates $\mathbf{q}_t$ and velocities $\dot{\mathbf{q}}_t$. For that, let us recall from Sect. 10.3.1 that we have

$$\begin{bmatrix}{}^{ij}\mathbf{t}_{ij}\\ \dot{\mathbf{q}}_{e_{ij}}\end{bmatrix} = \mathbf{J}_{ij}\dot{\mathbf{q}}_t \tag{11.16}$$

where the computation of the matrix $\mathbf{J}_{ij}$ is detailed in Sect. 10.3.2.

From the same section, we may also find that:

$$\mathbf{q}_{e_{ij}} = \mathbf{O}_{\mathbf{q}_{eij}}\mathbf{q}_t. \tag{11.17}$$

Introducing (11.16) and (11.17) into (11.14) and (11.15), we obtain

$$\begin{aligned} U_{et} &= \frac{1}{2}\sum_{i,j}\mathbf{q}_t^T\mathbf{O}_{\mathbf{q}_{eij}}^T\mathbf{K}_{ee_{ij}}\mathbf{O}_{\mathbf{q}_{eij}}\mathbf{q}_t \\ &= \frac{1}{2}\mathbf{q}_t^T\left(\sum_{i,j}\mathbf{O}_{\mathbf{q}_{eij}}^T\mathbf{K}_{ee_{ij}}\mathbf{O}_{\mathbf{q}_{eij}}\right)\mathbf{q}_t \end{aligned} \tag{11.18}$$

and

$$\begin{aligned} E_t &= \frac{1}{2}\sum_{i,j}\dot{\mathbf{q}}_t^T\mathbf{J}_{ij}^T\mathbf{M}_{ij0}\mathbf{J}_{ij}\dot{\mathbf{q}}_t \\ &= \frac{1}{2}\dot{\mathbf{q}}_t^T\left(\sum_{i,j}\mathbf{J}_{ij}^T\mathbf{M}_{ij0}\mathbf{J}_{ij}\right)\dot{\mathbf{q}}_t. \end{aligned} \tag{11.19}$$

11.2.3 Kinetic Energy of the Free Moving Platform

The kinetic energy of the free rigid platform is given from (6.8) by

$$E_p = \frac{1}{2}{}^0\mathbf{t}_p^T\,{}^0\mathbf{M}_p\,{}^0\mathbf{t}_p \tag{11.20}$$

where ${}^0\mathbf{M}_p$ is the inertia matrix of the platform expressed in the base frame $\mathscr{F}_0$ and is defined at (8.97).

As a result, the total kinetic and potential energies can be written as

$$\begin{aligned} E_t + E_p =& \frac{1}{2}\dot{\mathbf{q}}_t^T \left(\sum_{i,j} \mathbf{J}_{ij}^T \mathbf{M}_{ij0} \mathbf{J}_{ij} \right) \dot{\mathbf{q}}_t + \frac{1}{2} {}^0\mathbf{t}_p^{T\,0}\mathbf{M}_p {}^0\mathbf{t}_p \\ =& \frac{1}{2} \begin{bmatrix} \dot{\mathbf{q}}_t^T & {}^0\mathbf{t}_p^T \end{bmatrix} \begin{bmatrix} \sum_{i,j} \mathbf{J}_{ij}^T \mathbf{M}_{ij} \mathbf{J}_{ij} & \mathbf{0} \\ \mathbf{0} & {}^0\mathbf{M}_p \end{bmatrix} \begin{bmatrix} \dot{\mathbf{q}}_t \\ {}^0\mathbf{t}_p \end{bmatrix} \\ =& \frac{1}{2} \dot{\mathbf{q}}_{tot}^T \mathbf{M}_t \dot{\mathbf{q}}_{tot} \end{aligned} \tag{11.21}$$

with

- $\dot{\mathbf{q}}_{tot}^T = \begin{bmatrix} \dot{\mathbf{q}}_t^T & {}^0\mathbf{t}_p^T \end{bmatrix}$,
- $\mathbf{M}_t = \begin{bmatrix} \sum_{i,j} \mathbf{J}_{ij}^T \mathbf{M}_{ij} \mathbf{J}_{ij} & \mathbf{0} \\ \mathbf{0} & {}^0\mathbf{M}_p \end{bmatrix}$

and,

$$\begin{aligned} U_{et} =& \frac{1}{2} \begin{bmatrix} \mathbf{q}_t^T & \delta \mathbf{x}_p^T \end{bmatrix} \begin{bmatrix} \sum_{i,j} \mathbf{O}_{\mathbf{q}_{eij}}^T \mathbf{K}_{ee_{ij}} \mathbf{O}_{\mathbf{q}_{eij}} & \mathbf{0} \\ \mathbf{0} & \mathbf{0} \end{bmatrix} \begin{bmatrix} \mathbf{q}_t \\ \delta \mathbf{x}_p \end{bmatrix} \\ =& \frac{1}{2} \mathbf{q}_{tot}^T \mathbf{K}_{tot} \mathbf{q}_{tot} \end{aligned} \tag{11.22}$$

where

- $\delta \mathbf{x}_p$ is a small variation of the platform position,
- $\mathbf{q}_{tot}^T = \begin{bmatrix} \mathbf{q}_t^T & \delta \mathbf{x}_p^T \end{bmatrix}$,
- $\mathbf{K}_{tot} = \begin{bmatrix} \sum_{i,j} \mathbf{O}_{\mathbf{q}_{eij}}^T \mathbf{K}_{ee_{ij}} \mathbf{O}_{\mathbf{q}_{eij}} & \mathbf{0} \\ \mathbf{0} & \mathbf{0} \end{bmatrix}$ is the stiffness matrix of the virtual system.

11.2.4 Introducing the Actuator Inertia Effects

Adding the contributions of the motor inertia (from Sect. 8.2), the total kinetic energy becomes:

$$\begin{aligned} E_{tot} =& \frac{1}{2} \begin{bmatrix} \dot{\mathbf{q}}_t^T & {}^0\mathbf{t}_p^T \end{bmatrix} \left(\mathbf{M}_t + \begin{bmatrix} \mathbf{I}_t & \mathbf{0} \\ \mathbf{0} & \mathbf{0} \end{bmatrix} \right) \begin{bmatrix} \dot{\mathbf{q}}_t \\ {}^0\mathbf{t}_p \end{bmatrix} \\ =& \frac{1}{2} \dot{\mathbf{q}}_{tot}^T \mathbf{M}_{tot} \dot{\mathbf{q}}_{tot} \end{aligned} \tag{11.23}$$

where

- $\mathbf{I}_t$ is a diagonal matrix whose jth element corresponds to the value of the inertia of joint j (the jth element of $\mathbf{I}_t$ is equal to zero if the joint is passive or if it corresponds to an elastic coordinate),
- $\mathbf{M}_{tot} = \mathbf{M}_t + \begin{bmatrix} \mathbf{I}_t & \mathbf{0} \\ \mathbf{0} & \mathbf{0} \end{bmatrix}$ is the global inertia matrix of the virtual system.

11.3 Stiffness and Inertia Matrices of the *PKM*

As mentioned in Sect. 10.4, the model of the virtual tree structure and of the free moving platform does not consider the closed-loop kinematic chains and only a subset $\mathbf{q}$ in the variables $\mathbf{q}_{tot}$ are independent.

An efficient way to find this subset has been provided in Sect. 10.4, from which we have defined the relation:

$$\dot{\mathbf{q}}_{tot} = \mathbf{J}\dot{\mathbf{q}} \tag{11.24}$$

where the matrix $\mathbf{J}$ is defined in (10.114).

As a matter of fact, as the natural frequencies are evaluated around a nominal state for the robot, we also have

$$\mathbf{q}_{tot} = \mathbf{J}\mathbf{q} \tag{11.25}$$

in which the variables $\mathbf{q}_{tot}$ (and $\mathbf{q}$) represent indeed small displacements around the nominal configuration. Let us recall that $\mathbf{q}_{tot} = \left[\delta\mathbf{q}_a^T, \delta\mathbf{q}_d^T, \mathbf{q}_e^T, \delta\mathbf{x}_p^T\right]$.

Introducing (11.24) and (11.25) into (11.18) and (11.23), we finally obtain

$$U_{et} = \frac{1}{2}\mathbf{q}^T\mathbf{J}^T\mathbf{K}_{tot}\mathbf{J}\mathbf{q} = \frac{1}{2}\mathbf{q}^T\mathbf{K}\mathbf{q} = \frac{1}{2}\mathbf{q}_r^T\mathbf{K}_r\mathbf{q}_r \tag{11.26}$$

and

$$E_{tot} = \frac{1}{2}\dot{\mathbf{q}}^T\mathbf{J}^T\mathbf{M}_{tot}\mathbf{J}\dot{\mathbf{q}} = \frac{1}{2}\dot{\mathbf{q}}^T\mathbf{M}\dot{\mathbf{q}} = \frac{1}{2}\dot{\mathbf{q}}_r^T\mathbf{M}_r\dot{\mathbf{q}}_r. \tag{11.27}$$

where

- $\mathbf{q}_r = \mathbf{E}_q\mathbf{q}$, in which $\mathbf{E}_q$ is a matrix that makes it possible to sort vector $\mathbf{q}$ in such a manner that $\mathbf{q}_r^T = [\delta\mathbf{q}_a^T \; \mathbf{q}_c^T]$ in which the first n_a elements of $\mathbf{q}_r$ correspond to the vector $\delta\mathbf{q}_a$ of the actual active displacements, and the last $n_c = n_q - n_a$ elements of $\mathbf{q}_r$ correspond to the vector $\mathbf{q}_c$ of the actual non constrained variables (including the vectors $\delta\mathbf{q}_d$, $\delta\mathbf{x}_p$ and the parts of $\mathbf{q}_e$),
- $\mathbf{K}_r = \mathbf{E}_q^T\mathbf{J}^T\mathbf{K}_{tot}\mathbf{J}\mathbf{E}_q$ is the stiffness matrix of the *PKM*,
- $\mathbf{M}_r = \mathbf{E}_q^T\mathbf{J}^T\mathbf{M}_{tot}\mathbf{J}\mathbf{E}_q$ is the inertia matrix of the *PKM*.

Finally, we have

$$U_{et} = \frac{1}{2}\left[\delta\mathbf{q}_a^T\ \mathbf{q}_c^T\right]\begin{bmatrix}\mathbf{K}_{aa} & \mathbf{K}_{ac}\\ \mathbf{K}_{ac}^T & \mathbf{K}_{cc}\end{bmatrix}\begin{bmatrix}\delta\mathbf{q}_a\\ \mathbf{q}_c\end{bmatrix} \tag{11.28}$$

and

$$E_{tot} = \left[\dot{\mathbf{q}}_a^T\ \dot{\mathbf{q}}_c^T\right]\begin{bmatrix}\mathbf{M}_{aa} & \mathbf{M}_{ac}\\ \mathbf{M}_{ac}^T & \mathbf{M}_{cc}\end{bmatrix}\begin{bmatrix}\dot{\mathbf{q}}_a\\ \dot{\mathbf{q}}_c\end{bmatrix} \tag{11.29}$$

Since the natural frequencies are evaluated around an undeformed state (Wittbrodt et al. 2006), this means that the actuators are considered completely fixed, i.e. that $\delta\mathbf{q}_a = \mathbf{0}$ and $\dot{\mathbf{q}}_a = \mathbf{0}$. As a result, the equations (11.28) and (11.29) become

$$U_{et} = \frac{1}{2}\mathbf{q}_c^T\mathbf{K}_{cc}\mathbf{q}_c \tag{11.30}$$

and

$$E_{tot} = \frac{1}{2}\dot{\mathbf{q}}_c^T\mathbf{M}_{cc}\dot{\mathbf{q}}_c. \tag{11.31}$$

The Lagrangian of the system is thus:

$$L = E_{tot} - U_{et} = \frac{1}{2}\dot{\mathbf{q}}_c^T\mathbf{M}_{cc}\dot{\mathbf{q}}_c - \frac{1}{2}\mathbf{q}_c^T\mathbf{K}_{cc}\mathbf{q}_c. \tag{11.32}$$

The two $(n_c \times n_c)$ matrices $\mathbf{K}_{cc}$ and $\mathbf{M}_{cc}$

- are evaluated in the robot nondeflected configuration, namely, for $\mathbf{q}_e = \mathbf{0}$,
- depend on the robot configuration $\mathbf{q}_a$ and $\mathbf{q}_d$ but not on the variables $\mathbf{q}_c$ and $\dot{\mathbf{q}}_c$.

As a result, the Lagrange equations lead to

$$\frac{\mathrm{d}}{\mathrm{d}t}\left(\frac{\partial L}{\partial\dot{\mathbf{q}}_c}\right) - \frac{\partial L}{\partial\mathbf{q}_c} = \mathbf{M}_{cc}\ddot{\mathbf{q}}_c + \mathbf{K}_{cc}\mathbf{q}_c = \mathbf{0}. \tag{11.33}$$

A solution $\mathbf{q}_c^s$ of this equation satisfies:

$$\left(\omega_s^2\mathbf{M}_{cc} - \mathbf{K}_{cc}\right)\mathbf{q}_c^s = \mathbf{0} \tag{11.34}$$

where $\omega_s = 2\pi f_s$, f_s is the natural frequency associated with the sth natural mode of vibrations and $\mathbf{q}_c^s$ is its associated eigenvector.

Therefore, the natural frequencies of the parallel robot are evaluated by solving the following eigenvalue problem:

$$\det\left(\omega_s^2\mathbf{M}_{cc} - \mathbf{K}_{cc}\right) = 0. \tag{11.35}$$

11.4 Including the Actuator Elasticity

In case of actuator elasticity (see Sect. 10.5), Eqs. (11.29) and (11.29) become

$$U_{et} = \frac{1}{2}\left[\delta\mathbf{q}_M^T\ \delta\mathbf{q}_a^T\ \mathbf{q}_c^T\right]\begin{bmatrix}\mathbf{0} & \mathbf{0} & \mathbf{0}\\ \mathbf{0} & \mathbf{K}_{aa}+\mathbf{K}_a & \mathbf{K}_{ac}\\ \mathbf{0} & \mathbf{K}_{ac}^T & \mathbf{K}_{cc}\end{bmatrix}\begin{bmatrix}\delta\mathbf{q}_M\\ \delta\mathbf{q}_a\\ \mathbf{q}_c\end{bmatrix} \tag{11.36}$$

and

$$E_{tot} = \left[\dot{\mathbf{q}}_M^T\ \dot{\mathbf{q}}_a^T\ \dot{\mathbf{q}}_c^T\right]\begin{bmatrix}\mathbf{I}_a & \mathbf{0} & \mathbf{0}\\ \mathbf{0} & \mathbf{M}_{aa} & \mathbf{M}_{ac}\\ \mathbf{0} & \mathbf{M}_{ac}^T & \mathbf{M}_{cc}\end{bmatrix}\begin{bmatrix}\dot{\mathbf{q}}_M\\ \dot{\mathbf{q}}_a\\ \dot{\mathbf{q}}_c\end{bmatrix} \tag{11.37}$$

where

- $\delta\mathbf{q}_M$ represents the displacements of the motors,
- $\dot{\mathbf{q}}_M = \mathrm{d}(\delta\mathbf{q}_M)/\mathrm{d}t$,
- $\mathbf{I}_a = \mathrm{diag}([Ia_1 \ldots Ia_{n_a}])$ is the $(n_a \times n_a)$ diagonal matrix containing the inertia of the rotors, and
- $\mathbf{K}_a = \mathrm{diag}([ka_1 \ldots ka_{n_a}])$ is the $(n_a \times n_a)$ diagonal matrix containing the stiffness of the gearboxes.

Since the natural frequencies are evaluated for a given robot configuration around its undeformed state (Wittbrodt et al. 2006), the actuators are considered completely fixed, i.e. that $\delta\mathbf{q}_M = \mathbf{0}$ and $\dot{\mathbf{q}}_M = \mathbf{0}$ while now $\delta\mathbf{q}_a \neq \mathbf{0}$ and $\dot{\mathbf{q}}_a \neq \mathbf{0}$.

As a result, to compute the natural frequencies of the system, instead of considering only the matrices $\mathbf{K}_{cc}$ and $\mathbf{M}_{cc}$ in (11.35), we have to consider the new matrices

$$\mathbf{K}_r = \begin{bmatrix}\mathbf{K}_{aa}+\mathbf{K}_a & \mathbf{K}_{ac}\\ \mathbf{K}_{ac}^T & \mathbf{K}_{cc}\end{bmatrix} \tag{11.38}$$

and

$$\mathbf{M}_r = \begin{bmatrix}\mathbf{M}_{aa} & \mathbf{M}_{ac}\\ \mathbf{M}_{ac}^T & \mathbf{M}_{cc}\end{bmatrix} \tag{11.39}$$

associated to the generalized coordinates $\mathbf{q}_r^T = [\delta\mathbf{q}_a^T\ \mathbf{q}_c^T]$.

As a result, the Lagrange equations become

$$\frac{\mathrm{d}}{\mathrm{d}t}\left(\frac{\partial L}{\partial \dot{\mathbf{q}}_r}\right) - \frac{\partial L}{\partial \mathbf{q}_r} = \mathbf{M}_r\ddot{\mathbf{q}}_r + \mathbf{K}_r\mathbf{q}_r = \mathbf{0}. \tag{11.40}$$

A solution $\mathbf{q}_r^s$ of this equation satisfies:

$$\left(\omega_s^2 \mathbf{M}_r - \mathbf{K}_r\right) \mathbf{q}_r^s = \mathbf{0} \tag{11.41}$$

where $\omega_s = 2\pi f_s$, f_s is the natural frequency associated with the sth natural mode of vibrations and $\mathbf{q}_r^s$ is its associated eigenvector.

Therefore, the natural frequencies of the parallel robot are evaluated by solving the following eigenvalue problem:

$$\det\left(\omega_s^2 \mathbf{M}_r - \mathbf{K}_r\right) = 0. \tag{11.42}$$

11.5 Practical Implementation of the Algorithm

In order to finally obtain symbolic equations for the model with an optimized number of operations, the following method is used. First, the rigid kinematics of each element are modeled using the modified Denavit-Hartenberg notations (Sect. 4.2). If the body $\mathscr{B}_{ij}$ taken into consideration is flexible, N_{ij} supplementary elastic variables $\mathbf{q}_{e_{ij}}$ are introduced in combination with N_{ij} shape functions. Then, the previously developed equations are used in the following sequence:

- **Step 0:** *Initialization of the algorithm*
  Variables considered known: $\mathbf{q}_{tot}$, $\dot{\mathbf{q}}_{tot}$
  Computation of:
  - $\boldsymbol{\Phi}_{d_{il}}(O_{ij})$, $\boldsymbol{\Phi}_{r_{il}}(O_{ij})$, $\boldsymbol{\Phi}_{il}(O_{ij})$ from (10.8); ${}^{il}\mathbf{r}_{O_{il}O_{ij}}$ from (10.4);
  - $\mathbf{M}_{ij}$, $\mathbf{K}_{ee_{ij}}$ from (10.61) and (11.11);
  - ${}^{0}\mathbf{M}_p$ from (10.101);
  - ${}^{ij}\overline{\mathbf{R}}_{il}$, ${}^{ij}\overline{\mathbf{T}}_{il}$ from (10.97) and (10.98); ${}^{ij}\bar{\mathbf{a}}_{ij}$ from (10.75);
  - $\mathbf{A}_{ij}$, $\boldsymbol{\Phi}_{\mathbf{q}_{eij}}$ from (10.79);
- **Step 1:** *Forward recurrence (computation of the twist of each body, and computation of the Jacobian matrices required for applying the PVP on the virtual tree structure)*
  Computation of:
  - $\mathbf{J}_{t_{ij}}$ from (10.78); $\mathbf{J}_{ij}$ from (10.80);
- **Step 2:** *Computation of the global inertia and stiffness matrices of the virtual tree structure*
  Computation of:
  - $\mathbf{M}_t$ from (11.21);
  - $\mathbf{M}_{tot}$, $\mathbf{K}_{tot}$ from (11.22) and (11.23);

- **Step 3:** *Computation of the global inertia and stiffness matrices of the parallel robot*
  Computation of:
  - $\mathbf{J}_t, \mathbf{J}_p$ from (10.107);
  - $\mathbf{J}_t^f$ from (10.108); $\left(\mathbf{J}_t^f\right)^{-1}$ from (10.113);
  - $\mathbf{J}_d$ from (10.110); $\mathbf{J}$ from (10.114);
  - $\mathbf{M}_r, \mathbf{K}_r$ from (11.26) and (11.27);
- **Step 4:** *Solving the model*

The Mathematica algorithm for automatically computing the global inertia and stiffness matrices of the parallel robot is freely available on the webpage: http://www.irccyn.ec-nantes.fr/~briot/Books.html

11.6 Case Studies

11.6.1 Natural Frequencies of DualEMPS

Some experimental tests are carried out for measuring the natural frequencies of DualEMPS (see Sect. 10.7) using the setup presented in Fig. 11.1. The application of experimental modal testing to the DualEMPS is done through impact hammer excitation, a 3-D accelerometer response and data post-processing, conducted using the DataBox software developed at IRCCyN and commercialized by MITIS company. The impact point is near point P (Fig. 10.4) and the directions of excitation are contained in the horizontal plane in order to get the resonance frequencies that involves planar displacements of the robot. Piezoelectric triaxial accelerometers with a sensitivity of 1000 mV/g are used to measure the three acceleration responses. The acquisitions are performed for several robot configurations. However, the natural frequencies of the DualEMPS are near constant anywhere in the workspace, and all tested configurations lead to almost the same results. Therefore, only the results for the nominal configuration $q_{11} = q_{21} = 0$ will be presented. Each measurement resolution is equal to 1 Hz as the acquisition time and the sampling time are equal to 1 s and 40 μs, respectively.

The resonance frequencies are obtained with a fast Fourier transform of the signals given by the triaxial accelerometer. As a result, the measured resonance frequencies between 0 and 200 Hz are given in Table 11.1. It is noteworthy that the resonance frequencies of the DualEMPS amount to its natural frequencies as the damping is considered negligible. The obtained results show that the five first frequencies predicted with our model are very close to the measured ones. However, as the model is made of three beam elements only, the frequency prediction after the fifth mode is not correct anymore. This prediction could be improved by introducing a higher number of elements, but this will increase the computational time. Nevertheless,

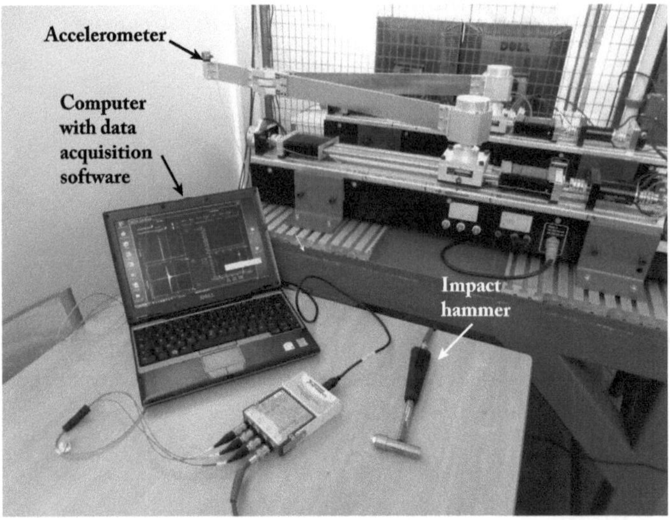

Fig. 11.1 Experimental setup for measuring natural frequencies

Table 11.1 Comparison of natural frequencies of the DualEMPS robot computed with the proposed model, computed with Adams and experimentally measured for $q_{11} = q_{21} = 0$

(Hz)	f_1	f_2	f_3	f_4	f_5	f_6
Adams	14.41	24.92	49.78	97.91	110.86	219.60
Model	14.22	25.09	48.22	92.37	109.08	423.39
Measured (±1Hz)	14.00	25.00	48.00	93.00	108.00	163.00

having a good prediction for all frequencies in the interval [14, 110] Hz is already a very good result. Note that the prediction of the Adams model mentioned in Sect. 10.7 gives almost the same result as our model for the five first frequencies and is not correct for the sixth frequency (even if it is closer to reality than our model).

11.6.2 Natural Frequencies of the NaVARo

11.6.2.1 Description of the NaVARo

The NaVARo (acronym for Nantes Variable Actuation Robot) is developed at IRCCyN and is shown in Fig. 11.2a. The NaVARo is a 3-*DOF* planar parallel manipulator composed of three identical legs and one moving platform made up of three segments E_1P, E_2P and E_3P rigidly linked at point P. The ith leg contains four links A_iB_i, B_iC_i, C_iE_i, A_iD_i (named link $i2$, link $i3$, link $i4$ and link $i1$, respectively) connected with five revolute joints in such a way that $A_iB_iC_iD_i$ is a parallelogram

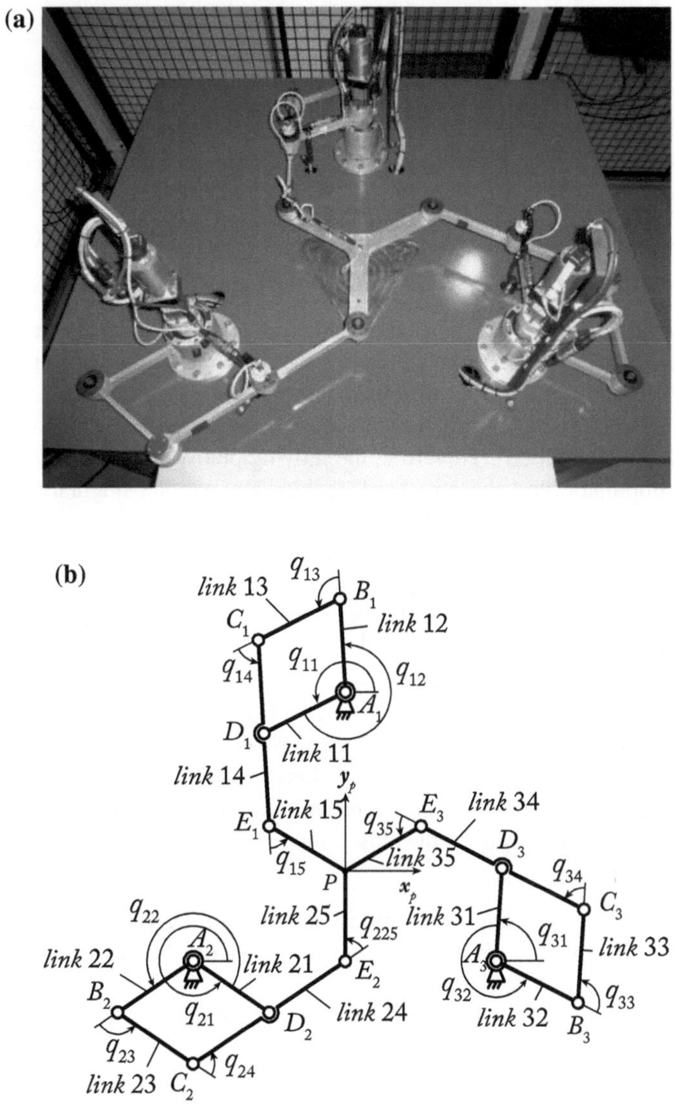

Fig. 11.2 The NaVARo. **a** Prototype of the NaVARo located at IRCCyN, Nantes, France. **b** Schematics of the NaVARo

linkage, $i = 1, 2, 3$. The base frame $\mathscr{F}_0\left(O, \boldsymbol{x}_0, \boldsymbol{y}_0, z_0\right)$ (not shown in Fig. 11.2b) is defined with point O being located at the geometric center of the equilateral triangle $A_1A_2A_3$. Frame $\mathscr{F}_p\left(P, \boldsymbol{x}_p, \boldsymbol{y}_p, z_p\right)$ is attached to the moving platform. In the home configuration shown in Fig. 11.2, $\mathscr{F}_0$ and $\mathscr{F}_p$ coincide. $\left(x_p, y_p\right)$ are the Cartesian coordinates of point P expressed in frame $\mathscr{F}_0$ and θ_p is the orientation angle of the moving platform, namely, the angle between $\boldsymbol{x}_0$ and $\boldsymbol{x}_p$.

Table 11.2 *MDH* parameters of the ith leg

ij	$a(ji)$	σ_{ij}	γ_{ij}	b_{ij}	α_{ij}	d_{ij}	θ_{ij}	r_{ij}
$i1$	0	0	γ_i	0	0	$d_1 = 0.4041\,\text{m}$	$q_{i1} - \gamma_i$	0
$i2$	0	0	γ_i	0	0	$d_1 = 0.4041\,\text{m}$	$q_{i2} - \gamma_i$	0
$i3$	$i2$	0	0	0	0	$d_3 = 0.2100\,\text{m}$	q_{3i}	0
$i4$	$i3$	0	0	0	0	$d_4 = 0.2100\,\text{m}$	q_{4i}	0
$i5$	$i4$	0	0	0	0	$d_5 = 0.4200\,\text{m}$	q_{5i}	0

q_{i1} denotes the angle between axis $\boldsymbol{x}_0$ and link $i1$. q_{i2} denotes the angle between link $i1$ and link $i2$. Three double clutches are mounted to the base and located at points A_i, $i = 1, 2, 3$, in order to actuate either angle q_{i1} or angle q_{i2}. As a consequence, the NaVARo has eight actuation modes as described in (Arakelian et al. 2008; Rakotomanga et al. 2008). Therefore, the moving platform can be moved throughout the manipulator workspace without reaching any parallel singularity thanks to a judicious actuation scheme.

The kinematics of the ith leg is described by the modified Denavit-Hartenberg parameters given in Table 11.2, in which $\gamma_i = \pi/2$ if $i = 1$, $\gamma_i = -5\pi/6$ if $i = 2$ and $\gamma_i = -\pi/6$ if $i = 3$. Besides, the circumradius of the moving-platform is equal to 0.2027 m, i.e., $l_{5i} = 0.2027$ m.

Each link of the rectangular c ross-section is made up of duraluminum alloy ($E = 74000\,\text{MPa}$, $G = 28900\,\text{MPa}$, $\rho = 2800\,\text{kg/m}^3$). Table 11.3 gives the cross-section area and the moments of inertia of the robot links.

In the experimental setup, the rotation of links $i1$ and $i2$ about point A_i, $i = 1, 2, 3$, is locked thanks to the double clutch mechanisms. The elasto-dynamic modeling of the NaVARo is complex due to the closed-loop chain in each leg and is obtained by following those three steps:

1. Computation of the mass and stiffness matrices of the virtual system assuming that the moving platform is cut at point P and the parallel linkages are opened at points D_i, $i = 1, 2, 3$;
2. Computation of the mass and stiffness matrices of the legs including the closed-loop chains;
3. Computation of the mass and stiffness matrices of the NaVARo.

A single 3D beam element is used to model links $i1$, $i2$, $i3$ and $i5$ ($i = 1, 2, 3$ – see Appendix F) while two 3D beam elements of equal lengths l ($l = l_{C_i D_i} = l_{D_i E_i}$) are used to model links $i4$. Links $i4$ are divided into two beam elements in order to close the loops as mentioned in step 2.

Table 11.3 Characteristics of the beam cross-sections

link	A_{ij} (m^2)	$I_{y_{ij}}$ (m^4)	$I_{z_{ij}}$ (m^4)	$I_{P_{ij}}$ (m^4)	$I_{O_{ij}}$ (m^4)
$i1, i2, i3, i4$	$2.4 \cdot 10^{-4}$	$1.152 \cdot 10^{-8}$	$2.000 \cdot 10^{-9}$	$1.352 \cdot 10^{-8}$	$5.902 \cdot 10^{-9}$
$i5$	$4 \cdot 10^{-4}$	$3.333 \cdot 10^{-8}$	$5.333 \cdot 10^{-8}$	$8.666 \cdot 10^{-8}$	$1.123 \cdot 10^{-8}$

Thus, the NaVARo is modeled as a spatial mechanism and its elasto-dynamic model contains 144 generalized coordinates: (i) 108 elastic coordinates; (ii) 12 passive joint coordinates, i.e., four passive joint angles per leg; (iii) 18 intermediary coordinates for the assembly of the legs; (iv) 6 coordinates for the moving-platform pose. From Sect. 11.3, it turns out that there are only 90 independent coordinates among those 144 coordinates.

11.6.2.2 Numerical Analysis

The model has been calculated using the proposed procedure and compiled into C code to obtain the robot's natural frequencies. The computation involves the use of 36183 '+' or '−' and 37341 '×' or '/' operators, while 21383 variables are defined. For reasons of comparison, the obtained frequencies are validated by means of an equivalent model developed using Cast3M software. Cast3M aims to determine the elastodynamic model of structures modeled with beams. Both models give the same values for the first 90 natural frequencies of the NaVARo. Table 11.4 gives the first five natural frequencies of the NaVARo for the eight robot postures shown in Fig. 11.3.

For the simulations, Cast3M gives the result after around 6 s of computation while our model send the results in around 0.01 s (for a Pentium 4 2.70 GHz, 8 Go of RAM).

The natural frequencies of the NaVARo are the same for poses 3, 5 and 7 (4, 6 and 8, resp.) as they correspond to a rotation of the robot base frame of ±120deg with respect to pose 3 (pose 4, resp.).

11.6.2.3 Experiments

Some experimental tests are carried out using the setup presented in Fig. 11.4. The application of experimental modal testing to the NaVARo was done through impact

Table 11.4 Comparison of the natural frequencies obtained with Cast3M and the Matlab model

(Hz)	Pose 1	Pose 2	Pose 3	Pose 4	Pose 5	Pose 6	Pose 7	Pose 8
f_1(Cast3M)	44.10	45.71	36.98	40.17	36.98	40.17	36.98	40.17
f_1(Matlab model)	44.10	45.71	36.98	40.17	36.98	40.17	36.98	40.17
f_2(Cast3M)	44.10	45.71	49.31	50.32	49.31	50.32	49.31	50.32
f_2(Matlab model)	44.10	45.71	49.31	50.32	49.31	50.32	49.31	50.32
f_3(Cast3M)	53.98	54.58	53.37	52.99	53.37	52.99	53.37	52.99
f_3(Matlab model)	53.98	54.58	53.37	52.99	53.37	52.99	53.37	52.99
f_4(Cast3M)	60.63	65.35	67.28	67.36	67.28	67.36	67.28	67.36
f_4(Matlab model)	60.63	65.35	67.28	67.36	67.28	67.36	67.28	67.36
f_5(Cast3M)	95.62	97.92	91.80	91.52	91.80	91.52	91.80	91.52
f_5(Matlab model)	95.62	97.92	91.80	91.52	91.80	91.52	91.80	91.52

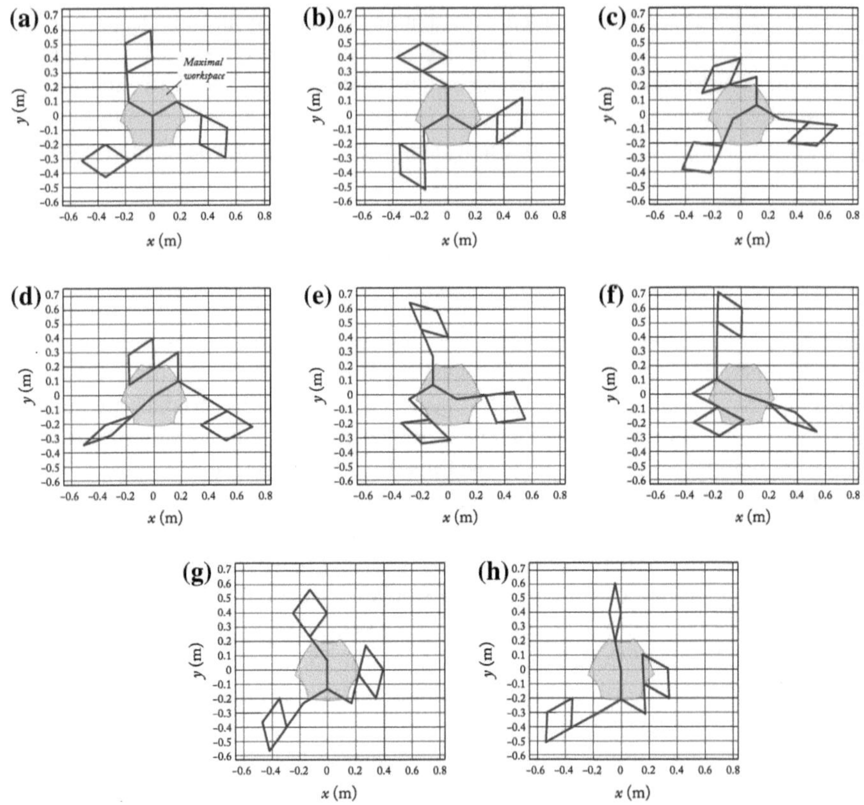

Fig. 11.3 The eight poses used for the experiments. **a** Pose 1 x = 0 m, y = 0 m, $\theta = 0$ rad. **b** Pose 2 x = 0 m, y = 0 m, $\theta = -\pi/3$ rad. **c** Pose 3 x = 0.117 m, y = 0.068 m, $\theta = -\pi/3$ rad. **d** Pose 4 x = 0.182 m, y = 0.105 m, $\theta = \pi/3$ rad. **e** Pose 5 x=.0.117 m, y= 0.068 m, $\theta = -\pi/3$ rad. **f** Pose 6 x = 0.182 m, y = 0.105 m, $\theta = -\pi/3$. **g** Pose 7 x = 0 m, y = .0.135 m, $\theta = -\pi/3$. **h** Pose 8 x = 0 m, y = 0.21 m, $\theta = -\pi/3$

hammer excitation, a 3-D accelerometer response and data post-processing, conducted using the DataBox software developed at IRCCyN and sold by MITIS company. The points and directions of excitation were chosen on points B_i and E_i of each leg along all axes in order to get the maximal number of resonance frequencies. Piezoelectric triaxial accelerometers with a sensitivity of 1000 mV/g were used to pick up the three acceleration responses. The acquisitions were performed for the eight robot postures shown in Fig. 11.3. Each measurement resolution is equal to 1 Hz as the acquisition time and the sampling times are equal to 1 s and 40 μs, respectively.

The resonance frequencies are obtained with a fast Fourier transform of the signals given by the triaxial accelerometer. As a result, the measured resonance frequencies between 0 and 80 Hz for poses 1 to 4 are given in Table 11.5. As the results for poses 3, 5 and 7 (poses 4, 6 and 8, resp.) are similar due to the manipulator symmetry, only

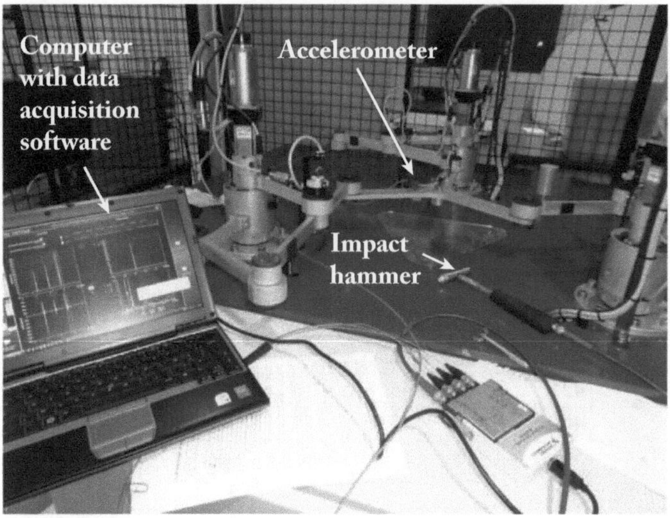

Fig. 11.4 Experimental setup: DataBox

the results for poses 3 and 4 are given in Table 11.5 and the redundant poses were used to highlight some resonance frequencies with low energy level.

It is noteworthy that the resonance frequencies of the NaVARo amount to its natural frequencies as the damping is supposed to be negligible.

It is apparent that the results given Table 11.5 do not match with those shown in Table 11.4. As a matter of fact, the elasticity of the clutches has not been modeled and the joint masses have been omitted with Cast3M software as the latter cannot model lumped masses. Thus, a refined Matlab model was written in order to consider joint masses (about 300 g per joint) and elasticities in clutches (about 2000 Nm/rad). The natural frequencies of the NaVARo computed with this refined model and the measured frequencies are gathered in Table 11.5 by comparing the computed mode shapes with the hammer impact direction and the direction of the vibration signals, the latter being measured by the triaxial accelerometer.

We can notice that there is a good correlation between the measured frequencies and the computed natural frequencies. Nevertheless, few predicted frequencies do not match with the measurements and *vice-versa*. Indeed, the theoretical and experimental results may differ due to the following reasons:

- The geometric parameters of the NaVARo have not been calibrated yet and there are some errors in the estimated moving platform pose;
- The passive joint elasticity has not been considered;
- The robot links are supposed to be coplanar in the theoretical model, whereas they are not in the prototype for collision avoidance;
- The robot links are not perfect beams as both ends are widened to insert ball bearings;
- The theoretical elastodynamic model does not consider any damping effect.

Table 11.5 NaVARo natural frequencies (measured and computed using refined model) between 0 and 80 Hz

(Hz)	Pose 1	Pose 2	Pose 3	Pose 4
f_1 meas.	22	19	17	18
f_1 calc.	19.25	19.46	17.91	18.44
f_2 meas.	24	21	19	20
f_2 calc.	20.43	20.49	19.71	19.26
f_3 meas.	32	–	23	22
f_3 calc.	40.25	41.88	20.91	21.28
f_4 meas.	–	44	27	33
f_4 calc.	43.16	45.55	–	36.88
f_5 meas.	42	45	32	43
f_5 calc.	44.10	47.05	36.88	40.60
f_6 meas.	50	53	43	44
f_6 calc.	–	–	41.86	46.13
f_7 meas.	52	54	46	50
f_7 calc.	–	56.37	45.61	55.29
f_8 meas.	62	56	48	56
f_8 calc.	67.94	–	50.52	57.81
f_9 meas.	66	60	57	58
f_9 calc.	68.81	63.10	55.45	62.27
f_{10} meas.	77	–	60	66
f_{10} calc.	79.79	–	61.04	–
f_{11} meas.	–	–	61	–
f_{11} calc.	–	–	–	–
f_{12} meas.	–	–	65	–
f_{12} calc.	–	–	65.00	–

However, from those experiments, we can claim that the theoretical model is satisfactory and the proposed modeling procedure is efficient for reproducing the real behavior of any parallel robot.

11.7 Conclusion

In this Part, we have introduced the dynamic modeling of flexible parallel robots. The goal of this Part was not to provide a complete lecture on mechanics of deformable bodies, but to show for people having some basic knowledge in this field how to obtain the dynamics of a flexible *PKM* starting from basic considerations in mechanics of deformable bodies. The reader requiring additional information on mechanics of deformable bodies is invited to read the reference books of Shabana (2005) and of Bauchau (2011).

We have proposed customized elastodynamic models for *PKM* able to minimize the number of operators, and thus to decrease the computational cost. Two types of models have been proposed:

- In Chap. 10, the full elastodynamic model of the *PKM* has been provided. In order to obtain it, it was necessary to investigate the dynamics of free flexible bodies and of flexible tree structure.
- In Chap. 11, the algorithm provided in Chap. 10 has been simplified in order to obtain the expressions of stiffness and inertia matrices of *PKM*, which are necessary for computation of the natural frequencies.

To go further in a decrease of the computational cost, it is also possible to combine our approach with the use of model reduction methods and with the definition of efficient truncated series of shape functions. The interested reader is referred to Blevins (2001), Briot et al. (2011), Craig (1981), Craig and Bampton (1968).

Note also that the elastodynamic model of parallel robots is not free of singularities which can be passed through a proper trajectory planning, such as defined in Chap. 9. The reader should read (Briot and Arakelian 2010, 2011) for more information.

Appendix A
Calculation of the Number of Degrees of Freedom of Robots with Closed Chains

A.1 Introduction

Let us recall first that the mobility, or number of *DOF*, of a robot is defined as the number of independent joint variables required to specify the location of all the links of the robot in space. It is equal to the minimal number of actuated joints to control the system.

The number of degrees of freedom N_{dof} of a robot is equal to the number of joints in the case of tree structure system L. In the case of a closed-loop mechanism, the calculation of the mobility N_{dof} can be expressed by the following relation:

$$N_{dof} = L - c \tag{A.1}$$

where L is the number of joints of the structure and c is the number of independent relationships (constraints) between the joint variables, i.e. the number of dependent joints.

Since 1854 with the work of P.A. Chebychev, several researchers have proposed different formulas that can be used to find the mobility of complex systems. Recently, Gogu (2008) has evaluated 35 methods that have been proposed to calculate the mobility of complex systems. He concluded that the majority of methods cannot properly calculate mobility for all mechanisms, and only those that require the construction of the kinematic constraint equations can give a good result.

In case of a single loop, c represents the number of independent kinematic constraint equations of the loop. Consequently $c \leq 6$ for a spatial loop and $c \leq 3$ for a planar loop. Consequently, N_{dof} gives the dimension of the space in which the situation of all the links belong. It is possible to calculate it by calculating the maximum rank of the Jacobian matrix of the serial structure constructed by cutting one link in the loop. This result can be interpreted by the fact that the open structure has L degrees of freedom, and since the c degrees of freedom of the terminal link will be lost when closing the loop, thus the number of remaining degrees of freedom is equal to $L - c$.

S. Briot and W. Khalil, *Dynamics of Parallel Robots*,
Mechanisms and Machine Science 35, DOI 10.1007/978-3-319-19788-3

In case of a system composed of B independent closed loops, the mobility of the system may be calculated by:

$$N_{dof} = L - \sum_{j=1}^{B} c_j. \tag{A.2}$$

This simple formula gives good results for most robot structures but it can yield bad results for certain complex systems and does not give information about the type of motion of the system. However, for some robots, the exact solution is obtained by analyzing the kinematic constraints and taking into account the coupling between the loops (Hervé 1978; Le Borzec and Lotterie 1975). In the following, we present two methods: the Morokine's method and the Gogu's method.

A.2 Moroskine's Method

The mobility can be calculated correctly using the rank of the matrix $\mathbf{J}_c$ defined in the following equation:

$$c = \max_{\mathbf{q}}(\text{rank}(\mathbf{J}_c(\mathbf{q}))) \tag{A.3}$$

where $\mathbf{J}_c$ is the Jacobian of the constraint equations between the joint variables such that:

$$\mathbf{J}_c(\mathbf{q})\dot{\mathbf{q}} = \mathbf{0}. \tag{A.4}$$

$\mathbf{J}_c(\mathbf{q})$ can be calculated by derivation of the geometric constraint equations of the loops or by constructing the constraint equations of the velocities through the loop. In fact, from this equation, we deduce that $\dot{\mathbf{q}}$ belongs to the null space of $\mathbf{J}_c$. Therefore, at a given configuration, the number of degrees of freedom is equal to the dimension of the null space of $\mathbf{J}_c$. Consequently:

$$N_{dof} = \min_{\mathbf{q}}(\dim(\mathscr{N}(\mathbf{J}_c(\mathbf{q})))) \tag{A.5}$$

where $\mathscr{N}(\mathbf{J}_c(\mathbf{q}))$ is the null space of the matrix $\mathbf{J}_c$.

In general, the rank of $\mathbf{J}_c(\mathbf{q})$ must be calculated for $\mathbf{q}$ satisfying the closure constraints of the loops. For a single closed loop, the rank can be calculated using random values of its joints.

This method yields the correct result, but it is significantly more difficult to execute. In order to find the rank of the Jacobian matrix defining a mechanism with closed loops, the kinematic constraint equations must be solved. The kinematic constraint equations display the relationship between the joint variables in the mechanism in order to ensure loop closure. In general it is difficult to solve these equations symbolically though it is possible to solve them numerically in order to obtain some random configurations satisfying the closure conditions.

A.3 Gogu's Method

In order to overcome the drawbacks of the previous method, Gogu proposed a method that does not require the construction of the kinematic constraint equations, but will yield the correct mobility for all mechanisms including complex parallel mechanisms. We present here how to use Gogu's method to calculate the mobility for:

- single loop kinematic chains,
- parallel mechanisms with simple legs, and
- parallel mechanisms with complex legs.

The proposed solution makes use of the mobility of the terminal link of simple open loop which is equal to the rank of the Jacobian matrix between the terminal link or the mobile platform in the case of a *PKM* and the base. In fact the mobility $M_{n/0}$ of the terminal link with respect to the base of an open loop chain is equal to the dimension of the task (Cartesian) space $\dim(E(\mathbf{x}))$.

$$M_{n/0} = M = \text{rank}(\mathbf{J}_n) = \dim(E(\mathbf{x})). \tag{A.6}$$

A.3.1 Mobility of Single Loop Kinematic Chains

A link must be opened to obtain an equivalent simple open loop. If the number of joints is equal to L, and the chain is opened around joint L such that the links $1, \ldots, L$ constitute a serial structure, then the rank of the Jacobian matrix J_L gives the number of joint variables that lose their independence after loop closure. Thus the mobility of the closed loop is given by:

$$N_{dof} = L - \text{rank}(\mathbf{J}_L) \tag{A.7}$$

where $\text{rank}(\mathbf{J}_L) = M_{L/0}$ is the mobility of link L w.r.t. link 0 in the open chain.

A second method to calculate the mobility is to open the structure around a joint $k < L$, in order to obtain two serial branches with n_1 and n_2 joints respectively.

Supposing $\text{rank}(\mathbf{J}_{n_i}) = \dim(E(\mathbf{x}_i))$ is the dimension of the task (Cartesian) space of branch i, with $i = 1, 2$, and $\dim(E(\mathbf{x}_1) \cap E(\mathbf{x}_2))$ gives the dimension of the common task space that the two branches share. Thus, the number c of joints losing their independence will be obtained as follows:

$$c = \text{rank}(\mathbf{J}_L) = \dim(E(\mathbf{x}_1)) + \dim(E(\mathbf{x}_2)) - \dim(E(\mathbf{x}_1) \cap E(\mathbf{x}_2)) \\ = \text{rank}(\mathbf{J}_{n_1}) + \text{rank}(\mathbf{J}_{n_2}) - \dim(E(\mathbf{x}_1) \cap E(\mathbf{x}_2)). \tag{A.8}$$

The mobility can be calculated using $N_{dof} = L - c$.

If there is no kinematic redundancies such that $\text{rank}(\mathbf{J}_{n_1}) = N_1$ and $\text{rank}(\mathbf{J}_{n_2}) = N_2$, the mobility of the mechanism is given by:

$$N_{dof} = L - \text{rank}(\mathbf{J}_L) = \dim(E(\mathbf{x}_1) \cap E(\mathbf{x}_2)). \tag{A.9}$$

In case of redundancy, the mobility will be:

$$N_{dof} = \dim(E(\mathbf{x}_1) \cap E(\mathbf{x}_2)) + \text{number of redundant joints in both branches.} \tag{A.10}$$

A.3.2 Mobility of Parallel Mechanisms with Serial Legs

Let us consider a parallel mechanism with a base platform and a mobile platform that are connected together with m simple open kinematic chains. The number of joints of each chain is denoted by n_i for $i = 1, \ldots, m$. The mobility of the platform M with respect to the base is given by the dimension of the common task spaces of the simple legs associated with the parallel mechanism, as seen in the following equations as long as there are no redundancy:

$$N_{P/0} = \dim(E(\mathbf{x}_1) \cap E(\mathbf{x}_2) \cdots \cap E(\mathbf{x}_m)). \tag{A.11}$$

The number of joints that lose their independence after loop closure is equal to the difference between the sum of mobilities of the terminal links of simple chains and the mobility of the platform:

$$c = \sum_{j=1}^{m} \dim(E(\mathbf{x}_j)) - \dim(E(\mathbf{x}_1) \cap E(\mathbf{x}_2) \cdots \cap E(\mathbf{x}_m)). \tag{A.12}$$

Thus the mobility of the structure is given by

$$N_{dof} = \sum_{j=1}^{m} n_j - \sum_{j=1}^{m} \dim(E(\mathbf{x}_j)) + \dim(E(\mathbf{x}_1) \cap E(\mathbf{x}_2) \ldots \cap E(\mathbf{x}_m)) \tag{A.13}$$

where $\sum_{j=1}^{m} n_j = L$, the number of total joints.

Note that in case of non-redundant legs, $n_j = \dim(E(\mathbf{x}_j))$, thus leading to mobility as the dimension of the common space between all the legs.

A.3.3 *Mobility of Parallel Mechanisms with Complex Legs*

A parallel mechanism with complex legs is a complex mechanism with L joints in which the mobile platform is connected to the base by $m \geq 2$ legs, of which at least one leg contains at least one closed loop. Theoretically in this case the platform is not uniquely defined, but in practice it is easy to select an appropriate one. The mobility of the mechanism in this case is calculated by:

$$N_{dof} = L - \sum_{j=1}^{m} \dim(E(\mathbf{x}_j)) + \dim(E(\mathbf{x}_1) \cap E(\mathbf{x}_2) \ldots \cap E(\mathbf{x}_m)) - C_{cl} \quad \text{(A.14)}$$

where $C_{cl} = \sum_{k=1}^{n_c} c_k$ is sum of the additional joint variables that lose their independence in the closed loops belonging to the complex legs. C_{cl} can be calculated using the previous cases in Sects. A.3.1 and A.3.2 depending on whether the leg contains loops connected serially or in parallel respectively.

A.4 Examples

In this section we calculate the mobility of some robot architectures treated in the different chapters of this book.

A.4.1 *The Planar Five-Bar Mechanism*

This structure is shown in Fig. 7.3. It consists of five revolute joints with parallel axes. The system ensures motion in the plane perpendicular to the joint axes. The mobility can be calculated using different interpretation:

1. The number of independent constraint equations around the loop is equal to 3, specifying the equality of x, y and the orientation ϕ. Thus, the mobility is equal to 2.
2. We open the loop at the terminal joint L, to obtain a serial architecture with five revolute joints. It is intuitive to deduce that the rank of this serial system is equal to the mobility of its terminal link, i.e. 3. Thus, connecting the terminal link with the base leads to a loss of these three degrees of freedom. Consequently, the mobility of the system is equal to 2.
3. We open the structure to obtain two serial chains connected with the base, such that one chain contains 3 joints and the other contains 2 joints. The dimension of the terminal link spaces of these chains are respectively 3 and 2. The motion of the first chain can be classified as: Tx (translation along $\boldsymbol{x}_0$), Ty (translation along $\boldsymbol{y}_0$) and Rz (rotation about z_0), whereas the motion of the terminal link of

the second chain, whose mobility is equal to 2, can be any two degrees of freedom among: Tx, Ty and Rz.

Consequently the common motion when connecting the two terminal links together will be of dimension 2, and can be represented by any two variables among (Tx, Ty, Rz). Thus the number of degrees of freedom of the terminal link (point) and the number of degrees of freedom of the system are respectively:

$$n_{dof} = 2$$

$$N_{dof} = L - \sum_{j=1}^{2} \dim(E(\mathbf{x}_j)) + \dim(E(\mathbf{x}_1) \cap E(\mathbf{x}_2)) = 2.$$

Consequently, two axes should be actuated. We selected the actuators near the base. In case of actuating more than two joints, the system will be redundantly actuated.

A.4.2 *The Planar 3–R$\underline{\text{P}}$R Parallel Robot*

This *PKM* is shown in Fig. 7.5. It has three legs, each being composed of three joints (*R*, *P*, and *R* joints) ensuring planar motions.

The mobility of each leg is composed of Tx, Ty and Rz.

Thus

$$E(\mathbf{x}_i) = \text{Tx, Ty, Rz for } i = 1, 2, 3$$

and thus the common space consists of: Tx, Ty and Rz.

Consequently, the mobility is equal to 3. The three prismatic joints are selected to be the actuated joints.

A.4.3 *The Orthoglide*

The Orthoglide (Fig. 7.7) is composed of three legs, each of them having a prismatic joint and a spatial parallelogram. The prismatic joints are perpendicular. The mobility of the Orthoglide will be analyzed using the equivalent kinematic chain of each leg, which is considered to be composed of PUU architecture, with U a universal joint represented by two intersecting revolute axes. Thus the motion of terminal link of each leg is composed of a prismatic joint, then two rotational and two translation degrees of freedom perpendicular on the prismatic axis of the first joint. The mobility of the three legs can be written respectively as:

$$E(\mathbf{x}_1) = \text{Tz, Rx, Ry, Tx, Ty}$$

$$E(\mathbf{x}_2) = \text{Tx, Ry, Rz, Ty, Tz}$$

$$E(\mathbf{x}_3) = \text{Ty, Rx, Ry, Tx, Tz.}$$

Thus

$$n_{dof} = \dim(E(\mathbf{x}_1) \cap E(\mathbf{x}_2) \cap E(\mathbf{x}_3)) = 3$$

$$E(\mathbf{x}_p) = \text{Tx, Ty, Tz}$$

$$N_{dof} = L - \sum_{j=1}^{3} \dim(E(\mathbf{x}_j)) + \dim(E(\mathbf{x}_1) \cap E(\mathbf{x}_2) \cap E(\mathbf{x}_3)) = 3.$$

We selected the three prismatic joints as actuators.

A.4.4 The Tripteron

This *PKM* is shown in Fig. 9.3. It is composed of three legs, each of them having a serial architecture with 4 joints. The first joint is prismatic and the other joints are revolute. All the *P* and *R* joints of the same leg have parallel axes.

We can easily deduce that:

$$E(\mathbf{x}_1) = \text{Tx, Ty, Tz, Rx}$$

$$E(\mathbf{x}_2) = \text{Tx, Ty, Tz, Ry}$$

$$E(\mathbf{x}_3) = \text{Tx, Ty, Tz, Rz.}$$

Thus

$$n_{dof} = \dim(E(\mathbf{x}_1) \cap E(\mathbf{x}_2) \cap E(\mathbf{x}_3)) = 3$$

$$E(\mathbf{x}_p) = \text{Tx, Ty, Tz}$$

$$N_{dof} = L - \sum_{j=1}^{3} \dim(E(\mathbf{x}_j)) + \dim(E(\mathbf{x}_1) \cap E(\mathbf{x}_2) \cap E(\mathbf{x}_3)) = 3.$$

We selected the three prismatic joints as actuators.

Remarks

Note that the simple relation of (A.2) can give the correct result for the first two examples (planar five-bar mechanism, and 3–*R*<u>*P*</u>*R*), but it does not give the correct result for the Orthoglide nor for the Tripteron.

Appendix B
Lagrange Equations with Multipliers

Let us consider a mechanical system whose Lagrangian L can computed by the knowledge of the generalized coordinates $\mathbf{q}$ (and $\dot{\mathbf{q}}$). Let us assume that $\mathbf{q}$ groups two sets of variables $\mathbf{q}_a$ and $\mathbf{q}_d$ ($\mathbf{q}^T = [\mathbf{q}_a^T \ \mathbf{q}_d^T]$ and $\dot{\mathbf{q}}^T = [\dot{\mathbf{q}}_a^T \ \dot{\mathbf{q}}_d^T]$) which are not independent and are related through the expressions:

$$\mathbf{h}(\mathbf{q}_a, \mathbf{q}_d) = \mathbf{0} \tag{B.1}$$

and:

$$\mathbf{A}(\mathbf{q}_a, \mathbf{q}_d)\dot{\mathbf{q}}_d + \mathbf{B}(\mathbf{q}_a, \mathbf{q}_d)\dot{\mathbf{q}}_a = \mathbf{0} \tag{B.2}$$

where $\mathbf{A}$ and $\mathbf{B}$ are two matrices depending on $\mathbf{q}_a$ and $\mathbf{q}_d$:

$$\mathbf{A}(\mathbf{q}_a, \mathbf{q}_d) = \left[\frac{\partial \mathbf{h}(\mathbf{q}_a, \mathbf{q}_d)}{\partial \mathbf{q}_d}\right] \tag{B.3}$$

and

$$\mathbf{B}(\mathbf{q}_a, \mathbf{q}_d) = \left[\frac{\partial \mathbf{h}(\mathbf{q}_a, \mathbf{q}_d)}{\partial \mathbf{q}_a}\right]. \tag{B.4}$$

Moreover, we consider that the variables $\mathbf{q}_a$ are some variables corresponding (in the frame of this book) to motor coordinates, motors which are exerting some input efforts $\boldsymbol{\tau}$ on the system. This is not the case for the coordinates $\mathbf{q}_d$.

The usual Lagrange equations (6.1) cannot be derived because all coordinates in $\mathbf{q}$ are not independent. In order to modify the Lagrange equations (6.1) to take into account the constraints (B.1) and (B.2), we must include some generalized constraint forces $\boldsymbol{\gamma}_a$ and $\boldsymbol{\gamma}_d$ such that:

S. Briot and W. Khalil, *Dynamics of Parallel Robots*,
Mechanisms and Machine Science 35, DOI 10.1007/978-3-319-19788-3

$$\boldsymbol{\tau} + \boldsymbol{\gamma}_a = \boldsymbol{\tau}_a, \text{ where } \boldsymbol{\tau}_a = \frac{\mathrm{d}}{\mathrm{d}t}\left(\frac{\partial L}{\partial \dot{\mathbf{q}}_a}\right)^T - \left(\frac{\partial L}{\partial \mathbf{q}_a}\right)^T$$
$$\boldsymbol{\gamma}_d = \boldsymbol{\tau}_d, \text{ where } \boldsymbol{\tau}_d = \frac{\mathrm{d}}{\mathrm{d}t}\left(\frac{\partial L}{\partial \dot{\mathbf{q}}_d}\right)^T - \left(\frac{\partial L}{\partial \mathbf{q}_d}\right)^T. \tag{B.5}$$

Of course, these generalized constraint forces $\boldsymbol{\gamma}_a$ and $\boldsymbol{\gamma}_d$ are internal to the system and produce no work, i.e. the *PVP* states that we have, for any arbitrary velocities $\dot{\mathbf{q}}_a^*$ and $\dot{\mathbf{q}}_d^*$,

$$\dot{\mathbf{q}}_d^{*T} \boldsymbol{\gamma}_d + \dot{\mathbf{q}}_a^{*T} \boldsymbol{\gamma}_a = 0. \tag{B.6}$$

Now, taking the transposed expression of (B.2), we obtain

$$\dot{\mathbf{q}}_d^T \mathbf{A}^T + \dot{\mathbf{q}}_a^T \mathbf{B}^T = \mathbf{0}. \tag{B.7}$$

This expression is also true if we right-multiply it by any arbitrary vector $\boldsymbol{\lambda}$:

$$\dot{\mathbf{q}}_d^T \mathbf{A}^T \boldsymbol{\lambda} + \dot{\mathbf{q}}_a^T \mathbf{B}^T \boldsymbol{\lambda} = 0. \tag{B.8}$$

By identification between (B.6) and (B.8), we can see that we have:

$$\boldsymbol{\gamma}_a = \mathbf{B}^T \boldsymbol{\lambda} \tag{B.9}$$

and

$$\boldsymbol{\gamma}_d = \mathbf{A}^T \boldsymbol{\lambda} \tag{B.10}$$

from which we obtain the new set of Lagrange equations, in which $\boldsymbol{\lambda}$ is called the vector of Lagrange multipliers:

$$\boldsymbol{\tau} + \mathbf{B}^T \boldsymbol{\lambda} = \boldsymbol{\tau}_a, \text{ where } \boldsymbol{\tau}_a = \frac{\mathrm{d}}{\mathrm{d}t}\left(\frac{\partial L}{\partial \dot{\mathbf{q}}_a}\right)^T - \left(\frac{\partial L}{\partial \mathbf{q}_a}\right)^T$$
$$\mathbf{A}^T \boldsymbol{\lambda} = \boldsymbol{\tau}_d, \text{ where } \boldsymbol{\tau}_d = \frac{\mathrm{d}}{\mathrm{d}t}\left(\frac{\partial L}{\partial \dot{\mathbf{q}}_d}\right)^T - \left(\frac{\partial L}{\partial \mathbf{q}_d}\right)^T. \tag{B.11}$$

Appendix C
Computation of Wrenches Reciprocal to a System of Twists

In this Appendix, we compute the actuation and constraint wrenches associated to some common *PKM* legs.

C.1 Definitions

The twist $\mathbf{t}$ of a body is parameterized by two vectors, the translational velocity $\mathbf{v}$ and the rotational velocity $\boldsymbol{\omega}$, such that we can define a vector of dimension 6:

$$\mathbf{t} = \begin{bmatrix} \mathbf{v} \\ \boldsymbol{\omega} \end{bmatrix}. \tag{C.1}$$

The twist $\mathbf{t}$ is also called the velocity screw. $\boldsymbol{\omega}$ is the resultant of the screw and $\mathbf{v}$ is its moment.

A wrench $\mathbf{w}$ is parameterized by two vectors, the force $\mathbf{f}$ and the moment $\mathbf{m}$, such that we can define a vector of dimension 6:

$$\mathbf{w} = \begin{bmatrix} \mathbf{f} \\ \mathbf{m} \end{bmatrix}. \tag{C.2}$$

The wrench $\mathbf{w}$ is a screw in which $\mathbf{f}$ is the resultant and $\mathbf{m}$ is the moment of the screw.

Let us define the screws:

- \$ which describes a unit twist; as a result, if
  - \$ characterizes a pure translation, it can be written as

$$\$ = \begin{bmatrix} \mathbf{u}_1 \\ \mathbf{0} \end{bmatrix} \tag{C.3}$$

  in which $\mathbf{u}_1$ represents the direction of the translation and $\|\mathbf{u}_1\| = 1$.

S. Briot and W. Khalil, *Dynamics of Parallel Robots*,
Mechanisms and Machine Science 35, DOI 10.1007/978-3-319-19788-3

- $ characterizes a pure rotation, it can be written as

$$\$ = \begin{bmatrix} \mathbf{u}_1 \times \mathbf{r} \\ \mathbf{u}_1 \end{bmatrix} \tag{C.4}$$

  in which $\mathbf{u}_1$ represents the axis of the rotation ($\|\mathbf{u}_1\| = 1$) and $\mathbf{r}$ is the vector defining the distance between the axis of rotation and the point at which $ is expressed.

- $\boldsymbol{\zeta}$ which describes a unit wrench.
  - $\boldsymbol{\zeta}$ characterizes a pure moment, it can be written as

$$\boldsymbol{\zeta} = \begin{bmatrix} \mathbf{0} \\ \mathbf{u}_2 \end{bmatrix} \tag{C.5}$$

    in which $\mathbf{u}_2$ represents the direction around which the moment is applied and $\|\mathbf{u}_2\| = 1$.
  - $\boldsymbol{\zeta}$ characterizes a pure force, it can be written as

$$\boldsymbol{\zeta} = \begin{bmatrix} \mathbf{u}_2 \\ \mathbf{u}_2 \times \mathbf{r} \end{bmatrix} \tag{C.6}$$

    in which $\mathbf{u}_2$ represents the direction along which the force is applied ($\|\mathbf{u}_2\| = 1$) and $\mathbf{r}$ is the vector defining the distance between the point on which the force is applied and the point at which $\boldsymbol{\zeta}$ is expressed.

C.2 Condition of Reciprocity

A twist $ is reciprocal to a wrench $\boldsymbol{\zeta}$ if their product is null, i.e. $\$^T \boldsymbol{\zeta} = \boldsymbol{\zeta}^T \$ = 0$. This means that the power developed by the wrench $\boldsymbol{\zeta}$ along the motion defined by $ is null.

From the definition of the unit twist and wrench $ and $\boldsymbol{\zeta}$, we can define the following rules (Zhao et al. 2009):

- for a revolute joint with axis along the direction $\boldsymbol{u}$, the reciprocal wrenches are:
  - forces coplanar to $\boldsymbol{u}$, i.e. forces directed along an axis either parallel to $\boldsymbol{u}$ or intersecting $\boldsymbol{u}$ at a point,
  - moments whose axes are orthogonal to $\boldsymbol{u}$.
- for a prismatic joint with axis aligned along $\boldsymbol{u}$, the reciprocal wrenches are:
  - forces whose directions are orthogonal to $\boldsymbol{u}$,
  - any moment.

These rules can be used to find the $(6-n)$ reciprocal wrenches for a system of n twists.

Similarly, we can define the twists reciprocal to a given wrench or system of wrenches. The interested reader could find more details in (Zhao et al. 2009).

C.3 Computation of Wrenches Reciprocal to a System of Twists Constrained in a Plane

This case where a system of twists is constrained in a plane appear for the *PPM* (planar parallel manipulators). These mechanisms are made of joints whose displacements are all constrained in the same plane. The most common legs of *PPM* are presented in Fig. C.1. They are all made of three joints, one of them being actuated (the joints in gray).

Let us consider that the plane of motion is the $(O_0, \boldsymbol{x}_0, \boldsymbol{y}_0)$ plane. As a result, any twist $\$_i$ associated to the motion of a joint i has the following form:

$$\$_i = \begin{bmatrix} v_{xi} \ v_{yi} \ 0 \ 0 \ 0 \ \omega_{zi} \end{bmatrix}^T . \tag{C.7}$$

This means that, for any twist system of dimension n (representing a planar leg composed of n active or passive joints $-n \in [1, +\infty[$) defined by $\$ = \begin{bmatrix} \$_1 \ \ldots \ \$_n \end{bmatrix}$, three constraint wrenches $\boldsymbol{\zeta}_{c1}$, $\boldsymbol{\zeta}_{c2}$ and $\boldsymbol{\zeta}_{c3}$ (i.e. the wrenches reciprocal to all twists representing the passive and active joint motions) can be easily defined as

$$\boldsymbol{\zeta}_{c1} = \begin{bmatrix} 0 \ 0 \ 1 \ 0 \ 0 \ 0 \end{bmatrix}^T \tag{C.8}$$

which represents a force along z_0,

$$\boldsymbol{\zeta}_{c2} = \begin{bmatrix} 0 \ 0 \ 0 \ 1 \ 0 \ 0 \end{bmatrix}^T \tag{C.9}$$

which represents a moment around $\boldsymbol{x}_0$, and

$$\boldsymbol{\zeta}_{c3} = \begin{bmatrix} 0 \ 0 \ 0 \ 0 \ 1 \ 0 \end{bmatrix}^T \tag{C.10}$$

which represents a moment around $\boldsymbol{y}_0$.

These three constraint wrenches prevent the translation along z_0 and the rotations around $\boldsymbol{x}_0$ and $\boldsymbol{y}_0$ of the body located at the leg extremity. All legs presented in the Fig. C.1 impose these three constraint wrenches to the platform.

Now, let us consider that one joint of the legs depicted in Fig. C.1 is actuated (the joints in gray). This is the case for the most usual *PPM*. Starting from this consideration, only two cases can appear:

1. the passive system is made of two *R* joints (Fig. C.2a),
2. the passive system is made of one *P* joint and one *R* joint (Fig. C.2b).

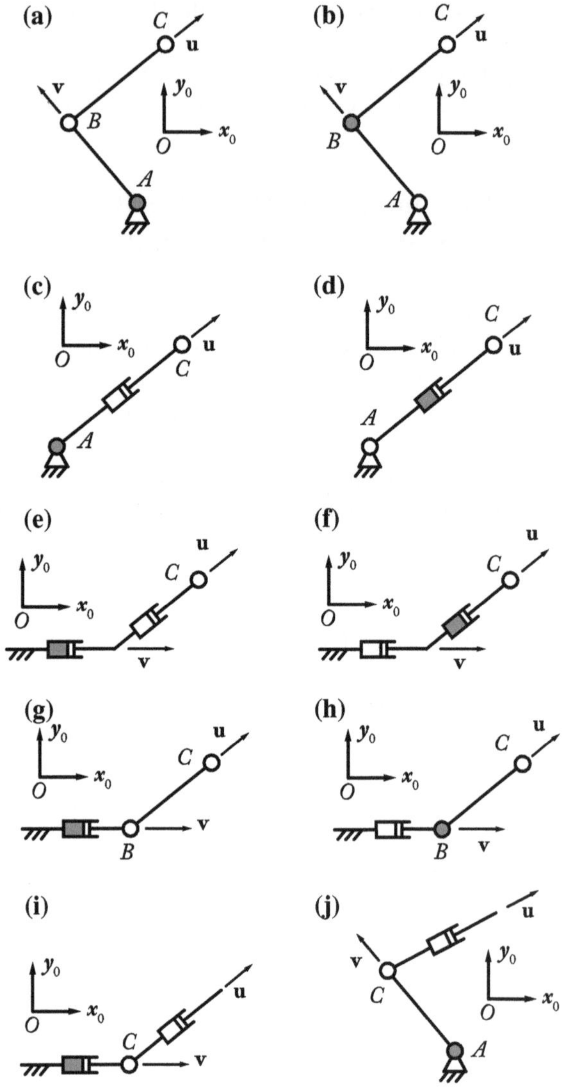

Fig. C.1 Usual legs for *PPM*. **a** *RRR* leg. **b** *RRR* leg. **c** *RPR* leg. **d** *RPR* leg. **e** *PPR* leg. **f** *PPR* leg. **g** *PRR*. **h** *PRR* leg. **i** *PRP* leg. **j** *RRR* leg

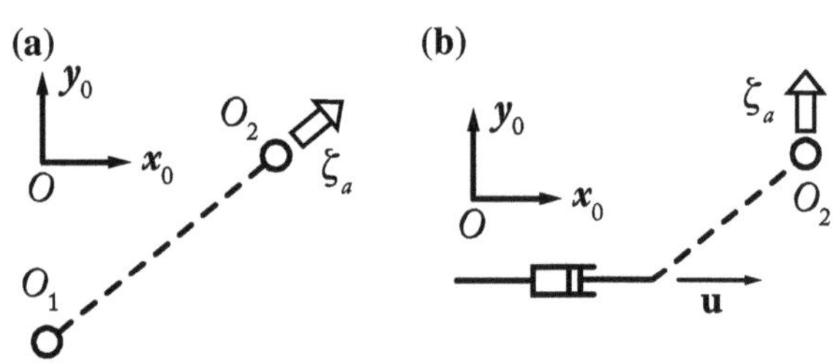

Fig. C.2 General passive systems for the usual legs of *PPM*. **a** *RR* passive system ($\boldsymbol{\zeta}_a$is aligned with $(O_1 O_2)$). **b** *PR* passive system ($\boldsymbol{\zeta}_a \perp \mathbf{u}$)

C.3.1 Computation of Wrenches Reciprocal to a Twist System Representing the Motion of Two Passive R Joints

For a system composed of two R joints (one located at point O_1 and the second one at point O_2, point O_i having the coordinates (x_i, y_i) in the plane $(O_0, \boldsymbol{x}_0, \boldsymbol{y}_0)$– Fig. C.2a, it is possible to define two twists $\mathbf{\$}_{R1}$ and $\mathbf{\$}_{R2}$ parameterized by (when expressed at point O_2):

$$^0\mathbf{\$}_{R1} = \left[-(y_2 - y_1)\; x_2 - x_1\; 0\; 0\; 0\; 1\right]^T \tag{C.11}$$

$$^0\mathbf{\$}_{R2} = \left[0\; 0\; 0\; 0\; 0\; 1\right]^T. \tag{C.12}$$

The constraint wrenches $\boldsymbol{\zeta}_{ci}$ $(i = 1, 2, 3)$ are reciprocal to $\mathbf{\$}_{R1}$ and $\mathbf{\$}_{R2}$ but one additional wrench, denoted as the actuation wrench $\boldsymbol{\zeta}_a$ is reciprocal to these twists. It can be easily seen from (Zhao et al. 2009) that

$$^0\boldsymbol{\zeta}_a = \left[x_2 - x_1\; y_2 - y_1\; 0\; 0\; 0\; 0\right]^T \tag{C.13}$$

i.e. it is a force passing through the centers of the two R joints (see Fig. C.2a). It should be noted that $\boldsymbol{\zeta}_a$ must not be reciprocal to the twist of the third (actuated) joint

C.3.2 Computation of Wrenches Reciprocal to a Twist System Representing the Motion of One P Joint and one R Joint

For a system composed of P joint of direction $\mathbf{u}_1 = [u_{1x}\; u_{1y}\; 0]^T$ and one R joint located at point O_2 (Fig. C.2b), it is possible to define two twists $\mathbf{\$}_{P1}$ and $\mathbf{\$}_{R2}$ parameterized by (when expressed at point O_2):

$$^0\mathbf{\$}_{P1} = \left[u_{1x}\; u_{1y}\; 0\; 0\; 0\; 0\right]^T \tag{C.14}$$

$$^0\mathbf{\$}_{R2} = \left[0\; 0\; 0\; 0\; 0\; 1\right]^T. \tag{C.15}$$

The constraint wrenches $\boldsymbol{\zeta}_{ci}$ $(i = 1, 2, 3)$ are reciprocal to $\mathbf{\$}_{P1}$ and $\mathbf{\$}_{R2}$ but one additional wrench, denoted as the actuation wrench $\boldsymbol{\zeta}_a$ is reciprocal to these twists ($\boldsymbol{\zeta}_a$ must not be reciprocal to the twist of the third (actuated) joint). It can be easily seen from (Zhao et al. 2009) that

$$^0\boldsymbol{\zeta}_a = \left[-u_{1y}\; u_{1x}\; 0\; 0\; 0\; 0\right]^T \tag{C.16}$$

i.e. it is a force lying in the plane $(O_0, \boldsymbol{x}_0, \boldsymbol{y}_0)$ passing through O_2 and orthogonal to the prismatic joint direction (see Fig. C.2b).

C.4 Computation of Wrenches Reciprocal to Other Types of Twist Systems

For *SPM*, other types of twist systems appear. Due to the large number of possible leg architectures for the existing *SPM*, it is impossible to deal with all possible twist systems in this Appendix. However, we will compute the wrenches reciprocal to the twist systems corresponding to:

- *U$\underline{P}$S* legs (legs of the Gough-Stewart platform—see Sect. 7.2.2.5)
- *U$\underline{P}$U* legs (legs of the Tsai mechanism (Tsai and Joshi 2000)),
- *$\underline{R}$US* legs (legs of Hexa-like robots, and most of Delta-like robots (Clavel 1989; Company et al. 2002; Pashkevich et al. 2006)).

C.4.1 Computation of Wrenches Reciprocal to a Twist System Representing the Motion of a U$\underline{P}$S *Leg*

Let us consider a *U$\underline{P}$S* leg composed of an actuated $\underline{P}$ joint of direction ${}^i\mathbf{u} = [u_x\ 0\ u_z]^T$ in the leg frame $\mathscr{F}_i$ (Fig. C.3), one passive *U* joint which can be represented as an assembly of two *R* joints whose axes $\mathbf{a}_1$ and $\mathbf{a}_2$ are orthogonal to the direction of the $\underline{P}$ joint, i.e. $\mathbf{a}_1 = [1\ 0\ 0]^T$ and $\mathbf{a}_2 = [0\ 1\ 0]^T$ in the leg frame $\mathscr{F}_i$, and one passive *S* joint allowing three independent rotations around three axes $\mathbf{a}_1 = [1\ 0\ 0]^T$ $\mathbf{a}_2 = [0\ 1\ 0]^T$ and $\mathbf{a}_3 = [0\ 0\ 1]^T$ without loss of generality in the leg frame $\mathscr{F}_i$ (Fig. C.3).

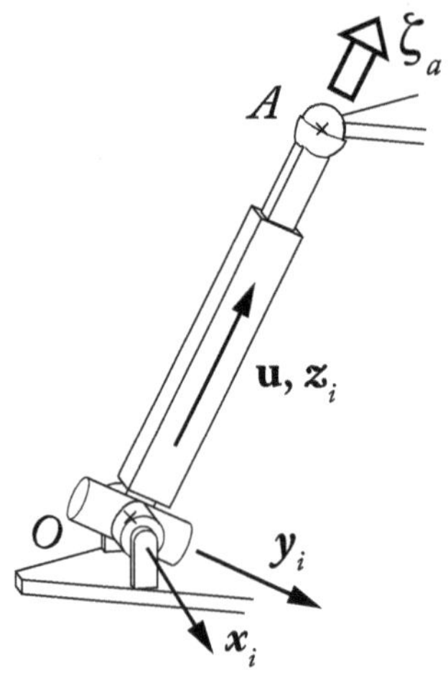

Fig. C.3 A *U$\underline{P}$S* leg (in this configuration, $\mathbf{u} \equiv z_i$, however, this is not the general case)

As a result, the twist system representing the motion of the leg in the frame $\mathcal{F}_i$ and expressed at the center A of the S joint is given by:

$$ {}^i\$_1 = \left[0\ -z_A\ 0\ 1\ 0\ 0\right]^T \tag{C.17}$$

$$ {}^i\$_2 = \left[z_A\ 0\ -x_A\ 0\ 1\ 0\right]^T \tag{C.18}$$

$$ {}^i\$_3 = \left[u_x\ 0\ u_z\ 0\ 0\ 0\right]^T \tag{C.19}$$

$$ {}^i\$_4 = \left[0\ 0\ 0\ 1\ 0\ 0\right]^T \tag{C.20}$$

$$ {}^i\$_5 = \left[0\ 0\ 0\ 0\ 1\ 0\right]^T \tag{C.21}$$

$$ {}^i\$_6 = \left[0\ 0\ 0\ 0\ 0\ 1\right]^T \tag{C.22}$$

where x_A and z_A are the coordinates of the point A along the axes $\boldsymbol{x}_i$ and $\boldsymbol{z}_i$ of the frame $\mathcal{F}_i$.

The total twist system $\$ = \left[\$_1 \ldots \$_6\right]$ is of rank 6. Therefore, there are no constraint wrenches.

Now, let us compute the actuation wrench when the $\underline{P}$ joint is considered actuated. We must thus consider the twist system $\$_d = \left[\$_1\ \$_2\ \$_4\ \$_5\ \$_6\right]$. It automatically gives that the actuation wrench $\boldsymbol{\zeta}_a$ is equal to, from (Zhao et al. 2009):

$$ {}^i\zeta_a = \left[u_x\ 0\ u_z\ 0\ 0\ 0\right]^T \tag{C.23}$$

i.e. it is a force directed along the direction connecting the centres of the U and S joints. We note that this force is not reciprocal to the actuated prismatic joint axis.

C.4.2 Computation of Wrenches Reciprocal to a Twist System Representing the Motion of a U$\underline{\text{P}}$U Leg

Let us consider a $U\underline{P}U$ leg composed of an actuated $\underline{P}$ joint of direction $\mathbf{u} = [u_x\ 0\ u_z^T$ in the leg frame $\mathcal{F}_i$, and two passive U joints. Each passive U joint can be represented as an assembly of two R joints whose axes $\mathbf{a}_1$ and $\mathbf{a}_2$ are orthogonal to the direction of the $\underline{P}$ joint, i.e. $\mathbf{a}_1 = [1\ 0\ 0]^T$ and $\mathbf{a}_2 = [0\ 1\ 0]^T$ in the leg frame $\mathcal{F}_i$ (Fig. C.4).

As a result, the twist system representing the motion of the leg in the frame $\mathcal{F}_i$ and expressed at the center A of the second U joint is given by:

$$ {}^i\$_1 = \left[0\ -z_A\ y_A\ 1\ 0\ 0\right]^T \tag{C.24}$$

$$ {}^i\$_2 = \left[z_A\ 0\ -x_A\ 0\ 1\ 0\right]^T \tag{C.25}$$

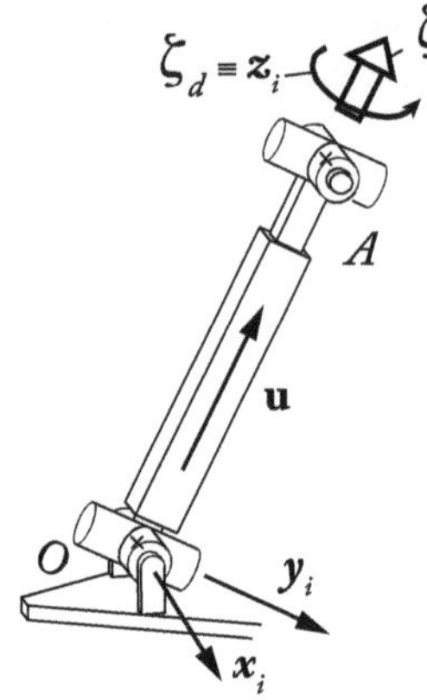

Fig. C.4 A $U\underline{P}U$ leg (in this configuration, $\mathbf{u} \equiv z_i$, however, this is not the general case)

$$^{i}\$_3 = \begin{bmatrix} u_x \; 0 \; u_z \; 0 \; 0 \; 0 \end{bmatrix}^T \quad \text{(C.26)}$$

$$^{i}\$_4 = \begin{bmatrix} 0 \; 0 \; 0 \; 1 \; 0 \; 0 \end{bmatrix}^T \quad \text{(C.27)}$$

$$^{i}\$_5 = \begin{bmatrix} 0 \; 0 \; 0 \; 0 \; 1 \; 0 \end{bmatrix}^T . \quad \text{(C.28)}$$

The total twist system $\$ = \begin{bmatrix} \$_1 \; \ldots \; \$_5 \end{bmatrix}$ is of rank 5. As a result, there is a constraint wrench given from (Zhao et al. 2009) by

$$^{i}\boldsymbol{\zeta}_c = \begin{bmatrix} 0 \; 0 \; 0 \; 0 \; 0 \; 1 \end{bmatrix}^T \quad \text{(C.29)}$$

i.e. it is a moment around the z_i axis (Fig. C.4).

Now, let us compute the actuation wrench when the $\underline{P}$ joint is considered actuated. We must thus consider the twist system $\$_d = \begin{bmatrix} \$_1 \; \$_2 \; \$_4 \; \$_5 \end{bmatrix}$. It automatically gives that the actuation wrench $\boldsymbol{\zeta}_a$ is equal to, from [Zhao *et al.*., 2009]:

$$^{i}\boldsymbol{\zeta}_a = \begin{bmatrix} u_x \; 0 \; u_z \; 0 \; 0 \; 0 \end{bmatrix}^T \quad \text{(C.30)}$$

i.e. it is a force directed along the direction of the prismatic joint (Fig. C.4).

C.4.3 Computation of Wrenches Reciprocal to a Twist System Representing the Motion of a $\underline{\text{R}}$US *Leg*

Let us consider a $\underline{R}US$ leg composed of an actuated $\underline{R}$ joint of direction $^0\mathbf{a}_1 = [1\ 0\ 0]^T$ in the base frame $\mathscr{F}_0$, one passive U joint which can be represented as an assembly of two R joints whose axes are $\mathbf{a}_1$ and $\mathbf{a}_2$, with $\mathbf{a}_1 = [1\ 0\ 0]^T$ and $\mathbf{a}_2 = [0\ a_y\ a_z]^T$ in the leg frame $\mathscr{F}_i$, and one passive S joint allowing three independent rotations around three axes $\mathbf{a}_1 = [1\ 0\ 0]^T$ $\mathbf{a}_3 = [0\ 1\ 0]^T$ and $\mathbf{a}_4 = [0\ 0\ 1]^T$ without loss of generality in the leg frame $\mathscr{F}_i$ (Fig. C.5).

Fig. C.5 A $\underline{R}US$ leg

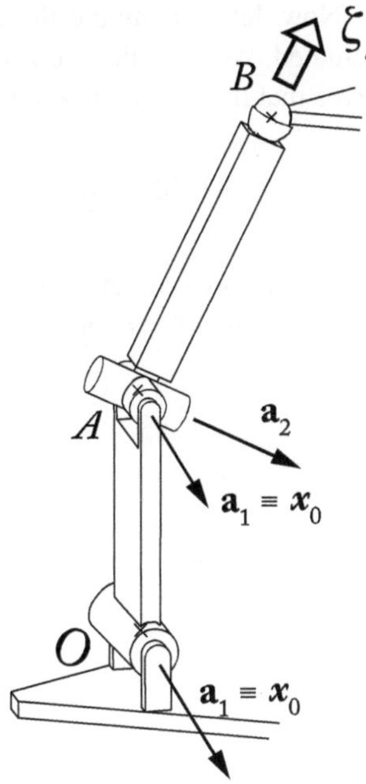

As a result, the twist system representing the motion of the leg in the frame $\mathscr{F}_0$ and expressed at the center B of the S joint (with coordinates in the base frame x_B, y_B and z_B) is given by:

$$^i\$_1 = \begin{bmatrix} 0 & -z_B & y_B & 1 & 0 & 0 \end{bmatrix}^T \tag{C.31}$$

$$^i\$_2 = \begin{bmatrix} 0 & -z_{AB} & y_{AB} & 1 & 0 & 0 \end{bmatrix}^T \tag{C.32}$$

$$^i\$_3 = \begin{bmatrix} a_y z_{AB} - a_z y_{AB} & a_z x_{AB} & -a_y x_{AB} & 0 & a_y & a_z \end{bmatrix}^T \tag{C.33}$$

$$^i\$_4 = \begin{bmatrix} 0 & 0 & 0 & 1 & 0 & 0 \end{bmatrix}^T \tag{C.34}$$

$$^i\$_5 = \begin{bmatrix} 0 & 0 & 0 & 0 & 1 & 0 \end{bmatrix}^T \tag{C.35}$$

$$^i\$_6 = \begin{bmatrix} 0 & 0 & 0 & 0 & 0 & 1 \end{bmatrix}^T \tag{C.36}$$

where x_{AB}, y_{AB} and z_{AB} are the coordinates of the vector $\overrightarrow{AB}$ along the axes of the frame $\mathscr{F}_0$. Note that $\overrightarrow{AB} \perp \mathbf{a}_2$, i.e. that $y_{AB} a_y + z_{AB} a_z = 0$.

The total twist system $\$ = \begin{bmatrix} \$_1 \ldots \$_6 \end{bmatrix}$ is of rank 6. Therefore, there are no constraint wrenches.

Now, let us compute the actuation wrench when the first $\underline{R}$ joint is considered actuated. We must thus consider the twist system $\$_d = \left[\$_2 \ldots \$_6\right]$. It automatically gives that the actuation wrench $\boldsymbol{\zeta}_a$ is equal to, from (Zhao et al. 2009):

$$^{i}\boldsymbol{\zeta}_a = \left[x_{AB}\ y_{AB}\ z_{AB}\ 0\ 0\ 0\right]^T / \sqrt{x_{AB}^2 + y_{AB}^2 + z_{AB}^2} \tag{C.37}$$

i.e. it is a force directed along the direction given by the vector $\overrightarrow{AB}$ (Fig. C.5). We can verify that $\boldsymbol{\zeta}_a$ is not orthogonal to the unit twist of the actuated joint.

Appendix D
Point-to-Point Trajectory Generation

Let us consider a robot displacement between an initial configuration A_0 parameterized by the Cartesian variables $\mathbf{x}_0$ and a final configuration A_f parameterized by the Cartesian variables $\mathbf{x}_f$. The trajectory between these two configurations can be defined by the functions

$$\mathbf{x}(t) = s(t)\left(\mathbf{x}_f - \mathbf{x}_0\right) + \mathbf{x}_0 \tag{D.1}$$

$$\dot{\mathbf{x}}(t) = \dot{s}(t)\left(\mathbf{x}_f - \mathbf{x}_0\right) \tag{D.2}$$

$$\ddot{\mathbf{x}}(t) = \ddot{s}(t)\left(\mathbf{x}_f - \mathbf{x}_0\right) \tag{D.3}$$

where:

- $t \in [0,\ t_f]$, where $t = 0$ s is the time at which the robot starts to move from the initial configuration A_0 and t_f is the time at which the robot arrives at the final configuration A_f
- $\mathbf{x}(t)$ denotes the robot Cartesian variables at the time t,
- $\dot{\mathbf{x}}(t)$ denotes the first derivative w.r.t. time of the robot Cartesian variables at the time t,
- $\ddot{\mathbf{x}}(t)$ denotes the second derivative w.r.t. time of the robot Cartesian variables at the time t,
- $s(t)$ is an interpolation function.

We can deduce from (D.1) that the path in the Cartesian space will be defined by a straight line. The boundary conditions for $s(t)$ are deduced as:

$$s(t = 0\ \text{s}) = 0 \tag{D.4}$$

$$s(t = t_f) = 1. \tag{D.5}$$

S. Briot and W. Khalil, *Dynamics of Parallel Robots*,
Mechanisms and Machine Science 35, DOI 10.1007/978-3-319-19788-3

Moreover, if at the initial and final configurations, the velocities and accelerations are null, we have

$$\dot{s}(t = 0\ \mathrm{s}) = 0 \tag{D.6}$$

$$\dot{s}(t = t_f) = 0 \tag{D.7}$$

$$\ddot{s}(t = 0\ \mathrm{s}) = 0 \tag{D.8}$$

$$\ddot{s}(t = t_f) = 0. \tag{D.9}$$

From these boundary conditions, and assuming that the interpolation function $s(t)$ is a polynomial of the form:

$$s(t) = \sum_{k=0}^{n} a_k\ t^k \tag{D.10}$$

we can find the coefficients a_k.

For high speed robots or when a robot is handling heavy or delicate loads, it is worth ensuring the continuity of the position, velocity, and accelerations as well, in order to avoid exciting resonances in the mechanics. The trajectory is said to be of class C^2. Thus, from (D.2) and (D.3), we must define the functions $\dot{s}(t)$ and $\ddot{s}(t)$, given by

$$\dot{s}(t) = \sum_{k=1}^{n} k\ a_k\ t^{k-1} \tag{D.11}$$

and

$$\ddot{s}(t) = \sum_{k=2}^{n} k\ (k-1)\ a_k\ t^{k-2}. \tag{D.12}$$

Since six constraints (D.4)–(D.9) have to be satisfied, the interpolation requires a polynomial of at least fifth degree (Binford et al. 1977).

Solving the six constraints yields the following interpolation function:

$$s(t) = 10\left(\frac{t}{t_f}\right)^3 - 6\left(\frac{t}{t_f}\right)^4 + 15\left(\frac{t}{t_f}\right)^5. \tag{D.13}$$

Obviously, if we increase the number of boundary conditions to take into account, the order of the polynomial will increase. For example, if n_c constraints have to be satisfied, the interpolation requires a polynomial of at least $n_c - 1$ degree.

Appendix E
Calculation of the Terms $\mathbf{f}_{acc_1}$, $\mathbf{f}_{acc_2}$ and $\mathbf{f}_{acc_3}$ in Chapter 10

E.1 Calculation of the Term $\mathbf{f}_{acc_1}$

From (10.12) and (10.15), we get that $\mathbf{f}_{acc_1}$ is given by

$$\begin{aligned}\mathbf{f}_{acc_1} &= \int_{\mathscr{B}_j} \dot{\mathbf{v}}_{M_j} \mathrm{d}m \\ &= \int_{\mathscr{B}_j} \left(\dot{\mathbf{v}}_j + \mathbf{\Phi}_{d_j}(M_{0j})\,\ddot{\mathbf{q}}_{e_j} + 2\boldsymbol{\omega}_j \times \mathbf{\Phi}_{d_j}(M_{0j})\dot{\mathbf{q}}_{e_j}\right) \mathrm{d}m \\ &+ \int_{\mathscr{B}_j} \left(\boldsymbol{\omega}_j \times \left(\boldsymbol{\omega}_j \times \mathbf{r}_{O_jM_j}\right) + \dot{\boldsymbol{\omega}}_j \times \mathbf{r}_{O_jM_j}\right) \mathrm{d}m \end{aligned} \tag{E.1}$$

which can be expanded to be rewritten as

$$\begin{aligned}\mathbf{f}_{acc_1} &= \int_{\mathscr{B}_j} \dot{\mathbf{v}}_j \mathrm{d}m + \int_{\mathscr{B}_j} \mathbf{\Phi}_{d_j}(M_{0j})\mathrm{d}m\ddot{\mathbf{q}}_{e_j} \\ &+ \dot{\boldsymbol{\omega}}_j \times \left(\int_{\mathscr{B}_j} \mathbf{r}_{O_jM_j}\mathrm{d}m\right) \\ &+ 2\boldsymbol{\omega}_j \times \left(\int_{\mathscr{B}_j} \mathbf{\Phi}_{d_j}(M_{0j})\mathrm{d}m\,\dot{\mathbf{q}}_{e_j}\right) \\ &+ \boldsymbol{\omega}_j \times \left(\boldsymbol{\omega}_j \times \left(\int_{\mathscr{B}_j} \mathbf{r}_{O_jM_j}\mathrm{d}m\right)\right). \end{aligned} \tag{E.2}$$

E.2 Calculation of the Term $\mathbf{f}_{acc_2}$

From (10.12) and (10.15), we get that $\mathbf{f}_{acc_2}$ is given by

S. Briot and W. Khalil, *Dynamics of Parallel Robots*,
Mechanisms and Machine Science 35, DOI 10.1007/978-3-319-19788-3

$$\begin{aligned}
\mathbf{f}_{acc_2} &= \int_{\mathscr{B}_j} \mathbf{r}_{O_jM_j} \times \dot{\mathbf{v}}_{M_j} \mathrm{d}m \\
&= \int_{\mathscr{B}_j} \mathbf{r}_{O_jM_j} \times \left(\dot{\mathbf{v}}_j + \mathbf{\Phi}_{d_j}(M_{0j})\ddot{\mathbf{q}}_{e_j} + 2\boldsymbol{\omega}_j \times \mathbf{\Phi}_{d_j}(M_{0j})\dot{\mathbf{q}}_{e_j} \right) \mathrm{d}m \\
&\quad + \int_{\mathscr{B}_j} \mathbf{r}_{O_jM_j} \times \left(\boldsymbol{\omega}_j \times \left(\boldsymbol{\omega}_j \times \mathbf{r}_{O_jM_j} \right) + \dot{\boldsymbol{\omega}}_j \times \mathbf{r}_{O_jM_j} \right) \mathrm{d}m
\end{aligned} \tag{E.3}$$

which can be expanded to be rewritten under a sum of five terms:

$$\mathbf{f}_{acc_2} = \sum_{k=1}^{5} \mathbf{a}_k \tag{E.4}$$

where:

$$\begin{aligned}
\mathbf{a}_1 &= \int_{\mathscr{B}_j} \mathbf{r}_{O_jM_j} \times \dot{\mathbf{v}}_j \mathrm{d}m \\
&= \left(\int_{\mathscr{B}_j} \mathbf{r}_{O_jM_j} \mathrm{d}m \right) \times \dot{\mathbf{v}}_j
\end{aligned} \tag{E.5}$$

$$\begin{aligned}
\mathbf{a}_2 &= \int_{\mathscr{B}_j} \mathbf{r}_{O_jM_j} \times \left(\mathbf{\Phi}_{d_j}(M_{0j})\ddot{\mathbf{q}}_{e_j} \right) \mathrm{d}m \\
&= \int_{\mathscr{B}_j} \hat{\mathbf{r}}_{O_jM_j} \mathbf{\Phi}_{d_j}(M_{0j})\ddot{\mathbf{q}}_{e_j} \mathrm{d}m \\
&= \left(\int_{\mathscr{B}_j} \hat{\mathbf{r}}_{O_jM_j} \mathbf{\Phi}_{d_j}(M_{0j}) \mathrm{d}m \right) \ddot{\mathbf{q}}_{e_j}
\end{aligned} \tag{E.6}$$

$$\begin{aligned}
\mathbf{a}_3 &= \int_{\mathscr{B}_j} \mathbf{r}_{O_jM_j} \times \left(\dot{\boldsymbol{\omega}}_j \times \mathbf{r}_{O_jM_j} \right) \mathrm{d}m \\
&= - \int_{\mathscr{B}_j} \mathbf{r}_{O_jM_j} \times \left(\mathbf{r}_{O_jM_j} \times \dot{\boldsymbol{\omega}}_j \right) \mathrm{d}m \\
&= - \int_{\mathscr{B}_j} \hat{\mathbf{r}}_{O_jM_j} \left(\hat{\mathbf{r}}_{O_jM_j} \dot{\boldsymbol{\omega}}_j \right) \mathrm{d}m \\
&= \int_{\mathscr{B}_j} \hat{\mathbf{r}}^T_{O_jM_j} \hat{\mathbf{r}}_{O_jM_j} \dot{\boldsymbol{\omega}}_j \mathrm{d}m \\
&= \int_{\mathscr{B}_j} \hat{\mathbf{r}}^T_{O_jM_j} \hat{\mathbf{r}}_{O_jM_j} \mathrm{d}m \; \dot{\boldsymbol{\omega}}_j
\end{aligned} \tag{E.7}$$

$$\mathbf{a}_4 = \int_{\mathscr{B}_j} \mathbf{r}_{O_jM_j} \times \left(\boldsymbol{\omega}_j \times \left(\boldsymbol{\omega}_j \times \mathbf{r}_{O_jM_j} \right) \right) \mathrm{d}m \tag{E.8}$$

$$\mathbf{a}_5 = 2 \int_{\mathscr{B}_j} \mathbf{r}_{O_jM_j} \times \left(\boldsymbol{\omega}_j \times \mathbf{\Phi}_{d_j}(M_{0j})\dot{\mathbf{q}}_{e_j} \right) \mathrm{d}m. \tag{E.9}$$

To simplify the term $\mathbf{a}_4$, let us recall the well-known identity for the double cross-product of three arbitrary vectors $\mathbf{u}$, $\mathbf{v}$ and $\mathbf{w}$:

$$\mathbf{u} \times (\mathbf{v} \times \mathbf{w}) + \mathbf{w} \times (\mathbf{u} \times \mathbf{v}) + \mathbf{v} \times (\mathbf{w} \times \mathbf{u}) = \mathbf{0}. \tag{E.10}$$

If $\mathbf{u} = \mathbf{r}_{O_jM_j}$, $\mathbf{v} = \boldsymbol{\omega}_j$, $\mathbf{w} = \boldsymbol{\omega}_j \times \mathbf{r}_{O_jM_j}$, so $\mathbf{w} \times (\mathbf{u} \times \mathbf{v}) = \mathbf{0}$ which leads to

$$\mathbf{u} \times (\mathbf{v} \times \mathbf{w}) = -\mathbf{v} \times (\mathbf{w} \times \mathbf{u}) \tag{E.11}$$

or also, replacing the vectors $\mathbf{u}$, $\mathbf{v}$ and $\mathbf{w}$ by their corresponding values

$$\mathbf{r}_{O_jM_j} \times \left(\boldsymbol{\omega}_j \times \left(\boldsymbol{\omega}_j \times \mathbf{r}_{O_jM_j}\right)\right) = -\boldsymbol{\omega}_j \times ((\boldsymbol{\omega}_j \times \mathbf{r}_{O_jM_j}) \times \mathbf{r}_{O_jM_j}). \tag{E.12}$$

Thus, $\mathbf{a}_4$ becomes

$$\begin{aligned}
\mathbf{a}_4 &= -\int_{\mathscr{B}_j} \boldsymbol{\omega}_j \times \left(\left(\boldsymbol{\omega}_j \times \mathbf{r}_{O_jM_j}\right) \times \mathbf{r}_{O_jM_j}\right) \mathrm{d}m \\
&= -\int_{\mathscr{B}_j} \boldsymbol{\omega}_j \times \left(\mathbf{r}_{O_jM_j} \times \left(\mathbf{r}_{O_jM_j} \times \boldsymbol{\omega}_j\right)\right) \mathrm{d}m \\
&= -\int_{\mathscr{B}_j} \boldsymbol{\omega}_j \times \left(\hat{\mathbf{r}}_{O_jM_j}\hat{\mathbf{r}}_{O_jM_j}\boldsymbol{\omega}_j\right) \mathrm{d}m \\
&= \boldsymbol{\omega}_j \times \left(\left(\int_{\mathscr{B}_j} \hat{\mathbf{r}}^T_{O_jM_j}\hat{\mathbf{r}}_{O_jM_j} \mathrm{d}m\right)\boldsymbol{\omega}_j\right).
\end{aligned} \tag{E.13}$$

Now, introducing (10.4) into (E.9), the expression of $\mathbf{a}_5$ becomes

$$\mathbf{a}_5 = 2(\mathbf{a}_{51} + \mathbf{a}_{52}) \tag{E.14}$$

where

$$\begin{aligned}
\mathbf{a}_{51} &= \int_{\mathscr{B}_j} \mathbf{r}_{O_jM_{0j}} \times \left(\boldsymbol{\omega}_j \times \boldsymbol{\Phi}_{d_j}(M_{0j})\dot{\mathbf{q}}_{e_j}\right) \mathrm{d}m \\
&= -\int_{\mathscr{B}_j} \mathbf{r}_{O_jM_{0j}} \times \left(\left(\boldsymbol{\Phi}_{d_j}(M_{0j})\dot{\mathbf{q}}_{e_j}\right) \times \boldsymbol{\omega}_j\right) \mathrm{d}m
\end{aligned} \tag{E.15}$$

$$\begin{aligned}
\mathbf{a}_{52} &= \int_{\mathscr{B}_j} \left(\boldsymbol{\Phi}_{d_j}(M_{0j})\mathbf{q}_{e_j}\right) \times \left(\boldsymbol{\omega}_j \times \boldsymbol{\Phi}_{d_j}(M_{0j})\dot{\mathbf{q}}_{e_j}\right) \mathrm{d}m \\
&= -\int_{\mathscr{B}_j} \left(\boldsymbol{\Phi}_{d_j}(M_{0j})\mathbf{q}_{e_j}\right) \times \left(\left(\boldsymbol{\Phi}_{d_j}(M_{0j})\dot{\mathbf{q}}_{e_j}\right) \times \boldsymbol{\omega}_j\right) \mathrm{d}m.
\end{aligned} \tag{E.16}$$

In order to simplify these two expressions, let us consider the fact that:

$$\mathbf{\Phi}_{d_j}(M_{0j})\mathbf{q}_{e_j} = \sum_{k=1}^{N_j} \mathbf{\Phi}_{dk_j}(M_{0j})q_{ek_j} \tag{E.17}$$

$$\mathbf{\Phi}_{d_j}(M_{0j})\dot{\mathbf{q}}_{e_j} = \sum_{k=1}^{N_j} \mathbf{\Phi}_{dk_j}(M_{0j})\dot{q}_{ek_j}. \tag{E.18}$$

Introducing (E.18) into (E.15) leads to:

$$\begin{aligned}
\mathbf{a}_{51} &= -\int_{\mathscr{B}_j} \mathbf{r}_{O_jM_{0j}} \times \left(\left(\sum_{k=1}^{N_j} \mathbf{\Phi}_{dk_j}(M_{0j})\dot{q}_{ek_j} \right) \times \boldsymbol{\omega}_j \right) \mathrm{d}m \\
&= -\sum_{k=1}^{N_j} \left(\int_{\mathscr{B}_j} \mathbf{r}_{O_jM_{0j}} \times \left(\left(\mathbf{\Phi}_{dk_j}(M_{0j})\dot{q}_{ek_j} \right) \times \boldsymbol{\omega}_j \right) \mathrm{d}m \right) \\
&= -\sum_{k=1}^{N_j} \left(\int_{\mathscr{B}_j} \mathbf{r}_{O_jM_{0j}} \times \left(\mathbf{\Phi}_{dk_j}(M_{0j}) \times \boldsymbol{\omega}_j \right) \mathrm{d}m \, \dot{q}_{ek_j} \right) \\
&= \sum_{k=1}^{N_j} \left(\int_{\mathscr{B}_j} \hat{\mathbf{r}}^T_{O_jM_{0j}} \hat{\mathbf{\Phi}}_{dk_j}(M_{0j}) \mathrm{d}m \right) \boldsymbol{\omega}_j \dot{q}_{ek_j}.
\end{aligned} \tag{E.19}$$

Then, introducing (E.17) and (E.18) into (E.16) leads to:

$$\begin{aligned}
\mathbf{a}_{52} &= -\int_{\mathscr{B}_j} \left(\sum_{i=1}^{N_j} \mathbf{\Phi}_{di_j}(M_{0j})q_{ei_j} \right) \times \left(\left(\sum_{k=1}^{N_j} \mathbf{\Phi}_{dk_j}(M_{0j})\dot{q}_{ek_j} \right) \times \boldsymbol{\omega}_j \right) \mathrm{d}m \\
&= -\sum_{i=1}^{N_j}\sum_{k=1}^{N_j} \left(\int_{\mathscr{B}_j} \left(\mathbf{\Phi}_{di_j}(M_{0j})q_{ei_j} \right) \times \left(\left(\mathbf{\Phi}_{dk_j}(M_{0j})\dot{q}_{ek_j} \right) \times \boldsymbol{\omega}_j \right) \mathrm{d}m \right) \\
&= -\sum_{i=1}^{N_j}\sum_{k=1}^{N_j} \left(\int_{\mathscr{B}_j} \mathbf{\Phi}_{di_j}(M_{0j}) \times \left(\mathbf{\Phi}_{dk_j}(M_{0j}) \times \boldsymbol{\omega}_j \right) \mathrm{d}m \, q_{ei_j}\dot{q}_{ek_j} \right) \\
&= -\sum_{i=1}^{N_j}\sum_{k=1}^{N_j} \left(\int_{\mathscr{B}_j} \hat{\mathbf{\Phi}}_{di_j}(M_{0j})\hat{\mathbf{\Phi}}_{dk_j}(M_{0j})\boldsymbol{\omega}_j \mathrm{d}m \, q_{ei_j}\dot{q}_{ek_j} \right) \\
&= \sum_{i=1}^{N_j}\sum_{k=1}^{N_j} \left(\int_{\mathscr{B}_j} \hat{\mathbf{\Phi}}^T_{di_j}(M_{0j})\hat{\mathbf{\Phi}}_{dk_j}(M_{0j}) \mathrm{d}m \right) \boldsymbol{\omega}_j q_{ei_j}\dot{q}_{ek_j}.
\end{aligned} \tag{E.20}$$

Finally, introducing expressions (E.5), (E.6), (E.7), (E.13), (E.14), (E.19), and (E.20) into (E.4), the expression (10.18) can be obtained.

E.3 Calculation of the Term $\mathbf{f}_{acc_3}$

From (10.12) and (10.15), we get that $\mathbf{f}_{acc_3}$ is given by

$$\begin{aligned}\mathbf{f}_{acc_3} &= \int_{\mathscr{B}_j} \mathbf{\Phi}_{d_j}^T(M_{0j})\dot{\mathbf{v}}_{M_j}\mathrm{d}m \\ &= \int_{\mathscr{B}_j} \mathbf{\Phi}_{d_j}^T(M_{0j})\left(\dot{\mathbf{v}}_j + \mathbf{\Phi}_{d_j}(M_{0j})\ddot{\mathbf{q}}_{e_j} + 2\boldsymbol{\omega}_j \times \mathbf{\Phi}_{d_j}(M_{0j})\dot{\mathbf{q}}_{e_j}\right)\mathrm{d}m \\ &+ \int_{\mathscr{B}_j} \mathbf{\Phi}_{d_j}^T(M_{0j})\left(\boldsymbol{\omega}_j \times \left(\boldsymbol{\omega}_j \times \mathbf{r}_{O_jM_j}\right) + \dot{\boldsymbol{\omega}}_j \times \mathbf{r}_{O_jM_j}\right)\mathrm{d}m \end{aligned} \quad \text{(E.21)}$$

which can be expanded to be rewritten under a sum of five terms:

$$\mathbf{f}_{acc_3} = \sum_{k=1}^{5} \mathbf{b}_k \quad \text{(E.22)}$$

where:

$$\begin{aligned}\mathbf{b}_1 &= \int_{\mathscr{B}_j} \mathbf{\Phi}_{d_j}(M_{0j})^T\dot{\mathbf{v}}_j\mathrm{d}m \\ &= \left(\int_{\mathscr{B}_j} \mathbf{\Phi}_{d_j}(M_{0j})\mathrm{d}m\right)^T \dot{\mathbf{v}}_j\end{aligned} \quad \text{(E.23)}$$

$$\begin{aligned}\mathbf{b}_2 &= \int_{\mathscr{B}_j} \mathbf{\Phi}_{d_j}^T(M_{0j})\left(\mathbf{\Phi}_{d_j}(M_{0j})\ddot{\mathbf{q}}_{e_j}\right)\mathrm{d}m \\ &= \left(\int_{\mathscr{B}_j} \mathbf{\Phi}_{d_j}^T(M_{0j})\mathbf{\Phi}_{d_j}(M_{0j})\mathrm{d}m\right)\ddot{\mathbf{q}}_{e_j}\end{aligned} \quad \text{(E.24)}$$

$$\begin{aligned}\mathbf{b}_3 &= \int_{\mathscr{B}_j} \mathbf{\Phi}_{d_j}^T(M_{0j})\left(\dot{\boldsymbol{\omega}}_j \times \mathbf{r}_{O_jM_j}\right)\mathrm{d}m \\ &= -\int_{\mathscr{B}_j} \mathbf{\Phi}_{d_j}^T(M_{0j})\left(\mathbf{r}_{O_jM_j} \times \dot{\boldsymbol{\omega}}_j\right)\mathrm{d}m \\ &= -\int_{\mathscr{B}_j} \mathbf{\Phi}_{d_j}^T(M_{0j})\hat{\mathbf{r}}_{O_jM_j}\dot{\boldsymbol{\omega}}_j\mathrm{d}m \\ &= \left(\int_{\mathscr{B}_j} \mathbf{\Phi}_{d_j}^T(M_{0j})\hat{\mathbf{r}}_{O_jM_j}^T\mathrm{d}m\right)\dot{\boldsymbol{\omega}}_j \\ &= \left(\int_{\mathscr{B}_j} \hat{\mathbf{r}}_{O_jM_j}\mathbf{\Phi}_{d_j}(M_{0j})\mathrm{d}m\right)^T \dot{\boldsymbol{\omega}}_j\end{aligned} \quad \text{(E.25)}$$

$$\mathbf{b}_4 = \int_{\mathscr{B}_j} \boldsymbol{\Phi}_{d_j}^T(M_{0j}) \left(\boldsymbol{\omega}_j \times \left(\boldsymbol{\omega}_j \times \mathbf{r}_{O_j M_j}\right)\right) \mathrm{d}m \tag{E.26}$$

$$\mathbf{b}_5 = 2\int_{\mathscr{B}_j} \boldsymbol{\Phi}_{d_j}^T(M_{0j}) \left(\boldsymbol{\omega}_j \times \boldsymbol{\Phi}_{d_j}(M_{0j})\dot{\mathbf{q}}_{e_j}\right) \mathrm{d}m. \tag{E.27}$$

Considering the ith component $\mathbf{b}_4|_i$ of the vector $\mathbf{b}_4$, we can rewrite (E.26) as

$$\mathbf{b}_4|_i = \int_{\mathscr{B}_j} \boldsymbol{\Phi}_{di_j}^T(M_{0j}) \left(\boldsymbol{\omega}_j \times \left(\boldsymbol{\omega}_j \times \mathbf{r}_{O_j M_j}\right)\right) \mathrm{d}m \tag{E.28}$$

in which $\boldsymbol{\Phi}_{di_j}(M_{0j})$ is the ith column of the matrix $\boldsymbol{\Phi}_{d_j}(M_{0j})$.

Now, introducing (10.4) into (E.28), the expression of $\mathbf{b}_4|_i$ becomes

$$\mathbf{b}_4|_i = b_{4i1} + b_{4i2} \tag{E.29}$$

where

$$\begin{aligned} b_{4i1} &= \int_{\mathscr{B}_j} \boldsymbol{\Phi}_{di_j}^T(M_{0j}) \left(\boldsymbol{\omega}_j \times \left(\boldsymbol{\omega}_j \times \mathbf{r}_{O_j M_{0j}}\right)\right) \mathrm{d}m \\ &= \int_{\mathscr{B}_j} \boldsymbol{\Phi}_{di_j}^T(M_{0j}) \left(\left(\mathbf{r}_{O_j M_{0j}} \times \boldsymbol{\omega}_j\right) \times \boldsymbol{\omega}_j\right) \mathrm{d}m \end{aligned} \tag{E.30}$$

$$b_{4i2} = \int_{\mathscr{B}_j} \boldsymbol{\Phi}_{di_j}^T(M_{0j}) \left(\boldsymbol{\omega}_j \times \left(\boldsymbol{\omega}_j \times \left(\boldsymbol{\Phi}_{d_j}\mathbf{q}_{e_j}\right)\right)\right) \mathrm{d}m. \tag{E.31}$$

To simplify the term b_{4i1}, let us recall the well-known identity for the triple product of three arbitrary vectors $\mathbf{u}$, $\mathbf{v}$ and $\mathbf{w}$:

$$\mathbf{u}^T(\mathbf{v} \times \mathbf{w}) = \mathbf{w}^T(\mathbf{u} \times \mathbf{v}). \tag{E.32}$$

Now, replacing $\mathbf{u}$ by $\boldsymbol{\Phi}_{di_j}(M_{0j})$, $\mathbf{v}$ by $\boldsymbol{\omega}_j \times \mathbf{r}_{O_j M_{0j}}$, and $\mathbf{w}$ by $\boldsymbol{\omega}_j$, we get

$$\begin{aligned} b_{4i1} &= \int_{\mathscr{B}_j} \boldsymbol{\omega}_j^T \left(\boldsymbol{\Phi}_{di_j}(M_{0j}) \times \left(\mathbf{r}_{O_j M_{0j}} \times \boldsymbol{\omega}_j\right)\right) \mathrm{d}m \\ &= \boldsymbol{\omega}_j^T \left(\int_{\mathscr{B}_j} \hat{\boldsymbol{\Phi}}_{di_j}(M_{0j}) \hat{\mathbf{r}}_{O_j M_{0j}} \mathrm{d}m\right) \boldsymbol{\omega}_j \\ &= -\,\boldsymbol{\omega}_j^T \left(\int_{\mathscr{B}_j} \hat{\boldsymbol{\Phi}}_{di_j}^T(M_{0j}) \hat{\mathbf{r}}_{O_j M_{0j}} \mathrm{d}m\right) \boldsymbol{\omega}_j \\ &= -\,\boldsymbol{\omega}_j^T \left(\int_{\mathscr{B}_j} \hat{\mathbf{r}}_{O_j M_{0j}}^T \hat{\boldsymbol{\Phi}}_{di_j}(M_{0j}) \mathrm{d}m\right)^T \boldsymbol{\omega}_j. \end{aligned} \tag{E.33}$$

Now, in order to simplify the term b_{4i2}, let us consider the fact that:

$$\boldsymbol{\Phi}_{d_j}(M_{0j})\mathbf{q}_{e_j} = \sum_{k=1}^{N_j} \boldsymbol{\Phi}_{dk_j}(M_{0j})q_{ek_j}. \tag{E.34}$$

Thus, b_{4i2} can be rewritten as

$$\begin{aligned}
b_{4i2} &= \int_{\mathscr{B}_j} \boldsymbol{\Phi}_{di_j}^T(M_{0j}) \left(\boldsymbol{\omega}_j \times \left(\boldsymbol{\omega}_j \times \left(\sum_{k=1}^{N_j} \boldsymbol{\Phi}_{dk_j}(M_{0j})q_{ek_j} \right)\right)\right) \mathrm{d}m \\
&= \sum_{k=1}^{N_j} \int_{\mathscr{B}_j} \boldsymbol{\Phi}_{di_j}^T(M_{0j}) \left(\boldsymbol{\omega}_j \times \left(\boldsymbol{\omega}_j \times \boldsymbol{\Phi}_{dk_j}(M_{0j})q_{ek_j}\right)\right) \mathrm{d}m \\
&= \sum_{k=1}^{N_j} \int_{\mathscr{B}_j} \boldsymbol{\Phi}_{di_j}^T(M_{0j}) \left(\boldsymbol{\omega}_j \times \left(\boldsymbol{\omega}_j \times \boldsymbol{\Phi}_{dk_j}(M_{0j})\right)\right) \mathrm{d}m \; q_{ek_j}.
\end{aligned} \tag{E.35}$$

Once again, using the identity (E.32) by replacing $\mathbf{u}$ by $\boldsymbol{\omega}_j$, $\mathbf{v}$ by $\boldsymbol{\omega}_j \times \boldsymbol{\Phi}_{dk_j}(M_{0j})$, and $\mathbf{w}$ by $\boldsymbol{\Phi}_{di_j}(M_{0j})$, we can obtain

$$\begin{aligned}
b_{4i2} &= \sum_{k=1}^{N_j} \int_{\mathscr{B}_j} \boldsymbol{\omega}_j^T \left(\left(\boldsymbol{\omega}_j \times \boldsymbol{\Phi}_{dk_j}(M_{0j})\right) \times \boldsymbol{\Phi}_{di_j}(M_{0j})\right) \mathrm{d}m \; q_{ek_j} \\
&= \sum_{k=1}^{N_j} \boldsymbol{\omega}_j^T \left(\int_{\mathscr{B}_j} \boldsymbol{\Phi}_{di_j}(M_{0j}) \times \left(\boldsymbol{\Phi}_{dk_j}(M_{0j}) \times \boldsymbol{\omega}_j\right) \mathrm{d}m \right) q_{ek_j} \\
&= \sum_{k=1}^{N_j} \boldsymbol{\omega}_j^T \left(\int_{\mathscr{B}_j} \hat{\boldsymbol{\Phi}}_{di_j}(M_{0j}) \hat{\boldsymbol{\Phi}}_{dk_j}(M_{0j}) \boldsymbol{\omega}_j \mathrm{d}m \right) q_{ek_j} \\
&= -\sum_{k=1}^{N_j} \boldsymbol{\omega}_j^T \left(\int_{\mathscr{B}_j} \hat{\boldsymbol{\Phi}}_{di_j}^T(M_{0j}) \hat{\boldsymbol{\Phi}}_{dk_j}(M_{0j}) \mathrm{d}m \right) \boldsymbol{\omega}_j q_{ek_j}.
\end{aligned} \tag{E.36}$$

Now, considering the ith component $\mathbf{b}_5|_i$ of the vector $\mathbf{b}_5$, we can rewrite (E.27) as

$$\mathbf{b}_5|_i = 2 \int_{\mathscr{B}_j} \boldsymbol{\Phi}_{di_j}^T(M_{0j}) \left(\boldsymbol{\omega}_j \times \boldsymbol{\Phi}_{d_j}(M_{0j})\dot{\mathbf{q}}_{e_j}\right) \mathrm{d}m. \tag{E.37}$$

In order to simplify the term $\mathbf{b}_5|_i$, let us consider the fact that:

$$\boldsymbol{\Phi}_{d_j}(M_{0j})\dot{\mathbf{q}}_{e_j} = \sum_{k=1}^{N_j} \boldsymbol{\Phi}_{dk_j}(M_{0j})\dot{q}_{ek_j}. \tag{E.38}$$

Thus, $\mathbf{b}_5|_i$ can be rewritten as

$$\begin{aligned}
\mathbf{b}_5|_i &= 2\int_{\mathscr{B}_j} \boldsymbol{\Phi}_{di_j}^T(M_{0j}) \left(\boldsymbol{\omega}_j \times \left(\sum_{k=1}^{N_j} \boldsymbol{\Phi}_{dk_j}(M_{0j})\dot{q}_{ek_j}\right)\right) \mathrm{d}m \\
&= 2\sum_{k=1}^{N_j}\left(\int_{\mathscr{B}_j} \boldsymbol{\Phi}_{di_j}^T(M_{0j}) \left(\boldsymbol{\omega}_j \times \boldsymbol{\Phi}_{dk_j}(M_{0j})\right) \mathrm{d}m\ \dot{q}_{ek_j}\right) \qquad \text{(E.39)} \\
&= -2\sum_{k=1}^{N_j}\left(\int_{\mathscr{B}_j} \boldsymbol{\Phi}_{di_j}^T(M_{0j}) \left(\boldsymbol{\Phi}_{dk_j}(M_{0j}) \times \boldsymbol{\omega}_j\right) \mathrm{d}m\ \dot{q}_{ek_j}\right).
\end{aligned}$$

Using the identity (E.32) by replacing $\mathbf{u}$ by $\boldsymbol{\Phi}_{di_j}(M_{0j})$, $\mathbf{v}$ by $\boldsymbol{\omega}_j$, and $\mathbf{w}$ by $\boldsymbol{\Phi}_{dk_j}(M_{0j})$, we can obtain

$$\begin{aligned}
\mathbf{b}_5|_i &= -2\sum_{k=1}^{N_j}\left(\int_{\mathscr{B}_j} \boldsymbol{\Phi}_{di_j}^T(M_{0j}) \left(\boldsymbol{\Phi}_{dk_j}(M_{0j}) \times \boldsymbol{\omega}_j\right) \mathrm{d}m\ \dot{q}_{ek_j}\right) \\
&= -2\sum_{k=1}^{N_j}\left(\int_{\mathscr{B}_j} \boldsymbol{\omega}_j^T \left(\boldsymbol{\Phi}_{di_j}^T(M_{0j}) \times \boldsymbol{\Phi}_{dk_j}(M_{0j})\right) \mathrm{d}m\ \dot{q}_{ek_j}\right) \\
&= 2\sum_{k=1}^{N_j}\left(\int_{\mathscr{B}_j} \boldsymbol{\omega}_j^T \left(\boldsymbol{\Phi}_{dk_j}(M_{0j}) \times \boldsymbol{\Phi}_{di_j}^T(M_{0j})\right) \mathrm{d}m\ \dot{q}_{ek_j}\right) \qquad \text{(E.40)} \\
&= 2\sum_{k=1}^{N_j}\left(\int_{\mathscr{B}_j} \boldsymbol{\Phi}_{dk_j}(M_{0j}) \times \boldsymbol{\Phi}_{di_j}^T(M_{0j})\mathrm{d}m\right)^T \boldsymbol{\omega}_j\dot{q}_{ek_j}.
\end{aligned}$$

Finally, taking the ith rows of expressions (E.23), (E.24) and (E.25) and summing them with the expressions (E.33), (E.36) and (E.40), the expression (10.19) can be obtained.

Appendix F
Dynamics Equations for a Clamped-Free Flexible Beam

F.1 Shape Functions for a Free Flexible Beam

Computation of the mass and stiffness matrices of 3D beams is useful for elastodynamic modeling of parallel manipulators.

The Bernoulli model describes beam deformation under the assumption that the shear effect is negligible, that the cross-sections remain perpendicular to the neutral axis and that the rotational inertia of sections is assumed to be zero Blevins (2001). With such a model, the 3D beam deformation $\mathbf{u}_{e_j}(M_{0j})$ (see Sect. 10.2.1) can be characterized with the six shape functions $\boldsymbol{\Phi}_{dx_j}$, $\boldsymbol{\Phi}_{dy_j}$, $\boldsymbol{\Phi}_{dz_j}$, $\boldsymbol{\Phi}_{rx_j}$, $\boldsymbol{\Phi}_{ry_j}$ and $\boldsymbol{\Phi}_{rz_j}$, i.e. $N_j = 6$, defined as:

$$\boldsymbol{\Phi}_{dx_j} = \left[\xi\ 0\ 0\ 0\ 0\ 0\right] \tag{F.1a}$$
$$\boldsymbol{\Phi}_{dy_j} = \left[0\ 3\xi^2 - 2\xi^3\ 0\ 0\ 0\ l_j\left(\xi^3 - \xi^2\right)\right] \tag{F.1b}$$
$$\boldsymbol{\Phi}_{dz_j} = \left[0\ 0\ 3\xi^2 - 2\xi^3\ 0\ -l_j\left(\xi^3 - \xi^2\right)\ 0\right] \tag{F.1c}$$
$$\boldsymbol{\Phi}_{rx_j} = \left[0\ 0\ 0\ \xi\ 0\ 0\right] \tag{F.1d}$$
$$\boldsymbol{\Phi}_{ry_j} = \left[0\ 0\ -6\left(\xi - \xi^2\right)/l_j\ 0\ 3\xi^2 - 2\xi\ 0\right] \tag{F.1e}$$
$$\boldsymbol{\Phi}_{rz_j} = \left[0\ 6\left(\xi - \xi^2\right)/l_j\ 0\ 0\ 0\ 3\xi^2 - 2\xi\right] \tag{F.1f}$$

where $\xi = x/l_j$ and l_j is the beam length.

x, y and z denote the Cartesian coordinates of point M_{0j} expressed in the local frame $\mathscr{F}_j$ and $\boldsymbol{\Phi}_{d_j}(M_{0j})$ defined at (10.3) is a (3×6) matrix that takes the form:

$$\boldsymbol{\Phi}_{d_j}(M_{0j}) = \begin{bmatrix} \boldsymbol{\Phi}_{dx_j} - y\boldsymbol{\Phi}_{rz_j} + z\boldsymbol{\Phi}_{ry_j} \\ \boldsymbol{\Phi}_{dy_j} - z\boldsymbol{\Phi}_{rx_j} \\ \boldsymbol{\Phi}_{dz_j} + y\boldsymbol{\Phi}_{rx_j} \end{bmatrix} \tag{F.2}$$

S. Briot and W. Khalil, *Dynamics of Parallel Robots*,
Mechanisms and Machine Science 35, DOI 10.1007/978-3-319-19788-3

while $\mathbf{\Phi}_{r_j}(M_{0j})$ defined at (10.5) is a (3×6) matrix equal to:

$$\mathbf{\Phi}_{r_j}(M_{0j}) = \begin{bmatrix} \mathbf{\Phi}_{rx_j} \\ \mathbf{\Phi}_{ry_j} \\ \mathbf{\Phi}_{rz_j} \end{bmatrix}. \tag{F.3}$$

F.2 Stiffness Matrix for a Free Flexible Beam

In the beam model, it is assumed that (Shabana 2005)

$$\sigma_{j22} = \sigma_{j33} = \sigma_{j23} = 0 \tag{F.4}$$

$$\varepsilon_{j22} = \varepsilon_{j33} = \varepsilon_{j23} = 0 \tag{F.5}$$

$$\sigma_{j11} = E_j \varepsilon_{j11} \tag{F.6}$$

$$\sigma_{j12} = G_j \varepsilon_{j12} \tag{F.7}$$

$$\sigma_{j13} = G_j \varepsilon_{j13} \tag{F.8}$$

where E_j is the Young modulus of body j and $G_j = E_j/(2(1+\nu_j))$ is its shear modulus, ν_j being the Poisson's coefficient.

Introducing (F.1a) to (F.8) into (10.35), the stiffness matrix of body $\mathscr{B}_j$ takes the form:

$$\mathbf{K}_{ee_j} = \frac{1}{l_j^3} \begin{bmatrix} E_j A_j l_j^2 & 0 & 0 & 0 & 0 & 0 \\ 0 & 12E_j I_{z_j} & 0 & 0 & 0 & -6E_j I_{z_j} l_j \\ 0 & 0 & 12E_j I_{y_j} & 0 & 6E_j I_{y_j} l_j & 0 \\ 0 & 0 & 0 & I_{0_j} G_j l_j^2 & 0 & 0 \\ 0 & 0 & 6E_j I_{y_j} l_j & 0 & 4E_j I_{y_j} l_j^2 & 0 \\ 0 & -6E_j I_{z_j} l_j & 0 & 0 & 0 & 4E_j I_{z_j} l_j^2 \end{bmatrix} \tag{F.9}$$

where A_j is the beam cross-section area, I_{y_j} and I_{z_j} are the second moments of area around axes y and z of the local frame, I_{0_j} is the torsion constant.

F.3 Evaluation of the Inertia Matrix of a Free Flexible 3D Bernoulli Beam for $\mathbf{q}_{e_j} = \mathbf{0}$

For $\mathbf{q}_{e_j} = \mathbf{0}$, the inertia matrix of the flexible 3D Bernoulli beam becomes, from Sect. 10.2.24:

$$\mathbf{M}_j = \begin{bmatrix} m_j \mathbf{1}_3 & \widehat{\mathbf{ms}}_{r_j}^T & \mathbf{MS}_{de_j} \\ \widehat{\mathbf{ms}}_{r_j} & \mathbf{I}_{rr_j} & \mathbf{MS}_{re_j}^r \\ \mathbf{MS}_{de_j}^T & \mathbf{MS}_{re_j}^{r\,T} & \mathbf{M}_{ee_j} \end{bmatrix} \tag{F.10}$$

where

$$\mathbf{MS}^{r}_{re_j} = \left[\boldsymbol{\beta}_{1_j} \; \cdots \; \boldsymbol{\beta}_{N_{j,j}}\right]. \tag{F.11}$$

After simplifications, we get that

$$\mathbf{MS}_{de_j} = \begin{bmatrix} \frac{m_j}{2} & 0 & 0 & 0 & 0 & 0 \\ 0 & \frac{m_j}{2} & 0 & 0 & 0 & \frac{m_j l_j}{12} \\ 0 & 0 & \frac{m_j}{2} & 0 & \frac{m_j l_j}{12} & 0 \end{bmatrix} \tag{F.12}$$

$$\mathbf{MS}^{r}_{re_j} = \begin{bmatrix} 0 & 0 & 0 & \frac{\rho_j l_j I_{p_j}}{2} & 0 & 0 \\ 0 & 0 & -\rho_j I_{y_j} - \frac{7 m_j l_j}{20} & 0 & -\frac{m_j l_j^2}{20} & 0 \\ 0 & \rho_j I_{z_j} + \frac{7 m_j l_j}{20} & 0 & 0 & 0 & -\frac{m_j l_j^2}{20} \end{bmatrix} \tag{F.13}$$

and

$$\mathbf{M}_{ee_j} = \begin{bmatrix} \frac{m_j}{3} & 0 & 0 & 0 & 0 & 0 \\ 0 & \frac{13 m_j}{35} + \frac{6 \rho_j I_{z_j}}{5 l_j} & 0 & 0 & 0 & -\frac{11 m_j l_j + 21 \rho_j I_{z_j}}{210} \\ 0 & 0 & \frac{13 m_j}{35} + \frac{6 \rho_j I_{y_j}}{5 l_j} & 0 & \frac{11 m_j l_j + 21 \rho_j I_{y_j}}{210} & 0 \\ 0 & 0 & 0 & \frac{\rho_j l_j I_{p_j}}{3} & 0 & 0 \\ 0 & 0 & \frac{11 m_j l_j + 21 \rho_j I_{y_j}}{210} & 0 & \frac{m_j l_j^2 + 14 \rho_j I_{y_j} l_j}{105} & 0 \\ 0 & -\frac{11 m_j l_j + 21 \rho_j I_{z_j}}{210} & 0 & 0 & 0 & \frac{m_j l_j^2 + 14 \rho_j I_{z_j} l_j}{105} \end{bmatrix} \tag{F.14}$$

where $I_{p_j} = I_{y_j} + I_{z_j}$ is the polar moment of inertia.

where

[illegible]

References

Abdellatif, H., & Heimann, B. (2009). Computational efficient inverse dynamics of 6-dof fully parallel manipulators by using the lagrangian formalism. *Mechanism and Machine Theory*, *44*(1), 192–207.

Afroun, M., Dequidt, A., & Vermeiren, L. (2012). Revisiting the inverse dynamics of the Gough-Stewart platform manipulator with special emphasis on universal-prismatic-spherical leg and internal singularity. *Proceedings of the Institution of Mechanical Engineers, Part C: Journal of Mechanical Engineering Science*, *226*(10), 2422–2439.

Ait-Ahmed, M. (1993). Contribution à la modélisation géométrique et dynamique des robots parallèles. Ph.D. thesis, LAAS, Toulouse.

Alizadeh, D. (2009). On the home posture of the McGill schoenflies motion generator. Master's thesis, McGill University, Montreal, QC, Canada.

Amine, S., Tale-Masouleh, M., Caro, S., Wenger, P., & Gosselin, C. (2011). Singularity analysis of the 4-RUU parallel manipulator using Grassmann-Cayley algebra. *Transactions of the Canadian Society for Mechanical Engineering*, *35*(5), 515–528.

Amine, S., Tale Masouleh, M., Caro, S., Wenger, P., & Gosselin, C. (2012a). Singularity analysis of 3T2R parallel mechanisms using Grassmann-Cayley algebra and Grassmann geometry. *Mechanism and Machine Theory*, *52*, 326–340.

Amine, S., Tale-Masouleh, M., Caro, S., Wenger, P., & Gosselin, C. (2012b). Singularity analysis of 3T2R parallel mechanisms using Grassmann-Cayley algebra and Grassmann line geometry. *Mechanism and Machine Theory*, *52*, 326–340.

Andreff, N., & Martinet, P. (2006). Vision-based kinematic modelling of some parallel manipulators for control purposes. In *Proceedings of EuCoMeS, the First European Conference on Mechanism Science*. Obergurgl, Austria.

Andreff, N., Marchadier, A., & Martinet, P. (2005). Vision-based control of a Gough-Stewart parallel mechanism using legs observation. In *Proceedings of the IEEE International Conference on Robotics and Automation, ICRA'05* (pp. 2546–2551). Barcelona, Spain.

Andreff, N., Dallej, T., & Martinet, P. (2007). Image-based visual servoing of gough-stewart parallel manipulators using legs observation. *International Journal of Robotics Research*, *26*(7), 677–687.

Angeles, J. (2003). *Fundamentals of robotic mechanical systems—theory, methods, and algorithms* (2nd ed.). New York: Springer.

Arai, T., & Cleary, K. et al. (1990). Design, analysis and construction of a prototype parallel link manipulator. In *Proceedings of the IEEE/RJS Conference on Intelligent Robots and Systems (IROS 1990)* (pp. 205–212). Ibaraki, Japan.

S. Briot and W. Khalil, *Dynamics of Parallel Robots*,
Mechanisms and Machine Science 35, DOI 10.1007/978-3-319-19788-3

Arakelian, V., Briot, S., & Glazunov, V. (2008). Increase of singularity-free zones in the workspace of parallel manipulators using mechanisms of variable structure. *Mechanism and Machine Theory*, *43*(9), 1129–1140.

Armstrong, B. (1988). *Dynamics for robot control: friction modeling and ensuring excitation during parameter identification*. Ph.D. thesis, Department of Electrical Engineering, Stanford University.

Armstrong, B. (1991). *Control of machines with frictions*. Dordrecht: Kluwer Academic.

Ayusawa, K., Venture, G., & Nakamura, Y. (2008). Identification of humanoid robots dynamics using floating-base motion dynamics. In *Proceedings of the IEEE/RSJ International Conference on Intelligent Robots and Systems (IROS 2008)*.

Baillieul, J. (1986). Avoiding obstacles and resolving kinematic redundancy. In *Proceedings of the 1986 IEEE International Conference on Robotics and Automation (ICRA 1986)* (pp. 1698–1704). San Francisco, CA, USA.

Baillieul, J., Hollerbach, J., & Brockett, R. (1984). Programming and control of kinematically redundant manipulators. In: *Proceedings of the 23rd IEEE Conference on Decision and Control (CDC 1984)* (pp. 768–774). Las Vegas.

Bauchau, O. (2011). *Flexible multibody dynamics*. New York: Springer.

Ben-Horin, P., & Shoham, M. (2006). Singularity analysis of a class of parallel robots based on Grassmann-Cayley algebra. *Mechanism and Machine Theory*, *41*(8), 958–970.

Ben-Horin, P., & Shoham, M. (2009). Application of Grassmann-Cayley algebra to geometrical interpretation of parallel robot singularities. *The International Journal of Robotics Research*, *28*(1), 127–141.

Ben-Horin, P., Shoham, M., Caro, S., Chablat, D., & Wenger, P. (2008). SINGULAB—a graphical user interface for the singularity analysis of parallel robots based on Grassmann-Cayley algebra. Advances in Robot Kinematics.

Bézout, E. (1764). *Recherches sur le degré des équations résultantes de l'éva-nouissement des inconnues*. Histoire de l'Académie Royale des Sciences.

Bhattacharya, S., Hatwal, H., & Ghosh, A. (1997). An on-line estimation scheme for generalized Stewart platform type parallel manipulators. *Mechanism and Machine Theory*, *32*(1), 79–89.

Bhattacharya, S., Nenchev, D., & Uchiyama, M. (1998). A recursive formula for the inverse of the inertia matrix of a parallel manipulator. *Mechanism and Machine Theory*, *33*(7), 957–964.

Binford, T., et al. (1977). Exploratory study of computer integrated assembly system—discussion of trajectory calculation methods. *Progress Report Memo AIM-285.4*. Stanford: Stanford Artificial Intelligence Laboratory.

Blevins, R. (2001). *Formulas for natural frequency and mode shape*. New York: Krieger Publishing Company.

Bonev, I. (2002). Geometric analysis of parallel mechanisms. Ph.D. thesis, Université Laval, QC, Canada.

Bonev, I. (2008). Direct kinematics of zero-torsion parallel mechanisms. In *Proceedings of the 2008 IEEE International Conference on Robotics and Automation (ICRA 2008)*.

Bonev, I., & Gosselin, C. (2005a). Singularity loci of spherical parallel mechanisms. In *Proceedings of the IEEE International Conference on Robotics and Automation (ICRA2005)* (pp. 2968–2973). Barcelona, Spain.

Bonev, I., & Gosselin, C. (2005b). Singularity loci of spherical parallel mechanisms. In *Proceedings of the 2005 IEEE International Conference on Robotics and Automation (ICRA 2005)* (pp. 2968–2973). Barcelona, Spain.

Bonev, I., & Gosselin, C. (2006). Analytical determination of the workspace of symmetrical spherical parallel mechanisms. *IEEE Transactions on Robotics*, *22*(5), 1011–1017.

Bonev, I., & Ryu, J. (1999). Orientation workspace analysis of 6-dof parallel manipulators. In *Proceedings of the ASME 1999 Design Engineering Technical Conferences (IDETC99)*. Las Vegas, NV, USA.

Bonev, I., Zlatanov, D., & Gosselin, C. (2002a). Advantages of the modified Euler angles in the design and control of PKMs. In *Proceedings of the Parallel Kinematic Machines International Conference* (pp. 171–188). Chemnitz, Germany.

Bonev, I., Zlatanov, D., Gosselin, C. (2002b). Advantages of the modified euler angles in the design and control of PKMs. In *Proceedings of the 2002 Parallel Kinematic Machines International Conference (PKS 2002)* (pp. 171–188). Chemnitz, Germany.

Bonev, I., Zlatanov, D., & Gosselin, C. (2003). Singularity analysis of 3-dof planar parallel mechanisms via screw theory. *ASME Journal of Mechanical Design, 125*(3), 573–581.

Bonev, I., Chablat, D., & Wenger, P. (2006). Working and assembly modes of the agile eye. In *Proceedings of the IEEE International Conference On Robotics And Automation (ICRA 1996)*. Orlando, FL, USA.

Borrel, P. (1986). *Contribution à la modélisation géométrique des robots-manipulateurs: Application à la conception assistée par ordinateur*. USTL, Montpellier, France: Thèse d'etat.

Bourbonais, F., Bigras, P., & Bonev, I. (2014). Minimum-time trajectory planning and control of a reconfigurable pick-and-place parallel robot. *IEEE/ASME Transactions on Mechatronics*.

Bouzgarrou, B. C., Ray, P., & Gogu, G. (2005). New approach for dynamic modelling of flexible manipulators. *Part K: Journal of Multi-body Dynamics, 219*(3), 285–298.

Boyer, F., & Coiffet, P. (1996). Symbolic modelling of a flexible manipulator via assembling of its generalized Newton-Euler model. *Mechanism and Machine Theory, 31*, 45–56.

Boyer, F., & Khalil, W. (1998). An efficient calculation of the flexible manipulator inverse dynamics. *International Journal of Robotics Research, 17*(3), 282–293.

Boyer, F., Khalil, W., Benosman, M., & LeVey, G. (2007). Robot manipulators. Modeling, performance, analysis and control: Modeling and control of flexible robots. *Control Systems, Robotics and Manufacturing Series* (pp. 337–394). ISTE

Briot, S., & Arakelian, V. (2008). Optimal force generation of parallel manipulators for passing through the singular positions. *International Journal of Robotics Research, 27*(8), 967–983.

Briot, S., & Arakelian, V. (2009). Complete shaking force and shaking moment balancing of the position-orientation decoupled PAMINSA manipulator. In *Proceedings of the IEEE/ASME International Conference on Advanced Intelligent Mechatronics (AIM2009)*. Singapore.

Briot, S., & Arakelian, V. (2010). On the dynamic properties of rigid-link flexible-joint parallel manipulators in the presence of type 2 singularities. *ASME Journal of Mechanisms and Robotics, 2*(2).

Briot, S., & Arakelian, V. (2011). On the dynamic properties of flexible parallel manipulators in the presence of Type 2 singularities. *ASME Journal of Mechanisms and Robotics, 3*(3).

Briot, S., & Bonev, I. (2008). Singularity loci of zero-torsion parallel mechanisms. In *Proceedings of the IEEE/RJS 11th International Conference on Intelligent Robots and Systems* (IROS 2008). Nice, France.

Briot, S., & Bonev, I. (2009a). Pantopteron: A new fully-decoupled 3-dof translational parallel robot for pick-and-place applications. *ASME Journal of Mechanisms and Robotics, 1*(2).

Briot, S., & Bonev, I. (2009b). Self motions of the pantopteron. In *Proceedings of the ASME 2009 International Design Engineering Technical Conferences and Computers and Information in Engineering Conference (IDETC/CIE 2009)*. San Diego, California, USA.

Briot, S., & Bonev, I. (2010). Pantopteron-4: A new 3T1R decoupled parallel manipulator for pick-and-place applications. *Mechanism and Machine Theory, 45*(5), 707–721.

Briot, S., & Gautier, M. (2012). Global identification of drive gains and dynamic parameters of parallel robots—part 1: Theory. In *Proceedings of the 19th CISM-IFToMM Symposium on Robot Design, Dynamics, and Control (RoManSy)*.

Briot, S., & Gautier, M. (2014). Global identification of joint drive gains and dynamic parameters of parallel robots. *Multibody System Dynamics* (In press).

Briot, S., & Khalil, W. (2014a). Recursive and symbolic calculation of the elastodynamic model of flexible parallel robots. *The International Journal of Robotics Research, 33*(3), 469–483.

Briot, S. & Khalil, W. (2014b). Recursive and symbolic calculation of the stiffness and mass matrices of parallel robots. In *Proceedings of the 20th CISM-IFToMM Symposium on Theory and Practice of Robots and Manipulators* (RoManSy 2014). Moscow, Russia.

Briot, S. & Martinet, P. (2013). Minimal representation for the control of Gough-Stewart platforms via leg observation considering a hidden robot model. In *Proceedings of the 2013 IEEE International Conference on Robotics and Automation (ICRA 2013)*. Karlsruhe, Germany.

Briot, S., Bonev, I., Chablat, D., Wenger, P., & Arakelian, V. (2008). Self motions of general 3-rpr planar parallel robots. *International Journal of Robotics Research, 27*(7), 855–866.

Briot, S., Pashkevich, A., & Chablat, D. (2009). On the optimal design of parallel robots taking into account their deformations and natural frequencies. In *Proceedings of the ASME 2009 International Design Engineering Technical Conferences & Computers and Information in Engineering Conference IDETC/CIE*. San Diego, California.

Briot, S., Pashkevich, A., & Chablat, D. (2010). Optimal technology-oriented design of parallel robots for high-speed machining applications. In *Proceedings of the 2010 IEEE International Conference on Robotics and Automation (ICRA 2010)*. Anchorage, Alaska, USA.

Briot, S., Pashkevich, A. & Chablat, D. (2011). Reduced elastodynamic modelling of parallel robots for the computation of their natural frequencies. In *World Congress in Mechanism and Machine Science*, 13.

Briot, S., Arakelian, V., & Le Baron, J. (2012a). Shaking force minimization of high-speed robots via centre of mass acceleration control. *Mechanism and Machine Theory, 57*, 1–12.

Briot, S., Caro, S., & Germain, C. (2012b). Robot Parallèle à Deux Degrés de Liberté Présentant Deux Chaînes Cinématiques dont la Raideur en Flexion est Maximisée. FR, FR2967603 - JP 2013–540316 - US 2014–0020500 - PCT/EP2011/070598.

Bruyninckx, H. (1998). Closed-form position kinematics of a (3-1-1-1) fully parallel manipulator. *IEEE Transactions on Robotics and Automation, 14*(2), 326–328.

Cammarata, A., Condorelli, D., & Sinatra, R. (2013). An algorithm to study the elastodynamics of parallel kinematic machines with lower kinematic pairs. *ASME Transactions Journal of Mechanisms and Robotics, 5*(1).

Campos, L., Bourbonais, F., Bonev, I. & Bigras, P. (2010). Development of a five-bar parallel robot with large workspace. In *Proceedings of the ASME 2010 International Design Engineering Technical Conferences*. Montréal, QC, Canada.

Canudas de Wit, C., & Aubin, A. (1990). Parameters identification of robots manipulators via sequential hybrid estimation algorithms. In *Proceedings IFAC Congress* (pp. 178–183).

Canudas de Wit, C., Noël, P., Aubin, A., Brogliato, B., & Drevet, P. (1989). Adaptive friction compensation in robot manipulators: Low-velocities. In *Proceedings of the IEEE International Conference on Robotics and Automation (ICRA 1989)* (pp. 1352–1357). Scottsdale.

Carricato, M., & Gosselin, C. (2009). On the modeling of leg constraints in the dynamic analysis of Gough/Stewart-type platforms. *ASME Journal of Computational and Nonlinear Dynamics, 4*(1), 1–8.

Carricato, M., & Parenti-Castelli, V. (2002). Singularity-free fully-isotropic translational parallel manipulators. *International Journal of Robotics Research, 21*(2), 161–174.

Chablat, D., & Wenger, P. (2003). Architecture optimization of a 3-dof parallel mechanism for machining applications, the Orthoglide. *IEEE Transactions on Robotics and Automation, 19*(3), 403–410.

Chablat, D., & Wenger, P. (2005). Design of a spherical wrist with parallel architecture: Application to vertebrae of an eel robot. In *Proceedings of the 2005 IEEE International Conference on Robotics and Automation (ICRA 2005)*. Barcelona, Spain.

Chablat, D., Moroz, G., & Wenger, P. (2011). Uniqueness domains and non singular assembly mode changing trajectories. In *Proceedings of the 2011 IEEE International Conference on Robotics and Automation (ICRA 2011)*. Shanghai, China.

Charentus, S., & Renaud, M. (1989). Modeling and control of a modular redundant robot manipulator. In *Proceedings of the First International Symposium on Experimental Robotics*.

Chedmail, P., & Gautier, M. (1990). Optimum choice of robot actuators. *ASME Journal of Engineering for Industry, 112*(4), 361–367.

Chedmail, P., Gautier, M., & Khalil, W. (1986). *Automatic modelling of robots including parameters of links and actuators* In *Proceedings of the IFAC Symposium on Theory of Robots* (pp. 295–299). Vienna, Austria

Clavel, R. (1989). Une nouvelle structure de manipulateur parallèle pour la robotique légère. *Automatique-Productique Informatique Industrielle*, *23*(6), 501–519.

Clavel, R. (1990). *Device for the movement and positioning of an element in space*. US, 4976582.

Codourey, A., & Burdet, E. (1997). A body oriented method for finding a linear form of the dynamic equatiojns of fully parallel robot. In *Proceedings of the 1997 IEEE International Conferenace on Robotics and Automation (ICRA 1997)* (pp. 1612–1619). Albuquerque, New Mexico, USA.

Company, O. & Pierrot, F. (1999). H4: A new family of 4-dof parallel robots. In *Proceedings of the IEEE/ASME International Conference on Advanced Intelligent Mechatronics (AIM1999)* (pp. 508–513). Atlanta, Georgia, USA.

Company, O., Pierrot, F., & Krut, S. (2002). Modelling and preliminary design issues of a four-axis parallel machine for heavy parts handling. *Proceedings of the Institution of Mechanical Engineers, Part K: Journal of Multi-body Dynamics*, *216*(1), 1–12.

Conconi, M., & Carricato, M. (2009). A new assessment of singularities of parallel kinematic chains. *IEEE Transactions on Robotics*, *25*(4), 757–770.

Craig, J. (1986). *Introduction to robotics: Mechanics and control*. Reading: Addison Wesley Publishing Company.

Craig, R. (1981). *Structural dynamics*. London: Wiley.

Craig, R., & Bampton, M. (1968). Coupling of substructures for dynamic analysis. *AIAA Journal*, *6*(7).

Crawford, N., Yamaguchi, G., & Dickman, C. (1999). A new technique for determining 3-D joint angles: The tilt/twist method. *Clinical Biomechanics*, *14*(3), 153–165.

Dahl, P. (1977). Measurements of solid friction parameters of ball bearings. In *Proceedings of the 6th Annual Symposium on Incremental Motion Control Systems and Devices*. University of Illinois, USA.

Dasgupta, B., & Choudhury, P. (1999). A general strategy based on the Newton-Euler approach for the dynamic formulation of parallel manipulators. *Mechanism and Machine Theory*, *34*(6), 801–824.

Dasgupta, B., & Mruthyunjaya, T. (1998a). Closed-form dynamic equations of the general Stewart platform through the Newton-Euler approach. *Mechanism and Machine Theory*, *33*(7), 993–1012.

Dasgupta, B., & Mruthyunjaya, T. (1998b). A Newton-Euler formulation for the inverse dynamics of the Stewart platform manipulator. *Mechanism and Machine Theory*, *33*(8), 1135–1152.

de Casteljau, P. (1987). *Les quaternions*. Paris, France: Hermès.

de Jalon, J. & Bayo, E. (1994). *Kinematic and dynamic simulations of multibody systems*. New York: Springer.

Dedieu, J., & Norton, G. (1990). Stewart varieties : A direct algebraic model for stewart platform. *SIGSAM Bulletin*, *24*(4), 42–59.

Denavit, J., & Hartenberg, R. (1955). A kinematic notation for lower pair mechanism based on matrices. *ASME Journal of Applied Mechanics*, *22*, 215–221.

Di Gregorio, R. (2003). Kinematics of the 3-UPU wrist. *Mechanism and Machine Theory*, *38*(3), 253–263.

Di Gregorio, R. (2004a). The 3-RRS wrist: A new, simple and non-overconstrained spherical parallel manipulator. *ASME Journal of Mechanical Design*, *126*(5), 850–855.

Di Gregorio, R. (2004b). Determination of singularities in Delta-like manipulators. *International Journal of Robotics Research*, *23*(1), 89–96.

Di Gregorio, R. (2005). Analytic form solution of the direct position analysis of a wide family of three-legged parallel manipulators. *ASME Journal of Mechanical Design*, *128*(1), 264–271.

Dwivedy, S. K. & Eberhard, P. (2006a). Dynamic analysis of flexible manipulators, a literature review. *Mechanism and Machine Theory*, 41 (7), 749–777. http://www.sciencedirect.com/science/article/B6V46-4JHMRPP-1/2/b2f90925df5e89d47b0aba9b6d67d06c.

Dwivedy, S., & Eberhard, P. (2006b). Dynamic analysis of flexible manipulators, a litterature review. *Mechanism and Machine Theory*, *41*(7), 749–777.

Ebrahimi, I., Carretero, J., & Boudreau, R. (2008). Kinematic analysis and path planning of a new kinematically redundant planar parallel manipulator. *Robotica*, *26*(3).

El Omri, J. (1996). Analyse géométrique et cinématique des mécanismes de type manipulateur. Ph.D. thesis, Université de Nantes et Ecole Centrale de Nantes, Nantes, France.

Espiau, B., Chaumette, F., & Rives, P. (1992). A new approach to visual servoing in robotics. *IEEE Transactions on Robotics and Automation*, *8*(3).

Faugère, J., & Lazard, D. (1995). Combinatorial classes of parallel manipulators. *Mechanism and Machine Theory*, *30*(6), 765–776.

Fichter, E. (1986). A Stewart platform based manipulator: General theory and practical construction. *The International Journal of Robotics Research*, *5*(2), 157–181.

Firmani, F., Zibil, A., Nokleby, S., & Podhorodeski, R. (2007). Force-moment capabilities of revolute-jointed planar parallel manipulators with additional actuated branches. *Transactions of the Canadian Society for Mechanical Engineering*, *31*(4), 469–481.

Forsythe, G., Malcom, M., & Moler, C. (1977). *Computer methods for mathematical computations*. Englewood Cliffs: Prentice-Hall.

Foucault, S., & Gosselin, C. (2004). Synthesis, design, and prototyping of a planar three degree-of-freedom re-actionless parallel mechanism. *ASME Journal of Mechanical Design*, *126*, 992–999.

Fournier, A. (1980). *Génération de mouvements en robotique : application des inverses généralisées et des pseudo-inverses*. USTL, Montpellier, France: Thèse d'etat.

Fu, S., Yao, Y., & Wu, Y. (2007). Comments on a Newton-Euler formulation for the inverse dynamics of the Stewart platform manipulator. *Mechanism and Machine Theory*, *42*(12), 1668–1671.

Gallardo, J., Rico, J., Frisoli, A., Checcacci, D., & Bergamasco, M. (2003). Dynamics of parallel manipulators by means of screw theory. *Mechanism and Machine Theory*, *38*(11), 1113–1131.

Gautier, M. (1986). Identification of robots dynamics. In *Proceedings IFAC Symposium on Theory of Robots* (pp. 351–356). Vienne, Austria

Gautier, M. (1991). Numerical calculation of the base inertial parameters. *Journal of Robotics Systems*, *8*(4), 485–506.

Gautier, M. (1997). Dynamic identification of robots with power model. *In Proceedings IEEE ICRA* (pp. 1922–1927). Albuquerque, USA

Gautier, M., & Briot, S. (2011a). New method for global identification of the joint drive gains of robots using a known inertial payload. In *Proceedings IEEE ECC CDC* (pp. 526–531). Orlando, FL, USA.

Gautier, M., & Briot, S. (2011b). New method for global identification of the joint drive gains of robots using a known payload mass. In *Proceedings IEEE IROS*. San Francisco, CA, USA.

Gautier, M. & Briot, S. (2012). Global identification of drive gains parameters of robots using a known payload. In *Proceedings of the 2012 International Conference on Robotics and Automation (ICRA 2012)*.

Gautier, M. & Briot, S. (2013). Dynamic parameter identification of a 6 dof industrial robot using power model. In: *Proceedings of the 2013 IEEE International Conference on Robotics and Automation (ICRA 2013)*. Karlsruhe, Germany.

Gautier, M., & Khalil, W. (1990). Direct calculation of minimum set of inertial parameters of serial robots. *IEEE Transactions on Robotics and Automation*, *6*(3), 368–373.

Gautier, M., Vandanjon, P., & Presse, C. (1994). Identification of inertial and drive gain parameters of robots. In *Proceedings IEEE CDC* (pp. 3764–3769). Lake Buena Vista, FL, USA

Gautier, M., Briot, S. & Venture, G. (2013). Identification of consistent standard dynamic parameters of industrial robots. In *Proceedings of the 2013 IEEE/ASME International Conference on Advanced Intelligent Mechatronics (AIM 2013)*. Wollongong, Australia.

Geng, Z., Haynes, S., Lee, J., & Carrol, R. (1992). On the dynamic model and kinematic analysis of a class of Stewart platforms. *Robotics and Autonomous Systems*, *9*, 237–254.

Germain, C., Caro, S., Briot, S. & Wenger, P. (2013). Optimal design of the IRSBot-2 based on an optimized test trajectory. In *Proceedings of the ASME 2011 International Design Engineering*

Technical Conferences and Computers and Information in Engineering Conference (IDETC/CIE 2013). Portland, Oregon, USA.

Gogu, G. (2004). Structural synthesis of fully-isotropic translational parallel robots via theory of linear transformations. *European Journal of Mechanics. A/Solids*, *23*(6), 1021–1039.

Gogu, G. (2008). *Structural synthesis of parallel robots*. New York: Springer.

Gogu, G. (2009). *Structural synthesis of parallel robots—part 2: Translational topologies with two and three degrees of freedom*. New York: Springer.

Gogu, G. (2010). *Structural synthesis of parallel robots—part 3: Topologies with planar motion of the moving platform*. New York: Springer.

Gogu, G. (2012). Structural synthesis of parallel robots—part 4: Other topologies with two and three degrees of freedom. New York: Springer.

Gogu, G. (2014). *Structural synthesis of parallel robots—part 5: Basic overconstrained topologies with schönflies motions*. New York: Springer.

Golub, G. & van Loan, C. (1983). Matrix computation (2nd ed.). *J. Hopkins*.

Gosselin, C. (1993). Parallel computational algorithms for the kinematics and dynamics of parallel manipulators. In *Proceedings of the 1993 IEEE International Conference on Robotics and Automation* (pp. 883–889). NY, USA.

Gosselin, C. (2009). Compact dynamic models for the tripteron and quadrupteron parallel manipulators. *Proceedings of the Institution of Mechanical Engineers, Part I: Journal of Systems and Control Engineering*, *223*(1), 1–11.

Gosselin, C., & Angeles, J. (1990). Singularity analysis of closed-loop kinematic chains. *IEEE Transactions on Robotics and Automation*, *6*(3), 281–290.

Gosselin, C., & Sefrioui, J. (1992). Determination of the singular loci of spherical 3 dof parallel manipulators. In *22nd Biennial Mechanisms Conference* (Vol. DE-45, pp. 329–336). Scottsdale.

Gosselin, C., Sefrioui, J., & Richard, M. (1992). Solution polynomiale au problème de la cinématique directe des manipulateurs parallèles plans à 3 degrés de liberté. *Mechanism and Machine Theory*, *27*(2), 107–119.

Gosselin, C., Saint-Pierre, E., & Gagné, M. (1996). On the development of the Agile Eye: Mechanical design, control issues and experimentation. *IEEE Robotics and Automation Society Magazine*, *3*(4), 29–37.

Gugliemetti, P., & Longchamp, R. (1994). A closed form inverse dynamic model ofthe Delta parallel robot. In *Proceedings of the 4th IFAC Symposium on Robot Control (SyRoCo 1994)* (pp. 51–56). Capri.

Hamon, P., Gautier, M., Garrec, P., & Janot, A. (2010). Dynamic identification of robot with a load-dependent joint friction model. In *Proceedings of the International Conference on Robotics, Automation and Mechatronics* (pp. 129–135). Singapore.

Hervé, J. (1978). Analyse structurelle des mécanismes par groupe de déplacement. *Mechanism and Machine Theory*, *13*(4), 437–450.

Hesselbach, J., Helm, M., & Soetebier, S. (2002). Connecting assembly modes for workspace enlargement. In *Advances in Robot Kinematics (ARK 2002)* (pp. 347–356).

Hesselbach, J., Wrege, J., Raatz, A., & Becker, O. (2004). Aspects on the design of high precision parallel robots. *Assembly Automation*, *24*(1), 49–57.

Hoffman, R., & Hoffman, M. (1979). Vibrational modes of an aircraft simulator motion system. In *Proceedings of the 5th World Congress on Theory of Machines and Mechanisms* (pp. 603–606). Montreal.

Hollerbach, J., Khalil, W., & Gautier, M. (2008). Handbook of robotics. *Model Identification* (pp. 321–344). New York: Springer.

Hollerbach, J., & Suh, K. (1985). Redundancy resolution of manipulators through torque optimization. In *Proceedings of the 1985 IEEE International Conference on Robotics and Automation (ICRA 1985)* (pp. 1016–1021). St Louis.

Horaud, R., Dornaika, F., & Espiau, B. (1998). Visually guided object grasping. *IEEE Transactions on Robotics and Automation*, *14*(4), 525–532.

Huang, T., Wang, J., & Whitehouse, D. (1999). Closed form solution to workspace of hexapod-based virtual axis machine tools. *ASME Journal of Mechanical Design, 121*(1), 26–31.

Hunt, K., & Primrose, E. (1993). Assembly configurations of some in-parallel actuated manipulators. *Mechanism and Machine Theory, 28*(1), 31–42.

Husain, M., & Waldron, K. (1994). Direct position kinematics of the 3-1-1-1 Stewart platform. *ASME Journal of Mechanical Design, 116*(4), 1102–1108.

Husty, M., & Karger, A. (2000). Architecture singular parallel manipulators and their self-motions. In *Advances in Robot Kinematics (ARK 2000)*. Piran.

Husty, M. (1996). An algorithm for solving the direct kinematics of general Stewart-Gough platforms. *Mechanism and Machine Theory, 31*(4), 365–379.

Ibrahim, O., & Khalil, W. (2010). Inverse and direct dynamic models of hybrid robots. *Mechanism and Machine Theory, 45*, 627–640.

Ider, S. K. (2005). Inverse dynamics of parallel manipulators in the presence of drive singularities. *Mechanism and Machine Theory, 40*, 33–44.

Innocenti, C. (1995). Direct kinematics in analytical form of the 6–4 fully parallel mechanism. *ASME Journal of Mechanical Design, 117*(1), 89–95.

Innocenti, C., & Parenti-Castelli, V. (1991a). Direct kinematics of the 6–4 fully parallel manipulator with position and orientation uncoupled. In *Proceedings of the European Robotics and Intelligent Systems Conference* (pp. 23–28). Corfou, Italy.

Innocenti, C., & Parenti-Castelli, V. (1991b). Direct kinematics of the reverse Stewart platform mechanism. In *Proceedings of the3rd IFAC/IFIP/IMACS Symposium on Robot Control (SyRoCo 91)* (pp. 75–80). Vienna, Austria.

Innocenti, C., & Parenti-Castelli, V. (1992). Analytical form solution of the direct kinematics of a 4–4 fully in-parallel actuated six degree-of-freedom mechanism. In *Proceedings of the 9th CISM-IFToMM SYMPOSIUM on Theory and Practice of Robots and Manipulators (RoManSy)* (pp. 41–50). Udine, Italy.

Innocenti, C., & Parenti-Castelli, V. (1993). Direct kinematics in analytical form of a general 5–4 fully-parallel manipulators. In *Computational Kinematics* (pp. 141–152).

Innocenti, C., & Parenti-Castelli, V. (1992). Reduction singularities in kinematics solution of the general geometry 6–6 Stewart platform. In *Proceedings of the IMACS/SICE International Symposium on Robotics, Mechatronics, and Manufacturing Systems* (pp. 451–458). Kobe, Japan.

Inoue, H., Tsusaka, Y., & Fukuizumi, T. (1985). Parallel manipulator. In *Proceedings of the 3rd ISRR* (pp. 321–327). Gouvieux, France.

Ji, Z. (1993). Study of the effect of leg inertia in Stewart platform. In *Proceedings of the 1993 IEEE International Conference on Robotics and Automation (ICRA 1993)* (pp. 121–126). Atlanta.

Joubair, A., Slamani, M., & Bonev, I. (2012). A novel XY-Theta precision table and a geometric procedure for its kinematic calibration. *Robotics and Computer-Integrated Manufacturing, 28*(1), 57–65.

Kanaan, D. (2008). Contribution à l'étude cinématique et dynamique des machines parallèles. Ph.D. thesis, École Centrale de Nantes.

Kanaan, D., Wenger, P., & Chablat, D. (2007). Kinematics analysis of the parallel module of the VERNE machine. In *Proceedings of the 12th World Congress in Mechanism and Machine Science*. Besançon, France.

Kanaan, D., Wenger, P., Caro, S., & Chablat, D. (2009). Singularity analysis of lower mobility parallel manipulators using Grassmann-Cayley algebra. *IEEE Transactions on Robotics, 25*(5), 995–1004.

Khalil, W., & Bennis, F. (1995). Symbolic calculation of the base inertial parameters of closed-loop robots. *The International Journal of Robotics Research, 14*(2), 112–128.

Khalil, W., & Creusot, D. (1997). Symoro+: A system for the symbolic modelling of robots. *Robotica, 15*, 153–161.

Khalil, W., & Dombre, E. (2002). *Modeling, Identification and Control of Robots*. London: Hermes Penton

Khalil, W. & Gautier, M. (2000). Modeling of mechanical systems with lumped elasticity. In *Proceedings of the IEEE International Conference on Robotics and Automation* (pp. 3965–3970). San Francisco, CA, USA.
Khalil, W., & Guégan, S. (2002). A novel solution for the dynamic modeling of gough-stewart manipulators. In *Proceedings of the IEEE International Conference on Robotics and Automation (ICRA 2002)*. Washington, DC.
Khalil, W., & Guégan, S. (2004). Inverse and direct dynamic modeling of Gough-Stewart robots. *IEEE Transactions on Robotics and Automation, 20*(4), 754–762.
Khalil, W., & Ibrahim, O. (2007). General solution for the dynamic modeling of parallel robots. *Journal of Intelligent and Robotic Systems, 49*(1), 19–37.
Khalil, W., & Kleinfinger, J. (1986). A new geometric notation for open and closed-loop robots. In *Proceedings of the IEEE International Conference on Robotics and Automation (ICRA1986)* (pp. 1174–1180). San Francisco, CA, USA.
Khalil, W., & Kleinfinger, J. (1987). Minimum operations and minimum parameters of the dynamic model of tree structure robots. *IEEE Journal of Robotics and Automation, 3*(6), 517–526.
Khalil, W., Kleinfinger, J., & Gautier, M. (1986). Reducing the computational burden of the dynamic model of robots. In *Proceedings of the IEEE International Conference on Robotics and Automation (ICRA 1986)* (pp. 525–531). San Francisco, CA, USA.
Khalil, W., Gallot, G. & Boyer, F. (2007a). Dynamic modeling and simulation of a 3D serial eel-like robot. *IEEE Transactions on Systems, Man and Cybernetics. Part C: Application and Reviews, 37*(6).
Khalil, W., Gautier, M., & Lemoine, P. (2007b). Identification of the payload inertial parameters of industrial manipulators. In *Proceedings IEEE ICRA* (pp. 4943–4948). Roma, Italy.
Khalil, W., Vijayalingam, A., Khomutenko, B., Mukhanov, I., Lemoine, P. and Ecorchard, G. (2014). OpenSYMORO: An open-source software package for symbolic modelling of robots. In *Proceedings of the 2014 IEEE/ASME International Conference on Advanced Intelligent Mechatronics (AIM 2014)* (pp. 1206–1211). Besançon, France.
Klein, C. (1984). Use of redundancy in the design of robotic systems. In *Proceedings of the 2nd International Symposium of Robotic Research* (pp. 58–65). Kyoto.
Kong, X., & Gosselin, C. (2002). A class of 3-dof translational parallel manipulators with linear input-output equations. In *Proceedings of the Workshop on Fundamental Issues and Future Research Directions for Parallel Mechanisms and Manipulators* (pp. 3–4). Québec City, QC, Canada.
Kong, X., & Gosselin, C. (2007). *Type synthesis of parallel mechanisms*. New York: Springer.
Kong, X., & Gosselin, C. (2011a). Forward displacement analysis of a quadratic 4-dof 3T1R parallel manipulator: The Quadrupteron. *Meccanica, 46*(1), 147–154.
Kong, X., & Gosselin, C. (2011b). Forward displacement analysis of a quadratic 4-dof 3T1R parallel manipulator: The Quadrupteron. *Meccanica, 46*(1), 147–154.
Korein, J. (1984). *A geometric investigation of reach*. Cambridge: MIT Press.
Kotlarski, J., Trung, D., Heimann, B., & Ortmaier, T. (2010). Optimization strategies for additional actuators of kinematically redundant parallel kinematic machines. In *Proceedings of the IEEE International Conference on Robotics and Automation (ICRA 2010)* (pp. 656–661).
Kruszewski, J., Gawronski, W., Wittbrodt, E., Najbar, F., & Grabowski, S. (1975). *The rigid finite element method*. Warszawa: Arkady.
Kurtz, R., & Hayward, V. (1992). Multiple-goal kinematic optimization of a parallel spherical mechanism with actuator redundancy. *IEEE Transactions on Robotics and Automation, 8*(5), 644–651.
Lazard, D. (1993). On the representation of rigid-body motions and its application to generalized platform manipulators. In J. Angeles, P. Kovacs & G. Hommel (Eds.), *Computational Kinematics* (pp. 175–182). Kluwer.
Le Borzec, R., & Lotterie, J. (1975). Principes de la théorie des mécanismes. Paris: Dunod.
le Rond d'Alembert, J. (1743). *Traité de dynamique*. Paris: David l'aine

Lebret, G., Liu, G., & Lewis, F. (1993). Dynamic analysis and control of a Stewart platform manipulator. *Journal of Robotic Systems*, *10*(5), 629–655.

Lee, K., & Shah, D. (1988). Dynamic analysis of a three-degrees-of-freedom in-parallel actuated manipulator. *IEEE Transactions on Robotics and Automation*, *4*(3), 361–368.

Leinonen, T. (1991). Terminology for the theory of machines and mechanisms. *Mechanism and Machine Theory*, 26.

Leonesio, M., & Bianchi, G. (2009). Self-locking analysis in closed kinematic chains. *Mechanism and Machine Theory*, *44*(11), 2038–2052.

Lin, W., Duffy, J., & Griffis, M. (1992). Forward displacement analysis of the 4–4 Stewart platform. *ASME Journal of Mechanical Design*, *114*(3), 444–450.

Lin, W., Crane, C., & Duffy, J. (1994). Closed-form forward displacement analysis of the 4–5 in-parallel platforms. *ASME Journal of Mechanical Design*, *116*(1), 47–53.

Liu M. J., Li, C. X., & Li, C. (2000). Dynamics analysis of the Gough-Stewart platform manipulator. *IEEE Transaction on Robotics and Automation*, *16*(1), 94–98.

Liu, X. J., Wang, J., & Pritschow, G. (2006). Performance atlases and optimum design of planar 5r symmetrical parallel mechanisms. *Mechanism and Machine Theory*, *41*(2), 119–144.

Llibre, M., Mampey, R., & Chrétien, J. (1983). Simulation de la dynamique des robots manipulateurs motorisés (pp. 197–207). Actes du Congrès AFCET: Productique et Robotique Intelligente. Besançon, France.

Long, P., Khalil, W., & Martinet, P. (2014). Dynamic modeling of parallel robots with flexible platforms. *Mechanism and Machine Theory*, *81*, 21–35.

Luh, J., Walker, M., & Paul, R. (1980). On-line computational scheme for mechanical manipulators. *ASME Journal of Dynamic Systems, Measurement and Control*, *102*(2), 69–76.

Maciejewski, A., & Klein, C. (1985). Obstacle avoidance for kinematically redundant manipulators in dynamically varying environments. *The International Journal of Robotics Research*, *4*(3), 109–117.

Martinet, P., Gallice, J., & Khadraoui, D. (1996). Vision based control law using 3D visual features. In *Proceedings of the World Automation Congress, WAC96, Robotics and Manufacturing Systems* (Vol. 3, pp. 497–502). Montpellier, France.

Mayeda, H., Yoshida, K., & Osuka, K. (1990). Base parameters of manipulator dynamic models. *IEEE Transactions on Robotics and Automation*, *6*(3), 312–321.

Merlet, J. (1989). Singular configurations of parallel manipulators and Grassmann geometry. *The International Journal of Robotics Research*, *8*(5), 45–56.

Merlet, J. (1997). Direct kinematics of planar parallel manipulators. In *Proceedings of the 1997 IEEE International Conference on Robotics and Automation (ICRA 1997)* (pp. 3744–3749). Minneapolis, USA

Merlet, J. (2004). Solving the forward kinematics of a Gough-type parallel manipulator with interval analysis. *The International Journal of Robotics Research*, *23*(3), 221–236.

Merlet, J. (2006a). Jacobian, manipulability, condition number, and accuracy of parallel robots. *ASME Transactions Journal of Mechanical Design*, *128*(1), 199–206.

Merlet, J. (2006b). *Parallel robots* (2nd ed.). New York: Springer

Merlet, J. (2014). www-sop.inria.fr/members/Jean-Pierre.Merlet/merlet.html.

Miller, K. (2004). Optimal design and modeling of spatial parallel manipulators. *The International Journal of Robotics Research*, *23*(2), 127–140.

Moon, F. (2007). Applied dynamics. J. Wiley and Sons.

Mourrain, B. (1993). The 40 generic positions of a parallel robot. In M. Bronstein (Ed.), *ISSAC'93* (pp. 173–182). Kiev, Ukraine: ACM Press.

Müller, A. (2005). Internal preload control of redundantly actuated parallel manipulators—its application to backlash avoiding control. *IEEE Transactions on Robotics*, *21*(4), 668–677.

Murphy, S., & Ting-Yung Wen, J. (1993). Analysis of active manipulator elements in space manipulation. *IEEE Transactions on Robotics and Automation*, *9*(5), 544–552.

Nabat, V. (2007). Parallel robot with articulated platforms: from concept to industrial solution for pick-and-place operations. Ph.D. thesis, Université de Montpellier.

Nahon, M., & Angeles, J. (1989). Force optimization in redundantly-actuated closed kinematic chains. In *Proceedings of the IEEE International Conference on Robotics and Automation (ICRA 1989).*

Nair, P. (1994). On the forward kinematics of parallel manipulators. *The International Journal of Robotics Research, 13*(2), 171–188.

Nanua, P., & Waldron, K. (1991). Direct kinematic solution of a Stewart platform. *IEEE Transactions on Robotics and Automation, 6*(4), 438–444.

Newman, W., & Sproull, R. (1979). *Principles of interactive computer graphics*. New York: McGraw Hill.

Notash, L., & Kamalzadeh, A. (2007). Inverse dynamics of wire-actuated parallel manipulators with a constraining linkage. *Mechanism and Machine Theory, 42*(9), 1103–1118.

Ozgur, E., Andreff, N., & Martinet, P. (2011). Dynamic control of the quattro robot by the leg edgels. In *Proceedings of the IEEE International Conference on Robotics and Automation, ICRA11*. Shanghai, China.

Özgür, E., Andreff, N., & Martinet, P. (2013). Linear dynamic modeling of parallel kinematic manipulators from observable kinematic elements. *Mechanism and Machine Theory*.

Pagis, G., Bouton, N., Briot, S., & Martinet, P. (2014). Design of a controller for enlarging parallel robots workspace through type 2 singularity crossing. *Proceedings of 2014 IEEE International Conference on Robotics and Automation (ICRA 2014)*. Hong Kong, China.

Pagis, G., Bouton, N., Briot, S., & Martinet, P. (2015). Enlarging parallel robot workspace through type-2 singularity crossing. *Control Engineering Practice, 39*, 1–11.

Park, F., Choi, J., & Ploen, S. (1999). Symbolic formulation of closed chain dynamics in independent coordinates. *Mechanism and Machine Theory, 34*(5), 731–751.

Pashkevich, A., Chablat, D., & Wenger, P. (2006). Kinematics and workspace analysis of a three-axis parallel manipulator: The Orthoglide. *Robotica, 24*(1), 39–49.

Paul, R. (1981). *Robot manipulators: Mathematics, programming and control*. Cambridge: MIT Press.

Pelaez, G., Pelaez, G., Perez, J., Vizan, A., & Bautista, E. (2005). Input shaping reference commands for trajectory following cartesian machines. *Control Engineering Practice, 13*(8), 941–958. Fault Detection, Supervision and Safety of Technical Processes (Safeprocess 2003).

Pfurner, M., & Husty, M. (2010). New trends in mechanism science. In *Implementation of a new and efficient algorithm for the inverse kinematics of serial 6R chains* (pp. 91–98). New York: Springer.

Pieper, D. (1968). The kinematics of manipulators under computer control. Ph.D. thesis, Stanford University.

Pierrot, F., Uchiyama, M., Dauchez, P., & Fournier, A. (1990). A new design of a 6-dof parallel robot. In *Proceedings of the 23rd International Symposium on Industrial Robots. Jounrla of Robotics and Mechatronics* (pp. 308–315).

Pierrot, F., Baradat, C., Nabat, V., Company, O., & Gouttefarde, S. K. K. (2009). Above 40g acceleration for pick-and-place with a new 2-dof pkm. In *Proceedings of the 2009 IEEE international Conference on Robotics and Automation (ICRA 2009)*. Japan: Kobe International Conference Center.

Potkonjak, V. (1986). Thermal criterion for the selection of DC drives for industrial robots. In *Proceedings of the 16th International Symposium on Industrial Robots. Bruxelles* (pp. 129–140). Belgium.

Raghavan, M. (1993). The Stewart platform of general geometry has 40 configurations. *ASME Journal of Mechanical Design, 115*(2), 277–282.

Rakotomanga, N., Chablat, D., & Caro, S. (2008). Kinetostatic performance of a planar parallel mechanism with variable actuation. *Advances in Robot Kinematics*.

Raucent, B., Bastin, G., Campion, G., Samin, J., & Willems, P. (1992). Identification of barycentric parameters of robotic manipulators from external measurements. *Automatica, 28*(5), 1011–1016.

Reboulet, C. and Berthomieu, T. (1991). Dynamic models of a six degree of freedom parallel manipulators. In *Proceedings of the International Conference on Advanced Robotics (ICAR 1991)* (pp. 1153–1157). Pisa, Italy.

Rizk, R., Munteanu, M., Fauroux, J., & Gogu, G. (2007). A semi-analytical stiffness model of parallel robots from the Isoglide family via the sub-structuring principle. In *Proceedings of the 12th IFToMM World Congress*. Besançon, France.

Roberts, L. (1965). Homogeneous matrix representation and manipulation of N dimensional constructs. Master's thesis, MIT Lincoln Lab.

Rognant, M., Courteille, E., & Maurine, P. (2010). A systematic procedure for the elastodynamic modeling and identification of robot manipulators. *IEEE Transactions on Robotics*, *26*(6), 1085–1093.

Ronga, F., & Vust, T. (1992). Stewart platforms without computer? In *Proceedings of the Conference on Real Analytic and Algebraic Geometry* (pp. 197–212). Trento, Italy.

Sciavicco, L., Siciliano, B., & Villani, L. (1994). On dynamic modelling of gear-driven rigid robot manipulators. In *Proceedings of the 4th IFAC Symposium on Robot Control (SYROCO 1994)* (pp. 543–549). Capri, Italy.

Seward, N., & Bonev, I. (2014). A new 6-dof parallel robot with simple kinematic model. In *Proceedings of the 2014 IEEE International Conference on Robotics and Automation (ICRA 2014)*. Hong Kong, China.

Shabana, A. (2005). *Dynamics of multibody systems*. Cambridge: Cambridge University Press.

Shabana, A. (1990). Dynamics of flexible bodies using generalized Newton-Euler equations. *Journal of Dynamic Systems, Measurement, and Control*, *112*, 496–503.

Shah, S., Saha, S., & Dutt, J. (2013). *Dynamics of tree-type robotic systems*. New York: Springer.

Sharf, I., & Damaren, C. (1992). Simulation of flexible-link manipulators: basis functions and nonlinear terms in the motion equations. In *Proceedings of the IEEE International Conference on Robotics and Automation (ICRA 1992)* (pp. 1956–1962). Nice, France.

Sheth, P., & Uicker, J. (1971). A generalized symbolic notation for mechanism. *ASME Journal of Engineering for Industry*, *93*, 102–112.

Singer, N. C., & Seering, W. P. (1988). Preshaping command inputs to reduce system vibration. Massachusetts Institute of Technology Artificial Intelligence Laboratory. A.I. Memo No. 1027.

Singh, T., & Singhose, W. E. (2002). Tutorial on input shaping/time delay control of maneuvering flexible structures. In *American Control Conference*.

Spong, M. (1987). Modeling and control of elastic joint robots. *ASME Journal of Dynamic Systems, Measurement, and Control*, *109*, 310–319.

Stachera, K., & Schumacher, W. (2008). Derivation and calculation of the dynamics of elastic parallel manipulators. *Automation and robotics*. I-Tech Education and Publishing.

Sugimoto, K. (1989). Computational scheme for dynamic analysis of parallel manipulators. *ASME Journal of Mechanics, Transmission and Automation in Design*, *111*, 29–33.

Tale-Masouleh, M., Gosselin, C., Husty, M., & Walter, D. (2011). Forward kinematic problem of 5-RPUR parallel mechanisms (3T2R) with identical limb structures. *Mechanism and Machine Theory*, *46*(7), 945–959.

Tsai, L. (2000). Solving the inverse dynamics of a Stewart-Gough manipulator by the principle of virtual work. *ASME Journal of Mechanical Design*, *122*, 3–9.

Tsai, L., & Joshi, S. (2000). Kinematics and optimization of a spatial 3-u Pu parallel manipulator. *ASME Journal of Mechanical Design*, *122*(4), 439–446.

Ur-Rehman, R., Caro, S., Chablat, D., & Wenger, P. (2009). Path placement optimization of manipulators based on energy consumption: Application to the Orthoglide 3-axis. *Transactions of the Canadian Society for Mechanical Engineering*, *33*(3).

Vakil, M., Pendar, H., & Zohoor, H. (2008). Comments on closed-form dynamic equations of the general Stewart platform through the Newton-Euler approach and a Newton-Euler formulation for the inverse dynamics of the Stewart platform manipulator. *Mechanism and Machine Theory*, *43*(10), 1349–1351.

van der Wijk, V., Krut, S., Pierrot, F., & Herder, J. (2011). Generic method for deriving the general shaking force balance conditions of parallel manipulators with application to a redundant planar 4-rrr parallel manipulator. In *Proceedings of the 13th World Congress in Mechanism and Machine Science*. Guanajuato, Mexico.

Voglewede, P., & Ebert-Uphoff, I. (2005). Overatching framework for measuring the closeness to singularities of parallel manipulators. *IEEE Transactions on Robotics*, *21*(6), 1037–1045.

Walker, M., & Orin, D. (1982). Efficient dynamic computer simulation of robotics mechanism. *ASME Journal of Dynamic Systems, Measurement, and Control*, *104*, 205–211.

Wampler, C. (1996). Forward displacement analysis of general six-in-parallel SPS (Stewart) platform manipulators using soma coordinates. *Mechanism and Machine Theory*, *31*(3), 331–337.

Wang, X. and Mills, J.K. (2006). Dynamic modeling of a flexible-link planar parallel platform using a substructuring approach. *Mechanism and Machine Theory*, *41*(6), 671–687. http://www.sciencedirect.com/science/article/B6V46-4HNSG0Y-1/2/81cd2e245e4875189413080286cf4c4c.

Wang, Y. (1999). Workspace analysis of a novel closed-chain manipulator. Ph.D. thesis, Case Western Reserve University, Cleveland, OH, USA.

Wei, W., & Simaan, N. (2010). Design of planar parallel robots with preloaded flexures for guaranteed backlash prevention. *ASME Journal of Mechanisms and Robotics*, *2*(1).

Wenger, P. (1992). A new general formalism for the kinematic analysis of all non redundant manipulators. In *Proceedings of the 1992 IEEE International Conference on Robotics and Automation (ICRA 1992)* (pp. 442–447). Nice, France.

Wenger, P., Chablat, D., & Zein, M. (2007). Degeneracy study of the forward kinematics of planar 3-r Pr parallel manipulators. *ASME Journal of Mechanical Design*, *129*(12), 1265–1268.

White, N. (2008). Grassmann-Cayley algebra and robotics applications. In *Handbook of geometric computing* (Vol. 8, pp. 629–656). New York: Springer

Whitney, D. (1969). Resolved motion rate control of manipulators and human prostheses. *IEEE Transactions on Man Machine Systems*, *10*(2), 47–53.

Wittbrodt, E., Adamiec-Wójcik, I., & Wojciech, S. (2006). *Dynamics of flexible multibody systems*. Berlin: Springer.

Yi, B., Freeman, R., & Tesar, D. (1994). Force and stiffness transmision in redundantly actuated mechanisms: The case for a spherical shoulder mechanism. *Robotics, Spatial Mechanisms, Mechanical Systems*, *45*, 163–172.

Yin, J., & Liang, C. (1994). The forward displacement analysis of a kind of special platform manipulator. *Mechanism and Machine Theory*, *29*(1), 1–9.

Yoshikawa, T. (1984). Analysis and control of robot manipulators with redundancy. In *The 1st International Symposium of Robotics Research* (pp. 735–748). Cambridge: MIT press.

Zein, M., Wenger, P., & Chablat, D. (2008). Non-singular assembly-mode changing motions for 3-RPR parallel manipulators. *Mechanism and Machine Theory*, *43*(4), 480–490.

Zhang, C., & Song, S. (1991). Forward kinematics of a class of parallel (Stewart) platforms with closed-form solutions. In *Proceedings of the 1991 IEEE International Conference on Robotics and Automation (ICRA 1991)* (pp. 2676–2681). Sacramento.

Zhao, J., Li, B., Yang, X., & Yu, H. (2009). Geometrical method to determine the reciprocal screws and applications to parallel manipulators. *Robotica*, *27*, 929–940.

Zlatanov, D., Fenton, R., & Benhabib, B. (1994a). Singularity analysis of mechanisms and robots via a motion-space model of the instantaneous kinematics. In *Proceedings of the 1994 IEEE International Conference on Robotics and Automation*. Ontario, Canada: Toronto University.

Zlatanov, D., Fenton, R., & Benhabib, B. (1994b). Singularity analysis of mechanisms and robots via a velocity-equation model of the instantaneous kinematics. In *Proceedings of the 1994 IEEE International Conference on Robotics and Automation*. Ontario, Canada: Toronto University.

Zlatanov, D., Bonev, I., & Gosselin, C. (2002). Constraint singularities of parallel mechanisms. In *Proceedings of the IEEE International Conference on Robotics and Automation (ICRA 2002).*

Index

S. Briot and W. Khalil, *Dynamics of Parallel Robots*,
Mechanisms and Machine Science 35, DOI 10.1007/978-3-319-19788-3

K

L

M

N

O

P

Q

R

S

MIX
Papier aus verantwortungsvollen Quellen
Paper from responsible sources
FSC® C105338

If you have any concerns about our products,
you can contact us on
ProductSafety@springernature.com

In case Publisher is established outside the EU,
the EU authorized representative is:
Springer Nature Customer Service Center GmbH
Europaplatz 3, 69115 Heidelberg, Germany

Printed by Libri Plureos GmbH
in Hamburg, Germany